Shawn Senk

MATHEMATICS: AN INTRODUCTION TO ITS SPIRIT AND USE

Readings from

SCIENTIFIC AMERICAN

MATHEMATICS: AN INTRODUCTION TO ITS SPIRIT AND USE

With Introductions by

Morris Kline

Courant Institute of Mathematical Sciences

New York University

W. H. Freeman and Company

San Francisco

Most of the SCIENTIFIC AMERICAN articles in *Mathematics: An Introduction to Its Spirit and Use* are available as separate Offprints. For a complete list of articles now available as Offprints, write to W. H. Freeman and Company, 660 Market Street, San Francisco, California 94104.

Library of Congress Cataloging in Publication Data

Main entry under title:

Mathematics: an introduction to its spirit and use.

Bibliography: p.
Includes index.
1. Mathematics—Addresses, essays, lectures.
I. Kline, Morris, 1908– II. Scientific American.
QA7.M3445 510 78-7878
ISBN 0-7167-0370-X
ISBN 0-7167-0369-6 pbk.

Printed in the United States of America

9 8 7 6 5 4 3 2 1

PREFACE

Mathematics is a multifaceted subject, limitless in extent and depth, vital for science and technology, and rich in cultural import. Unfortunately, our formal education does not and cannot encompass all that the subject has to offer. Elementary schools must teach the necessities. Because arithmetic is almost as indispensable as reading and writing, this aspect of mathematics receives the most attention at that level. Arithmetic is often dull and can prejudice us against mathematics. Who can enjoy multiplying one four-digit number by another? Secondary school education, though somewhat less restricted, is also tied to objectives that severely limit the presentation of many mathematical values. Educators must face the fact that teenagers do not know what careers they will pursue, and because technical mathematics is essential to so many of these careers, the practical procedure is to teach this technical knowledge—not very enticing to the student—so that students will indeed be prepared for any choice of career. To delay the preparation is to preclude choices.

At the college level, the curriculum can be more flexible and be adapted to a variety of student interests. Those students who will use mathematics in their life's work must continue to learn technical content; they may through years of close association with mathematics and some of its uses broaden and deepen their appreciation of the subject. But those who will not need mathematics in a career can devote themselves to an appreciation of its magnificent contributions to our civilization and culture. Nevertheless, one semester or one year of a liberal arts course in mathematics can only initiate students into this world of mathematics. Fortunately, there is now much literature available to the student who is inspired to read further. One source of such material is SCIENTIFIC AMERICAN.

The SCIENTIFIC AMERICAN articles selected for this Reader were written by scholars who excel in exposition, and I sincerely believe that the rewards for reading them will be handsome.

June 1978 Morris Kline

CONTENTS

IV STATISTICS AND PROBABILITY

V SYMBOLIC LOGIC AND COMPUTERS

VI APPLICATIONS

Note on cross-references: References to articles included in this book are noted by the title of the article and the page on which it begins; references to articles that are available as Offprints, but are not included here, are noted by the article's title and Offprint number; references to articles published by SCIENTIFIC AMERICAN, but which are not available as Offprints, are noted by the title of the article and the month and year of its publication.

MATHEMATICS: AN INTRODUCTION TO ITS SPIRIT AND USE

GENERAL INTRODUCTION

All of us readily appreciate the values that music, painting, history, literature, and many other disciplines offer to anyone interested in the liberal arts. But it is not at all evident that mathematics has much to offer to a person who does not intend to use it professionally. The values are there, values as great as any that human creation can offer. The effort to secure them may be more arduous than in literature, say, but the rewards are richer, just as the ascent of a high mountain, although more strenuous than climbing a low hill, offers a view that extends to far more distant horizons. Values there are in abundance; the only question one may raise is how to order them in importance. Here individual judgment, opinion, and taste enter.

Historically, the prime value of mathematics has been that it enables us to answer basic questions about our physical world, to comprehend the complicated operations of nature, and to dissipate much of the mystery that envelops life. The simplest arithmetic, algebra, and geometry suffice to determine the circumference of the earth, the distances to the moon and the planets, the speeds of sound and light, and the reasons for eclipses of the sun and moon. But the supreme value of mathematics, insofar as understanding the world about us is concerned, is that it reveals order and law where mere observation shows chaos. For example, among the phenomena of the world, motion is fundamental. And among motions, the most impressive are the motions of the heavenly bodies. But the appearances and disappearances of these bodies, their seemingly irregular wanderings, and the variations in their brightness and in the portions visible at various times of the month or year almost defy comprehension. However, law and order do prevail, and these are made manifest by mathematics. To use Alexander Pope's words, this mighty maze is not without a plan. The first truly great mathematical theory of planetary motion, a geocentric theory, was created by Claudius Ptolemy about 150 A.D. It was superseded in the sixteenth century by the heliocentric theory of Copernicus and Kepler. Moreover, by virtue of the work of Newton, we can deduce the laws of planetary motion from very simple laws that apply to all motions, from the motion of a ball thrown by a child to the motion of the massive planet Jupiter.

The understanding we have attained about the physical world is often credited to science. But mathematics provides the dies by which science is formed, and mathematics is the essence of our best scientific theories. Let us consider motion. In the scientific treatment of any motion, the force of gravity enters. Gravity "explains" why a ball that is dropped falls to earth and why the planets maintain their elliptical paths around the sun. But what is gravity and how does it act? We have no physical answers to these questions. All that we know about gravity is the mathematical law, called the inverse square law, and it is by means of this law and other mathematical laws of motion, commonly called Newton's laws, that we deduce by purely mathematical means the behavior of objects in motion. Of course, observation and experimentation are indispensable components of the scientific enterprise, but mathematical laws are the essence of our knowledge. Our world is to a large extent what mathematics says it is.

The ultimate goal of the search for understanding is knowledge about humanity itself. All people want to know the meaning of their own lives, and they seek the answer by attempting to understand the world in which they find themselves. Thus mathematics serves to mediate between man and nature, between man's inner and outer worlds. And the knowledge obtained about humanity and its place in this world is reflected in philosophy, literature, painting, and other branches of our culture.

More recently the search for understanding has extended to the social sciences—economics, political science, sociology, and education. The chief mathematical tools employed in this search, statistics and the theory of probability, though first conceived in the seventeenth century, have been expounded and exploited only during the past hundred years as social scientists became aware of their effectiveness. Because many of our articles will explain the nature of these subjects and their uses, we shall not dwell here on the importance of this value of mathematics.

Comprehension of the physical and social worlds and of our own role in these worlds satisfies man's deepest and strongest intellectual urge. But scientists discovered several centuries ago that knowledge yields power to improve material conditions, to ease pain, and even to prolong life. These values, proclaimed as the goal of mathematics and science by Francis Bacon three hundred and fifty years ago, have been realized to an extent that not even the most daring and imaginative person would have envisaged in Bacon's time. Perhaps one example will suffice.

Among the phenomena of nature that seemed most mysterious and that attracted the attention of scientists were electricity and magnetism. Long before mathematicians and scientists had much if any conception of what they could do with electricity and magnetism, they were exploring mathematical descriptions of these phenomena and succeeded in obtaining the relevant mathematical laws. The uses soon followed. Power to run factories, light for our homes, X rays for medical diagnosis, and radio and television are commonplace nowadays. Command of electrical phenomena coupled with the mathematical analysis of musical sounds has made possible still other conveniences and pleasures (see Section VI, "Applications"). Indeed, all of the benefits that Francis Bacon declared should be the objective of science and far more have been realized through mathematical analysis of phenomena, some as obscure and as physically incomprehensible as the action of gravity.

Most people should not expect to be able to repair their television sets, but as citizens they must have some understanding of what mathematics and science can achieve. Whether or not they care to have this understanding, they are called on to vote for people and measures that may or may not contribute to their own welfare. They are therefore obligated to obtain some knowledge of the role of mathematics and science in advancing technology.

To many mathematicians, the understanding gained by investigating nature and the technological advances made possible thereby are of relatively little interest. They seek beauty, and they find it in various types of mathematical activity. We are inclined to think that facts, observations, and experimental results dictate the mathematical theories of natural phenomena. But scientists have long since realized that the mathematical description of a physical phenomenon is not unique. Aesthetic considerations play a major role in the choice. The most notable example is the theory of planetary motions. We no longer believe that the heliocentric theory is true and the geocentric theory false. The former is preferred because it is mathematically more elegant. In the heliocentric theory the motions of all the planets around the sun are similar, whereas in the geocentric theory different schemes are required for each planet. Though Copernicus and Kepler believed that their creation was true and they exulted in the beauty of truth, the modern mathematician working on scientific problems opts for the truth of beauty. The magnificent theories of planetary motion and of electricity and magnetism—the unheard melodies of the mind—do exhibit elegant law, order, and uniformity in nature. Even in the very process of constructing a mathematical theory for a natural phenomenon, a choice must be made on the basis of what appeals to the mind, although, of course, the end product must conform to observation and experimentation. Reality is interpreted by man, and certainly a major consideration is what one of the most famous mathematicians of recent times, Henri Poincaré, called aesthetic sensibility.

Many mathematicians today find beauty in the theorems and proofs of mathematics proper. Among well-developed branches of mathematics, the theory of numbers and projective geometry are notable for the aesthetic satisfaction they furnish. We shall have the opportunity to examine some choice portions of these subjects in the ensuing articles.

Often inseparable from aesthetic appeal is still another value—intellectual challenge. Most people respond to some such challenges: Witness the popularity of crossword puzzles, chess, bridge, and even jigsaw puzzles. Mathematics offers far deeper, more varied, more intriguing, and more significant challenges. The tenacity with which mathematicians pursue problems is well known. The famous construction problems of geometry using straightedge and compass—squaring the circle, trisecting any angle, and constructing the side of a cube whose volume is double that of a given cube—were pursued by hundreds of people for over two thousand years. Unfortunately, the outcome of all these efforts was a proof that the constructions cannot be performed under the given conditions (although there were many serendipitous by-products of these searches). Unsolved problems in the theory of numbers, first attacked over three hundred years ago, are still engaging the best efforts of leading mathematicians.

Still another value has been attributed to the study of mathematics ever since Greek times—training the mind. The contention is that the study of mathematics makes a person a better and sharper thinker in all fields. It must be admitted that the contention is not supported by any hard and fast evidence. Moreover, deductive reasoning from explicit axioms, which is the distinctive feature of mathematics, is rarely applicable to political, social, and historical problems, because acceptable axioms are not available. Nevertheless, some forms of such reasoning do apply to other areas. We do learn in mathematics that if two triangles are congruent their corresponding angles are equal, but that the equality of the angles does not imply the congruence of the triangles. In general terms, if A implies B, it does not follow that B implies A. This principle and other laws of logic were, historically, abstracted from the thinking of mathematicians, and students of mathematics are called upon to apply them. An attentive student of mathematics will not argue that because good cars are expensive, expensive cars must be good.

Further, deductive reasoning is not the only form involved in mathematics. It is often overlooked that before mathematicians prove anything, they must first arrive at a conjecture and be reasonably confident that the conjecture is correct. Otherwise why spend weeks or months trying to prove it? What gives a mathematician confidence in a conjecture? Evidence, evidence of various sorts, depending on the problem. Also overlooked in the concentration on deductive reasoning is that mathematics includes the theory of probability, which yields no more than the probability that a conclusion is correct. And probability, too, is based on evidence—for example, statistical data. Certainly the student of mathematics must learn to look for evidence, and this habit should carry over to other fields. Even when an authoritative economist makes an assertion, one who has studied mathematics would almost surely ask for supporting evidence. We learn to think and reason by practicing thinking and reasoning. Hence it does seem likely that mathematical training inculcates important habits of thought.

While the extent to which the study of mathematics trains the mind is uncertain, there is no question that mathematics is humanity's most sustained, most enduring, and most powerful achievement. Comprehension of the physical world, intellectual satisfaction, technical utility, and aesthetic values (albeit more esoteric than those offered by music or painting) can be derived from mathematics. In addition, the achievements of mathematics demonstrate the capacity of the human mind, and this exhibition of what human reason can accomplish has given man the courage and confidence to tackle additional seemingly impenetrable mysteries of the cosmos, to seek cures for fatal diseases, and to question and improve the economic and political systems under which people live. Man's success in the difficult realm of mathematics provides one unquenchable source of hope for success in these other areas, whether or not mathematics itself plays a role in the undertakings. The many values described above attest to the assertion that mathematics is a fundamental and integral part of our culture.

No doubt many a student of mathematics has encountered and appreciated the values we have cited. For those who have, the examples in this Reader may add to their enjoyment and increase their admiration of the subject. Those who have not should look forward to acquiring new treasures. The articles selected from *Scientific American* were chosen to exhibit and substantiate the claims of mathematics for attention. The suggested readings at the end of this introduction and at the end of each section introduction offer additional material.

SUGGESTED READINGS

Courant, Richard, and Herbert Robbins. 1941. *What Is Mathematics?* Oxford University Press, New York.

Kasner, Edward, and James R. Newman. 1940. *Mathematics and the Imagination.* Simon and Schuster, New York.

Kline, Morris. 1953. *Mathematics in Western Culture.* Oxford University Press, New York. Also available in paperback.

Kline, Morris. 1962. *Mathematics: A Cultural Approach.* Addison-Wesley, Reading, Mass.

Newman, James R. 1956. *The World of Mathematics.* Simon and Schuster, New York. Also available in paperback.

Stein, Sherman. 1976. *Mathematics, the Man-Made Universe,* third edition. W. H. Freeman and Company, San Francisco.

Whitehead, Alfred N. 1926. *Science and the Modern World.* Cambridge University Press, London. Also available in paperback.

I

HISTORY

I HISTORY

INTRODUCTION

The history of mathematics is one of long, arduous struggles by many civilizations, which have culminated in the concepts and methodology that have given mathematics the commanding position, power, and values that we described earlier. The articles in this section do not attempt to relate this history; they do, however, call attention to several remarkable features.

Perhaps the most striking point is that although civilizations, primitive to be sure, existed tens of thousands of years ago, they created almost no mathematics. About ten thousand years ago, people turned to agriculture and animal husbandry, and even these more-developed civilizations produced no mathematics of consequence. The first civilizations to make significant contributions to our subject were the Babylonian, the Egyptian, and the Chinese. Although factual knowledge is sparse, we can be reasonably sure that these contributions date from about 4000 B.C. Of these three civilizations, the first two warrant more attention because they initiated the main line of mathematical development.

Though the mathematics of these two peoples differ somewhat—the Babylonians emphasized algebra and the Egyptians geometry—Newman's article on the Rhind papyrus, the best of the surviving Egyptian documents, gives a fine account of what this civilization produced. Both civilizations developed what is called empirical mathematics. That is, they arrived at procedures, rules, and formulas by trial and error and by generalizing on experiences with concrete objects and situations.

It was the civilization of ancient Greece that molded mathematics into the form that we regard as definitive today. Although the beginnings of this civilization can be traced as far back as 2800 B.C., its mathematical activity flourished during what is called the classical period, 600 to 300 B.C. The major contribution, Euclidean geometry, dwarfs Egyptian and Babylonian algebra and geometry. Even more important is the change that the classical Greeks introduced into the very concept of mathematics. Instead of accepting conclusions on the bases of trial and error and inferences from experience, the Greeks insisted on deductive proof from explicitly stated axioms—the methodology we usually learn in high school geometry.

When, under Alexander the Great, the Greeks expanded their empire to encompass Egypt and Babylonia, they shifted the center of their operations to Alexandria in Egypt. In this Alexandrian, also called Hellenistic, period (300 B.C.–600 A.D.), the Greeks absorbed and extended Babylonian algebra and initiated trigonometry. But surprisingly, in the domains of arithmetic and algebra, the Alexandrian Greeks succumbed to the empirical methods of the Egyptians and Babylonians. One of the anomalies of the history of mathematics is that although geometry remained deductive, arithmetic, algebra, and the branches built on algebra did not receive a deductive formulation until the end of the nineteenth century. Just how Greek mathematics reached Europe and the United States is a fascinating story that one can read in the histories by Struik and by Boyer listed in the suggested readings for this section.

Another facet of the history of mathematics is the lives of the people who created it. Since mathematics is humanity's finest intellectual achievement, curiosity about the kind of people who made contributions of this caliber is natural. Unfortunately, we know almost nothing about the lives of the ancient mathematicians, and so for biography we must turn to the Europeans.

The first great mathematician of the modern era was Jerome Cardan, and his book on algebra, *Ars Magna* [The Great Art] (1545), is the first European mathematical work of consequence. The scientific and humanistic leaders of the Renaissance possessed characteristics that we do not normally associate with intellectuals. They were indeed steeped in knowledge and endowed with genius, but they were also adventurers, swashbucklers, liars, cheats, and even murderers. Cardan was no exception. Newman's review of a biography of Cardan gives the highlights, and the book itself, by Oystein Ore, gives the details.

Though the men of the seventeenth century, which Alfred North Whitehead termed the century of genius, steered clear of the rascality of Cardan and his contemporaries and turned from a quest for encyclopedic knowledge to deep studies of nature, many of these men were marked by unusual qualities beyond intellectual brilliance, and their biographies are fascinating. One feature common to them all is the struggle to reconcile science and religion, and their resolutions of this conflict were remarkably different.

We must recall that the medieval European civilization was dominated by Catholic thought. God ruled the universe, and man's role was to serve and please Him and by so doing win salvation, whereupon the soul would live in an afterlife of joy and splendor. The conditions of life on this earth were immaterial, and hardship and suffering were to be tolerated and accepted as a test of faith in God. Understandably, interest in mathematics and science, which had been motivated in Greek times by the study of the physical world, was at a nadir. But interest in that world did reawaken, and with it a resurgence of mathematical and scientific activity took place. The reasons are complex, and much historical study has been devoted to pinpointing the causes of that resurgence. Certainly the geographical explorations around Africa to Asia, the discovery of America, and Magellan's circumnavigation of the globe broadened the physical horizon, and the encounter with totally different civilizations challenged the single-minded outlook of the Europeans. The Protestant Reformation defied and cast doubt on many Catholic doctrines. The rise within the feudal economy of Europe of a large class of artisans who made a living by their own independent handiwork motivated an interest in materials, skills, and, more broadly, technology. Whereas we normally think of technology as the by-product of science, in the seventeenth century the desire to improve technology gave rise to scientific problems. Still another blow to the medieval framework was the creation and rather rapid adoption of the heliocentric theory. Was it any longer reasonable to suppose that man, a mite on one of the many planets, could be the central figure in the universe and a major if not the chief concern of a God who ruled over the entire universe?

Such questioning of the prevailing doctrines disturbed the intellectuals but did not immediately cause them to drop beliefs that were inculcated during their own education and still held by the people at large and by secular and religious authorities. Rather, such questioning caused intellectuals to attempt to reconcile their newly acquired scientific knowledge with the long-established religious outlook. The resolutions given by Descartes, Newton, and Leibniz, all intensely religious men and yet equally attached to the certitudes about the physical world incorporated in mathematical laws and to the truths of mathematics proper, are different and absorbing. Each attempt warrants lengthy study because a definitive reconciliation is not available even today. The articles that follow at least introduce us to the thoughts that these men advanced.

The student of history should also seek to learn the specific problems that the seventeenth-century scientists tackled and that motivated still more mathematical creations in later centuries. One of our articles, Boyer's "The Invention of Analytical Geometry," describes the particular problems that Descartes sought to solve. In his *La Géométrie,* Descartes emphasizes attacking the construction problems of Euclidean geometry through algebra. However, both Descartes and the cocreator of analytic or coordinate geometry, Pierre de Fermat, had far broader objectives in mind.

In the seventeenth century motion was the major study—the motions of the planets as viewed under the new heliocentric theory, the motions of projectiles, and the motion of light. The study of motion involves the curves that describe motions; many new curves were introduced and new questions raised about known curves, such as the conic sections. In particular, the seventeenth-century men sought quantitative knowledge about the curves, whereas the Greeks had been content with qualitative knowledge. What was sorely needed, then, was a more effective method of studying curves and a method that would supply quantitative knowledge. Analytic geometry supplied both needs. Did Descartes and Fermat concern themselves with methods of studying curves? Descartes, as Boyer points out, concealed much of his thinking for various reasons, and Fermat published very little. Hence we must infer their motivations from their works. Descartes was a great scientist, and though Fermat did less with science, he produced the greatest advance of his times in the study of light, known as the principle of least time. It is hardly conceivable that Descartes and Fermat in their work on analytic geometry should not have had the prevailing scientific needs in mind.

The creation of analytic geometry illustrates another interesting feature of the history of mathematics. There is no question that Fermat and Descartes worked independently of each other, and yet both men arrived at the same result—the use of algebra to represent and study the properties of curves. This feature, independent work by two or more people producing the same theorem or theory, is displayed so often in the history of mathematics that one can safely say that people respond to the problems of their times; the times direct their thinking. Of course the converse is also true. Once made, the creations often alter the times radically.

Independent and practically simultaneous creation of the same work has had an unfortunate consequence that mars the history of mathematics. Mathematicians seek glory as avidly as any other human beings, and many of them, upon noting the appearance of a rival creation, immediately set about to establish priority and often accuse their rivals of stealing the ideas. The history is full of such quarrels, in which some of the greatest mathematicians behaved disgracefully. Best known is the quarrel between Newton and Leibniz about priority in creating the major ideas of the calculus. Several of the books in the suggested readings relate details.

Still another feature of the men whose biographies are included here is that although they produced mathematics as great as any created before or after their time, they were not primarily mathematicians. Pascal, Descartes, and Leibniz were primarily philosophers. Newton was a physicist. That mathematics should nevertheless figure so prominently in their work and that they devoted so much of their time and energy to the creation of mathematics substantiate the assertions that serious thought about our world must take into account what mathematics has to offer and that mathematics is inextricably involved in fashioning our culture.

One of the reasons for the great interest in the people who made mathematics—surely some of the finest intellects that any society has produced—is that originality and depth of thought are more readily recognized in mathematics than in other fields. Hence any attempt to study intelligence and creativity in particular should embrace a study of the mentalities of the great mathematicians. One question to which we have no definitive answer is whether there is a special intelligence or talent for mathematics.

Our historical articles deal only with sixteenth- and seventeenth-century men. But the biographies of numerous later mathematicians are available (see the suggested readings) and provide a fund of material that can be explored to study this vital problem. A study of the lives and a defense of whatever conclusions one can draw from it would take us far beyond the goals of this book. But the question of what marks intelligence and mathematical fertility can well be kept in mind as you read the few biographies included here.

SUGGESTED READINGS

Andrade, E. N. da C. 1954. *Isaac Newton.* Macmillan, New York.
Bell, Eric T. No date. *Men of Mathematics.* Dover, New York.
Boyer, Carl B. 1956. *History of Analytic Geometry.* Scripta Mathematica, New York.
Boyer, Carl B. 1968. *History of Mathematics.* Wiley, New York.
Carr, Herbert W. 1960. *Leibniz.* Dover, New York.
Dantzig, Tobias. 1954. *Number, the Language of Science,* fourth edition. Macmillan, New York.
Mahoney, Michael S. 1973. *The Mathematical Career of Pierre de Fermat, 1601–1665.* Princeton University Press, Princeton, N.J.
Manuel, Frank E. 1968. *A Portrait of Isaac Newton.* Harvard University Press, Cambridge, Mass.
Mortimer, Ernest. 1959. *Blaise Pascal: The Life and Works of a Realist.* Harper, New York.
Ore, Oystein. 1953. *Cardano, the Gambling Scholar.* Princeton University Press, Princeton, N.J.
Reé, Jonathan. 1974. *Descartes.* Allen Lane, London.
Scott, Joseph F. 1952. *The Scientific Work of René Descartes.* Taylor and Francis, London.
Smith, Alan G. R. 1972. *Science and Society in the 16th and 17th Centuries.* Science History Publications, New York.
Struik, Dirk J. 1947. *A Concise History of Mathematics.* Dover, New York.
Vrooman, Jack E. 1970. *René Descartes, a Biography.* Putnam's, New York.

1 The Rhind Papyrus

by James R. Newman
August 1952

In 1700 B.C. an Egyptian scribe named A'h-mosè set down his "knowledge of existing things all," a document which is now the principal source of what we know of Egyptian mathematics

IN THE WINTER of 1858 a young Scottish antiquary named A. Henry Rhind, sojourning in Egypt for his health, purchased at Luxor a rather large papyrus said to have been found in the ruins of a small ancient building at Thebes. Rhind died of tuberculosis five years later, and his papyrus was acquired by the British Museum. The document was not intact; evidently it had originally been a roll nearly 18 feet long and 13 inches high, but it was broken into two parts, with certain portions missing. By one of those curious chances that sometimes occur in archaeology, several fragments of the missing section turned up half a century later in the deposits of the New York Historical Society. They had been obtained, along with a noted medical papyrus, by the collector Edwin Smith. The fragments cleared up some points essential for understanding the whole work.

The scroll was a practical handbook of Egyptian mathematics, written about 1700 B.C. Soon after its discovery several scholars satisfied themselves that it was an antiquity of first importance, no less, as D'Arcy Thompson later said, than "one of the ancient monuments of learning." It remains to this day our principal source of knowledge as to how the Egyptians counted, reckoned and measured.

The Rhind was indited by a scribe named A'h-mosè (another, more sonorous form of his name is Aāh-mes) under a certain Hyksos king who reigned "somewhere between 1788 and 1580 B.C." A'h-mosè, a modest man, introduces his script with the notice that he copied the text "in likeness to writings of old made in the time of the King of Upper [and Lower] Egypt, [Ne-ma] 'et-[Rê]." The older document to which he refers dates back to the 12th Dynasty, 1849-1801 B.C. But there the trail ends, for one cannot tell whether the writing from which A'h-mosè copied was itself a copy of an even earlier work. Nor is it clear for what sort of audience the papyrus was intended, which is to say we do not know whether "it was a great work or a minor one, a compendium for the scholar, a manual for the clerk, or even a lesson book for the schoolboy."

The Egyptians, it has been said, made no great contributions to mathematical knowledge. They were practical men, not much given to speculative or abstract inquiries. Dreamers, as Thompson suggests, were rare among them, and mathematics is nourished by dreamers—as it nourishes them. Egyptian mathematics nonetheless is not a subject whose importance the historian or student of cultural development can afford to disparage. And the Rhind Papyrus, though elementary, is a respectable mathematical accomplishment, proffering problems some of which the average intelligent man of the modern world—38 centuries more intelligent, perhaps, than A'h-mosè—would have trouble solving.

Scholars disagree as to A'h-mosè's mathematical competence. There are mistakes in his manuscript, and it is hard to say whether he put them there

THE PAPYRUS was originally a roll 13 inches high and almost 18 feet long. This photograph shows a small section of it about 4 inches high and 10 inches wide. Hieratic script reads from right to left and top to bottom.

or copied them from the older document. But he wrote a "fine bold hand" in hieratic, a cursive form of hieroglyphic; altogether it seems unlikely that he was merely an ignorant copyist.

IT WOULD BE misleading to describe the Rhind as a treatise. It is a collection of mathematical exercises and practical examples, worked out in a syncopated, sometimes cryptic style. The first section presents a table of the division of 2 by odd numbers—from 2/3 to 2/101. This conversion was necessary because the Egyptians could operate only with unit fractions and had therefore to reduce all others to this form. With the exception of 2/3, for which the Egyptians had a special symbol, every fraction had to be expressed as the sum of a series of fractions having 1 as the numerator. For example, the fraction 3/4 was written as 1/2, 1/4 (note they did not use the plus sign), and 2/61 was expressed as 1/40, 1/244, 1/488, 1/610.

It is remarkable that the Egyptians, who attained so much skill in their arithmetic manipulations, were unable to devise a fresh notation and less cumbersome methods. We are forced to realize how little we understand the circumstances of cultural advance: why societies move—or is it perhaps jump—from one orbit to another of intellectual energy, why the science of Egypt "ran its course on narrow lines" and adhered so rigidly to its clumsy rules. Unit fractions continued in use, side by side with improved methods, even among Greek mathematicians. Archimedes, for instance, wrote 1/2, 1/4 for 3/4, and Hero, 1/2, 1/17, 1/34, 1/51 for 31/51. Indeed, as late as the 17th century certain Russian documents are said to have expressed 1/96 as a "half-half-half-half-half-third."

The Rhind Papyrus contains some 85 problems, exhibiting the use of fractions, the solution of simple equations and progressions, the mensuration of areas and volumes. The problems enable us to form a pretty clear notion of what the Egyptians were able to do with numbers. Their arithmetic was essentially additive, meaning that they reduced multiplication and division, as children and electronic computers do, to repeated additions and subtractions. The only multiplier they used, with rare exceptions, was 2. They did larger multiplications by successive duplications. Multiplying 19 by 6, for example, the Egyptians would double 19, double the result and add the two products, thus:

	1	19
\	2	38
\	4	76
Total	6	114

The symbol \ is used to designate the sub-multipliers that add up to the total multiplier, in this case 6. The problem 23 times 27 would, in the Rhind, look like this:

\	1	27
\	2	54
\	4	108
	8	216
\	16	432
Total	23	621

In division the doubling process had to be combined with the use of fractions. One of the problems in the papyrus is "the making of loaves 9 for man 10," meaning the division of 9 loaves among 10 men. This problem is not carried out without pain. Recall that except for 2/3 the Egyptians had to reduce all fractions to sums of fractions with the numerator 1. The Rhind explains:

"The doing as it occurs: Make thou the multiplication 2/3 1/5 1/30 times 10.

	1	2/3	1/5	1/30
\	2	1 2/3	1/10	1/30
	4	3 1/2	1/10	
\	8	7 1/5		

Total loaves 9; it, this is."

In other words, if one adds the fractions obtained by the indicated multiplications (2 + 8 = 10), he arrives at 9. The reader, understandably, may find the demonstration baffling. For one thing, the actual working of the problem is not given. If 10 men are to share 9 loaves, each man, says A'h-mosè, is to get 2/3, 1/5, 1/30 (*i.e.*, 27/30) times 10 loaves; but we have no idea how the figure for each share was arrived at. The answer to the problem (27/30, or 9/10) is given first and then verified, not explained. It may be, in truth, that the author had nothing to explain, that the problem was solved by trial and error—as, it has been suggested, the Egyptians solved all their mathematical problems.

AN often discussed problem in the Rhind is: "Loaves 100 for man 5, 1/7 of the 3 above to man 2 those below. What is the difference of share?" Freely translated this reads: "Divide 100 loaves among 5 men in such a way that the shares received shall be in arithmetical progression and that 1/7 of the sum of the largest three shares shall be equal to the sum of the smallest two. What is the difference of the shares?" This is not as easy to answer as its predecessors, especially when no algebraic symbols or processes are used. The Egyptian method was that of "false position"—a mixture of trial and error and arithmetic proportion. Let us look at the solution in some detail:

"Do it thus: Make the difference of the shares 5 1/2. Then the amounts that the five men receive will be 23, 17 1/2, 12, 6 1/2, 1: total 60."

Now the assumed difference 5 1/2, as we shall see, turns out to be correct. It is the key to the solution. But how did the author come to this disingenuously "assumed" figure? Probably by trial and error. Arnold Buffum Chace, in his definitive study *The Rhind Papyrus*—from which I have borrowed shamelessly—proposes the following ingenious reconstruction of the operation:

Suppose, as a starter, that the difference between the shares were 1. Then the terms of the progression would be 1, 2, 3, 4, 5; the sum of the smallest two would be 3, and 1/7 of the largest three shares would be 1 5/7 (1 1/2, 1/7, 1/14 Egyptian style). The difference between the two groups (3 minus 1 5/7) would be 1 2/7, or 1 1/4, 1/28. Next, trying 2 as the difference between the successive shares, the progression would be 1, 3, 5, 7, 9. The sum of the two smallest terms would be 4; 1/7 of the three largest terms would be 3, and the difference between the two sides, 1. The experimenter might then begin to notice that for each increase of 1 in the assumed common difference, the inequality between the two sides was reduced by 1/4,

PART OF TITLE PAGE of the papyrus is reproduced in facsimile. Here the hieratic script reads from top to bottom and right to left. It has been translated: "Accurate reckoning of entering into things, knowledge of existing things all, mysteries . . . secrets all. Now was copied book this in year 33, month four of the inundation season [under the majesty of the] King of [Upper and] Lower Egypt, 'A-user-Rê', endowed with life, in likeness of writings of old made in the time of the King of Upper [and Lower] Egypt, [Ne-ma] 'et-Rê'. Lo the scribe A'h-mosè writes copy this."

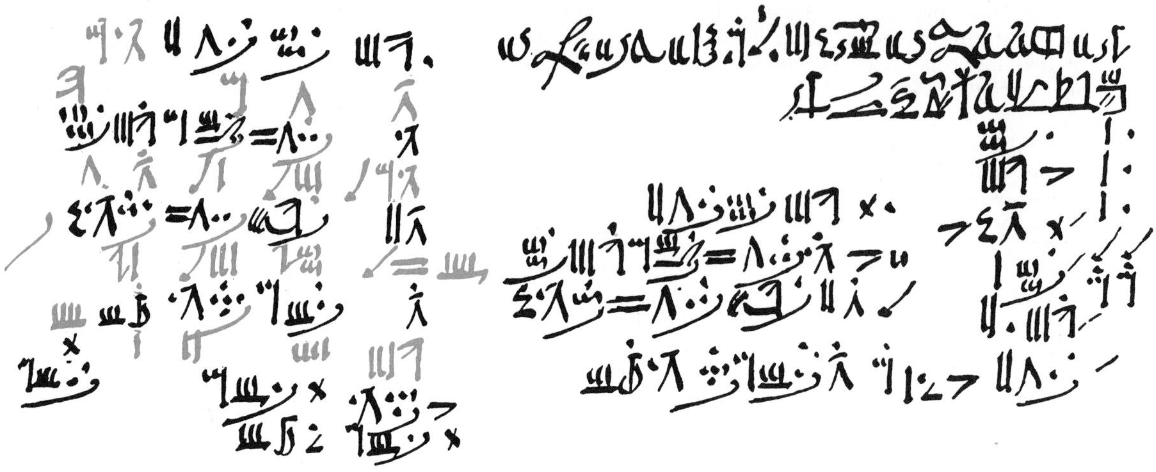

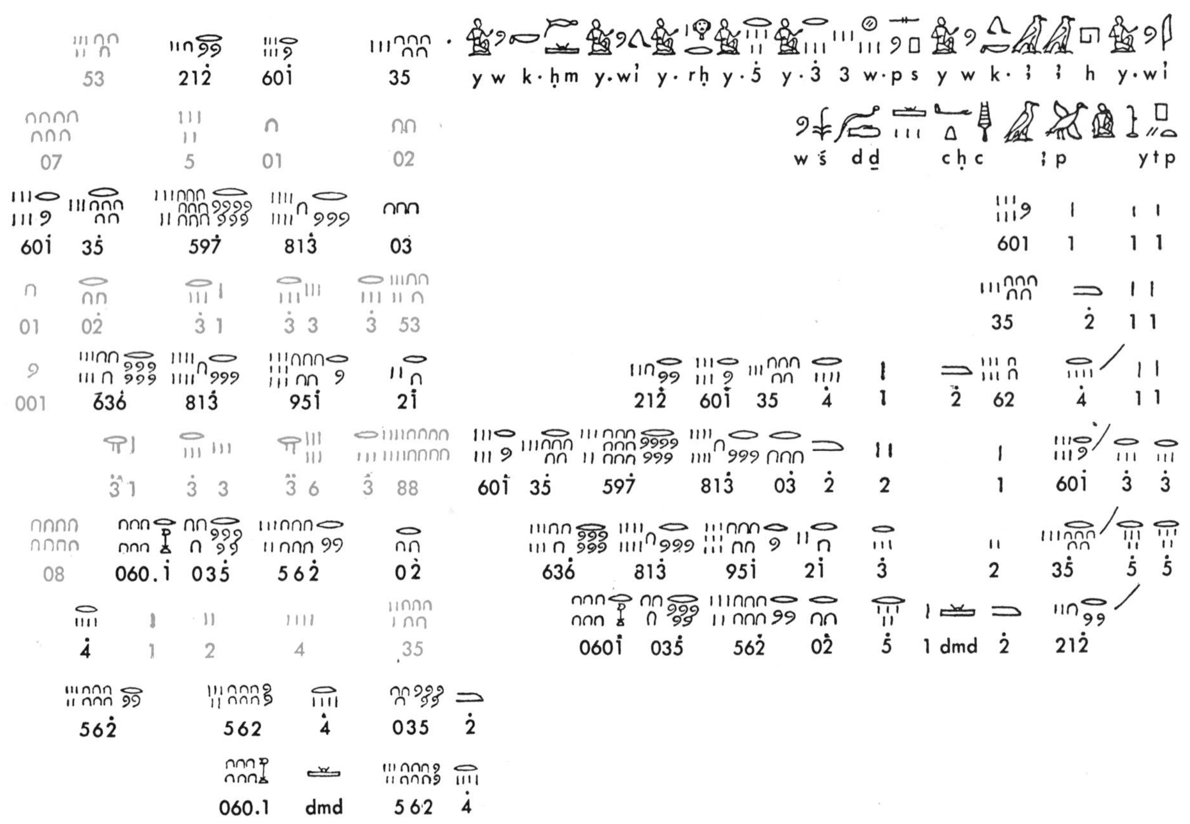

1 53 2̇12 60̇1 35 · y w k·ḥm y·wi y·rḥ y·5̇ y·3̇ 3 w·ps y w k·ꜣ ꜣ h y·wi

2 07 5 01 02 wś dd cḥc ꜣp ytp

3 60̇1 35̇ 59̇7 81̇3 03 60̇1 1 1 1

4 01 02̇ 3̇ 1 3̇ 3 3̇ 53 35 2̇ 1 1

5 001 6̇36 81̇3 95̇1 2̇1 21̇2 60̇1 35 4̇ 1 2̇ 62 4̇ 1 1

6 3̇1 3̇ 3 3̇ 6 3̇ 88 60̇1 35̇ 59̇7 81̇3 03̇ 2̇ 2 1 60̇1 3̇ 3̇

7 08 060.̇1 035̇ 562̇ 02̇ 6̇36 81̇3 95̇1 2̇1 3̇ 2 35̇ 5̇ 5̇

8 4̇ 1 2 4 35 060̇1 035̇ 562̇ 02̇ 5̇ 1 dmd 2̇ 21̇2

9 562̇ 562 4̇ 035 2̇

10 060.1 dmd 562 4̇

iw·y hꜣ·kwy sp·w 3 3̇ · y 5̇ · y ḥr· y iw · y mḥ · kwy pty pꜣ ꜥḥꜥ dd św

Go down I times 3, ⅓ of me, ⅕ of me is added to me; return I, filled am I. What is the quantity saying it?

1 1 1 4̇ 53 1̇06 2̇12
1 1 2 2̇ 30 3̇18 7̇95 5̇3 1̇06
1 1 3̇ 1̇2 1̇59 3̇18 6̇36
3̇ 3̇ 5̇ 20 2̇65 5̇30 1̇060
5̇ 5̇

1 106 53 1̇06 2̇12
2̇ 53 20 10 5 35
\ 4̇ 26 2̇ 30 3̇18 7̇95 5̇3 1̇06
\ 1̇06 1 35 3̇ 3 3̇ 1 3̇ 2̇0 10 70
\ 5̇3 2 1̇2 1̇59 3̇18 6̇36 2̇ 530
\ 2̇12 2̇ 88 3̇ 6 3̈ 3 3̇ 1 3̈ 100 4̇ 265
dmd 1 2̇0 2̇65 5̇30 1̇060 4̇ 265
Total 53 4 2 1 80 dmd 1060
4̇ 2̇65 Total

PROBLEM 36 of the papyrus begins: "Go down I times 3, 1/3 of me, 1/5 of me is added to me; return I, filled am I. What is the quantity saying it?" The problem is then solved by the Egyptian method. At the top of the opposite page is a facsimile of the problem as it appears in the papyrus. The hieratic script reads from right to left. The characters are reproduced in red and black, as they are written in the papyrus. In the middle of the page is a rendering in hieroglyphic script, which also reads from right to left. Beneath each line of hieroglyphs is a phonetic translation. The numbers are given in Arabic with the Egyptian notation. Each line of hieroglyphs and its translation is numbered to correspond to a line of the hieratic. At the bottom of the page the phonetic and numerical translation has been reversed to read from left to right. Beneath each phonetic expression is its English translation. A dot above a number indicates that it is a fraction with a numerator of one. Two dots above a 3 represent 2/3, the only Egyptian fraction with a numerator of more than one. Readers who would like to trace the entire solution are cautioned that the scribe made several mistakes that are preserved in the various translations.

1/28. Very well: to make the two sides equal, apparently he must multiply his increase 1 by as many times as 1/4, 1/28 is contained in 1 1/4, 1/28. That figure is 4 1/2. Added to the first assumed difference, 1, it gives 5 1/2 as the true common difference. "This process of reasoning is exactly in accordance with Egyptian methods," remarks Chace.

Having found the common difference, one must now determine whether the progression fulfills the second requirement of the problem: namely, that the number of loaves shall total 100. In other words, multiply the progression whose sum is 60 (see above) by a factor to convert it into 100; the factor, of course, is 1 2/3. This the papyrus does: "As many times as is necessary to multiply 60 to make 100, so many times must these terms be multiplied to make the true series." (Here we see the essence of the method of false position.) When multiplied by 1 2/3, 23 becomes 38 1/3, and the other shares, similarly, become 29 1/6, 20, 10 5/6 and 1 2/3. Thus one arrives at the prescribed division of the 100 loaves among 5 men.

THE AUTHOR of the papyrus computes the areas of triangles, trapezoids and rectangles and the volumes of cylinders and prisms, and of course the area of a circle. His geometrical results are even more impressive than his arithmetic solutions, though his methods, as far as one can tell, are quite unrelated to the discipline today called geometry. "A cylindrical granary of 9 diameter and height 6. What is the amount of grain that goes into it?" In solving this problem a rule is used for determining the area of a circle which comes to Area $= (8/9\,d)^2$, where d denotes the diameter. Matching this against the modern formula, Area $= \pi r^2$, gives a value for π of 3.16—a very close approximation to the correct value. The Rhind Papyrus gives the area of a triangle as 1/2 the base times the length of a line which may be the altitude of the triangle, but, on the other hand—Egyptologists are not sure—may be its side. In an isosceles triangle, tall and with a narrow base, the error resulting from using the side instead of the altitude in computing area would make little difference. The three triangle problems in the Rhind Papyrus involve triangles of this type, but it is clear that the author had only the haziest notion of what triangles were like. What he was thinking of was (as one expert conjectures) "a piece of land, of a certain width at one end and coming to a point, or at least narrower at the other end."

Egyptian geometry makes a very respectable impression if one considers the information derived not only from the Rhind but also from another Egyptian document known as the Moscow Papyrus and from lesser sources. Its attainments, besides those already mentioned, include the correct determination of the area of a hemisphere (some scholars, however, dispute this) and the formula for the volume of a truncated pyramid, $V = (h/3)\,(a^2 + ab + b^2)$, where a and b are the lengths of the side of the square and h is the height.

I should like to give one more example taken from the Rhind Papyrus, something by way of a historical oddity. Chace offers the following translation of the hard-to-translate Problem 79:

"Sum the geometrical progression of five terms, of which the first term is 7 and the multiplier 7.

"The sum according to the rule. Multiply 2801 by 7.

\ 1	2801
\ 2	5602
\ 4	11204
Total	19607

"The sum by addition

houses	7
cats	49
mice	343
spelt (wheat)	2401
hekat (half a peck)	16807
Total	19607"

This catalogue of miscellany provides a strange little prod to fancy. It has been interpreted thus: In each of 7 houses are 7 cats; each cat kills 7 mice; each mouse would have eaten 7 ears of spelt; each ear of spelt would have produced 7 hekat of grain. Query: How much grain is saved by the 7 houses' cats? (The author confounds us by not only giving the hekats of grain saved but by adding together the entire heterogeneous lot.) Observe the resemblance of this ancient puzzle to the 18th-century Mother Goose rhyme:

"As I was going to St. Ives
I met a man with seven wives.
Every wife had seven sacks,
Every sack had seven cats,
Every cat had seven kits.
Kits, cats, sacks and wives,
How many were there going
to St. Ives?"

(To this question, unlike the question in the papyrus, the correct answer is "one" or "none," depending on how it is interpreted.)

A CONSIDERABLE difference of opinion exists among students of ancient science as to the caliber of Egyptian mathematics. I am not impressed with the contention, based partly on comparison with the achievements of other ancient peoples, partly on the wisdom of hindsight, that the Egyptian contribution was negligible, that Egyptian mathematics was consistently primitive and clumsy. The Rhind Papyrus, though it demonstrates the inability of the Egyptians to generalize and their penchant for clinging to cumbersome calculating processes, proves that they were remarkably pertinacious in solving everyday problems of arithmetic and mensuration, that they were not devoid of imagination in contriving algebraic puzzles, and that they were uncommonly skillful in making do with the awkward methods they employed.

It seems to me that a sound appraisal of Egyptian mathematics depends upon a much broader and deeper understanding of human culture than either Egyptologists or historians of science are wont to recognize. As to the question how Egyptian mathematics compares with Babylonian or Mesopotamian or Greek mathematics, the answer is comparatively easy and comparatively unimportant. What is more to the point is to understand why the Egyptians produced their particular kind of mathematics, to what extent it offers a culture clue, how it can be related to their social and political institutions, to their religious beliefs, their economic practices, their habits of daily living. It is only in these terms that their mathematics can be judged fairly.

2 Cardano, The Gambling Scholar

by Oystein Ore

Reviewed by James R. Newman
June 1953

The fabulous life of Jerome Cardan, pioneer in the study of probability

THE LIFE of Jerome Cardan (Gerolamo Cardano) of Milan, physician, mathematician, astrologer, philosopher, gambler, is a melodrama too improbable even for the Italian Renaissance. His fortunes, from infancy to old age, fluctuated wildly, so that he was alternately poverty stricken and held in wretched contempt, or rich, influential and fervently admired. Untiring in his quest for knowledge, immensely industrious, restless in imagination and creativity, he was an incredibly prolific writer: his works fill 7,000 pages of folio, and no one knows how much more was lost. Cardan was an ambitious, dishonest, hot-tempered, quarrelsome, conceited and humorless man, but capable of generosity, kindliness and merciless self-revelation. During much of his life he suffered from ill health, physical and mental. His parents were an abominable pair; his favorite son was executed for murder; his other son was a scoundrel who managed to escape the gallows but brought Cardan nothing but unhappiness and disgrace. In his old age Cardan was stripped of his honors and high position and imprisoned for heresy; nevertheless he ended his days peacefully as a pensioner of the Pope. He was a genius, a fool and a charlatan who "embraced and amplified," as the English literary historian Henry Morley said, all the superstition of his age, and all its learning.

It is understandable that Cardan has not lacked biographers, yet none has succeeded in giving both a convincing portrait of the man and a dependable appraisal of his scientific achievements. Morley published in 1854 a biography of Cardan in the spacious two-volume Victorian tradition: moralistic but not unsympathetic, rich in details and background, amply stocked with quotations, elevated and flowing in style. It provides excellent reading, but it does not make its hero wholly believable nor is it the book to consult for an informed assessment of his labors in science. The less interesting but in some respects more disinterested appraisal written by W. G. Waters in 1898 also is unsatisfactory as regards Cardan's mathematical discoveries. The book I review here, by Oystein Ore of Yale University, is limited in scope, but within the area of its primary concern it is a first-rate contribution to the history of science.

Ore's account of Cardan's life is scarcely more than a pedestrian recital of the unpedestrian facts. His long analysis of Cardan's quarrel with the Italian mathematician Niccolò Tartaglia adds little to what has already been said on that subject. But the great merit of his book is in explaining Cardan's remarkable researches in probability. Here for the first time a mathematician has taken the trouble to disentangle and elucidate Cardan's obscure text on games of chance, a poorly printed work of 15 folio pages called *Liber de Ludo Aleae* (The Book on Games of Chance). Other biographers have scolded Cardan for being a gambler but have shown not the slightest comprehension of the fruits of his devotion to this naughty practice. Historians of mathematics also have failed to appreciate the full extent of his achievement in this sphere. Isaac Todhunter's *History of the Mathematical Theory of Probability* acknowledged Cardan's solution of certain simple problems of dice games but overlooked the principal insights of the *Liber de Ludo*. Moritz Cantor in his monumental *Geschichte der Mathematik* pointed out that Cardan was the first to answer correctly several basic questions in probability; but the examples Cantor gave were meager and unenlightening. Ore, with the help of S. H. Gould of Purdue University, a classical scholar trained in mathematics, has deciphered Cardan's little book and analyzed its contributions to the theory of probability—that

A contemporary picture of Cardan

great branch of mathematics, the "very guide of life," which owes its origins to the gaming table. Ore shows convincingly that Cardan's treatise formulated a number of fundamental probability principles, more than a century before Pascal and Fermat are supposed to have developed the theory of probability in their famous correspondence about the wagering problems of the French rake the Chevalier de Méré. Cardan's work gave the correct chances on dice, the so-called power law for the repetition of events (combined probabilities), a suggestion of the law of large numbers and several other important probability concepts. Ore's work in interpreting this first text on probability is of capital importance in the history of mathematics.

Cardan was born in 1501 in Pavia. His father was Fazio Cardano, a doctor in law and medicine and a geometrician. His mother was a widow, Chiara Michena, with three earlier children—a fat, dumpy woman with an ugly temper whose sole attraction for Fazio seems to have been that she was much younger than he. Fazio saw no reason to marry her, before or after Jerome's birth, but finally resigned himself to doing so shortly before his death at 80. Jerome's early years were filled with sickness and misery. His mother (whom he sentimentalized in his autobiography) spared no effort to make him feel unwanted, to remind him that he was an abiding source of shame and sorrow. When she grew tired of tormenting the child with words, she beat him—a sport in which her sister and Fazio were delighted to join. When Cardan became old enough to carry his father's bag through the streets of Milan, it was decided to give up beating him, to avoid any injury that might make it necessary to hire a servant to carry the bag.

At the age of 19 Jerome was permitted to enter the University of Pavia to study medicine. Fazio agreed to this step reluctantly only after Chiara, for once unaccountably supporting her son, raised a hideous commotion. Cardan completed his medical education at Padua and then applied for admission to the College of Physicians in Milan. When he was turned down—partly because he was a bastard, partly because his aggressiveness and other disagreeable traits had already gained him a crop of enemies—he settled as a country doctor in the little village of Sacco, a few miles outside Padua. It was a quiet, pleasant existence. "I gambled, played musical instruments, took walks, was of good cheer and studied only rarely. I had no pains, no fears . . . it was the springtime of my life." At 30 Cardan married Lucia Bandarini, the daughter of an innkeeper who was captain of the local Venetian militia. It was a reasonably happy union, although from his 20th to his 30th years Cardan had been impotent and convinced that he would remain so. Considering the character of the children he produced, continued impotence would not have been an unmitigated misfortune.

His writings during this period included an essay on palm-reading, a book called *The Method of Healing*, another on the plague, and two treatises, on "spittle" and on venereal diseases, which were destroyed, according to Morley's decorous euphemism, by "the misdeed of a cat." "*Hi libri corrupti sunt*," wrote Cardan, "*urina felis*."

Poverty forced Cardan to abandon country-doctoring and seek his fortune in Milan. At first matters went so badly that he and his family had to find shelter in a poorhouse; then suddenly his luck changed. He was appointed to a lectureship in mathematics once held by his father; he published several books, among them *The Practice of Arithmetic and Simple Mensuration* and *On the Bad Practice of Medicine in Common Use*; finally, with the help of prominent sponsors, he was elected a member of the College of Physicians.

Within a few years Cardan became the rector of the physicians' guild and the most prominent practitioner in Milan. "Before he was 50 years old he was second only to Vesalius among European physicians and was overwhelmed with flattering and magnificent offers for his services." Since a court physician was apt to keep his head only so long as the royal patient kept his health, Cardan generally stayed away from the highest places. But he could not resist a financially tempting call to go to Edinburgh to treat John Hamilton, the Archbishop of Scotland, who suffered from "suffocating attacks of asthma." This turned out to be a most profitable errand of mercy. After observing his patient carefully for a month, Cardan prescribed a sensible regimen of hygiene and the usual farrago of potions and ointments. In addition, he specifically recommended that the Archbishop substitute silk for feather mattresses and linen for leather pillow cases. The entire course of treatment would have made a modern allergist happy, and, more important, it actually made the Archbishop well. His health improved rapidly and in gratitude he lavished on Cardan an enormous fee of 1,800 gold crowns and various gifts.

On his journeys to and from Edinburgh Cardan passed through Lyons, Paris, London, Antwerp, Cologne and Strasbourg. In each of these cities the learned and the noble accorded him a festive reception. Henry II of France desired that he should "kiss hands" and accept court service, with a considerable pension. The 15-year-old Edward VI of England—whom Cardan described as "a marvelous boy"—requested his presence and conversed with him in polished Latin about various abstruse matters such as the motions of the stars and the causes of the rainbow. Edward also asked Cardan to calculate his horoscope. The great man was pleased to comply and prophesied, in some detail, a fairly long and sickly life for the delicate young monarch, including the specific prediction that at the age of 55 years, 3 months and 17 days "various diseases would fall to his lot." Edward was inconsiderate enough to die the following year, aged 16. Cardan was thereupon moved to write a dissertation wherein he explained (a) that he had been compelled to cast the horoscope against his better judgment, (b) that he had done so hastily and had therefore miscalculated, (c) that he had suspected the King had not long to live, (d) that the King had been poisoned.

Cardan was now at the summit of his fame as a physician, philosopher, astrologer and writer. His practice was prospering; he was in reasonably good health, and his favorite son, Giambatista, was completing his education for a medical degree. His books continued to pour from the presses at an incredible rate. In a single year, 1543, he published 53 separate works; altogether, as he himself reckoned, he published 131 books and left behind him in manuscript 111. He wrote on mathematics, astronomy, astrology, metoposcopy, physics, horoscopy, chess, gambling, the immortality of the soul, consolation, the uses of adversity, marvelous cures, dialectics, death, poisons, seven-month parturition, air, water, nourishment, dreams, urine, teeth, the plague, Galen, Hippocrates, Socrates, Nero, the Blessed Virgin, wisdom, morals and music. This list is representative, not inclusive. His popular scientific and philosophical books were widely read and many times reprinted, sometimes pirated by printers. Cardan's book on consolation was well known in England as the source of Hamlet's famous remarks on sleep and death.

The decline in Cardan's fortunes began with Giambatista's marriage to a young trollop named Brandonia Serono in December, 1557. The night before the messenger arrived with the bad news, Cardan had a premonition of evil: the house trembled "so that it was noticed even by the servants." He would not have the young couple in his house, but supported them financially. From the beginning Giambatista quarreled violently with his wife and her family. They exploited him shamelessly, and Brandonia bore three children in rapid succession, none of whom, she openly boasted, was genetically linked to her husband. Giambatista decided to settle the score by feeding his wife and several of her relatives a cake generously sprinkled with arsenic. The in-laws survived but Brandonia did not; nor did Giambatista. He was executed, after torture and the striking off of his left hand, on April 10, 1560.

Cardan never recovered from this

ghastly happening. "He reproached himself, wrote elegies to his son, brooded and relived the tragic events incessantly." The disgrace led him to relinquish his professorship at Milan, but in 1562 he yielded to his friends and took the chair in medicine at Bologna. Unfortunately he was accompanied by his son Aldo, who was, if anything, a scurvier specimen than his departed brother. Aldo gambled for outrageous stakes and managed to be jailed for sundry misdeeds no fewer than eight times within a couple of years. The father finally tired of rescuing him and had him banished from Bologna and its surrounding territories. But he continued to support Aldo and even provided for him in his will.

In 1570 Cardan was arrested and imprisoned for heresy. The causes of the indictment have never been adequately explained, but it is reasonable to suppose that he was a victim of the Counter-Reformation. He had expressed so many opinions on so many subjects that officials had no difficulty in proving that he was a heretic. Nevertheless, the inquisitors were disposed to be merciful: he was not tortured or required to recant publicly in the shirt of the penitent but was merely stricken from the university rolls and forbidden to lecture and to publish any more books. He journeyed to Rome seeking leniency. With the help of influential friends he gained membership in the College of Physicians and was permitted a limited consulting practice. The Pope, strangely enough, granted him a pension. Together with the income from his remaining personal means, it enabled him to live in comparative comfort. It was not to be expected that the interdict against publication would stop Cardan from writing. His last manuscript was an autobiography, *De Propria Vita,* in which he combined apologies, boasts, confessions, "sorrows and joys, successes and failures." Cardan died September 20, 1576, and was buried in the Church of the Augustins in San Marco.

Cardan's fame rests mainly on his mathematical inventions, and this is what Ore reassesses. His *Ars Magna,* a pioneer work on algebra, was published within two years of two other history-making books in science: Nicolaus Copernicus' *De Revolutionibus Orbium Coelestium* and Andreas Vesalius' *De Humani Corporis Fabrica.* The *Ars Magna* set forth the theory of equations, including the solution of the cubic and the biquadratic, not previously known. It established the author as a creative mathematician of the first rank. Yet Cardan is entitled to small credit for its two major innovations in the "cossick art," as algebra was long called in Europe—from the Italian word *cosa,* for "thing" or the unknown quantity x. The method of solving a biquadratic equation was discovered by a pupil of Cardan named Ferrari, and the solution of cubic equations by the unhappy stutterer Tartaglia.

The quarrel between Tartaglia and Cardan is one of the great Italian operas, rich in ridiculous and tragic interludes. In 1539 Cardan sent his bookseller to Venice to ask Tartaglia to disclose his rule for the solution of the cubic equation $x^3+bx=c$. Tartaglia refused. Cardan responded with an insulting letter. But after several furious, baiting exchanges, Tartaglia allowed himself to give up his precious discovery. He went to Milan and there extracted from Cardan a most solemn oath not to publish or divulge the secret; Cardan swore that he would write down Tartaglia's discovery in cipher so that even after his death no one could understand it. Having obtained the secret, Cardan proceeded to publish it in the *Ars Magna.* There were mitigating circumstances. For one thing, Cardan's book acknowledged the value of Tartaglia's work. For another, Cardan learned that Tartaglia's discovery had been anticipated by the Italian mathematician Scipio Del Ferro about 1515. Furthermore, Cardan carried the cubic problem well beyond Tartaglia's rule: he solved the third degree equation in its most general form where all the powers of the unknown are present; he used imaginary numbers to extract all the roots; he employed approximations in his solution and exhibited a grasp of the relations between the roots and the coefficients of the equation. These circumstances do not excuse his offense, but they lighten its hue.

Tartaglia did not take Cardan's betrayal lightly. The feud was long and acrid, marked by many challenges, appeals to the public conscience and mutual denunciations. Mathematics was a serious business in those days. Cardan delegated to his brilliant and hot-headed pupil Ferrari the responsibility for upholding his colors. The skill and ferocity of the pupil exceeded the master's expectations. Not only as a mathematician but as a name-caller Ferrari was unexcelled. The final act was a public mathematical joust between Tartaglia and Ferrari. The older man lost his temper and was defeated. The contest proved nothing, but it was a sad ending to an absurd dispute. Tartaglia is one of the outstanding victims of secrecy in science, a secrecy which he himself had imposed.

Cardan's treatise on probability did not appear in print until the publication of a 10-volume edition of his complete works in 1663, and even then it attracted scant attention. Yet the mathematical sections of the *Liber de Ludo* are perhaps even better proof of Cardan's creative abilities than his considerable achievements in algebra; moreover, here the originality of his ideas is without shadow. The *De Ludo* is first of all a gambling manual, based on the experiences of a veteran gambler. There are chapters on false dice, marked cards, strategic kibitzing, the card game Primera (resembling poker) and the board game Fritilla (resembling backgammon), and many other interesting items. Cardan admits gambling to be an evil but calls it "a natural evil." The law recognizes that it is a solace as well as an evil and permits gambling to those who are sick, "in prison and condemned to death." Cardan, giving advice on play, is skeptical of most gamblers' superstitions, but suggests that it is advisable to sit facing the rising moon.

In the technical sections of the book he discusses equiprobability, mathematical expectation (the correct amount to bet when a player has a probability p of winning some amount P), reasoning on the mean, frequency tables for dice probabilities, additive properties of probabilities and what a century later came to be known as "de Méré's problem": *i.e.,* how many trials are required to give a player an even chance to win, for example, to throw a given point in dice. In some of his investigations Cardan sets off in the wrong direction, arrives at an incorrect result but then a few pages later discovers his mistake and corrects it. An excellent example of this curious procedure is his treatment of compound probabilities. After several false starts he demonstrates his mathematical ability by establishing the power formula $p_n=p^n$ for obtaining n successes in n independent trials. Cardan did not have enough mathematics to express the so-called law of large numbers in a formula, but used the law in the following sense: "When the probability for an event is p, then by a large number n of repetitions the number of times it will occur does not lie far from the value $m=np$."

Even without his work in algebra, the *Liber de Ludo Aleae* would assure Cardan of a place in the company of the great mathematicians. The situation has its ironic side. Of the mountain of scribblings poor Cardan left, it turns out that this tiny gamblers' manual, which he did not consider worth publishing and a large portion of which he burned in manuscript, is his firmest claim to immortality. He said in his autobiography: "I gambled . . . at chess more than forty years, at dice about twenty-five; and not only every year, but—I say it with shame—every day, and with the loss at once of thought, of substance, and of time." Yet it is for a by-product of these misspent hours that Cardan deserves to be remembered.

Blaise Pascal: The Life and Work of a Realist

by Ernest Mortimer

Reviewed by James R. Newman
December 1959

The tortured life of Blaise Pascal, one of the great minds of all time

Did religion consume Blaise Pascal? Did it lead him to immolate the intellect, to forsake reason for faith? He stands out for his achievements even in the century of genius in which he lived; yet how much more, one wonders, might he have accomplished had he not wasted himself on theological sterilities and on religious quarrels? In one of Nietzsche's eloquent polemics against Christianity it is Pascal whom he holds up as a tragic victim of the ravages of religion: "What is it that we combat in Christianity? That it aims at destroying the strong, at breaking their spirit, at exploiting their moments of weariness and debility, at converting their proud assurance into anxiety and conscience-trouble; that it knows how to poison the noblest instincts and to infect them with disease, until their strength, their will to power, turns inwards, against themselves—until the strong perish through their excessive self-contempt and self-immolation: that gruesome way of perishing, of which Pascal is the most famous example."

No doubt Pascal's struggle with himself, his attempt to reconcile religious faith with the spirit of geometry, and his final justification of religion on a nonintellectual basis cost him dear. He was not a saint by nature. He had ferocious energy and a fiery temper (his sister Jacqueline tells us he had *une humeur bouillante*); he was a man of passion in his arguments, his friendships, his scientific exertions, his religious beliefs. He had a passion even for self-torture. But there are other facts which cast doubt on the opinion that Pascal sacrificed his magnificent creative powers to his God. They are set forth in this lucid, insightful biography by Ernest Mortimer, an English clergyman.

The subtitle of Mortimer's book is "The Life and Work of a Realist," and he endows the word realist, as a British reviewer has said, with its full breadth of meaning. Two of his major points should be mentioned at the outset. The first is that it is simply untrue that Pascal abjured science and society after his conversion in 1654. Over the years his religious interests deepened, but until the very end of his life he clung to his "free-thinking friends" and retained his scientific curiosity. The second point is that Pascal held facts to be no less sacred than faith. He believed in sense-data as a source of knowledge and truth; he recognized that the contradictory impressions of nature are no excuse for mystery-mongering; he respected the power of the intellect and reason to enlarge the "gates of perception" upon the universe. In all this he was a modern thinker. He was no less modern, it can be argued, in believing that the emotions are another gate of perception. It was he who said that the heart has its reasons which reason knows nothing of; modern psychology and psychoanalysis have also said it, but not so well.

Pascal was born at Clermont in Auvergne on June 19, 1623. His father, Etienne Pascal, was a well-to-do high public official (a judge and tax commissioner) who was to rise even higher. Blaise's mother, who came of a family of prosperous merchants, died at the age of 30, leaving him, aged three, an elder sister Gilberte and a younger sister Jacqueline.

The boy was sickly and precocious. The nature of his illness is not fully known (a witch was said to have been partly responsible), but an anatomical anomaly discovered after his death was probably a contributing factor. Of the two apertures in a baby's skull that should knit in the early weeks of life, one had never properly closed, and the other had clamped or overlapped so as to form a bony ridge. Pascal suffered from headaches all his life. His precocity served him better. His family was bound close by natural affection and by bereavement; they loved him and recognized his exceptional powers. Blaise was "encouraged rather than forbidden to stare."

On the death of his wife Etienne Pascal resolved to be father, mother and tutor to his children. They had their governess, but he alone gave them their formal education. His instruction, says Mortimer, was remarkable for its paternal self-denial, its eccentricity, rigor and patience; also for its fruits. He scorned "the sort of pedagogy which he had suffered himself"; he had his own theories of teaching and did not hesitate to practice them. His principal maxim, Gilberte tells us, "was to hold the child always above his work," not to rush him or overload him. Instruction was by easy conversation. The object was to quicken the child's interest, to let natural curiosity about language, about the things around him guide the direction and emphasis. It was time enough for a child to learn Latin when he was 12, when he could do it more easily. "By good fortune," Pascal wrote to a friend in later years, "for which I cannot be too grateful, I was taught on a peculiar plan and with more than fatherly care."

Etienne had a bent for mathematics, and decided to make that subject "the coping stone of his plan." Not until he was 16, after he had learned Latin, Greek, history and geography, was Blaise to be introduced to geometry. Meanwhile books on mathematics were not allowed in the house and the subject was not discussed in the daily conversational instructions. But the boy had secretly made his own way. One day when he was 12 his father came upon him surrounded with diagrams, "so absorbed that he still thought himself alone." He had been trying to work out for himself the principles of geometry, calling straight lines "bars" and circles "rounds," and he was now trying to prove that the three angles of a triangle add up to two right angles. (Another version of the story is that he had rediscovered the whole of Euclid up to the 32nd proposition of the first book, but there is no evidence that this is so.) Etienne forgot his theoretical scruples and was overcome with pride and joy.

Admirable though it was, Blaise's education had its gaps. History and literature were sketchily imparted; the natural sciences even more sketchily. Fortunately Blaise was not overburdened with theology; for the time being elementary religious teaching was enough. But he was soon to come out into the world and to glimpse new horizons of learning, including both science and theology.

When he was eight, the family had moved to Paris. Etienne had rented a house in the Rue de Tisseranderie (which, it is delightful to learn, was intersected by the Street of the Two Doors, the Street of the Bad Boys, Cock Street and the Street of the Devilish Wind) and had entered into the life of this marvelous city. He now had few professional duties and could give even more time to the education of his children. He was a sociable man, with a taste and talent for natural science, and he found friends with whom he could share his interests. Social groups in which the arts and sciences were discussed seriously "but outside academic settings" were a feature of the time. He was close to Guillaume Montory, the leading actor of the day, and to Charles Dalibray, the poet; the fashionable circles of literature and art were open to him. More important was his membership in the Abbé Bourdelot's group, which included, among others, the aging but still lively mathematician and physicist Père Mersenne ("a bold, capacious, untidy and extraordinarily vivacious mind"), who managed to reconcile his support both of Galileo and orthodox theology; the gifted but prosaic mathematician Gilles Personne de Roberval (who, it is told, saw a performance of *Le Cid* and complained that he could not see what it proved); the brilliant young engineer Pierre Petit; and Gérard Desargues, who invented projective geometry and introduced the method of perspective. Blaise was allowed, while still a child, to accompany his father to the séances of this group and was even encouraged to utter his own ideas. He was much influenced by Desargues, not only in mathematics but also in a concern, which the older man felt deeply, for improving the education and lightening the labors of plain workmen.

Pascal's first contribution to mathematics, composed at the age of 16, was his celebrated theorem. He had read Desargues' writings and had recognized as no one else had the worth of Desargues' method. His little *Essay on Conic Sections*, which acknowledged his debt to Desargues, described a property common to all conic curves that is invariant under projection and section: If any six points on the curve are joined by straight lines and the sides of the resulting hexagon are prolonged beyond the curve, the three pairs of opposite sides will intersect at three points that lie on one straight line.

The theorem was widely praised, but not by the great Descartes. In reply to Mersenne's enthusiastic letter he wrote: "He seems to have copied Desargues. I cannot pretend to be interested in the work of a boy." It is true Desargues' treatise suggested the property, but Pascal "isolated and proved it." It was a beautiful piece of work in its own right, and not merely a feat of precocity.

And now Pascal was no longer a child. The eight years in Paris had been "intensely formative." He was mastering Latin, beginning Greek, "absorbing paternal disquisitions on philosophy at mealtimes." He had met leading scientific thinkers; he had had his first taste of *la vie mondaine*. His energy matched the fecundity of his intellect; it is said, for example, that he followed up his theorem on conics with some 400 corollaries. Mathematics was unquestionably his great passion; he put his whole spirit into it. He reserved mathematics, to be sure, for his leisure moments; because it came easy to him he may have regarded it as a self-indulgence. But like the young Bertrand Russell he hoped to find in mathematics the solace and perfection of eternal truth.

In 1640 the Pascals moved to Rouen, Etienne having through the favor of Cardinal Richelieu been awarded the important but unpopular post of tax assessor for that city. There they remained for seven years. For Blaise it was a period of "much hard work, a religious awakening and the beginnings of a profound spiritual dilemma."

Etienne's job was backbreaking; in addition to various administrative chores he had the task of reassessing the taxes of 18 townships, which meant, among other things, copying endless columns of figures. "I never get to bed before two o'clock," he wrote Gilberte. The spectacle of his father's drudgery put an idea into Blaise's head. Why not make a machine to do the donkey calculations? During the first year at Rouen he conceived a mechanism that would perform all the operations of arithmetic; it would have a device for carrying digits from one column to the next, and another for recording the result. With the help of local craftsmen he worked on the idea. Some 50 experimental models were constructed over five years, and in 1645 he had his machine. It was the size of a glove box, simple in appearance, portable and workable. (It is believed that this very machine is now in private hands in the south of France.) Pascal hoped to get rich on the invention and staked a claim for a patent; by 1652 a standard model was in production and was placed on sale at 100 livres. His hopes were not realized. The manufacturing costs were too high, and the invention of logarithms cut the demand for the machine. Yet it was the first of all computers; as Gilberte writes in her life of her brother: "He reduced to mechanism a science which is wholly in the human mind." Now Blaise began to be referred to as *le grand M. Pascal.*

His concern with religion was growing, but it was still far from pervading his thoughts. The drama and poetry of the day did not appeal to him. He disliked and distrusted their emotional influence; he found them artificial and stilted. He read in philosophy and was attracted to the ancient Stoic writings and to Epictetus. What he admired was the value they placed on fortitude and duty, the high notion that man's fate is in his own power, that no outside force can crush him. But it was a "fatal flaw," he believed, to make man the complete master, to leave no place for God. Stoicism was too self-assured, too filled with self-pride, too apt "to greet the unseen with a sneer."

When his computer was finished, a new interest came to Pascal. On a visit to Rouen, Pierre Petit brought news of experiments in Florence two years before in which Evangelista Torricelli had demonstrated that we live at the bottom of a sea of air and that air has weight. Torricelli had taken a glass tube closed at one end, filled it with mercury and covered the open end with his finger; when he inverted the tube, placed the open end in a bowl of mercury and removed his finger, the liquid in the tube did not fall into the bowl but sank only part of the way down the tube and remained mysteriously suspended. "I assert," he wrote, "that the force [holding up the mercury in the tube] comes from without [and that the mercury stands] in a column high enough to make equilibrium with the weight of the external air which forces it up."

This was a remarkable and highly controversial conclusion because it contradicted what Aristotle, the Scholastics and Descartes had said about the impossibility of a vacuum. When the mercury that had originally filled it sank in the tube, what remained in the vacant space? Nothing, said Torricelli; but, said Descartes, nothing can be filled with Nothing. If any portion of the tube were truly empty, it would instantaneously collapse. Pascal, excited by Torricelli's discovery, decided to make his own experiments to extend it. Rouen had the best glassworks in Europe; Etienne had money to pay for the research; Petit had the engineering experience; Blaise had the enthusiasm. With glass tubes of different lengths, breadths and shapes (two were 46 feet long and were bound to ship's masts to strengthen them) he was able to show that the space vacated by the falling liquid was indeed a void. Neither rarefied air nor mercury vapor, as the "plenists" desperately asserted, was to be found in the vacant space. He varied the apparatus; he invented the plunger-type syringe to serve as a vacuum pump; he gave sensational public demonstrations with five-story-high siphons before 500 of the city's Eminences. All his ingenious experiments gave a consistent result: For a given liquid the height of the column in the tube was always the same; it was unaffected by the space at the top; it was not being sucked up from above but must be pressed up from below and from outside.

He now had in mind the writing of a treatise on the vacuum, and for this purpose other experiments were needed. Momentarily, however, he was diverted by a religious crisis.

It was occasioned, as Mortimer tells us, by an accident to Pascal's father. On a night in January, 1646, Etienne hastened out of doors to prevent a duel, slipped on ice and dislocated his hip. First aid was given by two "reformed characters," brothers named Deschamps who had been converted by a pious curé of a Rouen suburb and were devoting their lives to good works, "especially the good work of bone-setting, at which they were adept." The example of their charity and fervor awakened religious feelings in the whole Pascal family. Each was deeply stirred, and each responded in character. Etienne decided the time had come to look to his spiritual progress; Gilberte and her husband became "consistently devout"; Jacqueline moved closer to the decision of giving herself entirely to the religious life. Blaise's response was "more enigmatic." He immersed himself in theological tracts and began serious study of the Bible. His letters begin to show that he was "aflame with religious zeal." "Corporeal things," he wrote Gilberte, "are nothing but an image of the spiritual, and God has represented the invisible in the visible." Undoubtedly he was in conflict with himself over his scientific interests. They were "morally dangerous" in the eyes of the Deschamps brothers, "as leading to intellectual pride and distracting the attention from the quest of salvation" and of dubious value. But the realist and the lover of natural philosophy could not bring himself to renounce his experiments. He had undergone a crisis, and had turned in a direction that he would never abandon, yet the years 1648 to 1654 were to be his most fruitful and triumphant period of scientific discovery.

In 1647 he fell seriously ill, the diagnosis being "overwork." He was medically advised to avoid all intellectual exertion and to seek relaxation in society. (His physician was obviously a realist.) He retired to Paris with Jacqueline as his housekeeper and secretary. Soon he was back at his experiments and working on his *New Experiments Touching on Vacuum*. Descartes visited him in Paris and they discussed Torricelli's work. A couple of months later Pascal arranged to have his brother-in-law conduct an experiment on the Puy-de-Dôme, a mountain near Clermont, to determine whether a column of mercury would stand higher at the base of the mountain than at the summit. The results were conclusive, and the way was open for hydrostatics and hydrodynamics.

He sketched two treatises: *The Great Experiment on the Equilibrium of Liquids* and *On the Weight and Mass of Air*, neither of which was published till after his death. The pressure of the air was, he saw, a clue to a more general law of the behavior of "liquids" (including gases). His experiments led him to a startling principle. Pressure, it appeared, exerted at any place on a fluid in a closed vessel is transmitted undiminished throughout the fluid and acts at right angles to all surfaces. Suppose two tanks of equal size are filled with water and joined by a pipe at the base; obviously the water in the tanks will be in equilibrium, like equal weights in the pans of a scale. But suppose one tank is 100 times larger than the other? The water will still be in equilibrium. As Pascal explained, by putting into each tank a piston that fits it exactly, "a man pushing on the small piston will exert a force equal to that of one hundred men who are pushing on the piston which is one hundred times as large. . . . It may be added, for greater clearness, that the water is under equal pressure beneath the two pistons; for if one of them has one hundred times more weight than the other, it also touches one hundred times as many parts of the liquid." From this principle, which leads to anomalies and paradoxes that have puzzled students for years, came the hydraulic press.

"Nothing that has to do with faith can be the concern of the reason." This was the principle Blaise had learned in his boyhood. Some might regard it as even more anomalous than the principle of the behavior of fluids, but he lived by it and worked it into greater fullness. He held his religious beliefs with a growing fervor, but he refused to be "stampeded into a flight from reason." In one of the most famous of his *Pensées* he says: "It is the way of God, who does all things gently, to put religion into the mind by reason, and into the heart by grace."

The period from 1648 to 1653 was one of uncertainty and inner struggle in Pascal's life. His father died, his beloved sister Jacqueline, who had been his closest companion, finally renounced the outer world and entered a convent at Port-Royal. He contemplated marriage, but not very seriously. Bleak and lonely though his life had suddenly become, he was better off single; like Charles Lamb he was one of "nature's bachelors." He did not, however, cut himself off from the salons of Paris society. He was a man of science, but he was also a man of fashion and ambition, besides which he enjoyed "savoring human types." He formed an important friendship with a great nobleman who liked science better than women, the young Duc de Roannez; he lectured to a brilliant assembly in the drawing room of the Duchesse d'Aiguillon, Richelieu's niece, demonstrating his calculating machine and speaking on recent advances in hydrostatics; he sent Queen Christina of Sweden—who had recently finished off poor Descartes by requiring him to expound science and philosophy to her on cold mornings before breakfast—a gift of one of his calculating machines. This was accompanied by a celebrated letter which, though suitably gracious in tone, said in effect: "Your Majesty is a very great person by virtue of your sovereign rank. As we may agree, I am a still greater person by virtue of my sovereign intellect."

His social relaxations did not prevent him from reading and writing. At the very height of this period it is likely that he produced, in addition to scientific essays, a large part of the marvelous *Pensées*, which was not published until after his death. He was also immersed in Montaigne's *Essays*, which filled him with contradictory emotions. He cherished Montaigne's honesty, his hatred of cruelty, his scorn of dogmatism and love of intellectual freedom, his benevolent interest in human nature, "the stoical tranquility he practised and preached in an evil time"; and he admired the incomparable vitality of his style. But the *Essays* also shocked and horrified him. Montaigne had an ironical and detached view of religion. If religion was sublime, it was also celestial; it was heaven's affair rather than man's. Why should one imagine that municipal laws are the laws of the universe? Man is miserably weak, supremely foolish, moved by base passions and vanity. Wisdom and justice are unknown to him. His sages and philosophers are chatterers. He can rely neither upon his senses nor upon his sense. There was no vast difference, says Mortimer, between Montaigne's outlook and Hobbes's view that "the life of man is solitary, poor, nasty, brutish and short." Man, Montaigne said, is the prey of trivial chances. He wrote (in a translation of the time): "A gust of contrarie winds, the croking of a flight of Ravens, the false pase of a Horse, the casual flight of an Eagle, a dreame, a sodaine voyce, a false signe, a mornings mist, an evening fogge, are enough to overthrow, sufficient to overwhelme and pull him to the ground." And if it be urged in mitigation that, despite man's wretched condition and contemptible character, he has religion, the reply is yes, he has the religion that chance flung in his way: "Man cannot make a worm, but he makes Gods by the dozen."

In Sainte-Beuve's phrase, Montaigne became anchored in the soul of Pascal. He would not follow where Montaigne beckoned, but he could not turn away from him. This dilemma, centered on one man, was the epitome of the crisis of Pascal's entire outlook, of his doubting faith, his troubled convictions, his fluctuating commitments.

In 1653 Pascal made a trip to Poitou with the Duc de Roannez. One of the party was the Chevalier de Méré, a member of the great house of Condé, a soldier, scholar, skeptic and "cultivated libertine." To this man the world owes a debt made up of strangely incongruous parts. Méré was an elegant and skillful writer; Pascal's style, says Mortimer, attained "its complete freedom from affectation and verbiage" from Méré's teaching. The other part of the debt is more intriguing and important. Méré was a gambler, and the occasion arose, shortly after the journey to Poitou, for him to ask Pascal to solve two problems of practical use to gamblers. Pascal had for some months been disaffected and weary, preoccupied with his soul. On his return to Paris, however, he began to settle his estate, to buy property, to renew his efforts to sell his arithmetical machine, and before long his scientific interests, touched off by Méré, "awoke from a long sleep." The first problem was: "When one plays with two dice, what is the minimum number of throws on which one can advantageously bet that a double six will turn up?" This was easy for Pascal; 24 throws would be a bad bet; 25, a good one. The other question was harder.

Two players have agreed that the stakes will go to the one who first wins three games. Before this happens their play is interrupted. How, under different circumstances, are the stakes to be divided? Pascal found an ingenious and simple answer. Suppose A has won two games and B one, and the stake is 16 pistoles. In that case A should receive 12 pistoles and B four. Pascal reasoned as follows: Assume that one more game could be played. If A won it, he would be entitled to the full 16 pistoles; if B won it, at that point he could claim half the stakes. So when only three games have been played, A can say to B: "Win or lose the next game, I have at least eight pistoles due me. That leaves the other eight for the next game, in which the chances are equal, so if we cannot play the next game, we divide equally, and I get four." Pascal worked out the other cases. He sent his solution to Pierre de Fermat in Toulouse, who had got the same results by algebraic methods. Pascal was very pleased. "I see," he wrote, "that the truth is the same in Toulouse and in Paris." The independent work of both men laid the basis of the mathematical theory of probability. (Girolamo Cardano had anticipated some of their results in the preceding century, but his book on the subject lay unnoticed.)

Pascal's success in solving Méré's problems stimulated him to further study of the mathematical theory of chance. One of his brilliant ideas was the arithmetical triangle, a capsule of the calculus of probabilities. Pascal flung himself into this work with immense energy; it was his way when his interest had been roused to tackle a subject—physics, theology, mathematics, even business—in a frenzy, as if nothing else mattered, as if time were running out. His program was unbelievably ambitious. He proposed to write treatises on the theory of numbers, on the equilibrium of liquids, on the arithmetical triangle; to write papers on magic squares, on circles, on conic sections, on perspective; and "to reduce to an exact art, with the rigour of mathematical demonstration, the incertitude of chance, thus creating a new science which could justly claim the stupefying title: the geometry of hazard."

Of this fantastic plan only a small part came to fruition. Its grandiosity marked his desperation. He had to keep going, to drive himself, whether in high society or high science, so as not to be alone with his dark thoughts. He was "a made man and a celebrity," yet he was a tormented man. He doubted his God; he felt empty and lost.

On November 23, 1654, he experienced another intense religious episode, "a timeless eternal moment." Immediately after his vision he wrote at headlong speed an account which, together with a parchment copy, was found after his death, sewn into his doublet. His life from then on was changed. Prayer and sacred study were his main concerns. He never again attached his name to any of his writings ("an evident discipline against vanity"); he withdrew more and more from society. His health began to deteriorate. For a brief period he entered a retreat of the Port-Royalists. Doctrinal questions fevered him. We are "tied and bound with the chain of our sins," but does this mean there is no hope, that the truth is beyond us? He thought not; there is a function for the human will, a way to salvation; it is not as prideful as the way of Epictetus, nor as limited as the way of Montaigne. Out of the quarrel between the assignment of moral power and moral weakness Pascal drew a synthesis: "Faith teaches us to assign these two inclinations to different things: infirmity to human nature, but power to grace."

I shall not follow him in his career as a pamphleteer. His famous anti-Jesuit *Provincial Letters* are a mixture of tedious theological claptrap, moral grandeur, humor and irony. They are masterpieces of polemical literature; even Voltaire, who despised everything about Pascal, found the wit of the letters a match for Molière's finest comedies. The *Provincial Letters*, the background of the dispute that occasioned them, the history of Port-Royal, and Pascal's relation to

Jansenism are all well described in Mortimer's excellent book.

In the years that remained, Pascal neither entirely abandoned science nor renounced fashionable society, though religious questions were uppermost in his thoughts. He wrote his *Spirit of Geometry*, a philosophical essay that, in Mortimer's view, is Pascal's equivalent of Descartes' *Discourse on Method*. He greatly advanced knowledge of the mathematics of the cycloid. One night, awakened by a violent toothache, there came, as his sister Gilberte tells it, "uninvited into his mind some thought on the problem of the roulette [*i.e.*, cycloid]." A whole crowd of thoughts followed. He determined the area of a section produced by any line parallel to the base, the volume generated by it revolving around its base, and the centers of gravity of these volumes. Anonymously he sent out a challenge to all mathematicians to solve these problems; entries came from Christiaan Huygens, Christopher Wren, John Wallis and others. No one matched Pascal's solutions which, when published, caused a sensation, not only as an intellectual feat but also because he stirred up old and bitter scientific quarrels. He did not attach his name to the publications, but there was no doubt who was the author.

Now at last—this was 1659—he "bade a final farewell to the glories and quarrels of science." This may be the moment "at which he began to wear next his skin an iron belt with small spikes which he pressed at any temptation to pride." As his health deteriorated he became a solitary and an ascetic. He gave up his carriage and horses, his tapestries, even his books—"except the Bible . . . and (surely) Montaigne." He found it difficult to eat and could take only liquids.

He wished once more to meet with his friend Fermat, but he was not strong enough to make the journey even to a midway point. His letter to Fermat explaining why he could not come gives a moving self-portrait. "I find geometry the noblest exercise of the mind, yet I know it to be so useless that I see no difference between a geometer and a clever artisan. I call it the loveliest occupation in the world, but only an occupation. . . . A singular chance about a year or two ago did set me at mathematics, but having settled the matter I am not likely ever to touch the subject again."

He was dying. He had resigned himself and was waiting upon God, filling a page of the *Pensées* when he could. Yet in the midst of his departure came an almost comic interlude. Suddenly he immersed himself in the formation of the first omnibus company, which soon ran its first line from the Porte Saint-Antoine to the Luxembourg Palace. It was a far more profitable venture than his calculating machine. He gave the money to charity and bequeathed half his interest in the company to the hospitals of Clermont and Paris.

In June, 1663, he was attacked by terrible headaches. Convulsions followed. On August 18 he died, aged 39.

Mortimer has made Pascal understandable. But perhaps, after all, "realist" is not the best word for him. He had many gifts; he was many things. He was divided as experience itself is divided; he was uncertain in an uncertain world. His scientific achievements were extraordinary, yet it is in the *Pensées* that he achieved supremely. He had a profound sense of man's loneliness, of his terror of nothingness and longing to find his place in a vast, indifferent universe; and an exquisite way of transfixing truth, of making anguish stand still, of speaking the questions that men have asked since the beginning. No one, neither scientist nor philosopher, neither rationalist nor mystic, has transcended his insight into man's condition:

"For after all what is man in nature? A nothing in relation to infinity, all in relation to nothing, a central point between nothing and all, and infinitely from understanding either. The end of things and their beginnings are impregnably concealed from him in an impenetrable secret. He is equally incapable of seeing the nothingness out of which he was drawn and the infinite in which he is engulfed."

4 Descartes

by A. C. Crombie
October 1959

This extraordinary Frenchman is principally remembered for his invention of analytic geometry, but he attempted far more. His aim was nothing less than to reduce nature to mathematical law

"I should consider that I know nothing about physics if I were able to explain only how things *might* be, and were unable to demonstrate that they *could not be otherwise*. For, having reduced physics to mathematics, the demonstration is now possible, and I think that I can do it within the small compass of my knowledge."

With these words René Descartes declared the viewpoint that placed him among the principal revolutionaries in the 17th-century scientific revolution. Against the "forms" and "qualities" of Aristotelian physics, which had proved to be a blind alley, he asserted the "clear and fundamental idea" that the physical world was sheer mechanism and nothing else. Because the ultimate laws of nature were the laws of mechanics, everything in nature could ultimately be reduced to the rearrangement of particles moving according to these laws. In analytical geometry, perhaps Descartes' most enduring achievement, he created a technique for expressing these laws in algebraic equations. He thus put forward the ideal program of all theoretical science: to construct from the smallest number of principles a system to cover all the known facts and to lead to the discovery of new facts.

All subsequent theoretical physics has been aimed at the realization of this ideal of a single theoretical system in which the last details of observable regularities should be shown to be deducible from a minimum number of fundamental equations, written perhaps on a single page. Blaise Pascal and Isaac Newton may certainly be said to have carried on in the 17th century the Cartesian program of looking for the explanation of the physical world in terms of its mechanism. In this century we have witnessed attempts at universal theories by Albert Einstein and Werner Heisenberg, among others. In the vision of Descartes, however, his indisputable first principles—"nearly all so evident that it is only necessary to understand them in order to assent to them"—were not the end but the beginning of the search.

There can be no doubt of the revolutionary character and influence of Descartes' theoretical insights and program. The paradox is that he should have exercised so profound an influence over men who found his approach essentially distasteful and who rejected some of the most important of his fundamental assumptions and detailed conclusions. Christiaan Huygens, the great Dutch mathematician and astronomer whose father had been an intimate friend of Descartes, admitted late in life that he could no longer accept any but a small part of Cartesian physics. But he said that it was Descartes' *Principles of Philosophy* that first opened his eyes to science. Descartes, he said, had not only exposed the failure of the old philosophy but had offered "in its place causes which one could understand for all that exists in nature." As is so often the case with revolutionists, the legacy of Descartes was not only achievement but also prophecy and vision.

Descartes himself came to recognize that his purely deductive, mathematical ideal for science had failed in the face of the complexities of nature and the enigmas of matter. This failure was especially apparent in physiology, the field into which he ventured most daringly. Out of failure and compromise, however, Descartes extracted another contribution to scientific thinking in many ways as important as the original theoretical program itself. Forced to turn to experiment and hypothesis, he showed himself to be the first great master of the hypothetical model. This has become an essential tool in all scientific investigation. In his theoretical models of physiological processes Descartes displayed the most ingenious exercises of his imaginative and experimental genius.

René Descartes was born on March 31, 1596, at La Haye, a small and attractive town on the river Creuse in Touraine. His family were of the *petite-noblesse*, long in government service; his father was counselor to the *Parlement* of Brittany. From his mother, who died a month after his birth, he inherited "a dry cough and a pale complexion," which he kept until he was over 20. He also inherited property from her that gave him complete financial independence. Because he was a delicate child, it was thought that he would not live long. But he used his enforced inactivity to indulge an early passion for study.

When he was 10, his father sent him to the newly established Jesuit college of La Flèche, where he remained for eight and a half years and received an excellent education that embraced logic, moral philosophy, physics and metaphysics, classical geometry and modern algebra, as well as an acquaintance with the recent telescopic work of Galileo. All the main characteristics of his mind appeared precociously at La Flèche. Introduced to the classics, he fell in love with poetry. Far from being a "geometer who is only a geometer" (Pascal's description of him), Descartes himself wrote in an early essay, the *Olympica:* "There are sentences in the writings of the poets more serious than in those of the philosophers. The reason is that the poets wrote through enthusiasm and power of imagination. There are in us, as in a flint, seeds of knowledge. Philosophers adduce them through the reason;

poets strike them out from the imagination, and these are the brighter."

Mental facility was one of Descartes' most striking and perhaps more dangerous gifts. A fellow pupil described his prowess in argument. He would first get his opponents to agree on definitions and the meaning of accepted principles, and then he would build up a single deductive argument that was very difficult to shake. At La Flèche he also acquired a habit that persisted throughout his life. He was excused from certain work and allowed to lie late in bed. Here he found it possible to indulge most fully his natural inclination to solitary concentrated thought.

When he was 20, having graduated in law from the University of Poitiers, Descartes went to Paris. Here he became a young man of fashion somewhat

PORTRAIT OF DESCARTES by Frans Hals hangs in the Louvre. Among the fields that he worked in were physiology, psychology, optics and astronomy. Many consider him the father of modern philosophy. He died in 1650 while tutor to the Queen of Sweden.

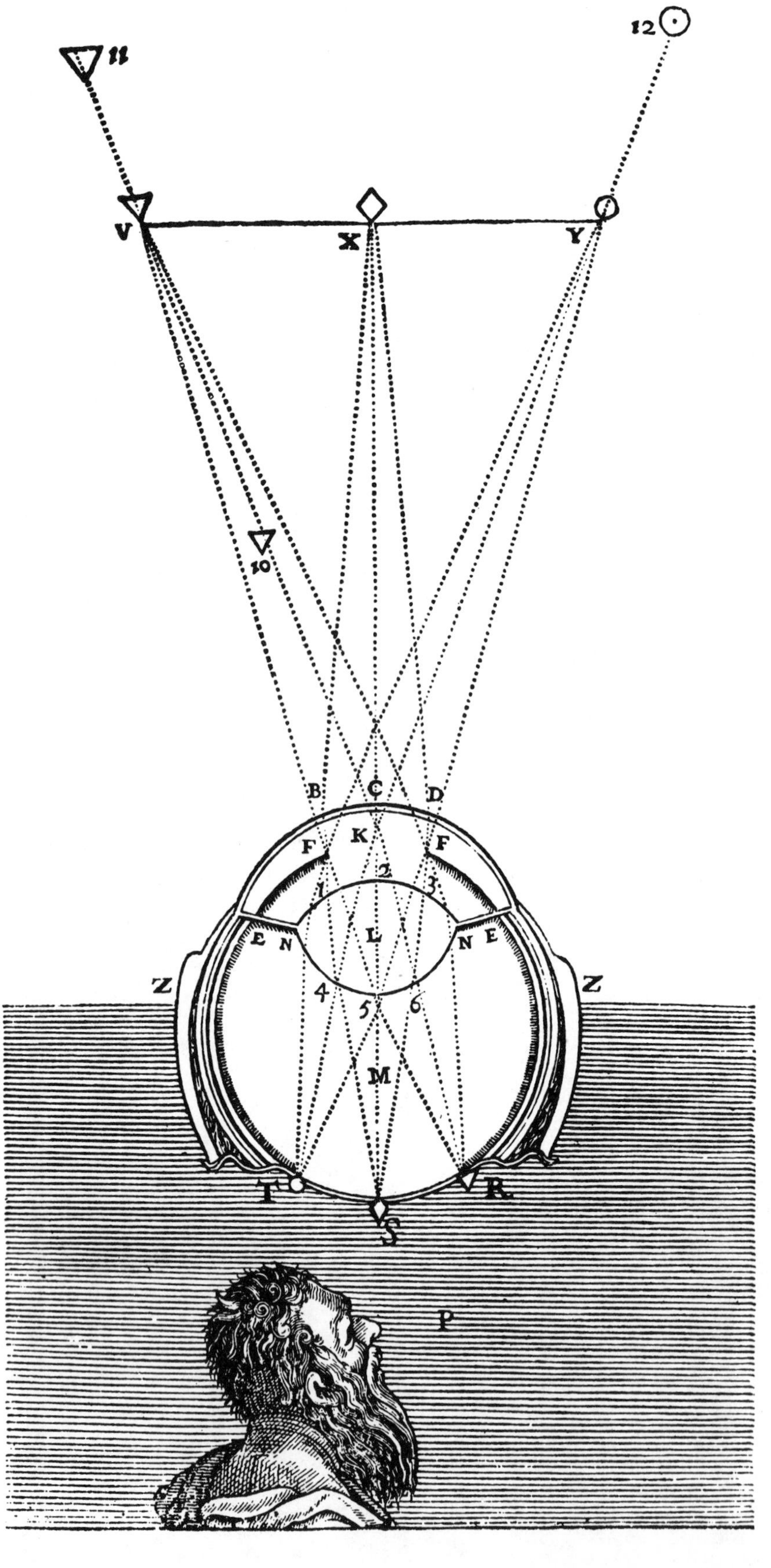

DESCARTES' EYE INVESTIGATIONS included removing the retina from the eye of an ox and replacing it with thin paper or eggshell so that he could study the image. This illustration is reproduced from Descartes' book *Dioptrics*, which was first published in 1637.

at a loose end. Soon, however, his thoughts returned to mathematics and philosophy. He was encouraged by his more serious friends, among them the Minim friar Marin Mersenne, whom Descartes had known at La Flèche. Mersenne was himself a competent mathematician and a skillful experimenter. His cell in the convent of the *Place Royale* was to become the meeting place of *savants*, an antecedent of the Academy of Sciences founded later in the century. Mersenne came to have a vast correspondence, of which only part has been published, and thus became a center of scientific intelligence in the days before there were any scientific journals. He also translated Galileo's *Dialogue* and *Discourses* into French, the former in 1634, the year after Galileo's condemnation. Until the end of his life Mersenne remained Descartes' principal friend, and after Descartes left France for good in 1628 Mersenne kept him posted with scientific news from Paris.

In 1618 Descartes joined the army of Prince Maurice of Nassau (later Prince of Orange) as a gentleman volunteer. He was sent to the garrison at Breda in the Netherlands, there being at that time a truce between the Franco-Dutch forces and the Spaniards, whose rule the Low Countries were throwing off. His scientific interests were such as were appropriate for an officer: ballistics, acoustics, perspective, military engineering, navigation.

One day—November 10, 1618—he came upon a group of people gathered about a notice pinned up in the street. It was in Flemish, and turning to someone in the group, Descartes asked him to translate it into Latin or French. The notice proved to be a challenge inviting all comers to solve the mathematical problem that it proposed. The man whom Descartes had asked to translate it was Isaac Beeckman, one of the country's leading mathematicians. Descartes solved the problem and presented his solution to Beeckman, who at once recognized the young man's mathematical genius and set out to revive his interest in theoretical problems. During that winter Beeckman proposed that Descartes should find the mathematical law of the acceleration of falling bodies. Neither knew that Galileo had in fact already solved this problem; his solution was to appear in his *Dialogue on the Two Principal Systems of the World* in 1632. Descartes produced solutions based on different assumptions. That none of them described the way bodies actually fall did not concern him. He had not yet learned to unite mathematical analysis with experiment.

We are indebted to Beeckman's journal, discovered in 1905, for a flood of light on this period of Descartes' life. It was a time of self-discovery; the young man's mind moved with incredible speed over a broad assortment of questions. Descartes now got on the track of the method by which he was to attempt the unification of human knowledge upon a single, central set of premises.

On March 26, 1619, Descartes reported to Beeckman "an entirely new science which will allow of a general solution of all problems that can be proposed in any and every kind of quantity, continuous or discontinuous, each in accordance with its nature . . . so that almost nothing will remain to be discovered in geometry." This was Descartes' announcement of his discovery of analytical geometry or, as Voltaire was to describe it, "the method of giving algebraic equations to curves." Descartes' 14th-century countryman Nicole Oresme may have contributed something toward this idea. In the 17th century Descartes' contemporary Pierre de Fermat was to make the same discovery quite independently, but he did not follow it up. Descartes did not publish his "new science" until 1637, when he included in his essay *Geometry* both an exposition of the principles and several particular applications. Its generality is there shown in Descartes' demonstration that the conic sections of Apollonius are all contained in a single set of quadratic equations. Since conic sections include the circles of the ancient astronomers, the ellipses of Johannes Kepler, and the parabola used by Galileo to describe the trajectory of a projectile, it is plain that Descartes' first invention placed a powerful tool in the hands of physicists. Without it Newton himself would have suffered a crippling handicap.

Exactly a year after his meeting with Beeckman, Descartes had a celebrated experience, perhaps the most important and certainly the most dramatic of his whole life. He had joined the army of the Duke of Bavaria, another of France's allies in the Thirty Years' War, and found himself in winter quarters at a remote place on the Danube. Much occupied with his thoughts, he spent the whole of November 10 shut up alone in the famous *poêle* (literally "stove," but actually an overheated room). In the course of the day he made two important decisions. First, he decided that he must methodically doubt everything he knew about physics and all other organized knowledge, and look for self-evident, certain starting points from which he could reconstruct all the sciences. Second, he decided that just as a perfect work of art or architecture is always the product of one master hand, so he must carry out the whole of this program himself.

That night, according to his 17th-century biographer Adrien Baillet, Descartes had three dreams. First he found himself in a street swept by a fierce wind. He was unable to stand because of a weakness in his right leg, but companions near him stood up firmly. He awoke, and fell asleep again; he was reawakened by dreaming that he had heard a clap of thunder and had found the room full of sparks. He fell asleep once more and dreamt that he had found a dictionary on his table. Then, in another book, his eye "fell upon the words *Quid vitae sectabor iter?* [What way of life shall I follow?]. At the same time a man he did not know presented him with some verses beginning with the words *Est et non*, which he recommended highly to him." These words Descartes recognized as the opening lines of two poems by Ausonius. Even before Descartes had finally awakened he had begun to interpret the first dream as a warning against past errors, the second as the descent of the spirit of truth to take possession of

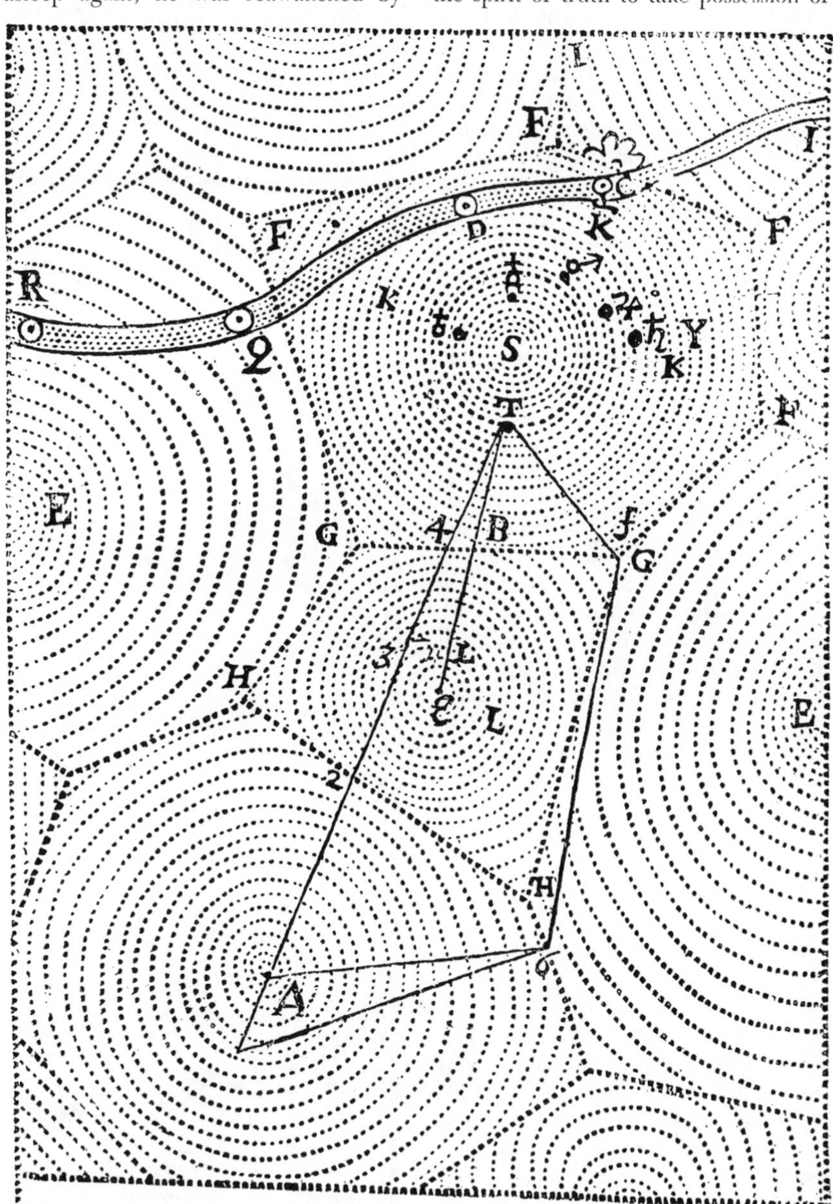

SYSTEM OF VORTEXES with which Descartes sought to account for the motions of the heavenly bodies consisted of whirlpools of "ether." In the case of the solar system the vortex carried the planets around the sun (S). Irregular path across top of the illustration is a comet, the motions of which Descartes believed could not be reduced to a uniform law.

him, and the third as the opening to him of the treasures of all the sciences and the path of true knowledge. However this incident may have been elaborated in the telling by Baillet, it stands as a symbol of Descartes' certainty in the rightness of his approach to true knowledge.

He went on soldiering until 1622, seeing action at the battle of Prague and the sieges of Pressburg and Neuhäusel. Then for a few years he was a traveler, ranging over Europe from Poland to Italy and returning at last to Paris in 1625. There he rejoined the circle round Mersenne, worked at his "universal mathematics," and engaged in speculations on many subjects from moral psychology to the prolongation of life. From such pursuits he was distracted, in the fashion of his leisured contemporaries, by the social whirlpool and by music, idle reading and gambling. His father expressed the opinion that he was "not good for anything but to be bound in buckskin."

Then an incident occurred that turned Descartes' vision into his life's mission. He found himself present, along with a fashionable and impressive audience including his friend Mersenne and the influential Cardinal de Bérulle, at the house of the papal nuncio to hear a certain Chandoux expound his "new philosophy." Descartes alone did not join in the applause. Pressed to give his opinion, he spoke at length, demonstrating how it was possible for a clever man to establish an apparently convincing case for a proposition and also for its opposite, and showing that by using what he called his "natural method" even mediocre thinkers could reach principles that were found to be true. His hearers were astonished. When Descartes visited Bérulle a few days later, the cardinal charged him to devote his life to working out the application of his method in philosophy and in "mechanics and medicine."

In October, 1628, Descartes left for the Netherlands, where he remained for the rest of his life except for three short visits to France and his last journey to Stockholm in 1649. He avoided the company of everyone but his intimate friends and disciples, and dedicated his time to the application of his principles in philosophy, science and mathematics and to the dissemination of his conclusions. Within a year of finally leaving the Netherlands at the invitation of Queen Christina of Sweden, he died in Stockholm in February, 1650.

Descartes may be described as a centrifugal thinker: he moved primarily outward from a firm central theoretical point, in diametrical contrast to thinkers like Francis Bacon or Isaac Newton. The French writer and amateur of science Bernard le Bovier de Fontenelle, in the well-known *Eloge de Newton* written on Newton's death, drew an eloquent contrast between the methods of Newton and Descartes:

"The two great men so placed in opposition had much in common. Both were geniuses of the first order, born to dominate the minds of others and to found empires. Both, being outstanding geometers, saw the need to carry geometry into physics. Both founded their physics on a geometry which they developed almost single-handed. But the one [Descartes] tried in one bold leap to put himself at the source of everything, to make himself master of the first principles by means of certain clear and fundamental ideas, so that he could then simply descend to the phenomena of Nature as to necessary consequences of these principles. The other [Newton], more timid or more modest, began his journey by leaning upon the phenomena in order to mount up unknown principles, resolved to admit them only in such a way that they could yield the chain of consequences. The one set out from what he knew clearly, in order to find the cause of what he saw. The other set out from what he saw, in order to find the cause."

The primary direction and movement of Descartes' philosophical and scientific enterprise are shown by the sequence in which he composed his major works. From 1618 to 1628, during the restless years of military life, travel and dissipation, he worked out his conception of

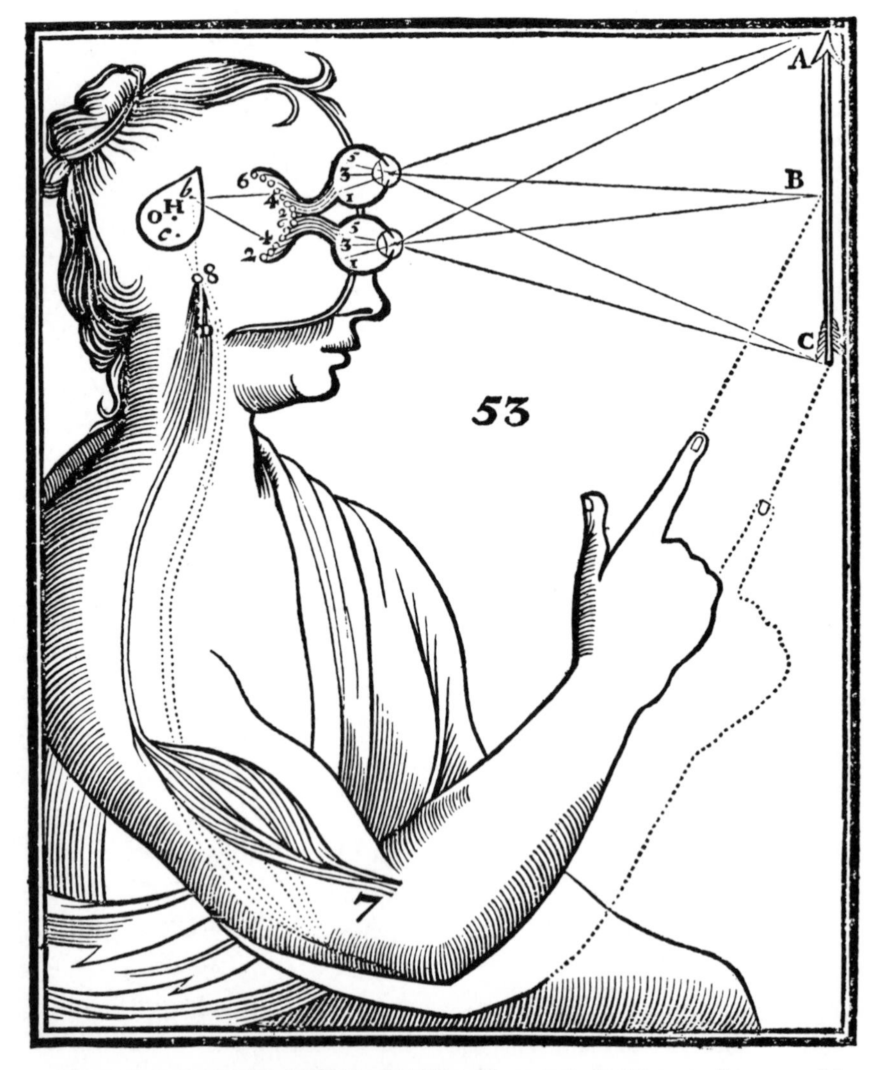

CENTRAL ROLE OF THE PINEAL GLAND in Descartes' physiology is diagrammed in *l'Homme*. Images fall on retinas (*5, 3, 1*) and are conveyed to the cerebral ventricles (*6, 4, 2*); these then form a single binocular image on the pineal gland (*H*), the site from which the soul controls the body. Stimulated by the image, the soul inclines the pineal gland, activating the "hydraulic system" of the nerves (*8*), causing a muscle to move (at 7).

true science and his highly rationalistic method for attaining it. These he described in his first work, *Rules for the Direction of the Mind*, finished in 1628 but published posthumously, and in the *Discourse on Method*, which he wrote after settling in the Netherlands. Before completing the latter he began work on the *Meteors*, the *Dioptrics* and the *Geometry*, which he presented as three illustrations of the power of the method applied to specific lines of investigation, when they were published with the *Discourse* in 1637. Meanwhile, by 1628, he had turned to the next stage of his investigations, the discovery of first principles. These he propounded in his *Meditations on First Philosophy*, published in 1641. From first principles he moved on quickly to the elaboration of his cosmology, which he completed in *Le Monde* in 1633, but withheld from publication upon the news of Galileo's condemnation. A revised version, with its Copernicanism mitigated by the idea that all motion is relative, was published in 1644 under the title *Principles of Philosophy*. At the same time Descartes was working out his conception of the relationship between the mind and the machinery of the body, and in his last work, *Passions of the Soul* (completed in 1649), he brought psychology within the compass of his system.

Perhaps the most revealing illustration of the power of his method is Descartes' *Dioptrics*. He characteristically announces at the outset that he intends to solve the problem of constructing a telescope on rational scientific principles. Accordingly he undertakes first an analysis of the nature of light: space is filled with fine contiguous globules of matter, forming a kind of "ether"; light is a mechanical phenomenon, an instantaneous pressure transmitted through this "ether" from a luminous source. Descartes then gives an elegant geometrical demonstration of the laws of reflection and refraction. Some years earlier the Dutch physicist Willebrord Snell had discovered the correct sine law of refraction, but he had not published it. Descartes' demonstration of what is now known as Snell's Law was almost certainly independent; he was the first to publish it.

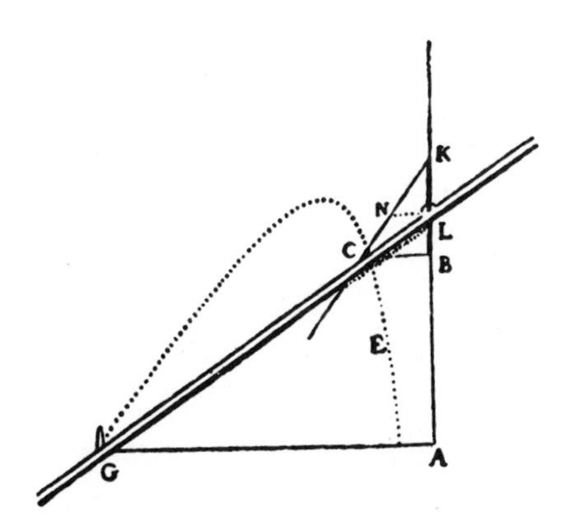

DESCARTES' GEOMETRY is, in Voltaire's words, "a method of giving algebraic equations to curves." Illustration is from a page in which equation for parabola is discussed.

Since the purpose of a telescope is to increase the power of vision, Descartes next makes a detailed analysis of the human eye in both its normal and its pathological states. For this, as his correspondence shows us, he conducted extensive studies and dissections. Repeating an experiment made by Christoph Scheiner, he removed the back of an ox's eye, replaced it with thin white paper or eggshell, and examined the reversed image cast upon it of an object placed in front of the eye. The whole investigation shows considerable anatomical knowledge and experimental skill; Descartes describes the functioning of the iris, the ciliary muscle, binocular vision, optical illusions and various forms of coordination and accommodation.

He now considered himself to be in a position, denied to Kepler and Galileo, to show scientifically what the curvatures of the lenses used in constructing a telescope should be. He concluded that their cross sections should be either hyperbolas or ellipses. He did not of course allow for chromatic aberration, a problem not then understood. Finally he gives a description of a machine designed to cut lenses on these scientific principles.

From a long correspondence between Descartes and a French spectacle-maker named Ferrier we know how this unfortunate man tried and failed to put Descartes' ideas into practice. In the end no actual telescope was constructed on Descartes' theoretical principles.

The essential structure and content of Descartes' physics and cosmology derive from the revolutionary conclusions at which he arrived soon after his retirement to the Netherlands in 1628. He found the basis for the possibility and certainty of knowledge in the fact of thought itself. This elemental fact, apprehended with "clarity and distinctness," became his criterion for determining whether or not anything else was true. The "qualities" of classical philosophy, apprehended by mere sensation, he found not to be clear and distinct. Thus he eliminated from the world outside everything but extension—the one measurable aspect of things and hence their true nature. This division of the world into the two mutually exclusive and collectively exhaustive realms of thought and extension enabled Descartes to offer what he regarded as a true science of nature. The task of science was now to deduce from these first principles the causes of everything that happens, just as a mathematics is deduced from its premises.

It was the very breadth of this program—which in effect declares that the whole of physical nature may ultimately be reduced to and comprehended by the laws of motion—that gave Descartes' work its revolutionary scientific importance. Descartes himself offered explanations, in terms of the motions of particles of various shapes and sizes, of chemical properties and combinations, taste and smell, heat, magnetism, light, the operation of the heart and nervous system as the source of action of the mechanism of the body, and many other phenomena that he investigated by sometimes rather naive experiments. The vastness of this program was its undoing; Descartes simply had no time to go into all these questions accurately and quantitatively. Coming from the author of a mathematizing program, Descartes' general physics and cosmology are surprisingly almost entirely qualitative. He was forced to fall back on speculation far beyond "the small compass of my knowledge," with the result that he came to fear that he had produced, to use his own phrase, nothing more than a beautiful "romance of nature."

His most disastrous failure occurred in fact at the very center of his program, in the laws of motion themselves. He had reached his conclusion that the essential property of matter was extension in space by a process of purely rational analysis. Since his method *a priori* ruled out other possibilities, it did not leave the question open to empirical test. From this supposedly firm basis he then proceeded to construct a system of mechanics that left out important facts, notably those included in what became the Newtonian notion of "mass." His mechanics certainly contains some valuable conclusions; for example, his account of the conservation of motion and his enunciation of an equivalent of the principle of inertia. But geometrically identical bodies of different masses do not behave identically when they collide or interact in other ways. Descartes' treatment of this subject was disastrously wrong because his antecedent analysis of matter into mere extension was itself mistaken.

In order to explain how the planets were kept in their orbits, Descartes put forward his famous vortex theory, according to which the fine matter of the "ether" forms great whirlpools or vortexes round the stars and the sun. The planets are carried about in the sun's vortex, rather like a set of children's boats in the celestial bathwater, and the moon is carried round the earth in the same way. The astonishing thing is that Descartes did not bother to check whether or not this very important part of his physical system agreed with the facts as expressed by Kepler's laws of planetary motion. It was Newton who destroyed Descartes' famous vortex theory. In fact, he may have chosen the title *Principia Mathematica* to give point to his polemic against Descartes' *Principia Philosophiae*. Newton treated the vortex theory as a serious problem of fluid dynamics and utterly demolished it.

Descartes' subsequent reputation as a mere speculator has been kept going largely by historians of mechanics writing under the influence of Newtonian polemics. But if we turn from Descartes' mechanics to his physiology we can observe him at work in a field where the qualitative hypotheses on which he had fallen back in dealing with other subjects yielded results more worthy of him.

Descartes is rightly ranked with William Harvey as a founder of modern physiology. Harvey was a master of experimental analysis, but Descartes introduced the master-hypothesis on which all subsequent physiology has been based. Having divided the world into extension and thought, Descartes was able to regard biology as a branch of mechanics and nothing else. In modern terms this view asserts that living organisms are in the last analysis explicable in terms of the physics and chemistry of their parts. In man, according to Descartes, the realm of thought makes contact with the extended body at a single point: the pineal gland in the brain.

Descartes' correspondence shows that during his long residence in the Netherlands he spent much time in making anatomical dissections. He found biology the most defeating of all the fields into which he tried to carry his explanations by means of mechanical principles. It is in this field that he found experiment most necessary, both to acquire information and to choose between different possible explanations of the same phenomenon. Although he accepted Harvey's discovery of the circulation of the blood, he engaged somewhat unsuccessfully in a controversy with him over the mechanism of the heart's action, each bringing forward a crucial experiment to establish his explanation. Descartes was wrong in fact, but he made the essential point that a full explanation of the heart's action cannot simply start with the fact that it is beating, but must try to account for this fact in terms of the underlying mechanism—ultimately in terms of the laws of motion common to all matter.

Although Descartes' mechanistic explanation of this still-obscure phenomenon now seems rather naive, the method by which he attacked it and the working of the machine of the body as a whole introduced one of the most powerful tools of all modern physiological research. This was the hypothetical model. Descartes' physiological writings contain many good observations and some brilliant mechanistic explanations of such phenomena as automatic actions like blinking and the coordination of different muscles in complicated movements such as walking. He was inclined to sacrifice real anatomy to the hypothetical anatomy demanded by his mechanism. But he always said explicitly that he was describing a hypothetical body to imitate the actions of the real body, just as a modern investigator will build an electronic machine to imitate the processes of the brain.

Descartes set out to produce a true science of nature in which everything would follow mathematically from self-evident first principles. Modern physicists, of course, reject the idea that the principles of physics can be self-evidently certain. Even in the 17th century Pascal and Huygens made the same criticism. They pointed out that there is an essential difference between physics and abstract mathematics in that the principles of physics, which explores the unknown in the concrete world of fact, are always exposed to complete or partial invalidation by the discovery of new facts.

Descartes, moving outward from his central principles, had himself come to appreciate the point made by Pascal and Huygens, and to realize that his mathematical ideal of unilinear deduction had broken down because of the difficulty of connecting abstract general principles with the particulars of fact. Yet as a positive scientific thinker he was perhaps not so different from his successors in our time. His search was for the causes and meaning of no less than everything that occurs.

The Mathematical Papers of Isaac Newton, Volume I: 1664–1666

5

Edited by D. T. Whiteside

Reviewed by I. Bernard Cohen
January 1968

Mathematics, the childhood of Isaac Newton's science

The inaugural volume of D. T. Whiteside's monumental edition of *The Mathematical Papers of Isaac Newton* (to be completed in eight volumes) deals with only three years in Newton's life: 1664–1666. But what golden years they were! In them Newton was first exposed to higher mathematics and firmly launched on his almost legendary scientific career. Guided by Whiteside's sure hand, we can now study Newton's early mathematical notes and first mathematical essays. We thus become privileged witnesses to the actual formation of one of the keenest minds of all time.

The basic task of editing Newton's mathematical papers is to place them in chronological order and transcribe them, but the editor must have the mathematical training and insight needed to understand and interpret them. Happily Whiteside brings to this assignment a rare combination of mathematical and linguistic ability and an abiding sense of history. He has had not only the patience to steep himself in the mathematical thought of another age but also the wit to discern the significance of each document in the light of Newton's development and currents of knowledge in the 17th century.

In this edition each document follows the form of the original as closely as type allows, and Whiteside has been careful to keep separate his critical emendations and Newton's own formulations. Copious notes form a continuous commentary to help the reader follow Newton's mathematical argument, which is often extremely difficult for the uninitiated modern reader. These notes, and the introductions to the several sections, constitute the best account in print of the main features of 17th-century mathematical thought and so go far beyond the major assignment of illuminating Newton's mathematical development.

Whiteside has spent 10 years studying the printed and manuscript sources, and he is already known to scholars for his earlier monograph *Patterns of Mathematical Thought in the Later Seventeenth Century*. With the publication of the Newton volume, with its riches of new and unexpected material and original interpretation, he has raised Newtonian scholarship to a new plane. It is difficult to think of any other work in recent years that surpasses it in importance in the entire field of the history of science. It profoundly illuminates the creative process of science in a manner completely worthy of its subject.

Newton, thrice great like the magical patron of alchemy Hermes Trismegistus, dominated the rise of three major sciences in modern times: rational mechanics, experimental optics and pure mathematics. Known in his day for achievements unsurpassed by any predecessor or contemporary, he had penetrated the mysteries of light and color (and had invented a new type of telescope), had found the law of universal gravitation (thus to explain at once why the planets move around the sun in accord with Kepler's three laws and why stones fall to earth as Galileo had found they do) and was an inventor of the "fluxional calculus," the new language of the exact sciences. With the kind of admiration usually granted to superheroes in war or athletics, Newton's contemporaries asked whether he was at all like ordinary men: did he eat and sleep like other mortals? They even compared him to the Nile (whose source had not yet been found), speculating that Newton was revealed to us only in the full stream of his mighty intellect—the source or sources being permanently hidden from our view.

What is it that makes a mind so creative that we still contemplate its achievement with awe some 300 years later? We cannot really say much more on this subject today than men of true insight in Newton's time could. We do have one advantage over our predecessors: thanks to the heroic labors of Whiteside we can see for ourselves the sources from which Newton's mathematical achievement sprang.

Newton's yeoman ancestors showed no special talent for science or mathematics. For that matter, Newton's early years of training and education are singularly lacking (for one who became so great) in auguries of a hidden genius straining to break the confines of its environment. Born on Christmas Day in 1642 (in the Julian calendar), Newton was educated in local country schools and the grammar school at Grantham in Lincolnshire. Like others in his time he learned Latin and Greek and eventually acquired enough Hebrew to write out texts in it. He must certainly have become acquainted with elementary arithmetic: addition, subtraction, multiplication, division, the reduction of fractions and the use of proportions. He does not seem to have read much, if anything, that could properly be called scientific. He did pay rather close attention to John Bate's *Mysteries of Nature and Art,* from which he got the idea of making water clocks, toy mills and other mechanical devices. Like many other scientists then and now he was a true gadgeteer, and he appears to have shown a certain proficiency in drawing. These talents must have stood him in good stead in later years, when he made his own telescopes (including grinding the mirrors) and the apparatus he needed for his experiments.

In 1947, when the old Newton homestead in Woolsthorpe was being renovated, a number of drawings and diagrams were uncovered, some carved into the stone and others scratched into the plaster. I can only agree with Whiteside's judgment: "It would need the blindness of maternal love to read into these sets of intersecting circles and scrawled line-figures either burgeoning artistic prowess or mathematical precocity."

When Newton went down to Cambridge from Lincolnshire, he had as yet shown no sign of the massive talent that was to change the world. A poor and rather young student, evidently shy and not gifted at making friends, he was in no way remarkable or worthy of notice from his contemporaries. Many notebooks survive from those undergraduate years to tell us in detail about his studies. We can turn the pages of notes in Greek on Aristotle's *Analytics* and *Ethics*, and then read comments in Latin taken from various 17th-century Aristotelians. We can see how the young Newton progressed from the customary topics of logic, rhetoric and ethics to physics; not yet the physics of his own time but the *Physics* of Aristotle, interpreted by the late-16th-century scholastic Johannes Magirus. Two extracts presented by Whiteside show the strongly scholastic or medieval quality of Newton's mathematical thinking at the time:

"Extension is related to places, as time to days. yeares &c. Place is ye principium individuationis of streight lines & of equall & like figures. . . ."

"If Extension is indefinite onely in greatness & not infinite y^n [then] a point is but indefinitely little & yet we cannot comprehend any thing lesse. To say y^t [that] extension is but indefinite . . . because we cannot perceive its limits, is as much as to say God is but indefinitely perfect because wee cannot apprehend his whole perfection."

So far all is traditional. Then, at the beginning of Newton's final year as an undergraduate (1664), his notebooks give us two very different signs. First of all, as Whiteside explains, Newton abruptly changed "his early, rather ornate handwriting for the simpler, less pretentious form which was to remain his throughout the rest of his life." Second, he gave up an exclusive diet of reading the ancients and their commentators and plunged into the moderns. From this year onward we can trace the genesis and flowering of Newton's genius as he came to know the scientific writings of the men of his own century. From them he took whatever tools or methods or concepts he could use in forging his own, and soon we see him exercising his newly acquired powers to produce a novel Newtonian science and Newtonian mathematics and to surpass one by one those who had been his mentors. It is for this reason that the documents that are printed in the first volume of Whiteside's edition are so precious; they come from those crucial years 1664–1666, when Newton first developed his mathematical prowess.

Newton's notebooks show us at close range how he learned the "new science." Among the major authors he read were Walter Charleton, an expert on ancient and modern atomism and on the physics of Galileo and Pierre Gassendi, and Kenelm Digby, whose presentation included the physics and philosophy of Descartes along with the ideas of Galileo. Newton also read philosophical and scientific writings by Joseph Glanville, Thomas Hobbes and Henry More, and he devoured books by Robert Boyle and Robert Hooke. He took careful notes on his reading in the *Philosophical Transactions* of the young Royal Society of London. He also read and made notes on Galileo's *Dialogues on the Two Great Systems of the World* and Descartes's *Principles of Philosophy;* from Descartes he copied out an English version of the principle of inertia, which was later to become Newton's first law of motion. As we turn the pages of his notebooks we can see his mind leap from summaries of his reading to his own new principles and results. An outstanding example is his formulating the law of "centrifugal force" (that the force when a body moves uniformly along a circle is at once directly proportional to the square of the speed and inversely proportional to the radius) years before it was published by Christiaan Huygens, who is usually credited with the discovery. He also began to think of gravity as a force extending as far as the moon.

In mathematics as in physics and astronomy Newton turned to the new authors of his century. According to the mathematician Abraham De Moivre, who knew him well, Newton's introduction to higher mathematics (higher, that is, than the simple arithmetic he knew on entering college) began when he "bought a book of Astrology" in 1663 "out of a curiosity to see what there was in it." He found he could not understand "a figure of the heavens" for "want of being acquainted with Trigonometry." He therefore got himself a book of trigonometry "but was not able to understand the Demonstrations," and so he got "Euclid to fit himself for understanding the ground of Trigonometry." De Moivre then records the following about Newton: "Read only the titles of the propositions, which he found so easy to understand that he wondered how any body would amuse themselves to write any demonstrations of them. Began to change his mind when he read that Parallelograms upon the same base & between the same Parallels are equal, & that other proposition that in a right angled Triangle the square of the Hypothenuse is equal to the squares of the two other sides."

De Moivre continues: "Began to read Euclid with more attention than he had done before & went through it." Next Newton read William Oughtred's *Clavis*. Then:

"Took Descartes's Geometry in hand, tho he had been told it would be very difficult, read some ten pages in it, then stopt, began again, went a little farther than the first time, stopt again, went back again to the beginning, read on till by degrees he made himself master of the whole, to that degree that he understood Descartes's Geometry better than he had done Euclid.

"Read Euclid again & then Descartes's Geometry for a second time. Read next Dr. [John] Wallis's Arithmetica Infinitorum, & . . . found that admirable Theorem for raising a Binomial to a power given. . . ."

The last statement is a reference to one of the best known of Newton's early mathematical discoveries, the series expansion of a binomial to any power n, that is, $(1 + a)^n$.

Whiteside shows that De Moivre's story conforms to the evidence of Newton's notebooks and certain books from Newton's library, happily preserved at Trinity College, Cambridge. Among these books Whiteside finds a "well-thumbed and marginally annotated copy of Euclid's *Elements* (in Barrow's 1655 edition)," which bears witness to Newton's examination of this work. As for Descartes's *Géométrie*, Whiteside tells us that Newton did not read it in the original French but in the second Latin edition of 1659–1661, and that he studied not only Descartes's original text but also an extensive commentary by Frans van Schooten. In support of De Moivre's account Whiteside also adduces testimony from Newton himself, in the form of certain autobiographical notes:

"July 4th 1699. By consulting an accompt of my expenses at Cambridge in the years 1663 & 1664 I find that in ye year 1664 a little before Christmas I being then senior Sophister [an undergraduate] I bought Schooten's Miscellanies & Cartes's Geometry (having read this Geometry & Oughtreds Clavis above half a year before) & borrowed Wallis's works and by consequence made these Annotations out of Schooten & Wallis in [the] winter between the years 1664 & 1665. At w^ch [which] time I found the method of Infinite series. And in summer

1665 being forced from Cambridge by the Plague I computed ye area of ye Hyperbola at Boothby in Lincolnshire to two & fifty figures by the same method."

From these beginnings in 1664, in Whiteside's judgment, "Newton over the next two years (and not only of course in mathematics) was to develop a remarkable series of researches formidable in technical content and effervescent with still untested creative thoughts.... Their detailed systematization, carried through by a typically stubborn perseverance and massive power of mental concentration, was to take most of the rest of his life."

Assembling Newton's mathematical papers and putting them in chronological order was in itself a formidable task. Although the bulk of the manuscripts is in Cambridge (primarily in the University Library but also in the libraries of Trinity College and King's College), other documents are scattered far and wide. The original order in which Newton had arranged his manuscripts had been violated by later librarians and cataloguers. Scraps and fragments had to be identified, some of them belonging to the same document but differing in style or physical appearance because they were earlier or later versions or states. Dating the documents was altogether a vexing problem. In some cases a date could be assigned on internal evidence or by an ancillary autobiographical document, but in the end Whiteside became so steeped in Newton's written words that he could in most cases date documents closely by the form of the handwriting, which varied slightly from year to year. I can speak from personal experience in saying that dates Whiteside has assigned in this way have repeatedly been confirmed by independent evidence.

Whiteside received very little help from earlier works; only a few of the documents in his collection had been published before. Although this situation has been acute with respect to Newton's mathematical work, it applies in Newtonian scholarship generally. In spite of the fact that the bulk of Newton's scientific manuscripts have been available for almost a century, little use has been made of them by scholars, who have largely based their studies of Newton on published material rather than on manuscripts. In recent years, however, the situation has been changing as it has become increasingly clear that a full comprehension of the development of Newton's thought requires close study of all his writings. Among those who have now intensively mined the manuscript archives are the late Herbert W. Turnbull (the editor of Newton's *Correspondence*) and J. F. Scott (the current editor), the late Alexandre Koyré, A. Rupert Hall and Marie Boas Hall, John Herivel, Joseph Lohne, J. E. McGuire, R. S. Westfall and myself.

In his first volume Whiteside presents Newton's texts in three groups: the first mathematical annotations (1664–1665), from Oughtred, Descartes, Schooten and Huygens; then Newton's manuscript records of his research in analytic geometry and calculus (1664–1666), and finally some miscellaneous early researches (1664–1666). From the evidence presented there can be little doubt of the correctness of Whiteside's judgment that "beyond reasonable doubt" Newton was "self-taught in mathematics, deriving his factual knowledge from the books he bought or borrowed, with little or no outside help." There appears to be no ground whatever for the "pleasant story of [Isaac] Barrow's tutorial guidance," which is enshrined in one of Walter Savage Landor's *Imaginary Conversations*.

Newton's notes, observes Whiteside, "are not mere inferior, copied images but have their own life, revealing a young mathematician at work stretching his mind, shaping what he read and recording his own impressions." At times it is difficult to draw a hard and fast line between the end of a summary and "the following wave of new ideas" as it becomes "a piece of original research." The transformation of a youth who knew no more mathematics than simple arithmetic, and who could not read a treatise on astrology for want of trigonometry, into a profound creator of higher mathematics is marvelous to follow. The first annotation presented, taken from a small notebook in which Newton had begun to make a Hebrew-English dictionary, lists the properties "of right angled triangles," or the Pythagorean theorem, in answer to the problem, "Any two leggs given to find y^e^ other." The solution is given in three equations, with a reference to Euclid's "lib. 1. pr: 47":

$$1.\ bq + cq = hq.$$
$$2.\ r{:}hq - bq{:} = c.$$
$$3.\ r{:}hq - cq[{:}] = b.$$

In these equations q stands for *quadratus*, or "squared," so that the first equation merely states that the sum of the squares of the two sides b and c equals the square of the hypotenuse h. In the second and third equations c is given in terms of the root (r) of $h^2 - b^2$ and b in terms of the root of $h^2 - c^2$. Newton has gone beyond addition and multiplication, but not too far. Before long, however, he is deep in the geometry of conic sections. Then he becomes fascinated by the problems of permutations and combinations, for example, the number of conjunctions that can be made of the seven planets, taken in any combination. He concludes that the "7 Planets may be conjoyned 120 divers ways," mistakenly copying 120 for 127. He also finds that "$1 \times 2 \times 3 \times 4 \times 5 \times 6. = 720.$ are y^e^ number of changes [that can be rung] in six bells." This is 1664, but before long he has gone deeply into number theory and algebra and shows himself to have mastered "the arithmetical symbolisms from Oughtred" and the "algebraical from Descartes." This was the basis of his own symbolism, with many modifications and innovations added.

The rise of Newton as a mathematician can be seen most clearly in his notes on the writings of John Wallis. Here we find him gaining an acquaintance with the theory of "indivisibles" (an immediate precursor of the calculus) and acquiring from Wallis a sure grasp of the methods of infinite series and continued products. By 1665 he had thus advanced from the algebra of Cartesian geometry and from simple number theory to the realm of higher mathematics, even as we still understand the expression. For instance, we find Newton using at this stage a logarithmic series for $\log(1 + x)$ to "square the Hyperbola," that is, to find the area of a segment:

$$\text{"}x - \frac{x^2}{2} + \frac{x^3}{3} - \frac{x^4}{4} + \frac{x^5}{5} - \frac{x^6}{6} + \frac{x^7}{7} - \frac{x^8}{[8]} + \frac{x^9}{9} - \frac{x^{10}}{10}.\ \&\text{c."}$$

He calculated the value for x as $1/10$, but Whiteside finds his results in part "vitiated by two small numerical slips, one of addition and one of subtraction." So much for those who would believe Newton was a marvelous calculator! To judge by a number of different manuscripts, Newton was particularly fascinated by the problem of determining the areas of hyperbolas, and it was in the course of investigating this topic that he found the general binomial expansion theorem.

On the basis of his study of the manuscripts Whiteside observes: "Already by mid-1665, one short crowded year after his first beginnings, the urge to learn from the work of others was largely abated. The indication of his rapid rise to mathematical maturity is telling. It was time for him to go his own way in earnest and thereafter, though he continued to draw in detail on the ideas of others, Newton took his real inspiration from the workings of his own fertile mind." This is amply confirmed by the documents—short essays or tracts rather than annotations—that contain "the written record of his first researches in the interlocking structures of Cartesian coordinate geometry and infinitesimal analysis," which comprise the second part of this volume, accounting for more than half of the whole.

It is on these papers on analytical geometry and calculus, Whiteside concludes, that "Newton laid the foundations of his mature work in mathematics, revealing for the first time the true magnitude of his genius." Whiteside then presents an admirable summary of Newton's meditations and analyses, which deserves to be reproduced here:

"As in so many other fields—and though we should not underestimate the impact of Wallis' work upon him—in the creation of his calculus of fluxions Newton's chief mentor was Descartes. From the latter's *Geometrie,* a veritable mathematical bible in Schooten's richly annotated second Latin edition, he drew a continuous inspiration over the two years from the summer of 1664. Beginning with the Cartesian technique for constructing the subnormal to an algebraic curve, Newton swiftly soaked up the algorithmic facility of Hudde's rule *de maximis et minimis* but quickly came to appreciate its hidden riches, applying it to the construction of the subtangent and of the circle of curvature at a general point on an algebraic curve and ultimately formulating a differentiation procedure founded on the concept of an indefinitely small, vanishing increment. On that basis and little afterwards he was able to set down the standard differential algorithms in the generality with which they were to be expounded by Leibniz two decades later. Along with this a parallel stream of researches, built on Wallis' work in the theory of the algebraic integral and on Heuraet's general rectification procedure, culminated about the same time (mid-1665) in a limited mastery of the quadrature problem and in geometrical insight into the inverse problem of tangents. In the summer and early autumn of that year, away from books in Lincolnshire and with time for unhurried thought, Newton recast the theoretical basis of his new-found calculus techniques, rejecting the concept of the indefinitely small increment in favour of that of the fluxion, a finite instantaneous speed defined with regard to an independent dimension of time and on the geometrical model of the line-segment. Soon after, in the autumn of 1665, he was led to restudy the tangent-problem by the Robervallian method of combining limit-motions of a point defined in a suitable co-ordinate-system. After an initial crisis in his construction of the quadratrix-tangent he was able correctly to generalize the method, giving in May 1666 a comprehensive treatment of tangents by limit-motion analysis and extending its area of application to include the construction of inflexion points. In the autumn of 1666, lastly, and as a not unintended finale, almost all these researches were ordered and condensed in a short, unfinished and till recently unpublished work to which he gave no title but which, following his own practice in later reference, we may name the 'October 1666' tract.

"In those two years a mathematician was born: a man, certainly, still capable of profound error but with a depth of mathematical genius which by late 1666 had made him the peer of Huygens and James Gregory and probably the superior of his other contemporaries. His only earnest regret must have been that he had yet found no outlet for communicating his achievement to others. The papers printed in the following pages throb with energy and imagination but yet convey the claustrophobic air of a man completely wrapped up in himself, whose only real contact with the external world was through his books. That was to change somewhat in years to come, but it was Newton's continuing tragedy that he was never to find a collaborator of his own mental stature."

One final observation can be made. In the papers presented by Whiteside, Newton shows himself to have been extraordinarily fertile in the invention of mathematical symbols as they were needed. In dealing with the problems of tangents and of curvature he introduced a kind of dot notation for partial differentiation but not for ordinary differentiation. This is surprising because Newton's fluxional calculus is generally known for the use of "pricked," or dotted, letters for ordinary differentials, for example $\dot{x}$ for dx/dt and $\ddot{x}$ for d^2x/dt^2. He also employed a small square to indicate integration. The most interesting symbol is a kind of script X he seems to have devised around 1666 for a general algebraic function of two variables $f(x,y)$. Then, by placing one or more dots to the left or right, he could indicate xf_x, yf_y, x^2f_{xx}, xyf_{xy} or y^2f_{yy}. Whiteside finds, however, that he did not invent dotted letters for ordinary differentials until 1691, which provides an answer for those who are puzzled because he did not use his dotted letters in the *Principia.* It was published in 1687.

Leibniz

6

by Frederick C. Kreiling
May 1968

The theme of his career was the search for a universal language and algebra of reasoning. He perfected the calculus by establishing its fundamental notation, and he pointed the way toward symbolic logic

> ...*there would be no more need of dispute between two philosophers than between two accountants. It would suffice for them to take their pencils in their hands, sit down to their slates, and say to each other...: "Let us calculate."*
>
> —GOTTFRIED WILHELM LEIBNIZ

Such was Leibniz' dream: to develop a generalized symbolic language, and an algebra to go with it, so that the truth of any proposition in any field of human inquiry could be determined by simple calculation. His quest was unsuccessful, but he did invent the calculus, a mathematical way of dealing with change and motion, and he did devise and promote much of modern mathematical notation. Perhaps even more important, his vision pointed the way toward modern symbolic logic, as such 20th-century mathematicians and logicians as Bertrand Russell, Kurt Gödel and Alfred Tarski have recognized. Indeed, Norbert Wiener suggested that Leibniz might be considered the patron saint of communication theory and control theory, twin mathematical foundations of much contemporary technology, because his thought centered on "two closely related concepts, that of a universal symbolism and that of a calculus of reasoning. From these are descended the mathematical notation and the symbolic logic of the present day."

The same passion for universality that prompted Leibniz' search for a mathematical *lingua generalis* informed his entire varied career. He functioned at one time or another, and often simultaneously, as political theorist, diplomat, engineer, inventor of gadgets and house historian for German princely families. He aimed at binding into a consistent whole all the tangled threads of 17th-century thought, and in his philosophy he did manage to link mathematics, physics, metaphysics, psychology and theology. In the last analysis, however, his effort at synthesis failed. His life is interesting more for his extraordinary versatility and his sharp perceptions, many of them foreshadowing modern developments, than for a unified and systematic body of thought.

PORTRAIT OF LEIBNIZ on the following page was painted in about 1695, when he was head of the ducal library and an adviser to the court in Braunschweig (Brunswick). The portrait, now in the Duke Anton Ulrich Museum there, was once attributed to Andreas Scheits but is now thought to have been done by Christoph Bernhard Francke.

Gottfried Wilhelm Leibniz was born in Leipzig in 1646, the son of a professor of moral philosophy at the university. At eight he taught himself Latin; soon he was reading Greek and at 14 he was immersed in Aristotle. He immediately found himself questioning the master's formal system. The 10 Aristotelian "categories"—substance, quantity, quality, relation, place, time, position, possession, action, affection—had been held for centuries to be necessary elements of all thought, scientific or otherwise; for the scientific revolution of the 17th century to advance, their hold on the imagination of learned people had to be loosened. Encountering them as a boy, Leibniz wondered whether perhaps the Aristotelian categories had subsets, whether they could fit into larger categories, whether a "regular passage" might not be found among them. (Eventually just such questions were to be dealt with in the 19th century by George Boole and John Venn.)

Leibniz' philosophy of mathematics turned out to be much like Aristotle's in that its metaphysical and logical theories were closely related. Aristotle had held that in logic every proposition could be reduced to a subject-predicate form; this directly paralleled his metaphysical doctrine that the world is made up of "substances" with "attributes." Leibniz held that the predicate of every proposition is "contained in" the subject, and this paralleled his metaphysical doctrine that the world consists of self-contained points, the "monads," that operate in preestablished harmony. The metaphysics, in turn, was inextricably bound up with mathematics. "There are," he wrote in the *Monadology*, the final summary of his philosophy, "two kinds of truth, those of reasoning and those of fact. Truths of reasoning are necessary and their opposite is impossible; truths of fact are contingent [on definition or on perception, for example] and their opposite is possible." Truths of reasoning, for Leibniz, included all mathematical axioms, postulates, definitions and theorems, since their opposites involve contradiction.

Leibniz agreed in theory with Plato that diagrams, geometric figures and means of notation in general were mere aids to mathematical thinking, but he placed great emphasis on their importance in practice. He called them the "thread of Ariadne" that could guide the mind, and he was always looking for "methods of forming and arranging characters and signs, so that they represent thoughts, that is to say, that they are related to each other as the corresponding thoughts." Unlike most mathematicians of his day, he made an extended study of notation, in the course of which he corresponded with many of the leading mathematicians he knew: the Bernoullis in Switzerland, John Wallis in England, Christiaan Huygens in Paris. He was responsible for introducing more signs and symbols than any other mathematician with the exception of Leonhard Euler, and he promoted the use of still

more. He recommended the use of parentheses to set apart portions of algebraic expressions instead of the vinculum, a line above the terms, which had long been in use. He proposed the dot to indicate multiplication (because the St. Andrew's cross was too readily confused with the letter X), the decimal point, the equal sign, the colon for division and ratio. He also introduced numerical superscripts as exponents for the letter terms in algebra instead of merely repeating the letter.

The notation of the calculus as we know it today is in large part due to Leibniz. For the process of integration, which had been developing slowly ever since the time of Archimedes, he proposed the familiar symbol $\int$, an elongated *s* signifying "summation." Perhaps the best-known of his innovations is the *d* for the differential. Isaac Newton, who had invented the calculus independently, used dots and dashes above the letters to indicate what he called "fluxions" and "fluents," and they were difficult to read and to print. It is generally agreed that the development of the calculus in England was hindered until well into the 19th century because English mathematicians remained loyal to Newton's notation while their Continental colleagues moved ahead into new areas with Leibniz' more expressive system. Leibniz sought to make the form of a symbol reflect its content. "In signs," he wrote, "one sees an advantage for discovery that is greatest when they express the exact nature of a thing briefly and, as it were, picture it; then, indeed, the labor of thought is wonderfully diminished."

Leibniz had a touch of the mystic, and sometimes he saw mathematical notation as a reflection of a higher order of

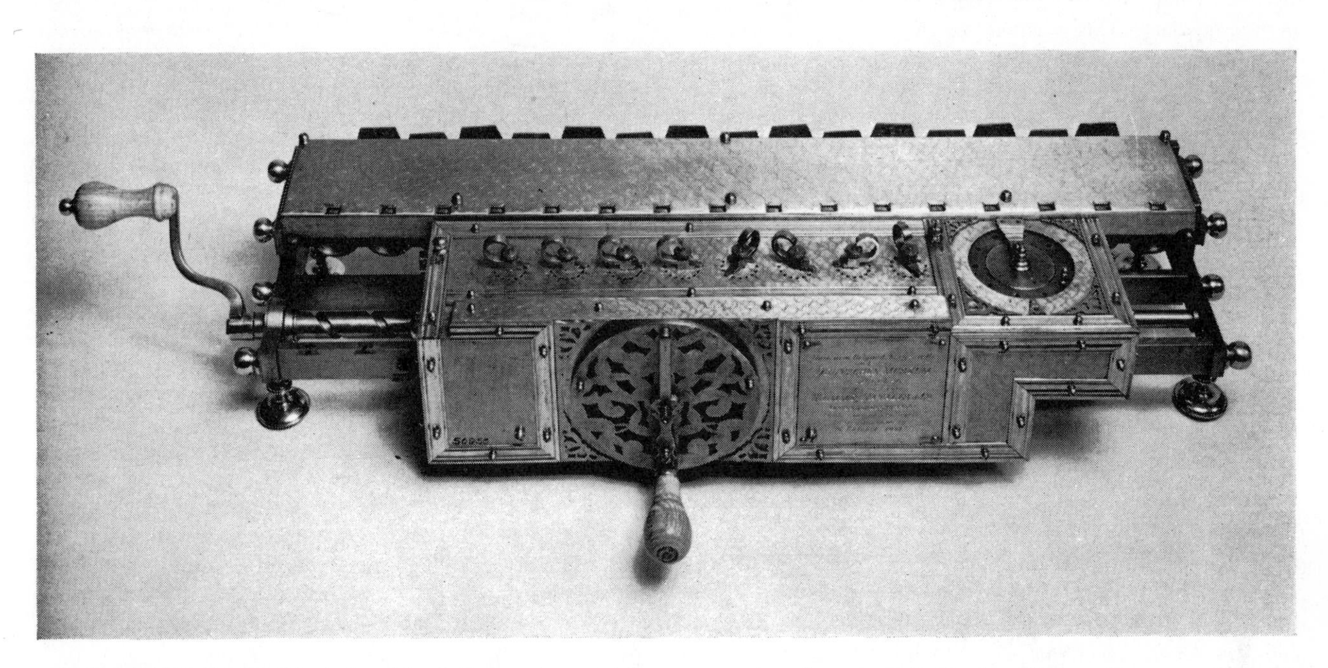

CALCULATING MACHINE, designed by Leibniz to do multiplication and division by repeated addition and subtraction, may never have been built in his day. This version was made to his specifications in 1923 and is in the Deutsches Museum in Munich.

MACHINE is seen from the bottom. The key elements are the eight "stepped cylinders" with teeth of different lengths. As each turns it advances a totalizer gear by an amount that depends on the number of teeth engaged, which in turn depends on the cylinder's position.

reality. Having devised a system of binary arithmetic, for example, he came to see its symbols as analogues of the ancient Chinese magic symbols called trigrams, the meaning of which he thereupon tried to fathom. His interest in Chinese culture had a more practical aspect, however. The fact that Chinese ideograms symbolize concepts rather than sounds, and can therefore represent the same things in different dialects, suggested to him that he might find in Chinese a means of constructing his universal symbolic language. He strove to advance cultural contact between Europe and China, with Russia serving as an intermediary. This was one of the objectives he proposed to Peter the Great when in 1711 he discussed with the czar the formation of a scientific society in Russia.

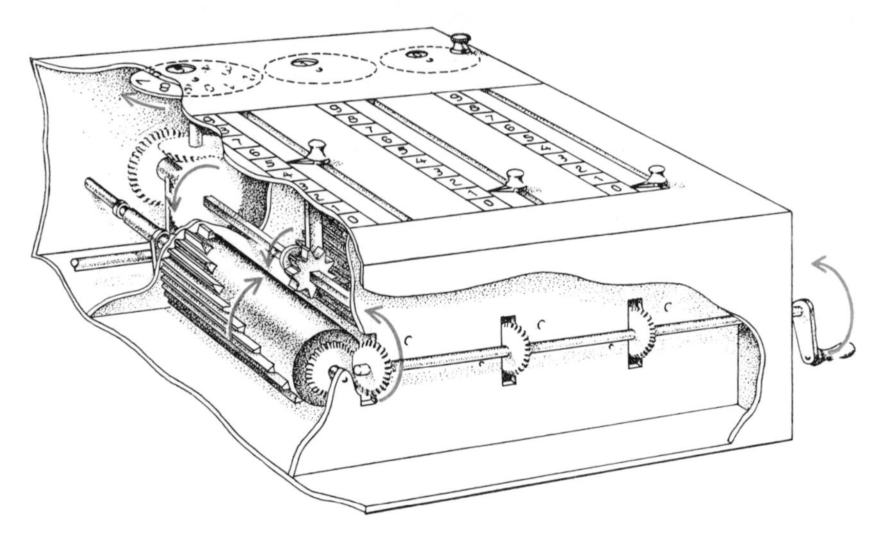

PRINCIPLES of the Leibniz calculating machine are illustrated in a diagram based on one from a 19th-century book. In this version a crank (*right*) turns a set of stepped cylinders. The totalizer gears are positioned by knobs to represent the multiplicand, in this case 510; a turn of the cylinder at left, for example, therefore engages five teeth of its totalizer, turning the square shaft and, through bevel gears, advancing the counter (*top*) to 5. Thus each turn of the crank adds the multiplicand one more time. As a counter goes from 9 to 0 an arm attached to it causes the next cylinder to the left to advance one unit to effect the carrying function.

If Peter were to send a mission to China, Leibniz suggested, a suitable diplomatic present might well be Leibniz' own calculating machine [*see illustrations on previous page*]. Blaise Pascal had earlier constructed several devices for addition and subtraction; Leibniz extended Pascal's principles to accomplish multiplication and division by repeated addition or subtraction. The key elements in Leibniz' machine were the "stepped cylinders," in effect long gears with nine teeth, each a different length. Smaller gears were set above them, each representing a digit of the multiplicand and placed so as to be engaged by that number of the long gears' teeth [*see illustration on this page*]. Each complete turn of the set of long gears therefore registered the multiplicand once; the multiplier was expressed by the number of times the long gears were turned. When the first commercial calculating machines were made in the early 19th century by Charles Xavier Thomas in Alsace, they incorporated the Leibniz stepped wheel.

Leibniz' considerable flair for engineering was exhibited primarily in the large number of designs he left, most of them unexecuted: wagon wheels that would plow through mud, improved ship hulls and smokestacks, even a new type of nail with tiny spurs to fix it more firmly in the wood. To some extent his technological ability was put to the service of the various dukes and princes by whom he was employed as a kind of all-purpose civil servant. There was, for example, the incident of the silver mines. Leibniz' patron Duke John Frederick of Hanover owned mines in the Harz mountains; their yield was low, but they helped to maintain the duke's balance of payments. Thus it was a serious matter when a drought lowered the level of streams at the surface, rendering inoperative the waterwheels that drove pumps to clear the mine passages of seeping ground water. Leibniz was given a contract to set up a windmill-operated pump and work it for one year; if the pilot project was successful, Leibniz would receive 1,200 talers a year, a considerable income. The inevitable friction between science adviser and bureaucrat ensued: the mine directors protested that Leibniz was a theorist who lacked practical experience; he accused them of incompetence. His plan—to use windmills to pump the mine water above the surface and then let it flow down an inclined plane, turning a waterwheel that would help to pump more water—was technically valid. It ran into all kinds of trouble, however, at least partly because the directors had given him the most difficult mine to work with. Leibniz persisted, and after four years the system finally seemed to be functioning. A delegation from the duke arrived for an inspection. That day the wind failed, and Leibniz' grant was promptly terminated. He persisted for two years with his own money, but finally gave up, fearing that his reputation would be ruined.

Typically the trips to the Harz mountains developed new interests for Leibniz: geology and paleontology. In these areas his ideas were essentially those of most other 17th-century students of natural history, but he asked some interesting questions. He saw that the traditional views could not account for the age of the earth, which, it seemed to him, must be immensely old. The whole question of chronology, both evolutionary and historical, was a burning issue in those years. Newton was interested in it. Robert Boyle and the Italian scholar Lodovico Antonio Muratori spent years trying to establish precise dates for historical events, and chronology was a major concern in Leibniz' ventures into historiography.

In 1685, as the windmill project was abandoned, Duke Ernst August of Hanover commissioned Leibniz to write a full-length history of his house. The duke wanted a document that would justify his political aspirations; Leibniz looked on the project as an intellectual challenge. He grounded the history in a number of geological observations and then burrowed into old documents, copying out royal edicts and diplomatic letters. After several years he brought out one thick volume, and then two more, all stuffed with quotations from original sources. His readers had expected something a little more digestible. When some of them told him so, he replied that what he had attempted was a unique kind of work that would put the events of the past into a clear new light; among other things, he felt that an authentic history could establish a new code of international relations.

Leibniz was in some respects rather conservative. The disunity of the remains of the Holy Roman Empire pained him; unlike most political theorists of his day, he was for holding it together in the name of a united Europe rather than seeing it break up into nation states. His role as a diplomat was often that of a mediator between opposing political factions. The same was true in religion: he spent years trying to bring about unity between the Catholic and the Protestant church. In position papers and correspondence he often used the pen name "Pacidius."

If he was a peacemaker in diplomacy, Leibniz was an impassioned partisan in physics, mathematics and philosophy. Through correspondence and publication he engaged in a series of strenuous debates, largely with English scholars. It all began when he and Newton became embroiled in their bitter struggle over which of them had first devised the calculus. Neither appears to have been exactly forthright; each unnecessarily denigrated the contributions of the other. It is now fairly certain that they each discovered it independently. Newton wrote on his method of fluxions for dealing with velocities and change as early as 1665 and continued to do so for the next decade, but he did not publish on the subject until 1687, three years after Leibniz had published in the journal *Acta Eruditorum* a brief and cryptic essay: "A new method for maxima and minima, as well as tangents, . . . and a curious type of calculus for it."

Leibniz had come to the calculus by way of combinatorial analysis. (It was in his *De arte combinatoria,* written at the age of 20, that he first proposed "a general method in which all truths of the reason would be reduced to a kind of calculation.") At first it was the logical and the occult that interested him most; he studied the number diagrams of the 13th-century Spanish mystic Ramón Lull, for example. It was only when he visited Paris and London in the 1670's that he became acquainted with leading mathematicians, including Huygens, and learned about contemporary advances in algebra and geometry. Soon he discovered the fundamental principle of the calculus: that differentiation, the means of studying limits and rates, is the inverse of integration. Neither Leibniz nor Newton was ever able to establish a rigorous basis for the calculus, but both had overcome the prime obstacle set up by the mathematics of the ancients: the belief, inspired by Plato, that scientific treatment of variability was impossible because of the unchanging nature of true reality.

Leibniz' dispute with Newton spread into other areas, notably the nature of gravitation. The controversy that surrounded Newton's ideas is easier to understand for us—who have been exposed to relativity and quantum theory and to the intellectual resistance to them—than it was for 19th-century scholars, whose belief in the absoluteness of Newton's universe had been repeatedly confirmed.

What is important to realize is that to Leibniz—and to Huygens and other leading natural philosophers—Newton's theory seemed a regression to medieval notions that had been dispelled with great effort. Natural philosophers interested in physics had gradually freed themselves from the old scholastic concepts of qualities and powers and from animistic ideas in general. They had come to see every force as the effect of the motion of material particles; they admitted of no way for bodies to affect one another except through the force of impact exerted when they came into contact. They had gone as far as to develop complex systems for explaining the motions of the planets and the behavior of heavy bodies on the earth by means of the motion of corpuscles. To accept the theory of gravitation uncritically would be to forsake these ideas for an explanation that apparently depended on a mysterious force exerted on each other by two separate bodies in empty space, without benefit of a medium between them. It meant a return to "action at a distance," which even scholastic philosophers had rejected.

In short, gravitation seemed to Leibniz to contradict the mechanistic principles to which he was fully committed. He began his critique in 1690, after reading Newton's *Principia,* and kept it up until he died. Some of his ideas on the subject, developed in correspondence with Samuel Clarke, an English philosopher and friend of Newton's, had a prophetic edge. Space and time are not independent, absolute entities as René Descartes and Newton held, he wrote, but rather systems of relations and order among things. "As for my own opinion," he wrote in his third letter to Clarke, "I have said more than once that I hold space to be something purely relative, as time is; that I hold it to be an order of coexistences, as time is an order of successions. For space denotes, in terms of possibility, an order of things which exist at the same time, considered as existing together without inquiring into their manner of existing." Again, in his fifth letter he observed that "it is sufficient to consider these . . . relations and the rules of their changes without needing to fancy any absolute reality out of the things whose situation we consider."

As Albert Einstein noted, Leibniz' criticisms were justified. Einstein pointed out, however, that "had they won out at that time, it hardly would have been a boon to physics, for the empirical and theoretical foundations necessary to follow up his idea were not available in the 17th century" [see "On the Generalized Theory of Gravitation," by Albert Einstein; SCIENTIFIC AMERICAN, April, 1950].

INTEGRATION SIGN, the $\int$ (for "summation"), was proposed by Leibniz in a letter to Henry Oldenburg, secretary of the Royal Society, written in 1675. "It will be useful," he suggests, "to write $\int$ for omn.," which had been used to indicate the product of integration.

Together with Huygens, Leibniz also contributed to the first clear formulation of the principle of the conservation of mechanical energy, the energy of motion and position. Here again his ideas emerged in the course of a long dispute, this time with the Cartesians. Descartes had believed the quantity of rest and motion in the universe was constant and the "efficiency" of objects was equal, in the case of colliding bodies, to the product of their mass and velocity (mv). Leibniz held that the velocity should be squared—that the measure was mv^2. It is easy to resolve the dispute in contemporary terms: what the Carte-

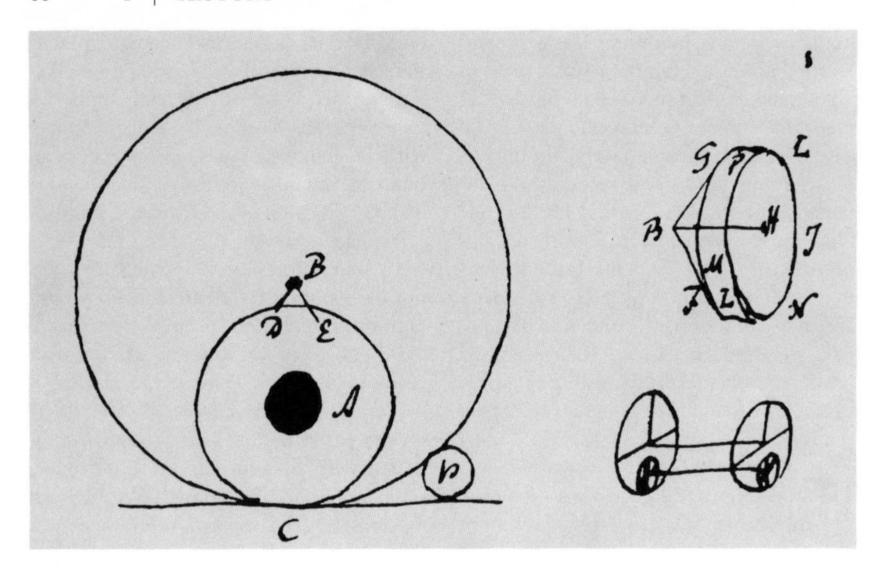

SKETCH BY LEIBNIZ illustrates his design of a wheel to move through mud or over obstacles. An eccentric flywheel (*A*) rides in the rim of a wagon wheel (*B*), storing momentum to help surmount a stone (*p*) in the road. The spokes could be bowed or the wheel made of two disks (*top right*) to accommodate the flywheel. Complete assembly is at bottom right.

sians were talking about is what we call momentum, whereas Leibniz was talking about kinetic energy.

When Leibniz spoke of energy, however, he had more in mind than physical laws. What was real for Leibniz was not "extension," as Descartes had held, but "activity" proportional to degree of "sensitivity." The fundamental unit in his metaphysics was the monad, and the simplest monad was a kind of dimensionless elementary particle of energy; then there was a series of gradually more sensitive, and consequently more active, units including the human psyche and ultimately God. At certain of the higher levels the monad was something like a point of view—a substance capable of perception. Here Leibniz distinguished between animals and men. Animals have perceptions and, through memory, a "sort of consecutiveness" that as he describes it is very like the modern concept of a conditioned reflex. Men "act in like manner as animals" most of the time but also have "the knowledge of eternal and necessary truths" that "gives us reason and the sciences, thus raising us to a knowledge of ourselves and of God."

Leibniz' psychological ideas were more fully developed in a long unpublished work he wrote to refute John Locke's *Essay concerning Human Understanding*. He respected Locke but he could not accept the English philosopher's doctrine that the mind amounts to a blank tablet on which experience writes. It was not that Leibniz believed in a concrete factual memory that was present in the brain at birth, but he did insist that experiences are registered in the mind in certain distinct patterns. Moreover, he did not agree with Locke and Descartes that conscious awareness makes up the whole of mental activity. In effect, he proposed the existence of an unconscious. He observed that dreams often awaken previous thoughts; he even conceived of a theory of psychic trauma, noting that "when we are stunned by some blow, fall, symptom, or other accident, an infinite number of minute confused sensations take form within us."

In this remarkable man's body of thought, then, one finds the seeds of symbolic logic and computer design, hints of the special theory of relativity and some anticipation of Freudian psychology. Many scholars have held that Leibniz could have been a greater mathematician if he had stuck to mathematics—but he simply did not want to stick to mathematics. Scholars deplore the fact that he wasted many years as an adviser on trivial matters to German princes—but he chose that career deliberately, in preference to what he considered a dull life in a university town, because routine made him uneasy. He enjoyed meeting people, being involved in the tumultuous public affairs of the day, feeling that he was helping to set things right. He enjoyed traveling, and as he bounced in stagecoaches from capital to capital, engaged in petty diplomatic intrigues, he filled hundreds of sheets of paper with speculations on scores of different subjects. The papers were found when he died, neglected and quite alone, in 1716. They have yet to be completely sorted and analyzed.

The Invention of Analytic Geometry

by Carl B. Boyer
January 1949

It is generally attributed to the great Descartes, but its development goes back as far as attempts to solve the famous riddle of the oracle at Delos

ORNATE DECORATION appears on the first page of Pierre de Fermat's *Introduction to plane and solid loci*. This presents Fermat's independently developed principles of analytic geometry.

ANALYTIC geometry is usually defined as the combination of algebra and geometry through the method of analysis, and its invention is credited to the great 17th-century French philosopher René Descartes. Both these notions, as it happens, are historically inadequate. The term analytic geometry is perhaps as difficult to define as is the word mathematics, and the origin of the one is not more easily indicated than that of the other. The story of the evolution of analytic geometry, indeed, is the history not of a single discovery but of a famous problem and its slow solution, of the essential unity of mathematics and of the growth of mathematical ideas.

The realms of arithmetic and geometry have of course never been entirely independent. The very idea of the measurement of lengths, areas and volumes implies the application of number to geometrical configuration, and this general concept may be taken as the source from which analytic geometry arose. The Babylonians of some 4,000 years ago measured the sizes of rectangular figures with great accuracy, and they made a start on the geometry of the circle. They were well aware of the fact that the sum of the squares of the legs of a right triangle is equal to the square of the hypotenuse—which history has misnamed the Pythagorean theorem. They also took the initial steps toward the idea of a coordinate system. Coordinates are simply magnitudes or distances that determine the position of one point or object with respect to certain other fixed points, lines or objects. Babylonian astronomers, for example, determined the position of a planet at a given time by specifying its angular distances from certain fixed stars or stellar configurations. Egyptian surveyors similarly located points in the Nile Valley by means of a coordinate frame not unlike that formed by a network of city streets and avenues.

Inasmuch as analytic geometry is known also as coordinate geometry, one might be tempted to assume that its invention was a direct consequence of such pre-Hellenic astronomy and geography. This, however, was not the case. Analytic geometry did not arise out of practical problems; it was instead the outgrowth of questions of a purely theoretical and speculative nature.

The source from which the invention arose was one of the classical problems of the age of Pericles. The story goes that Athens was afflicted by a plague, and its citizens, upon appealing to the oracle of Apollo at Delos, were instructed to double the size of the cubical altar. It was generally understood that the cube was to be doubled or "duplicated" exactly, using only compasses and an unmarked straightedge. The people carefully doubled each dimension of the altar, but the plague continued; doubling each edge had increased its volume eightfold rather than twofold. The plague ran its natural course. The Athenians nevertheless continued their attempts to solve the "Delian problem." Not until some 2,000 years later was it recognized that the oracle had sardonically proposed an unsolvable problem.

The Delian problem amounts simply to solving the equation $x^3=2$, but this could not be accomplished by the geometry of the line and circle alone without recourse to arithmetic, as the original restriction dictated. The Greeks were prevented from attempting an algebraic solution, in any case, by a disconcerting discovery which is said to have cost the discoverer his life by shipwreck. A member of the Pythagorean school, one Hippasus, had proved by rigorous reasoning that there is no number that will measure exactly the diagonal of a unit square. The answer to this problem, which is expressed in algebraic terms by the equation $x^2=2$, is an irrational number. To the Greeks this meant that geometrical problems are not to be solved through arithmetic, and hence they banned from their mathematics the notions of a variable and of an arithmetical continuum —a number system having the same continuity as a line. Consequently the Greeks never developed an algebra appropriate to the methods of coordinate geometry.

After many fruitless efforts to solve the cubic equation $x^3=2$, the Greeks finally decided to seek a solution by relaxing the rules, permitting curves other than the circle and straight line to be used. But here a peculiar difficulty was encountered. The Greeks were aesthetically one of the most gifted people of all time, yet the only curves that they had observed in the heavens and on the earth were circles and straight lines. The straight and the round seem to have possessed for the Greeks a peculiar fascination, and upon them they sought to build their astronomy and mechanics, as well as most of their mathematics.

Of all curves seen in routine experience, the most common, with the exception of the straight line, is not the circle but the ellipse. Wheels and other circular objects, when viewed obliquely, appear as ellipses, and the shadows cast by circles are practically always elliptical. Yet there appears to be no evidence that the Greeks noticed this ubiquitous curve until they began their relentless search for a geometrical solution of the equation $x^3=2$. It is reported that the ellipse was discovered, together with the hyperbola and parabola, toward the middle of the fourth century B.C. by Menaechmus, tutor of Alexander the Great. Menaechmus is said to have advised his impatient pupil, "There is no royal road to geometry"— all unaware that the three curves he had discovered were to be crucial in the development of analytic methods, a path to geometry far easier than the one he was expounding.

The equations of two of the curves he obtained, the parabola and the hyperbola, would now be written as $x^2=y$ and $xy=2$, respectively. If y is eliminated from these two equations, the result is $x^3=2$. The Delian problem can therefore be solved by finding the point of intersection of these two curves on a coordinate graph (*see diagram on page 43*).

Menaechmus was aware in a general way of the properties expressed by the equations of these curves, but he did not use analytic geometry in the modern sense. Greek study of the so-called conic sections was entirely in the language of pure geometry, without reference to algebraic considerations. The writings of Menaechmus have been lost, so we do not know the precise manner in which he was led to the curves. It seems likely, however, that he discovered them through a consideration of familiar geometrical solids. He may well have obtained the three conic sections by cutting each of three right circular cones —one acute-angled, one right-angled and one obtuse-angled—by a plane perpendicular to one element of the cone. The Greek mathematician Apollonius showed later in his famous *Conics* that by varying the angle of the cutting plane, all three types of curves can be obtained from a single right or oblique circular cone. It was Apollonius who gave the curves their present names—ellipse, parabola and hyperbola, the literal meanings of which are "less than," "the same as," and "more than," respectively. (Three figures of speech— ellipsis, parable and hyperbole—are derived from the same source.)

The word "curve" is difficult to define precisely, for, as the German mathematician Felix Klein wrote, "Everyone knows what a curve is, until he has studied enough mathematics to become confused through the countless number of possible exceptions." Nevertheless, it is sufficient for most purposes to define a curve in a plane as the locus, or totality, of points in the plane which satisfy a given condition—*i.e.*, it is the path of a point which moves according to a given law. The circle, for example, is the locus of all points at a given distance from a given fixed point. The Greeks knew that the locus of a point which moves so that its distance from a given point is in a fixed ratio to its distance from a given line is either an ellipse, a parabola, or a hyperbola, according as the ratio is less than, equal to, or greater than one. Lacking modern algebra, however, the ancients experienced great difficulty in attacking some questions

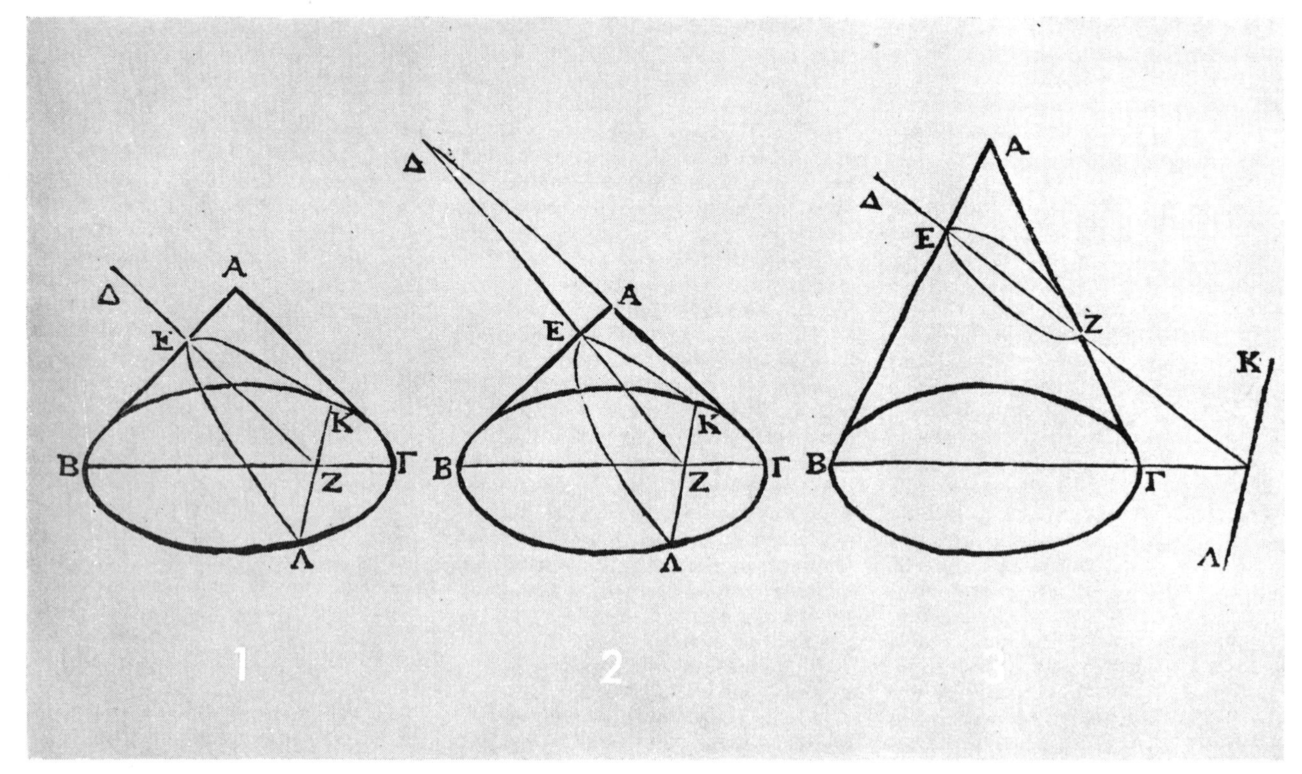

CONIC SECTIONS, which are curves obtained by cutting a circular cone with a plane, were probably first described by Menaechmus, tutor of Alexander the Great. The three sections are the parabola (1), the hyperbola (2) and the ellipse (3). These illustrations of conic sections are from an edition of the Greek mathematician Apollonius' *Conics*. The edition was brought out in 1710 at Oxford by the English astronomer Edmund Halley.

VIÈTE, a 16th-century counsellor to the King of France, was in the direct line of intellectual descent that led to Descartes. He suggested that vowels be used as the symbols for unknown quantities, and consonants for known.

FERMAT, a contemporary of Descartes, expressed basic principle of analytic geometry: "Whenever in a final equation two unknown quantities are found, we have a locus, the extremity of one . . . describing a line."

involving loci which today even a beginning student of analytic geometry handles with ease.

One of their troublesome problems was the following: Given four fixed lines, to find the locus of a point which moves so that the product of its distances from two of the lines shall be in a fixed ratio to the product of its distances from the other two lines. The locus is in all cases one of the conic sections. The problem has become known as "the problem of Pappus," after the Greek mathematician who called attention to it in the fourth century A.D. This problem and its generalization led Descartes to the development of analytic geometry some 1,300 years later.

THE essential ingredient that made this possible was the development of algebra. Pappus had failed to solve the generalization of his problem for want of algebra, and it was in algebra that the Hindus and Arabs of the medieval period, and the European scholars of the Renaissance or early modern period, were strongest. They took less seriously the distinction that the Greeks had made between the discreteness of number and the continuity of geometrical magnitudes. Algebraic symbols, introduced freely into arithmetic and geometry, were used indifferently to designate either numbers or lines. The Persian poet Omar Khayyám, for example, devised an algebra in which he gave both numerical and geometrical solutions for linear and quadratic equations. It was Descartes who first combined an algebraic study of such equations with the geometrical problem of Pappus, and the result of the combination was analytic geometry. But Descartes was partly anticipated by the medieval Latin theologian Nicole Oresme, who had been thinking along entirely different lines.

Oresme's discovery had to do with something called the "latitude of forms," derived from a study of physical variables. In the 14th century, Scholastics at Oxford and Paris, following up Aristotle's generative idea of change, concerned themselves with problems of varying acceleration, density, thermal content, and intensity of illumination. They distinguished not only between uniform and non-uniform rates of change, but also subdivided the latter according as the rate of change of the rate of change was or was not constant. In their "latitude of forms" concept, a "form" was any variable quantity in nature (such as motion, heat or light), and its "latitude" was the value of this quantity corresponding to a given value of the "longitude" or independent variable (generally time or space). The latitude of forms illustrates the important mathematical concepts of "variable" and "function"—the dependence of one quantity upon another so that a change in the one implies a change in the other.

At first this useful idea was not related to graphical methods. But in the middle of the 14th century Oresme, director of the Collège de Navarre in Paris and later Bishop of Lisieux, hit upon the important idea of clarifying functional relationships through reference to geometrical figures. If, for example, the velocity of an object was to be represented as a function of time, the time units were measured along a horizontal base line (longitudes), and the corresponding velocities were illustrated by lines (latitudes) drawn perpendicular to the base line. The totality of the latitudes or velocity lines constituted the graph of the function. For example, the graph of uniformly accelerated motion is a straight line, as shown in manuscripts describing Oresme's ideas.

Oresme's system of latitudes and longitudes appears to be the earliest use of coordinates in the graphical representation of arbitrary functions. Oresme consequently has been hailed by some historians as the inventor of analytic geometry. The usefulness of his ideas, however, was sharply limited. Although he handled linear graphs correctly, he was prevented by deficiencies in geometrical knowledge and algebraic technique from extending his novel idea to curvilinear figures. There is in his work no systematic association of algebra and geometry in which an equation in two variables determines a specific curve, and conversely.

Because Descartes carefully avoided any reference to his predecessors, one cannot say with assurance that he was familiar with the work of Oresme. It seems quite probable that he was. Yet the differences between Descartes' system, called

DESCARTES, a philosopher who utilized mathematics as a basis for rational thought, apparently was not fully aware of the importance of analytic geometry. His concept appeared as appendix to philosophical work.

EULER, a mathematician who was born in Switzerland, lived in Germany and died in Russia, published the first textbook of analytic geometry more than a century after the appearance of Descartes' original principles.

Cartesian geometry, and the graphical representation of the latitude of forms are so great as to make questionable any decisive influence. While Descartes was perhaps slightly in debt to Oresme, the immediate inspiration for his work came not from the 14th-century Bishop of Lisieux, but from a 16th-century counsellor to Henry IV of France named François Viète.

Viète's contribution was the simplification and systematization of algebra by the invention of certain symbols and ideas. The signs + and — had already been substituted for the words plus and minus at the time of Columbus, and half a century later the symbol = for equality had been introduced in England. But convenient notations for the quantities entering into an equation were lacking. Viète suggested that unknown quantities be designated by vowels, and that quantities assumed as known be represented by consonants. This made algebra more than arithmetical legerdemain, for it became a systematic study of types or forms. Equations with numerical coefficients gave way to equations with literal coefficients. Where previously attention had been centered upon the construction of the roots of a particular cubic equation, such as $x^3=2$, Viète showed that the solutions of *all* cubic equations were reducible to the Delian problem of duplicating the cube, or to the trisection of an angle. Geometric problems could be expressed in the language of algebra; and, after algebraic simplification, the roots of the resulting equations could be constructed geometrically.

The application of algebra simplified geometry to a great extent, yet Viète did not discover the royal road for which Alexander had hoped. Oresme had missed the invention of analytic geometry because he had had no algebra adequate to the study of graphical representation; Viète, however, failed to associate his algebraic geometry with a coordinate system.

BUT the stage was set for someone to fuse algebra and geometry through the crucial idea of coordinates, and the result was one of the many cases in science and mathematics of simultaneous discovery. Analytic geometry was the independent invention of two Frenchmen who were the greatest mathematicians of their day, yet neither of whom was a professional in the field. Pierre de Fermat was a lawyer with an absorbing interest in the geometrical works of antiquity; René Descartes was a philosopher who found in mathematics a basis for rational thought. Both men began where Viète had left off, but they continued in somewhat different directions. Fermat retained the notation of Viète, but applied it in a new connection: the study of loci; Descartes adopted the aim of Viète—the geometric solution of algebraic equations—but extended it, in conjunction with modern symbolism, to equations of higher degree.

Fermat composed in Latin only a very short treatise on analytic geometry—*Introduction to plane and solid loci.* It is a work of but eight folio pages, devoted to the line, circle and conic sections. It opens with the statement that although the ancients studied loci, they must have found them difficult, judging from the fact that they often failed to state the problem in general form. Fermat proposes to submit the theory of loci to an analyis that is appropriate to such problems and that will open the way for a general study of them. Without further introduction, he then states in clear and precise language the fundamental principle of analytic geometry:

> *Whenever in a final equation two unknown quantities are found, we have a locus, the extremity of one of these describing a line, straight or curved.*

This brief sentence represents one of the most significant statements in the history of mathematics. It introduces not only analytic geometry, but also the immensely useful idea of an algebraic variable. The vowels in Viète's terminology previously had represented unknown, but nevertheless fixed or determinate, magnitudes. Fermat's point of view gave meaning to indeterminate equations in two unknowns —which previously had been rejected in geometry—by permitting one of the vowels to take on successive line values (corresponding to Oresme's longitudes), and plotting the values of the other as perpendicular lines (latitudes. Oresme would have called them). Thus Fermat rediscovered the graphical representation of variables, and this time there was an algebra at hand with which to exploit the idea.

Fermat showed that equations of the first degree, which today are expressed in general terms as $ax+by+c=0$, represent straight lines; equations of the second degree of the form $x^2+y^2+ax+by+c=0$ represent circles; other equations of second degree represent ellipses, parabolas, and hyperbolas. As the "crowning point" of his treatise, Fermat gave the following proposition:

> *Given any number of fixed lines, the locus of a point which moves so that the sum of the squares of the segments drawn at given angles from the point to the lines shall be constant is a conic section.*

Fermat's analytic geometry was not published during his lifetime, so it is difficult to determine the extent of its influence. The *Introduction* appeared in print for the first time in 1679, 14 years after the death of its author, 40 years after the publication of the corresponding work of Descartes, and half a century after the treatise had been composed. Manuscript works of that day often enjoyed a wide circulation among scholars, but readers of Fermat's work, unaware of the early date of its composition, missed its significance as evidence of his independent invention of analytic geometry. Hence analytic geometry came to be known as "Cartesian geometry." This designation is perhaps not entirely unwarranted, for it was largely through the influence of Descartes, rather than of Fermat, that the new geometry took root.

Descartes' only work on the subject, *La géométrie*, appeared somewhat unobtrusively in 1637, as an appendix to the longer and better-known philosophical treatise, *Discours de la méthode pour bien conduire sa raison, et chercher la vérité dans les sciences.* The whole was published without the author's name, although the authorship was generally known.

The theme of *La géométrie* is set by the opening sentence:

> *Any problem in geometry can easily be reduced* [algebraically by means of coordinates] *to such terms that a knowledge of the lengths of certain lines is sufficient for its construction.*

Descartes was concerned primarily with the geometric solution of equations; indeed, he was a direct descendent of those who, more than 2,000 years before, had attempted to appease the oracle of Apollo at Delos. This concern was reflected in his treatise. Book I is on "problems the construction of which requires only straight lines and circles"—the original Greek limitation on constructions. The goal of the treatise—the third and last book—is on the geometric solution of equations of degree higher than two. The really important, modern part of his work, Book II, which deals with "The nature of curved lines," was dismissed by Descartes as a preliminary to Book III. It is paradoxical that Descartes, from whom the world derived coordinate geometry, showed little interest in this basic principle; he used coordinates simply as an aid to the solution of geometrical problems, and was so indifferent to the theory of curves that he never fully understood the significance of negative coordinates.

Descartes had been much impressed by the power of his method in dealing with the locus of Pappus, and this problem runs like a thread of Ariadne through the three books of *La géométrie*. His fundamental principle of analytic geometry is enunciated in Book II in this way:

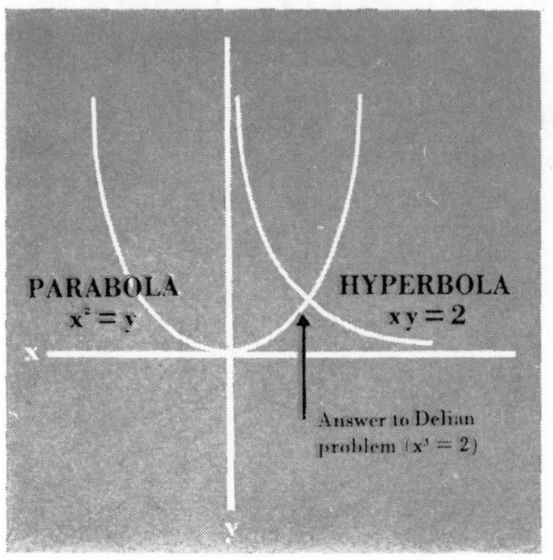

COORDINATES of analytic geometry make it possible to solve the problem stated by oracle of Delos.

> *For the solution of any one of these problems of loci is nothing more than the finding of a point for whose complete determination one condition is wanting . . . In every such case an equation can be obtained containing two unknown quantities.*

This crucial statement means that an equation in two unknowns in general represents a curve. For example, the equation $xy=x^3-2x^2-x+2$, which appears repeatedly, represents what Isaac Newton later called the "Cartesian parabola" or trident.

The loci in Pappus' problem for the case of four lines lead to equations of the first and second degree. Descartes showed that for five, six, seven or eight lines the locus is a curve of degree three or four; and in general if the number of lines does not exceed 2n, the degree of the locus will be not more than n. Descartes was not much concerned, however, with the shape of these curves or loci; he wished to use them to solve, graphically, algebraic equations in a single unknown. But here Descartes made a bad blunder. The Pythagoreans knew that equations of degree one or two could be solved by straight lines and circles alone; Menaechmus, Omar Khayyám and Viète knew that equations of degree three or four could be solved by conic sections, *i.e.*, by curves of order two. Descartes extrapolated too rapidly and asserted that equations of degree 2n or 2n—1 require, for their graphical solution, curves of order n. This would mean, for example, that an equation of degree nine, such as $x^9+x+1=0$, called for curves of order five, whereas actually the cubic curves $y=x^3$ and $y^3+x+1=0$ suffice, through the elimination of y, to solve the equation. As Fermat and others pointed out, the correct rule is that equations of degree not exceeding n^2 are solvable by means of curves of order not greater than n.

Estimates of the relative merit of the works of Fermat and Descartes differ widely, partly because of differences encountered in notation, emphasis and point of view. The algebraic notation of Descartes was far more appropriate than that of Viète and Fermat. To Descartes we owe the use of letters near the end of the alphabet—such as x, y and z—to represent unknowns, and of letters near the beginning of the alphabet to represent known quantities. The immensely convenient notation of exponents for powers (x^2, y^3, etc.) was also introduced by Descartes. On the other hand, the fundamental idea of the equation of a curve is more clearly set forth by Fermat. The work of Descartes is more general in scope, that of Fermat being limited to equations of first and second degree; but the expository treatment of Fermat's *Introduction* is more systematic than that of Descartes' *La géométrie*. One gets the impression that Descartes wrote his geometry to boast rather than to explain. He built it about a difficult problem (the Pappus locus), and he did not go into detail to make his argument clear. In concluding the work he justifies this inadequacy of exposition by the remark that he has left much unsaid in order not to rob the reader of the joy of discovery. In so doing, however, he deprived many a student of the milder pleasure of comprehension. Descartes might better have followed his own advice: "When you have to deal with transcendent questions, you must be transcendently clear."

THE geometry of Descartes and Fermat did not bring about a rapid transformation of mathematics. For one thing, the haughty attitude of Descartes and the indifference to fame of Fermat made the new subject accessible almost exclusively to professional geometers of marked ability. (The incomparable Newton quickly mastered analytic geometry and extended it to include the general theory of curves.) So, although analytic geometry had been twice invented before 1637, it was not until more than a hundred years later, in 1748, that what may be called the first

MATHEMATICAL FUNCTIONS, magnitudes which relate one changing quantity to another, were first represented graphically in a book written in 14th century either by the theologian Nicole Oresme or one of his students.

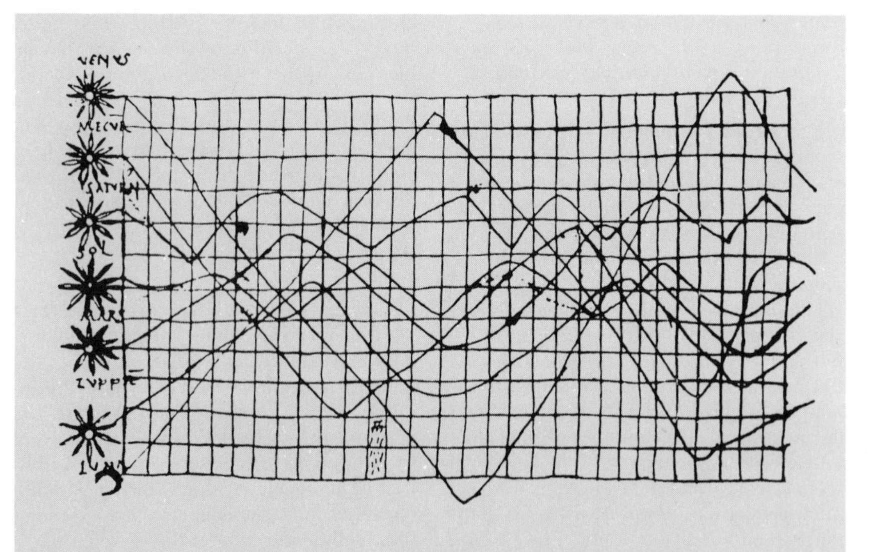

USE OF COORDINATES, one of the fundamental concepts of analytic geometry, is of great antiquity. This sketch of the paths of the sun, moon and planets was drawn in the 10th century. Symbols of these bodies are at the left.

textbook on the subject appeared. This treatise, the *Introduction to infinitesimal analysis* of Leonhard Euler, was written in Latin by a mathematician who was born in Switzerland, lived in Germany and died in Russia. The work remains a classic, yet it has never been translated into English. To the contributions of Fermat, Descartes and Newton, Euler added a significant achievement of his own: he extended Cartesian geometry to the space of three dimensions.

In the end, it was a practicing revolutionist who brought analytic geometry into general use. This was Gaspard Monge, a Frenchman who participated in three "revolutions." As an ardent republican, he took an active part in the affairs of the French revolutionary government, and later was closely associated with Napoleon. As an experimental chemist, he shared the credit for the discovery of the composition of water, a keystone in the chemical revolution. But primarily he was a mathematician, the foremost specialist of his day in geometry, and it was he who saw, more clearly than others, that analytic geometry differs from synthetic geometry as one language does from another. Both say the same thing in different ways. Pure or synthetic geometry is expressed by means of diagrams and constructions; analytic geometry, in terms of algebra and equations. Monge therefore suggested to his students at the famed École Polytechnique that the elementary geometry of lines and planes, of circles and spheres, is quite as appropriately studied by analytic means as are the more advanced properties of the conic sections. This view resulted in a transformation of Cartesian geometry that was every bit as striking in its way as was the new chemistry of Antoine Lavoisier; hence the phrase "analytical revolution" may aptly be applied to the movement initiated by Monge.

There is considerable justification for the proposition that analytic geometry is primarily a French contribution. Its two most important modern precursors, Oresme and Viète, were French; so were its two inventors, Fermat and Descartes, as well as Monge, the man who together with his disciples did most to fashion Cartesian geometry into modern form. On the other hand, the multitudinous developments of the past hundred years have been contributed by men of all nationalities. Indeed, the most noteworthy accounts of the development of analytic geometry are not by Frenchmen but by an Italian, two Germans and an American. These four scholars—Gino Loria, Johannes Tropfke, Heinrich Wieleitner and Julian Coolidge—are symbols of the fact that in mathematics the one-world ideal is not a mirage.

II

NUMBER AND ALGEBRA

II NUMBER AND ALGEBRA

INTRODUCTION

Mathematics proper deals with number and geometrical figures, and abstractions built on these foundation stones. Of the two, geometry would seem to be more important. Objects in this world, including our own bodies, could not exist without shape, and this can be described without calling upon number. However, any civilization that proposes to utilize scientific knowledge for prediction or for engineering must obtain quantitative information, whereas geometrical knowledge is qualitative. Moreover, the history of the development of geometry shows us (see Boyer's article in the previous section) that to make progress even in that subject algebra is immensely useful. Hence, in our Western civilization, arithmetic, algebra (which is essentially generalized arithmetic), and subjects such as the calculus that are built on number have taken precedence over geometry.

The articles in this section are in the main concerned with numbers as such. The scientific uses of number will be far more evident in the later sections on "Statistics and Probability" and "Applications." (See also the book by Gamow in the suggested readings.) In particular, many of the present articles are concerned with whole numbers. Why this concentration on the whole numbers? Of course, all types of numbers—fractions, irrationals (such as $\sqrt{2}$ and π), and complex numbers—are built up from whole numbers. But the whole numbers also have had a special significance. The importance of these numbers was emphasized by the Pythagoreans, the first major group of Greek mathematicians, who were active in the sixth and fifth centuries B.C. The Pythagoreans thought that all matter was composed of indecomposable particles and that the variety of shapes found in the universe could be described not only by the number of particles in the shapes but by the arrangement of the particles. Therefore they classified whole numbers in accordance with the shapes that could be formed with the corresponding number of particles. For example, the numbers 1, 3, 6, 10, 15, etc., were called triangular numbers, because the dots could be arranged in the form of triangles, and thus these numbers could describe triangularly shaped objects (see Figure 1). The Pythagoreans also introduced terms such as square, pentagonal, and hexagonal for whole numbers, each of which described shapes that could be formed with that species of number. Thus the Pythagoreans initiated a scientific program in which the

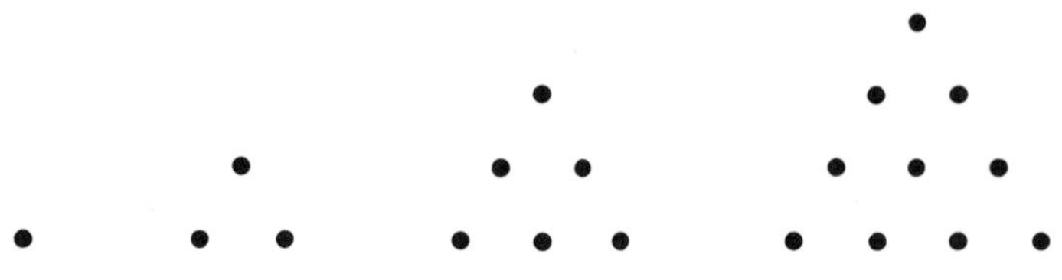

Figure 1

whole numbers would be the key to (or, as they put it, the measure of) all things.

The Pythagorean scientific philosophy was abandoned later in the classical Greek period in favor of geometry, but the study of the properties of the whole numbers for the sake of the intellectual challenge and, to many, the aesthetic interest was continued by the Greeks and became a major and flourishing branch of mathematics from the seventeenth century onward. It is now called the theory of numbers.

What properties of the whole numbers can be so intriguing as to cause people to devote substantial parts of their lives to the subject? Many of the ensuing articles will answer this question. But the mathematics of number does not end with the theory of numbers. Historically, the important additional types—irrational numbers, negative numbers, and complex numbers—entered in due course. Though centuries had to pass before these strange types became accepted and respectable—time does soothe disquieted minds—these types are now staid members of our number system. Another category, the hypernumbers, entered mathematics during the nineteenth century, and these are described in Davis's article. The most recent entry, made in the late nineteenth century, is transfinite numbers—numbers that represent the quantity of objects in infinite or nonfinite collections. Transfinite numbers are still a subject of controversy. Nevertheless, they are here to stay. Gardner's article on transfinite numbers explains their nature. The realm of number is now vast, and Davis's article surveys its full extent.

One of the ingenious ideas of mathematics is the use of letters to represent classes of numbers. Thus the letter a might stand for any one of the rational numbers (positive and negative whole numbers and fractions), or it might stand for any real number (rationals and irrationals). In the equation $ax^2 + bx + c = 0$, a, b, and c might each stand for any whole number or any real number. By using letters to represent any one of a class of numbers, an idea that seems simple now but which was introduced only in the late sixteenth century by François Vieta, mathematics advanced from number to algebra in a real sense. For the first time, assertions could be made and proved about classes of numbers, classes of equations, classes of matrices, or classes of any other concepts relating to number. As the variety of types of numbers and algebraic expressions increased, it became apparent that these collections possessed different properties. Thus if we limit ourselves to the positive and negative whole numbers we can add and subtract them, but we cannot divide 5 by 6, say, and still stay within the class of whole numbers. However, in the class of rational numbers we may add, subtract, multiply, and divide any two numbers (excluding, of course, division by zero) and still end up with a rational number. For any real or complex numbers a and b, we may assert that $ab = ba$; this is called the commutative property of multiplication. But this property does not hold for hypernumbers and matrices.

To classify the types of algebras—the algebra of whole numbers, the algebra of complex numbers, the algebra of matrices, etc.—mathematicians undertook the study of algebraic structures. This new and very active branch of mathematics is called abstract algebra, and its nature is more fully described in Davis's article and in Sawyer's book, which is listed in the suggested readings. One of the surprises encountered in mathematics is that ideas created and developed for one purpose prove to be applicable to situations or phenomena totally different from this original purpose. In the case of abstract algebra, the notion of group has proved to be broadly applicable to geometry and even to physics. Although an investigation of these applications would take us far beyond the content that we can hope to encompass in this book, the interested reader can get a fine presentation of the application of group theory to physics in Freeman Dyson's article on "Mathematics in the Physical Sciences," *Scientific American,* September 1964.

SUGGESTED READINGS

Archibald, Ralph G. 1970. *An Introduction to the Theory of Numbers.* Merrill, Columbus, Ohio.

Cantor, Georg. No date. *Contributions to the Founding of the Theory of Transfinite Numbers.* Dover, New York.

Dantzig, Tobias. 1954. *Number, the Language of Science,* fourth edition. Macmillan, New York.

Davis, Philip J. 1961. *The Lore of Large Numbers.* Random House, New York.

Gamow, George. 1947. *One, Two, Three . . . Infinity.* Viking, New York. Also available in paperback.

Ore, Oystein. 1948. *Number Theory and Its History.* McGraw-Hill, New York.

Ore, Oystein. 1969. *Invitation to Number Theory.* Random House, New York.

Sawyer, W. W. 1959 *A Concrete Approach to Abstract Algebra.* W. H. Freeman and Company, San Francisco.

Vinogradov, I. M. 1954. *Elements of Number Theory.* Dover, New York.

Zippin, Leo. 1962. *Uses of Infinity.* Random House, New York.

The Remarkable Lore of the Prime Numbers

8

by Martin Gardner
March 1964

No branch of number theory is more saturated with mystery and elegance than the study of prime numbers: those exasperating, unruly integers that refuse to be divided evenly by any integers except themselves and 1. Some problems concerning primes are so simple that a child can understand them and yet so deep and far from solved that many mathematicians now suspect they *have* no solution. Perhaps they are "undecidable." Perhaps number theory, like quantum mechanics, has its own uncertainty principle that makes it necessary, in certain areas, to abandon exactness for probabilistic formulations.

The central difficulty is that the primes are scattered along the series of integers in a pattern that clearly is not random and yet defies all attempts at precise description. What is the 100th prime? The only way a mathematician can answer is by obtaining a list of primes and counting to the 100th. How is such a list obtained initially? The simplest method is to go through the integers and cross out all the composite (not prime) numbers. Of course a computer can do this with great speed, but it still must use essentially the same simple-minded procedure that Eratosthenes, the Alexandrian geographer-astronomer and friend of Archimedes, devised 2,000 years ago.

There is no better way to become familiar with the primes than by using Eratosthenes' Sieve (as his procedure is called) for sifting out all primes under 100. Kenneth P. Swallow of Monterey, Calif., has proposed an efficient way to do this. Write the numbers from 1 to 100 in the rectangular array shown at the right. Cross out all multiples of 2, except 2 itself, by drawing vertical lines down the second, fourth and sixth columns. Eliminate the remaining multiples of 3 by drawing a line down the third column. The next integer not crossed out is 5. Multiples of 5 are removed by a series of diagonal lines running down and to the left. Remaining multiples of 7 are eliminated by lines sloping the other way. The integers 8, 9 and 10 are composite: their multiples have already been crossed out. Our job is now finished because the next prime, 11, is larger than the square root of 100, the highest number in the table. Had the table been longer, larger multiples of 11 would have been removed by diagonal lines of steeper slope.

All but 26 numbers (shown in color) have fallen through the sieve. These are the first 26 primes. Mathematicians prefer to say 25 primes, because various important theorems are simpler to express if 1 is not called a prime. For example, the "fundamental theorem of arithmetic" states that every integer greater than 1 can be factored into a unique set of prime numbers. Thus 100 is the product of four primes: $2 \times 2 \times 5 \times 5$. No other set of positive primes has a product of 100. If 1 were called a prime, we could not say this. There would be an infinite number of different sets of prime factors, such as $2 \times 2 \times 5 \times 5 \times 1 \times 1$.

Much can be learned about the primes by studying the array on this page. You see at once that all primes greater than 3 are either one less or one more than a multiple of 6. Also, it is clear why there are so many "twin primes": pairs of primes that have a difference of 2, such as 71 and 73, 209,267 and 209,-269, or 1,000,000,009,649 and 1,000,-000,009,651. After eliminating multiples of 2 and 3, *all* remaining numbers are twin-paired. Subsequent sievings simply remove one or both partners of a pair, but they leave many untouched. Twin primes get scarcer as the numbers get bigger. It is conjectured that an infinity of them continue to sift through the sieve, but no one knows for certain. The chart also shows at a glance that 3, 5, 7 is the only possible triplet of primes.

If the integers are differently placed, the primes will of course form a different geometrical pattern. Last fall Stanislaw M. Ulam of the Los Alamos Scientific Laboratory, attended a scientific meeting at which he found himself listening to what he describes as a "long and very boring paper." To pass the time he doodled a grid of horizontal and vertical lines on a sheet of paper. His first impulse was to compose some chess problems, then he changed his mind and began to number the intersections, starting near the center with 1 and moving out in a counterclockwise spiral. With no spe-

1	2	3	4	5	6
7	8	9	10	11	12
13	14	15	16	17	18
19	20	21	22	23	24
25	26	27	28	29	30
31	32	33	34	35	36
37	38	39	40	41	42
43	44	45	46	47	48
49	50	51	52	53	54
55	56	57	58	59	60
61	62	63	64	65	66
67	68	69	70	71	72
73	74	75	76	77	78
79	80	81	82	83	84
85	86	87	88	89	90
91	92	93	94	95	96
97	98	99	100		

The Sieve of Eratosthenes

cial end in view, he began circling all the prime numbers. To his surprise the primes seemed to have an uncanny tendency to crowd into straight lines. The illustration at the bottom of page 51 shows how the primes appeared on the spiral grid from 1 to 100. (For clarity the numbers are shown inside cells instead of on intersections.)

Near the center of the spiral the lining up of primes is to be expected because of the great "density" of primes and the fact that all primes except 2 are odd. Number the squares of a checkerboard in spiral fashion and you will discover that all odd-numbered squares are the same color. If you take 17 checkers (to represent the 17 odd primes under 64) and place them at random on the 32 odd-numbered squares, you will find that they form diagonal lines. But in the higher, less dense areas of the number series one would not expect many such lines to form. How would the grid look, Ulam wondered, if it was extended to thousands of primes?

The computer division at Los Alamos has a magnetic tape on which 90 million prime numbers are recorded. Ulam, together with Myron L. Stein and Mark B. Wells, programed the MANIAC computer to display the primes on a spiral of consecutive integers from 1 to about 65,000. The picture of the grid presented by the computer is shown at the right below. Note that even near the picture's outer limits the primes continue to fall obediently into line!

The eye first sees the diagonally compact lines, where odd-number cells are adjacent, but there is also a marked tendency for primes to crowd into vertical and horizontal lines on which the odd numbers mark every other cell. Straight lines in all directions (once they have been extended beyond the consecutive numbers on a segment of the spiral) bear numbers that are the values of quadratic expressions beginning with $4x^2$. For example, in the illustration on the cover the diagonal sequence of primes 5, 19, 41, 71 is given by the expression $4x^2 + 10x + 5$ as x takes the values 0 through 3. The grid on the cover suggests that throughout the entire number series expressions of this form are likely to vary markedly from those "poor" in primes to those that are "rich," and that on the rich lines an unusual amount of clumping occurs.

By starting the spiral with numbers higher than 1 other quadratic expressions form the lines. Consider a grid formed by starting the spiral with 17 [*see illustration at left top, page 51*]. Numbers in the main diagonal running northeast by southwest are generated by $4x^2 + 2x + 17$. Plugging positive integers into x gives the diagonal's lower half; plugging negative integers gives the upper half. If we consider the entire diagonal, rearranging the numbers in order of increasing size, we find—pleasantly enough—that all the numbers are generated by the simpler formula $x^2 + x + 17$. This is one of many "prime-rich" formulas discovered by Leonhard Euler, the 18th-century Swiss mathematician. It generates primes for all values of x from 0 through 15. This means that if we continue the spiral shown in the illustration until it fills a 16-by-16 square, the entire diagonal will be solid with primes.

Euler's most famous prime generator, $x^2 + x + 41$, can be diagramed similarly on a spiral grid that starts with 41 [*see illustration at right top, p. 51*]. This produces an unbroken sequence of 40 primes, filling the diagonal of a 40-by-40 square! It has long been known that of the first 2,398 numbers generated by this formula, exactly half are prime. After testing all such numbers below 10,000,000, Ulam, Stein and Wells found the proportion of primes to be .475... Mathematicians would like to discover a formula expressing a function of n that would give a different prime for *every* integral value of n. It has been proved that no polynomial formula of this type exists. There are many nonpolynomial formulas that *will* generate only primes, but they are of such a nature that they are of no use in computing primes because the sequence of primes must be known in order to operate with the formulas. (See "History of a Formula for Primes," by Underwood Dudley, *The American Mathematical Monthly*, January, 1969.)

Ulam's spiral grids have added a touch of fantasy to speculations about the enigmatic blend of order and haphazardry in the distribution of primes. Are there grid lines that contain an infinity of primes? What is the maximum prime

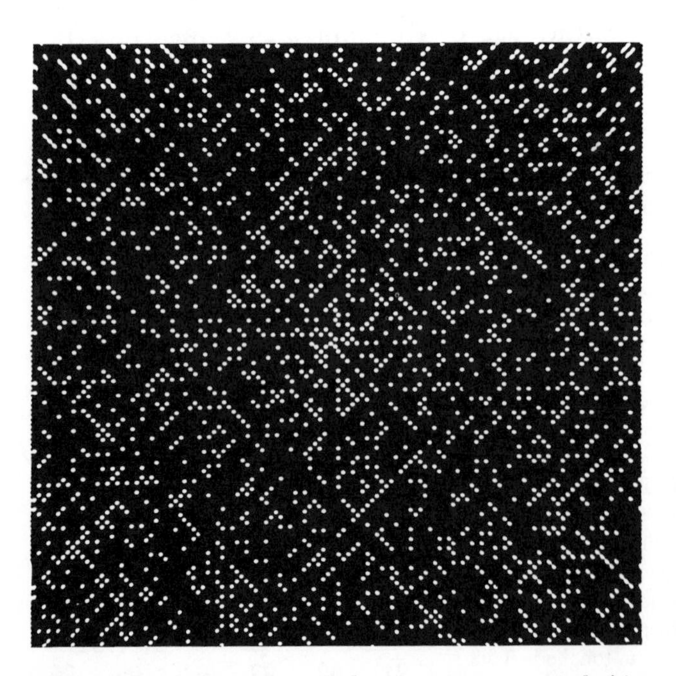

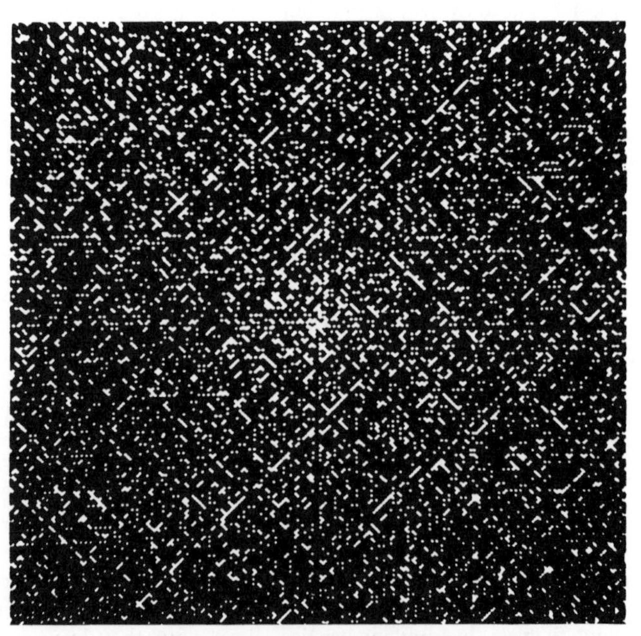

***Photographs of a computer grid showing primes as a spiral of integers from 1 to about 10,000* (left) *and from 1 to about 65,000* (right)**

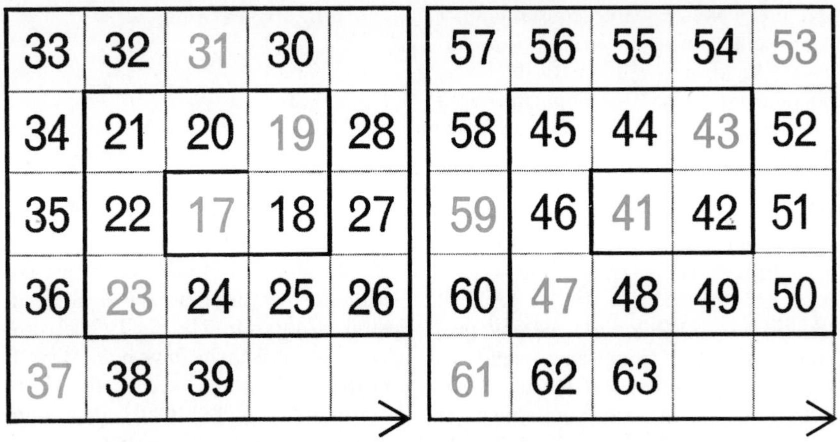

33	32	31	30	
34	21	20	19	28
35	22	17	18	27
36	23	24	25	26
37	38	39		

57	56	55	54	53
58	45	44	43	52
59	46	41	42	51
60	47	48	49	50
61	62	63		

Diagonals generated by the formula $x^2 + x + 17$ **(left)** ***and*** $x^2 + x + 41$ **(right)**

density of a line? On infinite grids are there density variations between top and bottom halves, left and right, the four quarters? Ulam's doodlings in the twilight zone of mathematics are not to be taken lightly. It was he who made the suggestion that led him and Edward Teller to think of the "idea" that made possible the first thermonuclear bomb.

Although primes grow steadily rarer as numbers increase, there is no highest prime. The infinity of primes was concisely and beautifully proved by Euclid. One is tempted to think, because of the rigidly ordered procedure of the sieve, that it would be easy to find a formula for the exact number of primes within any given interval on the number scale. No such formula is known. Early-19th-century mathematicians made an empirical guess that the number of primes under a certain number n is approximately n/natural log of n, and that the approximation approaches a limit of exactness as n approaches infinity. This astonishing theorem, known as the "prime-number theorem," was rigorously proved in 1896. (See David Hawkins' article "Mathematical Sieves," beginning on page 55, for a discussion of this theorem and its application to other types of numbers, including the "lucky numbers" invented by Ulam.)

It is not easy to find the mammoth primes isolated in the vast deserts of composite numbers that blanket ever larger areas of the number series. At the moment the largest known prime is $2^{19937} - 1$, a number of 6,002 digits. It was discovered in 1971 by Bryant Tuckerman, at IBM's research center, Yorktown Heights, New York. Before the advent of modern computers, testing a number of only six or seven digits could take weeks of dreary calculation. Euler once announced that 1,000,009 was prime, but he later discovered that it is the product of two primes: 293 and 3,413. This was a considerable feat at the time, considering that Euler was 70 and blind. Pierre Fermat was once asked in a letter if 100,895,598,169 is prime. He shot back that it is the product of primes 898,423 and 112,303. Feats such as these have led some to think that the old masters may have had secret and now-lost methods of factoring. As late as 1874 W. Stanley Jevons could ask, in his *Principles of Science:* "Can the reader say what two numbers multiplied together will produce the number 8,616,460,799? I think it unlikely that anyone but myself will ever know; for they are two large prime numbers." Jevons, who himself invented a mechanical logic machine, should have known better than to imply a limit on future computer speeds. Today a computer can find his two primes (96,079 and 89,681) faster than he could multiply them together.

Numbers of the form $2^p - 1$, where p is prime, are called Mersenne numbers after Marin Mersenne, a 17th-century Parisian friar (he belonged to a humble order known as the Minims—an appropriate order for a mathematician), who was the first to point out that many numbers of this type are prime. For some 200 years the Mersenne number $2^{67} - 1$ was suspected of being prime. Eric Temple Bell, in his book *Mathematics, Queen and Servant of Science,* recalls a meeting in New York of the American Mathematical Society in October, 1903, at which Frank Nelson Cole, a Columbia University professor, rose to give a paper. "Cole—who was always a man of very few words—walked to the board and, saying nothing, proceeded to chalk up the arithmetic for raising 2 to the sixty-seventh power. Then he carefully subtracted 1. Without a word he moved over to a clear space on the board and

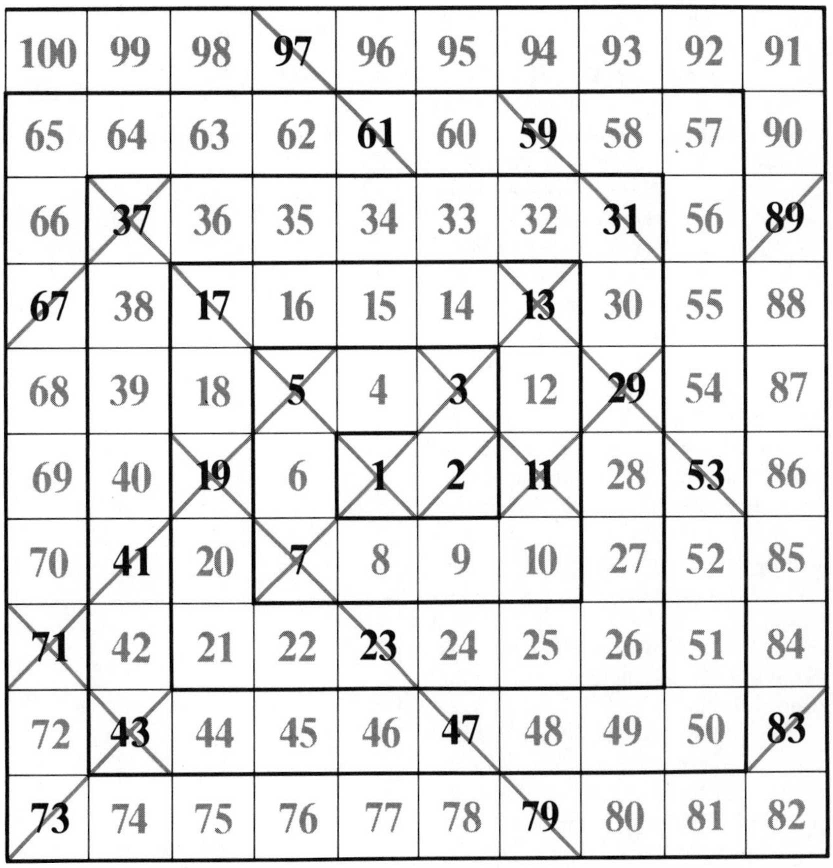

100	99	98	97	96	95	94	93	92	91
65	64	63	62	61	60	59	58	57	90
66	37	36	35	34	33	32	31	56	89
67	38	17	16	15	14	13	30	55	88
68	39	18	5	4	3	12	29	54	87
69	40	19	6	1	2	11	28	53	86
70	41	20	7	8	9	10	27	52	85
71	42	21	22	23	24	25	26	51	84
72	43	44	45	46	47	48	49	50	83
73	74	75	76	77	78	79	80	81	82

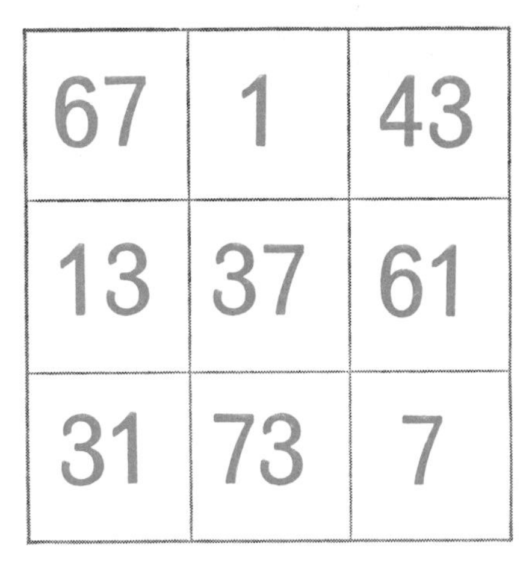

67	1	43
13	37	61
31	73	7

Prime magic square with lowest constant

multiplied out, by longhand,

193,707,721 × 761,838,257,287.

The two calculations agreed.... For the first and only time on record, an audience of the American Mathematical Society vigorously applauded the author of a paper delivered before it. Cole took his seat without having uttered a word. Nobody asked him a question." Years later, when Bell asked Cole how long it took him to crack the number, he replied, "Three years of Sundays."

The British puzzle expert Henry Ernest Dudeney, in his first puzzle book (*The Canterbury Puzzles*, 1907), pointed out that 11 was the only known prime consisting entirely of 1's. (Of course, a number formed by repeating any other digit would be composite.) He was able to show that all such "repunit" numbers, from 3 through 18 units, are composite. Are any larger "repunit" chains prime? Oscar Hoppe, a New York City reader of Dudeney's book, took up the challenge and actually managed to prove, in 1918, that the 19-"repunit" number 1,111,-111,111,111,111,111 is prime. Later it was discovered that 23 repeated 1's is also prime. There the matter rests. No one knows if the "repunit" primes are infinite, or even if there are more than three. It is easy to see that no repunit number is prime unless the number of its units is prime. (For example, if its number of digits has a factor of, say, 13, then clearly it is divisible by a repunit of 13 digits.) As of 1970 repunits have been tested through 373 digits without finding a fourth prime.

Can a magic square be constructed solely of different primes? Yes; Dudeney was the first to do it. The top illustration on this page shows such a square. It sums in all directions to the "repunit" number 111: the lowest possible constant for a prime square of order 3. (Curiously, an order-4 square is possible with the lower magic constant of 102. See Dudeney's, *Amusements in Mathematics;* New York: Dover, 1917; problem 408.)

Can a magic square be made with consecutive odd primes? (The even prime, 2, must be left out because it would make the odd or even parity of its rows and columns different from the parity of all other rows and columns, thereby preventing the array from being magic.) In 1913 J. N. Muncey of Jessup, Iowa, proved that the smallest magic square of this type is one of order 12. This remarkable curiosity is so little known that I reproduce it in the illustration at the bottom of this page. Its cells hold the first 144 consecutive odd primes, starting with 1. All rows, columns and the two main diagonals sum to 4,514.

Readers may test their familiarity with primes by answering the following elementary questions:

1. Identify the four primes among the following six numbers. (NOTE: The second number is the first five digits in the decimal of pi.)

10,001
14,159
76,543
77,377
123,456,789
909,090,909,090,909,090,-
909,090,909,091

2. Two gear wheels, each marked with an arrow, mesh as shown in the illustration opposite. The small wheel turns clockwise until the arrows point directly toward each other once more. If the large wheel has 181 teeth, how many times will the small wheel have rotated? (Contributed by Burris Smith of Greenville, Miss.)

3. Using each of the nine digits once, and only once, form a set of three primes that have the lowest possible sum. For example, the set 941, 827 and 653 sum to 2,421, but this is far from minimal.

4. Find the one composite number in the following set:

31 331 3331 33331 333331 3333331 33333331 333333331

1	823	821	809	811	797	19	29	313	31	23	37
89	83	211	79	641	631	619	709	617	53	43	739
97	227	103	107	193	557	719	727	607	139	757	281
223	653	499	197	109	113	563	479	173	761	587	157
367	379	521	383	241	467	257	263	269	167	601	599
349	359	353	647	389	331	317	311	409	307	293	449
503	523	233	337	547	397	421	17	401	271	431	433
229	491	373	487	461	251	443	463	137	439	457	283
509	199	73	541	347	191	181	569	577	571	163	593
661	101	643	239	691	701	127	131	179	613	277	151
659	673	677	683	71	67	61	47	59	743	733	41
827	3	7	5	13	11	787	769	773	419	149	751

Smallest possible magic square of consecutive odd primes

5. Find a sequence of a million consecutive integers that contain not a single prime.

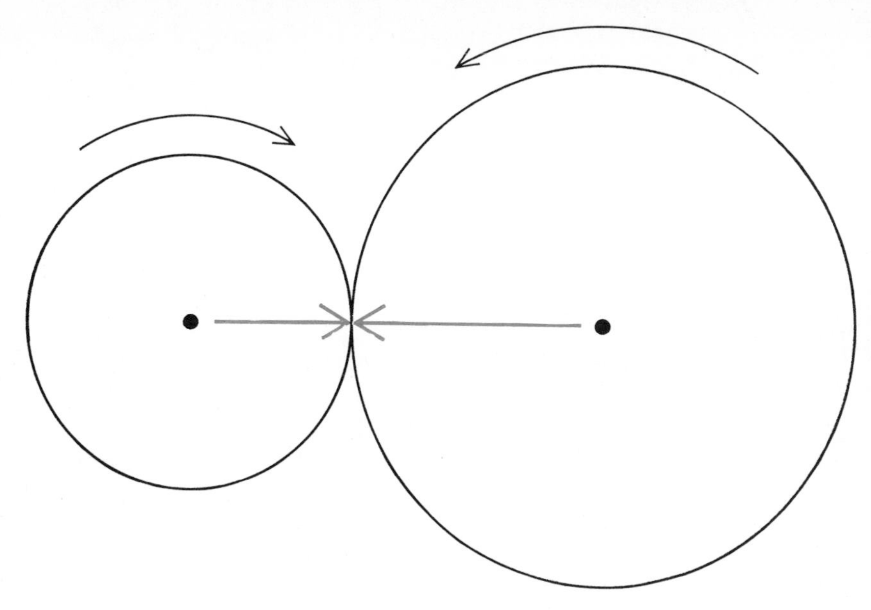

A problem in primes

Addendum

Many *Scientific American* readers experimented with triangular and hexangular arrays of integers and found that the primes cluster along straight lines in the same manner as in Stanislaw Ulam's square spiral grids. Laurence M. Klauber of San Diego, California, sent me a copy of a paper he had read to a meeting of the Mathematical Association of America in 1932, discussing his search for prime-rich polynomial formulas in such an array. Ulam has also used the Los Alamos computer for investigating a variety of other types of grid, including the triangular, and in every case he found that significant departures from random distributions of primes were at once apparent. This is hardly surprising, because any orderly arrangement of consecutive integers in a grid will have straight lines that are generated by polynomial expressions. If the expression is factorable, the line cannot contain primes; this fact alone can account for a concentration of primes along certain other lines.

All diagonals of even numbers are obviously prime-empty, and other lines are empty because they are factorable by other numbers. Many readers noticed that the diagonal line extending down and to the right from 1 on Ulam's spiral grid contains in sequence the squares of odd integers, and the diagonal line extending up and to the left from 4 gives the squares of even integers. Both diagonals are, of course, prime-empty. Conversely, other lines are prime-rich because they are generated by formulas that act as sieves, removing numbers that are multiples of low primes. The significance of Ulam's spiral grids lies not in the discovery that primes are nonrandomly distributed, which is to be expected in any orderly arrangement of integers, but in the use of a computer and scope to extend such grids quickly so that photographs provide, so to speak, a bird's-eye view of the pattern from which hints can be obtained that may lead to new theorems.

Several readers called my attention to W. H. Mills's formula in the *Bulletin of the American Mathematical Society;* June, 1947, page 604, which contains an irrational number between 1 and 2. When positive integers are substituted for n in the formula, the expression gives prime values; but since the irrational number is not known, the formula is of no value in computing primes. In fact, it is easy to write irrational numbers that generate every prime in sequence, for example .20305070110130170190230. . . . To be sure, one has to know the sequence of primes before computing the number. There are many ways of writing complicated functions of n so that integral values of n produce distinct primes, but the catch is that the function itself requires the introduction of the prime-number sequence, making the formula valueless for finding primes. Readers interested in formulas of this type will find a nontechnical discussion of them in Oystein Ore's excellent book *Number Theory and Its History* (New York: McGraw-Hill, 1948).

Mathematical Sieves

by David Hawkins
December 1958

They sift out prime numbers and similar series of integers. Recent research into their properties suggests that a kind of uncertainty principle may exist even in pure mathematics

It is no accident that the theories of probability and statistics are among the most rapidly growing branches of modern mathematics. Science demands them. Faced with problems too complex, or too little understood, to solve exactly, it falls back on laws or facts that are true only probably, or on the average. And from physics, considered the most exact of sciences, we learn that at bottom nature is inescapably uncertain and chancy.

But if we must settle for a gambler's view of the real world, can we not console ourselves with the thought that in the abstract realm of mathematics certainty is always possible? As this article will indicate, the answer is by no means clear. Some provinces of mathematics are so difficult that, for the present at least, they must make do with rules which are only probably true. Even in mathematics there may be an uncertainty principle not utterly unlike the uncertainty principle of physics.

The text of this sermon derives not from some new and exotic kind of mathematics but from arithmetic. We shall discuss the classical problem of prime numbers. These are the positive integers—2, 3, 5, 7, 11 and so on—which cannot be represented by multiplying two smaller numbers. (Numbers which can be represented by such multiplication—4, 6, 8, 9, 10, 12 and so on—are called composite numbers.) Prime numbers have fascinated mathematicians for centuries. It was Euclid who proved there is an infinite number of them. Since then many brilliant minds have turned to primes and have discovered a number of remarkable theorems concerning them. Even more remarkable is what has not been discovered. For example, what is the 34th prime number? What is the billionth? The *n*th? To this day there is no general formula to answer these questions. The only way to find the billionth prime would be to write down all of the first billion and take the last. As another example, consider the famous twin-prime problem. Pairs of primes such as 11 and 13 or 29 and 31, which are separated by only

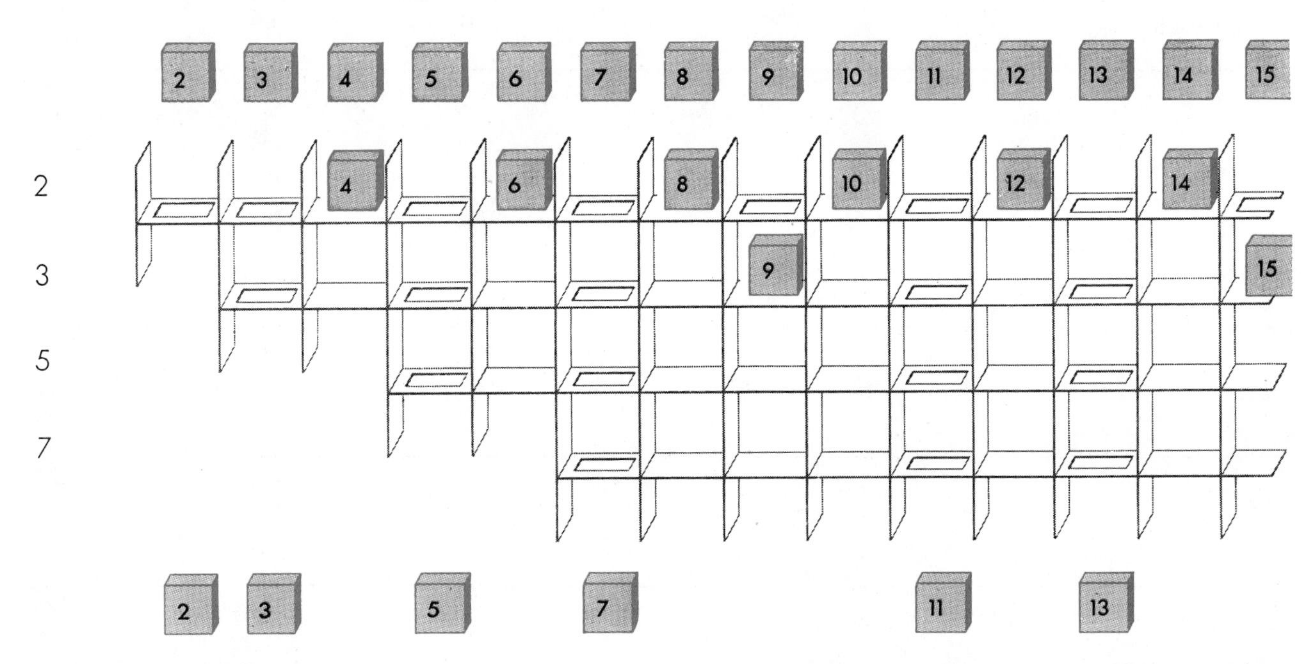

SIEVE OF ERATOSTHENES, a small part of which is shown here, was devised more than 2,000 years ago to separate prime and composite numbers. The first "layer" of the sieve screens out multiples of 2 from the series of integers at the top. Since 3 passes through this layer, it screens out its own multiples in the next layer. Numbers at the bottom are primes which have passed through all previous layers; they will become screening numbers in their turn. No simpler method of deriving primes has yet been devised.

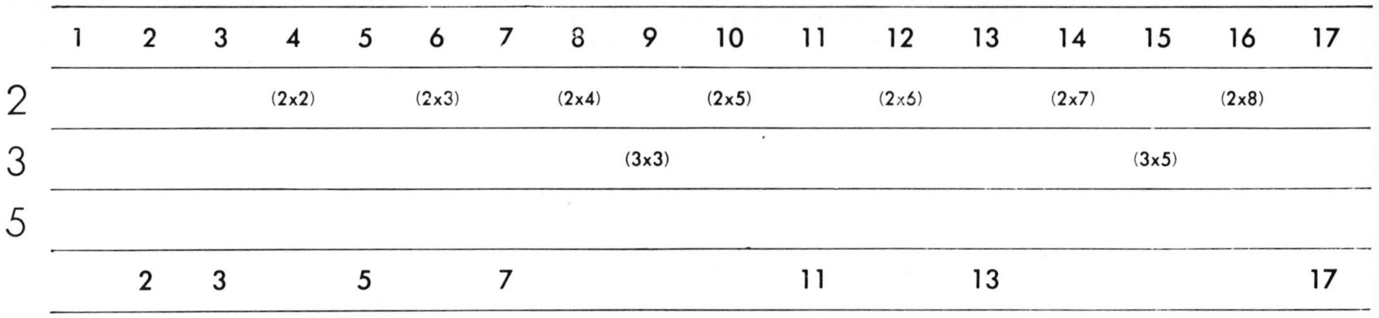

	1	2	3	4	5	6	7	8	9	10	11	12	13	14	15	16	17
2				(2x2)		(2x3)		(2x4)		(2x5)		(2x6)		(2x7)		(2x8)	
3									(3x3)						(3x5)		
5																	
		2	3		5		7				11		13				17

PRIME-NUMBER SIEVE shown here is a larger portion of the sieve on the preceding page. Primes appear on the bottom line. Each prime in turn becomes a sieving number which eliminates its own multiples, beginning with its square (lower multiples have

one other number, are known as twins. They keep turning up in the longest series of primes that have yet been listed. Will they continue to recur indefinitely? Is their number infinite? It seems probable, but no one has been able to prove it.

The study of prime numbers has been quite literally as much an experimental as a theoretical investigation. Most of the facts that have been proved began as conjectures, based on the inspection of an actual series of primes. Many conjectures remain, seeming more or less probably true. Thus an indispensable tool of the number theorist is a long list of primes.

One of the best known, now found in every well-equipped mathematics library, was compiled by D. N. Lehmer of the University of California in 1914. The volume contains a table of the 664,580 prime numbers smaller than 10,000,000, plus a few more to fill the last column, ending with the prime 10,006,721. Lehmer's work was completed before the age of automatic computation; today there are even longer lists, the longest being "published" only on magnetic tape.

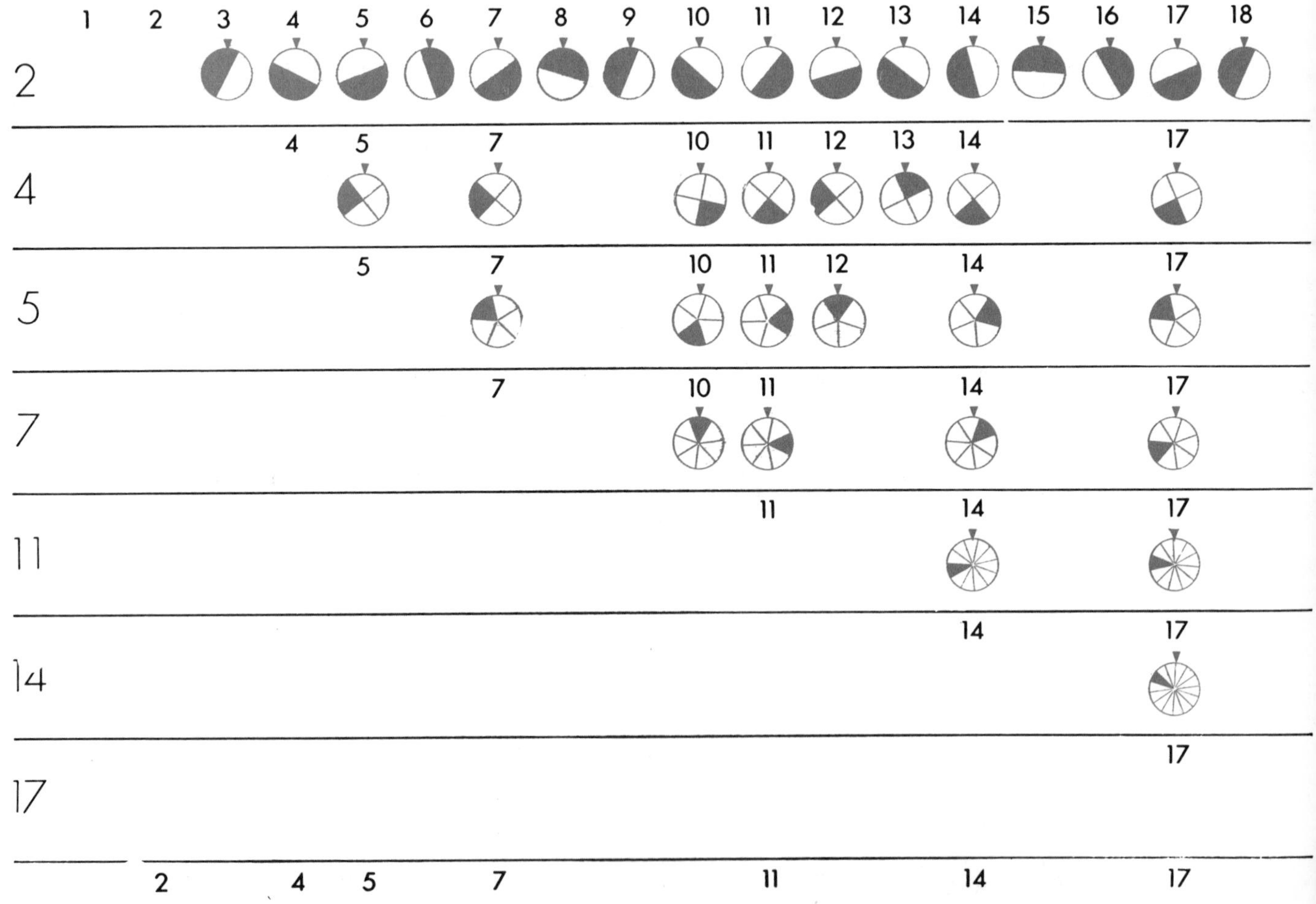

	1	2	3	4	5	6	7	8	9	10	11	12	13	14	15	16	17	18
2																		
4				4	5		7			10	11	12	13	14			17	
5					5		7			10	11	12		14			17	
7							7			10	11			14			17	
11											11			14			17	
14														14			17	
17																	17	
		2		4	5		7				11			14			17	

RANDOM-NUMBER SIEVE is statistically similar to the prime sieve but differs from it in detail. In both cases numbers not previously eliminated become sieving numbers; these screen out a proportion of the remaining numbers equal to their reciprocals.

18	19	20	21	22	23	24	25	26	27	28	29	30	31	32	33	34
(2x9)		(2x10)		(2x11)		(2x12)		(2x13)		(2x14)		(2x15)		(2x16)		(2x17)
			(3x7)						(3x9)						(3x11)	
							(5x5)									
	19				23						29		31			

already been removed by lower primes). Thus each prime eliminates a proportion of the remaining numbers equal to its reciprocal (*e.g.*, 3 removes 1/3, 5 removes 1/5). The steps shown here in part yield all primes up to 49, the square of the next sieving number.

Modern tables of primes are prepared by a method, essentially unaltered for 2,000 years, which is called the sieve of Eratosthenes. Its inventor was one of those great figures of the Hellenistic Age who seem today, across the intervening centuries, so clairvoyant of the spirit of modern science. Eratosthenes of Alexandria is best known for his feat of measuring the size of the earth. But he was a man of universal learning who wrote also on geometry, the measurement of time, and the drama. In his own day he was nicknamed "Beta" because, it was said, he stood at least second in every field. Modern electronic computers can make far longer lists of primes than Eratosthenes could have, but his principle of computation has not been much improved.

The method is almost obvious [*see illustration on page 55*]. Simply write down a series of positive integers and proceed systematically to eliminate all the composite numbers. The numbers that remain—that fall through the "sieve"—are primes. We begin by knocking out the even numbers, which are multiples of the first prime number: 2. (One is not usually called a prime.) When we have

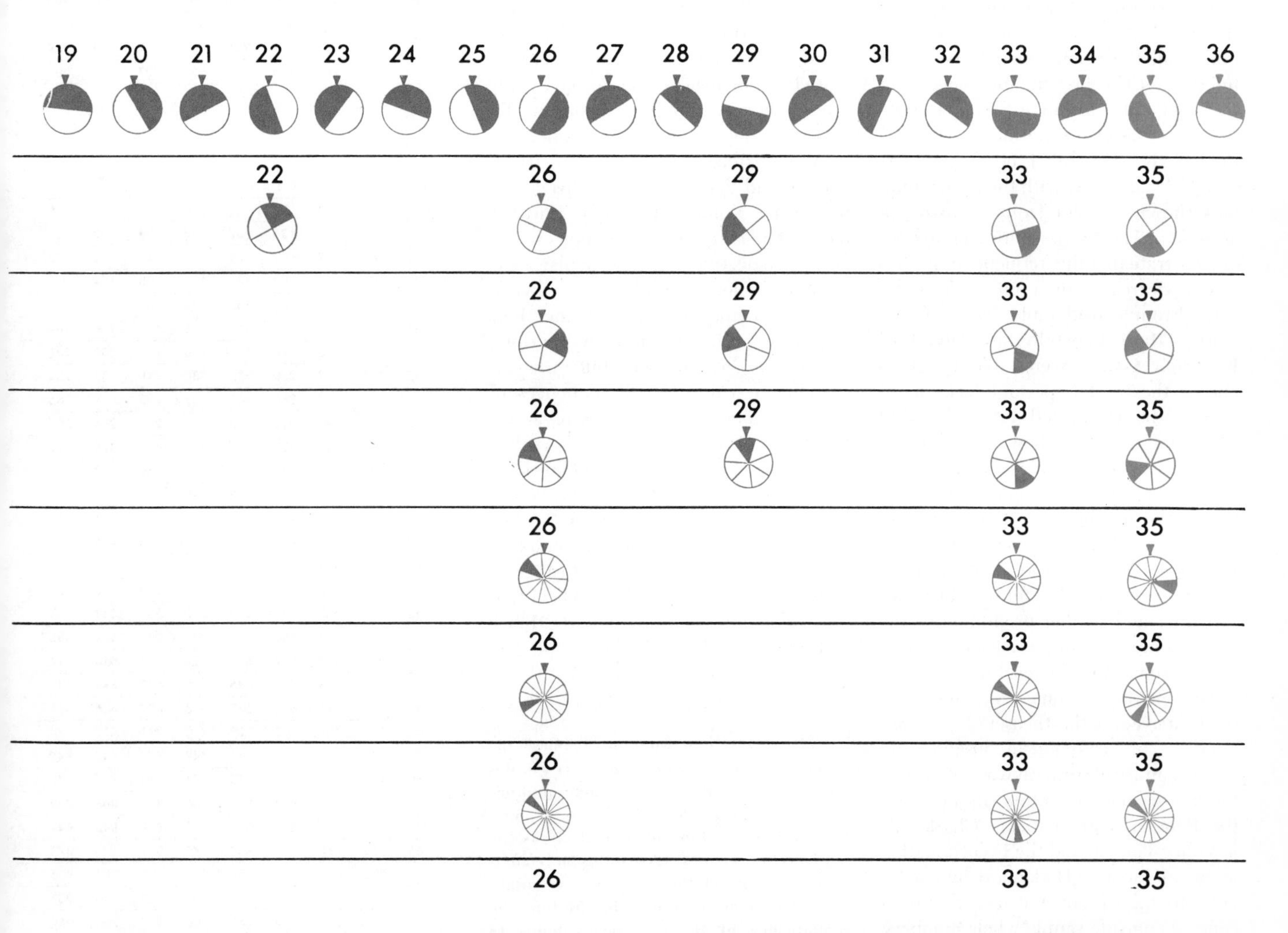

In the random sieve, however, the specific numbers to be eliminated are chosen by a random process symbolized by the colored wheels. Thus the random sieve produces a different set of numbers each time it is used, while the set of prime numbers is invariant.

done this, the smallest of the remaining numbers is the second prime: 3. Now we eliminate the multiples of 3 from the numbers which survived the first sieving operation. Five is the next number remaining, so its multiples drop out next; then the multiples of 7, and so on.

The reader may wish to try a somewhat longer version of the sieve than the one shown in the illustration, where 7 is the largest sieving number. In number theory the distance from the obvious to the profound is sometimes very short, and any amateur willing to play the game is on the verge of some first-class mysteries. At any rate, a little manipulation of the sieve will make clear some of its properties. Every sieving number is a prime. The first number sieved out by each one is its own square: the first number eliminated by 2 is 4; by 3, 9; by 5, 25 and so on. In addition, the fraction of the remaining integers eliminated by each sieving number is its own reciprocal: 2 sieves out half of the remaining numbers, 3 sieves out a third, 5 sieves out a fifth.

By carrying out the sieving operation through the prime number 31, we can obtain all the primes in the first 1,368 integers. (The first number sieved out by 37, the next prime, is 37^2, or 1,369.) For purposes of illustration we have arranged the first 1,024 of the integers in a 32×32 array, with the prime numbers shown in color [*upper illustration at right*]. The list is short, but it does demonstrate that the frequency of primes slowly decreases in a rather irregular way. From considerably longer tables Adrien Marie Legendre, and later Karl Friedrich Gauss, were able to guess one of the most important facts about primes—the celebrated Prime Number Theorem. This tells how many primes we may expect to find by carrying the list out to any given number. It states that if the number is n, then there are about n divided by the logarithm of n ($n/\log n$) primes before it. As n grows larger, the error in the formula becomes a smaller and smaller proportion of the exact number of primes. Gauss, whose skill in computing belied the myth that mathematicians cannot add and subtract, arrived at the theorem by a combination of arithmetical insight and purely empirical study. It was not proved for almost another century. In the 1890s the Belgian mathematician Charles de la Vallée Poussin and the French mathematician Jacques Hadamard independently found a proof, but it made use of concepts outside simple whole numbers. It was not until 1950 that the Norwegian mathematician Atle Selberg discovered a purely arithmetical proof. In the quaint vocabulary of number theory his proof is called elementary, but it is not easy.

The difficulties of the Prime Number Theorem are connected with the puzzlingly irregular way in which the primes are distributed. Indeed, the theorem itself does no more than state a statistical average. Outrageous as it may seem, the sequence of primes is just as "random" as many of the natural phenomena on which we make bets. Sometimes we think that if we knew enough about the individual events of which such phenomena are composed, we could predict their outcome with certainty. This is surely true of the primes. The sieve will eventually tell us about the primality of any given number. But it cannot tell us about all numbers, because the sequence is itself an infinite, unending process.

From the time of Gauss mathematicians have talked, perhaps rather shamefacedly, about the "probable" behavior of primes, and this kind of reasoning has been very helpful. No mathematician, however, seems to have gone the whole way and made a purely statistical model of the prime-number distribution. Recently I was led to try it, and I found that the model helps clarify the Prime Number Theorem. Furthermore, it places the whole subject in a new perspective. In particular, the theorem no longer appears as a special fact about the sequence of numbers which cannot be produced by multiplying two smaller numbers, but rather as a common feature of all sequences of numbers generated by sieves of a certain type.

The model is called the random sieve, and it works like this [*see illustration at bottom of preceding two pages*]. Start with 2 as the first sieving number, just as in the method of Eratosthenes. Now make a kind of roulette wheel that is divided into two equal parts, black and white. Go down the list of integers

PRIME SERIES

1	2	3	4	5	6	7	8	9
33	34	35	36	37	38	39	40	41
65	66	67	68	69	70	71	72	73
97	98	99	100	101	102	103	104	105
129	130	131	132	133	134	135	136	137
161	162	163	164	165	166	167	168	169
193	194	195	196	197	198	199	200	201
225	226	227	228	229	230	231	232	233
257	258	259	260	261	262	263	264	265
289	290	291	292	293	294	295	296	297
321	322	323	324	325	326	327	328	329
353	354	355	356	357	358	359	360	361
385	386	387	388	389	390	391	392	393
417	418	419	420	421	422	423	424	425
449	450	451	452	453	454	455	456	457
481	482	483	484	485	486	487	488	489
513	514	515	516	517	518	519	520	521
545	546	547	548	549	550	551	552	553
577	578	579	580	581	582	583	584	585
609	610	611	612	613	614	615	616	617
641	642	643	644	645	646	647	648	649
673	674	675	676	677	678	679	680	681
705	706	707	708	709	710	711	712	713
737	738	739	740	741	742	743	744	745
769	770	771	772	773	774	775	776	777
801	802	803	804	805	806	807	808	809
833	834	835	836	837	838	839	840	841
865	866	867	868	869	870	871	872	873
897	898	899	900	901	902	903	904	905
929	930	931	932	933	934	935	936	937
961	962	963	964	965	966	967	968	969
993	994	995	996	997	998	999	1,000	1,001

RANDOM SERIES

1	2	3	4	5	6	7	8	9
33	34	35	36	37	38	39	40	41
65	66	67	68	69	70	71	72	73
97	98	99	100	101	102	103	104	105
129	130	131	132	133	134	135	136	137
161	162	163	164	165	166	167	168	169
193	194	195	196	197	198	199	200	201
225	226	227	228	229	230	231	232	233
257	258	259	260	261	262	263	264	265
289	290	291	292	293	294	295	296	297
321	322	323	324	325	326	327	328	329
353	354	355	356	357	358	359	360	361
385	386	387	388	389	390	391	392	393
417	418	419	420	421	422	423	424	425
449	450	451	452	453	454	455	456	457
481	482	483	484	485	486	487	488	489
513	514	515	516	517	518	519	520	521
545	546	547	548	549	550	551	552	553
577	578	579	580	581	582	583	584	585
609	610	611	612	613	614	615	616	617
641	642	643	644	645	646	647	648	649
673	674	675	676	677	678	679	680	681
705	706	707	708	709	710	711	712	713
737	738	739	740	741	742	743	744	745
769	770	771	772	773	774	775	776	777
801	802	803	804	805	806	807	808	809
833	834	835	836	837	838	839	840	841
865	866	867	868	869	870	871	872	873
897	898	899	900	901	902	903	904	905
929	930	931	932	933	934	935	936	937
961	962	963	964	965	966	967	968	969
993	994	995	996	997	998	999	1,000	1,001

DISTRIBUTION of random "primes" between 1 and 1,024 (*lower table*) resembles that of true primes in the same number sequence (*upper table*). Both sets of numbers (*in color*) thin out irregularly as the sequence progresses (*see totals at right*). Another "run" of the random sieve might yield an even more similar distribution. The resemblance of the two series tends to intensify as they are increased in length.

																							ROW TOTAL	CUMULATIVE TOTAL
10	11	12	13	14	15	16	17	18	19	20	21	22	23	24	25	26	27	28	29	30	31	32	11	11
42	43	44	45	46	47	48	49	50	51	52	53	54	55	56	57	58	59	60	61	62	63	64	7	18
74	75	76	77	78	79	80	81	82	83	84	85	86	87	88	89	90	91	92	93	94	95	96	6	24
106	107	108	109	110	111	112	113	114	115	116	117	118	119	120	121	122	123	124	125	126	127	128	7	31
138	139	140	141	142	143	144	145	146	147	148	149	150	151	152	153	154	155	156	157	158	159	160	6	37
170	171	172	173	174	175	176	177	178	179	180	181	182	183	184	185	186	187	188	189	190	191	192	6	43
202	203	204	205	206	207	208	209	210	211	212	213	214	215	216	217	218	219	220	221	222	223	224	5	48
234	235	236	237	238	239	240	241	242	243	244	245	246	247	248	249	250	251	252	253	254	255	256	6	54
266	267	268	269	270	271	272	273	274	275	276	277	278	279	280	281	282	283	284	285	286	287	288	7	61
298	299	300	301	302	303	304	305	306	307	308	309	310	311	312	313	314	315	316	317	318	319	320	5	66
330	331	332	333	334	335	336	337	338	339	340	341	342	343	344	345	346	347	348	349	350	351	352	4	70
362	363	364	365	366	367	368	369	370	371	372	373	374	375	376	377	378	379	380	381	382	383	384	6	76
394	395	396	397	398	399	400	401	402	403	404	405	406	407	408	409	410	411	412	413	414	415	416	4	80
426	427	428	429	430	431	432	433	434	435	436	437	438	439	440	441	442	443	444	445	446	447	448	6	86
458	459	460	461	462	463	464	465	466	467	468	469	470	471	472	473	474	475	476	477	478	479	480	6	92
490	491	492	493	494	495	496	497	498	499	500	501	502	503	504	505	506	507	508	509	510	511	512	5	97
522	523	524	525	526	527	528	529	530	531	532	533	534	535	536	537	538	539	540	541	542	543	544	3	100
554	555	556	557	558	559	560	561	562	563	564	565	566	567	568	569	570	571	572	573	574	575	576	5	105
586	587	588	589	590	591	592	593	594	595	596	597	598	599	600	601	602	603	604	605	606	607	608	6	111
618	619	620	621	622	623	624	625	626	627	628	629	630	631	632	633	634	635	636	637	638	639	640	4	115
650	651	652	653	654	655	656	657	658	659	660	661	662	663	664	665	666	667	668	669	670	671	672	6	121
682	683	684	685	686	687	688	689	690	691	692	693	694	695	696	697	698	699	700	701	702	703	704	5	126
714	715	716	717	718	719	720	721	722	723	724	725	726	727	728	729	730	731	732	733	734	735	736	4	130
746	747	748	749	750	751	752	753	754	755	756	757	758	759	760	761	762	763	764	765	766	767	768	5	135
778	779	780	781	782	783	784	785	786	787	788	789	790	791	792	793	794	795	796	797	798	799	800	4	139
810	811	812	813	814	815	816	817	818	819	820	821	822	823	824	825	826	827	828	829	830	831	832	6	145
842	843	844	845	846	847	848	849	850	851	852	853	854	855	856	857	858	859	860	861	862	863	864	5	150
874	875	876	877	878	879	880	881	882	883	884	885	886	887	888	889	890	891	892	893	894	895	896	4	154
906	907	908	909	910	911	912	913	914	915	916	917	918	919	920	921	922	923	924	925	926	927	928	3	157
938	939	940	941	942	943	944	945	946	947	948	949	950	951	952	953	954	955	956	957	958	959	960	5	162
970	971	972	973	974	975	976	977	978	979	980	981	982	983	984	985	986	987	988	989	990	991	992	5	167
1,002	1,003	1,004	1,005	1,006	1,007	1,008	1,009	1,010	1,011	1,012	1,013	1,014	1,015	1,016	1,017	1,018	1,019	1,020	1,021	1,022	1,023	1,024	5	172

																							ROW TOTAL	CUMULATIVE TOTAL
10	11	12	13	14	15	16	17	18	19	20	21	22	23	24	25	26	27	28	29	30	31	32	8	8
42	43	44	45	46	47	48	49	50	51	52	53	54	55	56	57	58	59	60	61	62	63	64	6	14
74	75	76	77	78	79	80	81	82	83	84	85	86	87	88	89	90	91	92	93	94	95	96	7	21
106	107	108	109	110	111	112	113	114	115	116	117	118	119	120	121	122	123	124	125	126	127	128	8	29
138	139	140	141	142	143	144	145	146	147	148	149	150	151	152	153	154	155	156	157	158	159	160	7	36
170	171	172	173	174	175	176	177	178	179	180	181	182	183	184	185	186	187	188	189	190	191	192	6	42
202	203	204	205	206	207	208	209	210	211	212	213	214	215	216	217	218	219	220	221	222	223	224	4	46
234	235	236	237	238	239	240	241	242	243	244	245	246	247	248	249	250	251	252	253	254	255	256	4	50
266	267	268	269	270	271	272	273	274	275	276	277	278	279	280	281	282	283	284	285	286	287	288	7	57
298	299	300	301	302	303	304	305	306	307	308	309	310	311	312	313	314	315	316	317	318	319	320	6	63
330	331	332	333	334	335	336	337	338	339	340	341	342	343	344	345	346	347	348	349	350	351	352	3	66
362	363	364	365	366	367	368	369	370	371	372	373	374	375	376	377	378	379	380	381	382	383	384	4	70
394	395	396	397	398	399	400	401	402	403	404	405	406	407	408	409	410	411	412	413	414	415	416	3	73
426	427	428	429	430	431	432	433	434	435	436	437	438	439	440	441	442	443	444	445	446	447	448	5	78
458	459	460	461	462	463	464	465	466	467	468	469	470	471	472	473	474	475	476	477	478	479	480	4	82
490	491	492	493	494	495	496	497	498	499	500	501	502	503	504	505	506	507	508	509	510	511	512	3	85
522	523	524	525	526	527	528	529	530	531	532	533	534	535	536	537	538	539	540	541	542	543	544	3	88
554	555	556	557	558	559	560	561	562	563	564	565	566	567	568	569	570	571	572	573	574	575	576	4	92
586	587	588	589	590	591	592	593	594	595	596	597	598	599	600	601	602	603	604	605	606	607	608	3	95
618	619	620	621	622	623	624	625	626	627	628	629	630	631	632	633	634	635	636	637	638	639	640	8	103
650	651	652	653	654	655	656	657	658	659	660	661	662	663	664	665	666	667	668	669	670	671	672	3	106
682	683	684	685	686	687	688	689	690	691	692	693	694	695	696	697	698	699	700	701	702	703	704	4	110
714	715	716	717	718	719	720	721	722	723	724	725	726	727	728	729	730	731	732	733	734	735	736	5	115
746	747	748	749	750	751	752	753	754	755	756	757	758	759	760	761	762	763	764	765	766	767	768	5	120
778	779	780	781	782	783	784	785	786	787	788	789	790	791	792	793	794	795	796	797	798	799	800	4	124
810	811	812	813	814	815	816	817	818	819	820	821	822	823	824	825	826	827	828	829	830	831	832	3	127
842	843	844	845	846	847	848	849	850	851	852	853	854	855	856	857	858	859	860	861	862	863	864	5	132
874	875	876	877	878	879	880	881	882	883	884	885	886	887	888	889	890	891	892	893	894	895	896	3	135
906	907	908	909	910	911	912	913	914	915	916	917	918	919	920	921	922	923	924	925	926	927	928	0	135
938	939	940	941	942	943	944	945	946	947	948	949	950	951	952	953	954	955	956	957	958	959	960	1	136
970	971	972	973	974	975	976	977	978	979	980	981	982	983	984	985	986	987	988	989	990	991	992	5	141
1,002	1,003	1,004	1,005	1,006	1,007	1,008	1,009	1,010	1,011	1,012	1,013	1,014	1,015	1,016	1,017	1,018	1,019	1,020	1,021	1,022	1,023	1,024	4	145

DERIVATION OF THE PRIME NUMBER THEOREM FOR THE RANDOM SIEVE

Let us consider the fate of any two consecutive numbers, say 127 and 128, on a run through the random-sieving operation. We shall compare their probabilities of getting through the sieve; *i.e.*, of becoming sieving numbers or "random primes" themselves.

Call these probabilities P_{127} and P_{128}. Now it is obvious that 128 runs the same risk of being eliminated by previous sieving numbers as does 127, except for one possibility. If 127 becomes a sieving number, it can eliminate 128, but not *vice versa*. The probability that 127 is a sieving number is P_{127}. If it is a sieving number, the probability that it will eliminate 128 (or any other following number) is 1/127. The chance that the two events will occur and that 127 will eliminate 128 is the product of their probabilities: $P_{127} \times 1/127$. The probability that this will not happen is $1 - P_{127}/127$. Except for this factor the chance of survival for 128 is the same as that for 127. Its net probability is therefore the product of the two: $P_{128} = P_{127}(1 - P_{127}/127)$.

At this point it will be more convenient to shift from the probabilities to their reciprocals. The reciprocal of a probability has itself a clear statistical meaning: it gives the average interval, or range, between two events. (Instead of saying that the probability of double six in dice is 1/36, we can as well say that the average interval between throws of double six is 36.) Denote the reciprocal of P_{127} by X_{127}, and of P_{128} by X_{128}. X_{127} measures the average interval between sieving numbers in the neighborhood of 127 and X_{128} measures the same interval in the slightly shifted neighborhood of 128.

By a little algebra we can show that if $P_{128} = P_{127}\ (1 - P_{127}/127)$, then $X_{128} = X_{127} + 1/127 + r$, where r is a negligibly small remainder. For practical purposes we can say that $X_{128} = X_{127} + 1/127$. Now a similar argument would show that $X_{127} = X_{126} + 1/126$, and so on. Eventually we arrive at the result that $X_{128} = 1 + 1/2 + 1/3 + 1/4 \ldots + 1/127$, or, in general, $X_n = 1 + 1/2 + 1/3 + 1/4 \ldots + 1/n$, with a remainder that is still negligibly small. In calculus books we discover that the series $1 + 1/2 + 1/3 + 1/4 \ldots + 1/n$ is nearly equal to log n for fairly long series. The difference can be made as small as we like by making n large enough. Therefore we can say that, in the long run, $X_n = \log n$, or $P_n = 1/\log n$.

The graph on the opposite page shows the values of $1/\log n$ (and, for comparison, the reciprocal of the actual values of the series $1 + 1/2 + 1/3 + 1/4 \ldots + 1/n$). Thus the curve is also a graph of P_n. Suppose we now want to know how many random primes, on the average, there should be before any number n. We simply add the probabilities that each smaller number becomes a sieving number. Graphically this is the same as taking the area under the curve. But if n is very large, then the difference between the area under the curve and the area of the shaded rectangle, which is $n \times P_n$, is negligible. Hence we can say that the average number of random primes out to n is $n \times P_n$. But $P_n = 1/\log n$, so the number becomes $n/\log n$. And this is the Prime Number Theorem!

Having completed the proof, we may reexamine our reasoning to see why the result is plausible. The essential step was to find that $X_n + 1 = X_n + 1/n$. This equation says that on the average, over many repetitions of the sieve, any number n removes enough of the numbers following to lengthen the interval between them by $1/n$. Take a specific example. Suppose that P_{127} is 1/5 and X_{127} is 5. Then 127 will be a sieving number 1/5 of the time. When it is, it will eliminate about 1/127 of the remaining numbers, lengthening the average interval between them from 5 to 5 + 5/127. Since it only does this about one time out of every five trials of the sieve, its average effect will be to lengthen the interval from 5 to 5 + 1/127.

The same chain of reasoning is plausible for the prime-number sieve.

following 2, and for each one spin the wheel. If the black part of the wheel stops at the pointer, strike the integer out; if white stops at the pointer, leave the integer in. Note what you have accomplished. In the long run you will have sieved out half of the integers, just as the first step in the prime-number sieve does. But just which ones go out is a matter of chance, and the list will be different each time you try it.

Next take the first number that was not removed. Suppose it was 4. Make a new wheel of which a fourth is black and three fourths is white. Spin the wheel for each succeeding number left after the first sieving. When black comes up, strike the number out; when white comes up, leave the number in. This time you have removed a fourth of the remaining numbers. Proceed again to the first number not removed—say 5. Repeat the procedure using a sieving probability of 1/5, and so on. After any number of steps you will be left with a series of integers which might be called "random primes."

If you want to try the sieve yourself, you need not actually make roulette wheels. A table of random numbers or, failing that, a telephone book will do. Express each sieving probability as a four-digit decimal (*e.g.*, 1/4 = .2500). For each "spin of the wheel" read successive telephone numbers. If your probability is 1/4, then any number whose last four digits are 2499 or less tells you to eliminate the integer in question; 2500 or more means to leave it in.

One run of the random sieve for the first 1,024 integers is summarized in the table on the preceding two pages. Comparing the distribution of these random primes with the actual ones, we can see that our sieve acts something like the sieve of Eratosthenes. This is partly in spite of the random element, but partly because of it. For a much longer series the general statistical similarity would be even closer.

It may seem paradoxical that we can take a statistical model, involving an infinity of random choices, as *ersatz* for the straightforward and perfectly defined sieve of Eratosthenes. The paradox is the same as the one which underlies statistical mechanics: the average behavior of an assembly of molecules is easier to describe than the actual behavior of any one of them. Of course the random sieve preserves only the general features of the prime-number sieve. The eccentricities of the latter are averaged out by randomizing them. In either case any number not sieved out becomes in turn a sieving number. It starts a process by which a proportion of later numbers is removed, equal to the reciprocal of that sieving number. Every wave of sieving in the prime-number sieve, except the first, is determined strictly by the result of previous waves. At every corresponding point the random sieve makes probability choices, partly determined by its own earlier statistical behavior.

How closely the random sieve actually approximates the sieve of Eratosthenes is demonstrated by the fact that the Prime Number Theorem holds for random primes. This can be proved by some elementary mathematics, which in this case is also fairly easy [*see box on opposite page*].

Perhaps the parallel between the two sieves is not so surprising. We might say, indeed, that the prime-number sieve would have to be remarkably abnormal in its detailed behavior not to lead to the same general result as the random sieve. This statement implies that the random sieve can be taken as a criterion of normality.

If so, there must be other sieves—in fact, an infinite number of other sieves—that have the same general characteristics as those of the sieve of Eratosthenes, but which differ somewhat in the details of their definition. They will not yield the prime numbers in general, but numbers having some other special property. In 1956, as it happens, Stanislas M. Ulam and his associates at the Los Alamos Scientific Laboratory published some results of a new type of sieve which yields what they called "lucky" numbers. Their sieve begins by removing the multiples of 2, leaving 3 as the first number not sieved out. Instead of removing next the multiples of 3, the Ulam sieve removes every third remaining number. Since 5 is the third number in the list of remaining numbers, it drops out, but 7 remains. Hence in the next wave every seventh number of those still remaining is eliminated, and so on [*see illustration at top of next page*]. The numbers that escape are "lucky." It has been proved that the analogue of the Prime Number Theorem holds for lucky numbers. Thus the random sieve is a model for the lucky numbers as well as for the primes.

So far the random sieve has only duplicated results that can be obtained independently and rigorously for the sieves of Eratosthenes and of Ulam. The mathematics of it, however, is mostly easier. Therefore many additional theorems can be obtained from the random sieve and conjectured to be true of the other two. Such conjectures are not proofs, but we can say that unless the prime number and lucky sieves are vastly abnormal, the results must hold for them.

Let us look at a couple of examples. As we go to larger and larger numbers in the table of integers, the spacing between successive primes (or luckies or random primes) grows greater in an irregular way. In the neighborhood of any number, n, the average interval is about the logarithm of n. What is the greatest interval? We do not know the answer for primes or luckies. But for the random sieve we can prove that, with only a finite number of exceptions, the interval is never greater than the square of the logarithm of n, that is, $(\log n)^2$. The chance that there will be any further exceptions can be made as small as we please by taking a sufficiently large n. No upper boundary to the interval between successive primes or successive luckies has been found which is anywhere nearly as small, although from the existing tables it looks as though the formula should hold for them too.

Another example is the twin-prime problem mentioned earlier. In the random sieve there is almost certainly an infinite number of twins. Indeed the average interval between twins ought to be about $(\log n)^2$, and the maximum interval between them, with only a finite number of exceptions, ought to be $(\log n)^3$. Again the tables suggest that these results are also true for primes and luckies, but no one has any idea how to prove such results.

Although the random sieve does not solve any classical problems concerning primes, it does enable us to reformulate such problems. We may ask: "Are the prime numbers normal in such and such a respect?" The random sieve, or certain modifications of it, defines what we mean by normality. If the properties we are talking about depend on the exact fine structure of the sequence of primes, the answer will obviously be no. Thus all primes except the number 2 are odd, while this is infinitely improbable in the sequences of random primes. But average properties such as those we have discussed do not seem to depend on the fine structure, and those may be presumed to be normal for primes or luckies. Can anyone find a major abnormal property, in this sense, of the sequence of primes? Or the sequence of luckies?

In the opinion of the author the concept of normality raises some very deep questions about numbers and the theory of numbers. Sieves as a class are a type of feedback mechanism: the output of one stage of the process determines the input of the next stage. Now in any such mechanism the nature of the coupling between output and input is crucial; the result may be stable and predictable for one type of coupling and unstable for another. So far as the outcome of the random sieve is concerned, it is in one respect extremely stable. If by chance there are relatively few sieving numbers in the early stages, they will remove relatively few later on, and so there will be an increase in the later stages to compensate for the initial deficit. The sieves of primes and luckies share this characteristic. But this is a statistical stability.

When we look at other aspects of the prime or the lucky sieve, however, we find elements of instability. The detailed ordering of primes or luckies depends upon the individual sieving numbers that precede them, and this involves a growth of complexity without apparent limit. Some easily defined properties of normal sequences, for example the two described, may depend strongly enough on

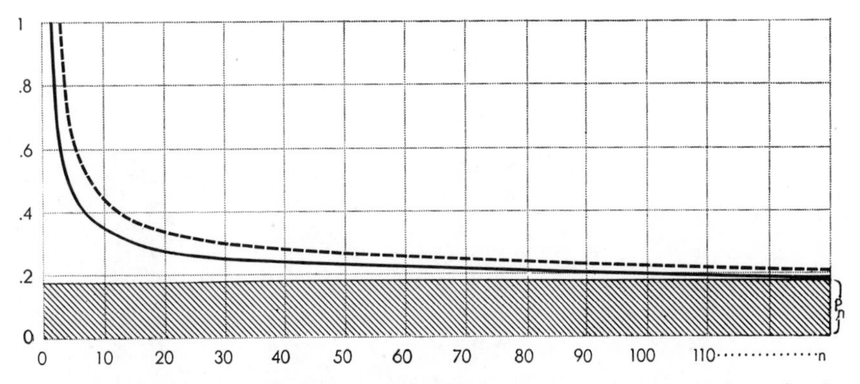

AVERAGE NUMBER OF RANDOM "PRIMES" in the first n integers is shown by the area under the solid curve, roughly approximated by the hatched rectangle (drawn here for $n = 130$). The area under the broken curve gives the approximate number of true primes. Since the two curves approach each other as n increases, the two sets of primes are very like.

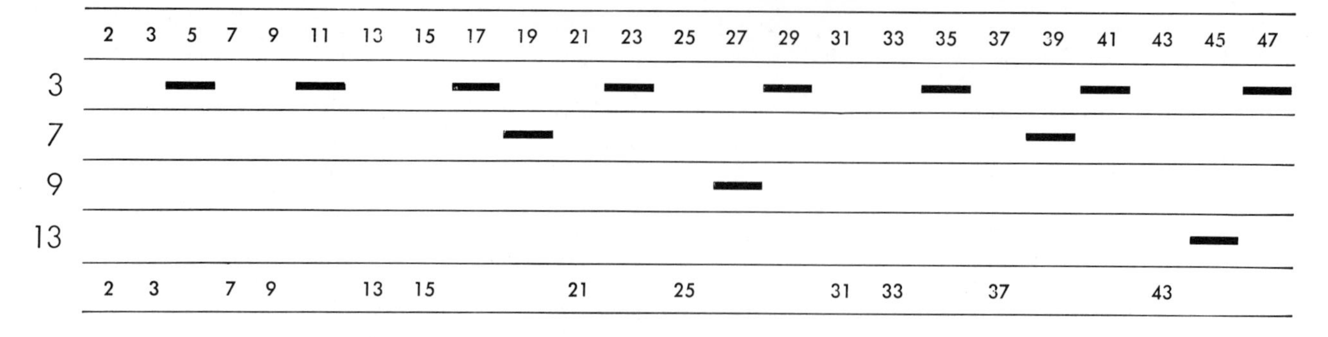

"LUCKY" NUMBER SIEVE resembles the prime and random sieves already described. Here, also, numbers which are not eliminated become sieving numbers and remove a proportion of the remaining numbers equal to their reciprocals. Elimination is by counting: thus 3 removes every third remaining number, 7 every seventh. Like primes, the "lucky" numbers form an invariant series.

2	3	7	9	13	15	21	25	31 ★	33	37	43	49	51	63 ★	67	69	73	75	79	87	93 ★	99
105	111	115	127 ★	129	133	135	141	151	159 ★	163	169	171	189 ★	193	195	201	205	211	219	223 ★	231	235
237	241 ★	259	261	267	273	283	285 ★	289	297	303	307	319 ★	321	327	331	339	349 ★	357	361	367 ★	385	391
393	399	409	415 ★	421	427	429	433 ★	451	463	475	477 ★	483	487	489	495	511 ★	517	519	529	535	537	541 ★
553	559 ★	577	579	583	591	601 ★	613	615	619	621	631	639 ★	643	645	651	655 ★	673	679	685	693	699 ★	717
723	727	729	735 ★	739	741	745 ★	769	777	781	787 ★	801	805	819	823	831 ★	841	855 ★	867	873	883	885	895 ★
897	903	925	927 ★	931	933	937	957 ★	961	975	979	981	991 ★	993	997	1009	1011	1021	1023 ★				

DISTRIBUTION OF "LUCKY" numbers between 1 and 1,024 resembles that of primes and random primes, thinning out gradually but irregularly as the list increases. This table shows only the "luckies"; the intervening numbers are omitted. Stars set off luckies within successive series of 32 integers; each of these groups corresponds to a single line of the tables on pages 58 and 59.

this complexity to make it impossible, in a finite number of steps, to prove that they hold. Here is the analogy, if it be one, with the uncertainty principle of physics: An infinite complexity requires infinite time to resolve it. If our suggestions have substance, we will have examples of mathematical statements which are almost certain, but which cannot, in principle, be proved. Examples of undecidable propositions are known in modern arithmetic [see "Gödel's Proof," by Ernest Nagel and James R. Newman; SCIENTIFIC AMERICAN, June, 1956], but so far none of the unproved conjectures about prime numbers has been shown to be undecidable. Perhaps none of them is. If any are, however, the random sieve will be a model for the primes in a deeper sense than any we have exploited in this article. We cannot distinguish an infinitely complex order from a random one, and so we might be forced to admit that there is a certain background of noise even among the eternal verities.

10

A Short Treatise on the Useless Elegance of Perfect Numbers and Amicable Pairs

by Martin Gardner
March 1968

One would be hard put to find a set of whole numbers with a more fascinating history and more elegant properties, surrounded by greater depths of mystery—and more totally useless—than the perfect numbers and their close relatives the amicable (or friendly) numbers.

A perfect number is simply a number that equals the sum of its divisors (including 1 but not the number itself). The smallest such number is 6, which equals the sum of its three divisors, 1, 2 and 3. The next is 28, the sum of $1 + 2 + 4 + 7 + 14$. Early commentators on the Old Testament, both Jewish and Christian, were much impressed by the perfection of those two numbers. Was not the world created in six days and does not the moon circle the earth in 28? In *The City of God*, Book 11, Chapter 30, St. Augustine argues that although God could easily have created the world in an instant, he preferred to take six days because the perfection of 6 signifies the perfection of the universe. (Similar views had been advanced earlier by the first-century Jewish philosopher Philo Judaeus in the third chapter of his *Creation of the World*.) "Therefore," St. Augustine concludes, "we must not despise the science of numbers, which, in many passages of Holy Scripture, is found to be of eminent service to the careful interpreter."

The first great achievement in perfect number theory was Euclid's ingenious proof that the formula $2^{n-1}(2^n - 1)$ always gives an even perfect number if the parenthetical expression is a prime. (It is never a prime unless the exponent n also is prime, although if n is prime, $2^n - 1$ need not be, indeed rarely is, prime.) It was not until 2,000 years later that Leonhard Euler proved that this formula gives *all* even perfects. In what follows, "perfect number" will mean "even perfect number" because no odd perfects are known and they probably do not exist.

To get an intuitive grasp of Euclid's remarkable formula and see how closely it ties perfect numbers to the familiar doubling series 1, 2, 4, 8, 16..., consider the legendary story about the Persian king who was so delighted with the game of chess that he told its originator he could have any gift he wanted. The man made what seemed to be a modest request: he asked for a single grain of wheat on the first square of the chessboard, two grains on the second square, four on the third and so on up the powers of 2 to the 64th square. It turns out that the last square would require 9,223,372,036,854,775,808 grains. The total of all the grains is twice that number minus 1, or a few thousand times the world's annual wheat crop!

If each square of a chessboard is marked according to the number of grains it would hold [*see illustration below*], then, if we remove one grain from a square, the number remaining on that square will fit the parenthetical expression in Euclid's formula. If that number is a prime, multiply it by the number of grains on the preceding square, the

2^{0}	2^{1}	2^{2}	2^{3}	2^{4}	2^{5}	2^{6}	2^{7}
2^{8}	2^{9}	2^{10}	2^{11}	2^{12}	2^{13}	2^{14}	2^{15}
2^{16}	2^{17}	2^{18}	2^{19}	2^{20}	2^{21}	2^{22}	2^{23}
2^{24}	2^{25}	2^{26}	2^{27}	2^{28}	2^{29}	2^{30}	2^{31}
2^{32}	2^{33}	2^{34}	2^{35}	2^{36}	2^{37}	2^{38}	2^{39}
2^{40}	2^{41}	2^{42}	2^{43}	2^{44}	2^{45}	2^{46}	2^{47}
2^{48}	2^{49}	2^{50}	2^{51}	2^{52}	2^{53}	2^{54}	2^{55}
2^{56}	2^{57}	2^{58}	2^{59}	2^{60}	2^{61}	2^{62}	2^{63}

Powers of 2 on a chessboard. Colored squares yield Mersenne primes

2^{n-1} of the formula. *Voilà,* we have a perfect number! Primes of the form $2^n - 1$ are now called Mersenne primes after the 17th-century French mathematician who studied them. The colored squares in the illustration mark the cells that become Mersenne primes after losing one grain and that consequently provide the first nine perfect numbers.

From Euclid's formula it is not difficult to prove all kinds of weird and beautiful properties of perfect numbers. For example, all perfects are triangular. This means that a perfect number of grains can always be arranged to form an equilateral triangle like the 10 bowling pins or the 15 pool balls. Put another way, every perfect number is a partial sum of the series $1 + 2 + 3 + 4 + \ldots$. It is also easy to show that every perfect number except 6 is a partial sum of the series of consecutive odd cubes: $1^3 + 3^3 + 5^3 + \ldots$. Still more surprising, the sum of the reciprocals of the divisors of a perfect number, including the number itself as a divisor, is always 2. For example, the reciprocals of the divisors of 28:

$$\frac{1}{1}+\frac{1}{2}+\frac{1}{4}+\frac{1}{7}+\frac{1}{14}+\frac{1}{28}=2.$$

The digital root of every perfect number (except 6) is 1. (To obtain the digital root add the digits, then add the digits of the result, and continue until only one digit remains. This is the same as casting out nines, and thus to say that a number has a digital root of 1 is equivalent to saying that the number has a remainder of 1 when divided by 9.) The proof involves showing that Euclid's formula gives a number with a digital root of 1 whenever n is odd, and since all primes except 2 are odd, perfect numbers belong to this class. The one even prime, 2, provides the only perfect number, 6, that does not have 1 as its digital root.

Because perfect numbers are so intimately related to the powers of 2, one might expect them to have some kind of striking pattern when expressed in the binary system. This proves to be correct. Indeed, given the Euclidean formula for a perfect number, one can instantly write down the number's binary form. Readers are invited first to determine the rule by which this can be done and then to see if they can show, before looking at the simple proof in the solution section, that the rule always works.

The two greatest unanswered questions about perfects are: Is there an odd perfect number? Is there a largest even perfect number? No odd perfect has yet been found; nor, although it looks so easy, has anyone proved that such a number cannot exist. The second question hinges, of course, on whether there is an infinity of Mersenne primes, since every such prime immediately leads to a perfect number. When each of the first four Mersenne primes (3, 7, 31 and 127) is substituted for n in the formula $2^n - 1$, the formula gives a higher Mersenne prime. For more than 70 years mathematicians hoped this procedure would define an infinite set of Mersenne primes, but the next possibility, $n = 2^{13} - 1 = 8{,}191$, let them down: in 1953 a computer found that $2^{8,191} - 1$ was not a prime. No one knows whether the series of Mersenne primes continues forever or has a highest member.

Oystein Ore, in his *Number Theory and Its History,* quotes an amusing prediction from Peter Barlow's 1811 book *Theory of Numbers.* After giving the ninth perfect, Barlow adds that it "is the greatest that will ever be discovered, for, as they are merely curious without being useful, it is not likely that any person will attempt to find one beyond it." In 1876 the French mathematician Édouard Lucas, who wrote a classic four-volume work on recreational mathematics, announced the next perfect to be discovered, $2^{126}(2^{127} - 1)$. The 12th Mersenne prime, on which it is based, is one less than the number of grains on the last square of a *second* chessboard, if the doubling plan is carried over to another board. Years later Lucas had doubts about this number, but eventually its primality was established. It is the largest Mersenne prime to have been found without the aid of modern computers. The illustration at the left lists formulas for the 23 known perfects and the number of digits in each, and gives the numbers themselves until they get too large. The last perfect—which has 22,425 divisors—came to light in 1963 when a computer at the University of Illinois discovered the 23rd Mersenne prime. (The university's mathematics department is so proud of this that its post-

	FORMULA	NUMBER	NUMBER OF DIGITS
1	$2^1 (2^2 - 1)$	6	1
2	$2^2 (2^3 - 1)$	28	2
3	$2^4 (2^5 - 1)$	496	3
4	$2^6 (2^7 - 1)$	8.128	4
5	$2^{12} (2^{13} - 1)$	33.550.336	8
6	$2^{16} (2^{17} - 1)$	8.589.869.056	10
7	$2^{18} (2^{19} - 1)$	137.438.691.328	12
8	$2^{30} (2^{31} - 1)$	2.305.843.008.139.952.128	19
9	$2^{60} (2^{61} - 1)$		37
10	$2^{88} (2^{89} - 1)$		54
11	$2^{106} (2^{107} - 1)$		65
12	$2^{126} (2^{127} - 1)$		77
13	$2^{520} (2^{521} - 1)$		314
14	$2^{606} (2^{607} - 1)$		366
15	$2^{1.278} (2^{1.279} - 1)$		770
16	$2^{2.202} (2^{2.203} - 1)$		1,327
17	$2^{2.280} (2^{2.281} - 1)$		1,373
18	$2^{3.216} (2^{3.217} - 1)$		1,937
19	$2^{4.252} (2^{4.253} - 1)$		2,561
20	$2^{4.422} (2^{4.423} - 1)$		2,663
21	$2^{9.688} (2^{9.689} - 1)$		5,834
22	$2^{9.940} (2^{9.941} - 1)$		5,985
23	$2^{11.212} (2^{11.213} - 1)$		6,751

The 23 known perfect numbers

```
EXECUTION STARTED ON 25/09/67   AT   17.17.29  WITH  THE INSTRUCTION COUNTER READING      248

                                        2 81411 20136 97373 13339 31529 75842 58419 18186 62382 01360 07878 92419 34934 55151 76682
27631 38107 15094 74563 32570 74198 78930 85350 71537 34244 50164 18881 80178 93905 48709 41439 18572 57571 56575 87064
78418 35674 70706 74633 49718 80530 50875 41682 16243 25680 55582 60711 10691 94660 74608 73056 96536 08305 71590 24277
49342 26866 18396 63091 85433 46251 45374 84258 65598 23862 35046 02922 75078 01410 90716 33484 39547 78109 33972 60096
90967 70918 43944 55575 42211 15477 34376 02069 79650 06708 78849 93478 01297 72778 78532 80743 22345 54020 93157 18023
10429 92316 75884 32457 03610 41108 50960 43976 90384 50365 51402 23496 25383 66575 12071 69661 69735 27322 36111 92684
64547 51701 73452 70113 79148 17510 78208 21297 62894 67956 31098 96076 74922 50494 83425 40733 34414 12162 78339 39461
53921 25289 32010 72613 66892 93688 81566 54916 71395 17471 04526 63709 17575 36037 74156 85576 65153 13827 61372 72816
96692 63352 96663 63787 28653 97699 41609 10777 71835 93336 00268 01245 17633 45149 04395 98324 82383 64572 51219 40639
14326 35639 22560 45560 42396 00430 77993 61927 37990 05864 00420 76309 23208 13392 26249 29420 76312 93326 80338 18471
55525 58206 39308 88994 86655 70202 40381 58563 13578 94977 97670 46261 84532 79567 25767 28920 52623 11752 01478 62478
13331 83401 50844 75386 76052 66122 17340 57972 12374 14485 80372 53554 63022 00953 63010 08145 86752 47046 04618 86203
90935 55206 19532 82409 51895 10704 07932 84825 09546 25301 51872 82399 71717 64140 66331 58043 09008 61194 25783 80931
06474 89915 94407 47632 84377 85848 82542 39211 70614 93829 40294 83257 16297 92993 88940 69587 73754 48948 08110 83452
93394 32780 84527 29789 83413 51401 93912 41966 17994 88795 21032 82381 12742 21870 06345 41149 74365 72872 32843 42636
93488 04878 99347 19624 03393 96785 76761 50371 60019 66502 52168 25011 77931 78488 01200 05054 22821 36255 05205 09209
72445 98958 52366 82747 78516 19190 50325 48531 15029 40313 21789 89005 19575 11943 01340 27728 27303 90683 65112 05878
95060 19875 31218 82187 78865 70240 07291 78418 65185 89977 78851 03067 43945 89610 86452 58766 41569 28256 64174 47061
61533 05144 85227 38845 49635 05925 54106 06458 42732 38641 09506 68763 63144 47514 26909 49329 53219 92421 25946 95157
65500 91585 21173 42092 32758 82063 32762 54086 17963 03296 20335 72563 55360 40560 97832 11154 75359 08988 43381 69197
47615 81716 16066 20557 30700 03771 94730 01343 18155 60750 15902 78421 64901 42254 45712 24546 93679 32349 70894 95466
84254 36412 34778 53761 94310 03013 90805 68383 42077 26286 18722 64610 97075 06566 92810 28000 33941 70434 39919 62002
05979 45655 27774 91388 32377 56792 72006 55437 68640 79217 74415 59278 27235 08230 92843 68353 43966 79150 22967 61018
34243 78782 04200 87274 02861 72126 84576 38873 36057 69491 22410 98665 92577 36066 62414 67280 15898 86055 23486 34588
08822 27855 50570 63092 76349 41503 45476 77180 61829 63528 66263 00550 92222 54318 45976 81941 26727 60304 74603 44175
58102 92983 20171 22635 52344 39676 81630 99191 27574 20633 48077 19021 87541 38915 80871 52904 91878 29308 41213 34009
10419 75631 30215 40478 43660 41784 46757 73899 86320 83586 20799 22340 85162 63437 54067 71169 70732 32139 88284 94377
91221 71985 95360 58979 02291 78176 82865 48287 87818 04150 60635 46004 71641 04095 48377 72017 37468 87332 40685 50430
69582 62103 04316 33638 53113 84093 49002 13323 72463 46337 39774 27405 89667 38275 44203 12857 48745 81960 33523 20056
37229 31959 23692 88171 37527 67022 60450 91173 50695 04025 01666 77552 14932 07364 36541 99488 47701 03639 09372 00575
78999 89580 77577 51266 21113 05790 57174 49417 22201 60705 30243 91611 67059 90451 30425 62063 18289 29773 83030 95152
43054 97722 39514 96482 16018 38628 86144 63019 36017 71054 67775 03109 26303 09947 47397 61857 62073 73447 72544 14271
35362 42636 08636 69327 15763 59830 45447 97181 67188 01639 86954 75251 46305 65557 18437 17916 87566 91403 20724 97856
85867 18527 58660 24396 02335 28351 39449 80064 32703 02781 04224 14497 18836 80541 68978 47962 67391 47608 76963 92191

END OF JOB.
```

The 23rd Mersenne prime, $2^{11,213} - 1$

age meter has been stamping on envelopes a rectangle bearing the statement: "$2^{11213} - 1$ is prime.") The printout of this largest prime known [*see illustration above*] is from John McKay, who used a computer at the Atlas Computer Laboratory of the British Science Research Council and a fast program devised by Fred Lunnon of the University of Manchester.

The end digits of perfect numbers present another tantalizing mystery, appropriately discussed in the year 1968 since it concerns the digits 6 and 8. It is quite easy to prove from Euclid's formula that any even perfect must end in 6 or 8. (If it ends in 8, the preceding digit is 2; if it ends in 6, the preceding digit must be 1, 3, 5 or 7 except in the cases of 6 and 496.) The ancients knew the first four perfects—6, 28, 496 and 8,128—and from them rashly concluded that the 6's and 8's alternated as the series continued. Scores of mathematicians from ancient times through the Renaissance repeated this dogmatically, without proof, particularly after the fifth perfect number (first correctly given in an anonymous 15th-century manuscript) turned out to end in 6. Alas, so does the seventh. The series of terminal digits for the 23 known perfects is

6, 8, 6, 8, 6, 6, 8, 8, 6, 6, 8, 8,
6, 8, 8, 8, 6, 6, 6, 8, 6, 6, 6.

The sequence contains infuriating hints of order. The first four digits alternate 6 and 8, then 66 and 88 alternate for eight digits, followed by a meaningless 6, then 888 and the palindromic 6668666. Are the digits trying to tell us something or is all this accidental? If we partition the series into triplets, starting at the left, no triplet contains three of a kind. If this pattern continues, the next perfect should end in 28, but so far no one has found a reliable rule for predicting the last digit of the next, undiscovered perfect. It is easy, of course, to determine the terminal digit of any perfect number if you know its Euclidean formula. Can the reader work out a simple rule?

Amicable numbers derive from an obvious generalization of the perfects. Suppose we start with any number, add its divisors to obtain a second number, then add the divisors of *that* number and continue the chain in the hope of eventually getting back to the original number. If the first step immediately restores the original number, the chain has only one link and the number is perfect. If the chain has two links, the two numbers are said to be amicable. Each is equal to the sum of the divisors of the *other*. The smallest such numbers, 220 and 284, were known to the Pythagoreans, who considered them symbols of friendship. Biblical commentators spotted 220, the lesser of the pair, in Genesis 32:14 as the number of goats given Esau by Jacob. A wise choice, the commentators said, because 220, being one of the amicable pair, expressed Jacob's great love for Esau. During the Middle Ages this pair of numbers played a role in horoscope casting, and talismans inscribed with 220 and 284 were believed to promote love. One poor Arab of the 11th century recorded that he once tested the erotic effect of *eating* something labeled with 284, at the same time having someone else swallow 220, but he failed to add how the experiment worked out.

It was not until 1636 that another pair of amicable numbers, 17,296 and 18,416, were discovered by the great Pierre de Fermat. He and René Descartes independently rediscovered a rule for constructing certain types of amicable pairs—a rule they did not know had previously been given by a ninth-century Arabian astronomer. Using this rule, Descartes found a third pair: 9,363,584 and 9,437,056. In the 18th century Euler drew up a list of 64 amicable pairs (two of which were later shown to be unfriendly). Adrien Marie Legendre found another pair in 1830. Then in 1867 a 16-year-old Italian, B. Nicolò I. Paganini, startled the mathematical world by announcing that 1,184 and 1,210 were friendly. It was the second lowest pair and had been completely overlooked until then! Although the boy probably found it by trial and error, the discovery put his name permanently into the history of number theory.

Today more than 600 amicable pairs are known, many of the numbers running to more than 30 digits. J. Alanen, Oystein Ore and J. Stemple, in their pa-

Postage-meter stamp honoring the largest prime

1	220	284
2	1,184	1,210
3	2,620	2,924
4	5,020	5,564
5	6,232	6,368
6	10,744	10,856
7	12,285	14,595
8	17,296	18,416
9	63,020	76,084
10	66,928	66,992
11	67,095	71,145
12	69,615	87,633
13	79,750	88,730

Amicable pairs with five or fewer digits

per "Systematic Computations on Amicable Numbers" (*Mathematics of Computation*, Vol. 21, No. 98, pages 242-245; April, 1967), list the 42 pairs of amicables in which the smaller number is less than 1,000,000. The 66 amicables between 1,000,000 and 10,000,000 were tabulated last year by Paul Bratley and John McKay with a computer at the University of Edinburgh, but their list (which includes 38 new pairs) has not yet been published. The list at the bottom of the next page is of all pairs smaller than 100,000. The last pair was a 1964 computer discovery by Howard L. Rolf of Baylor University.

All known amicable pairs consist of two even numbers or (more rarely) two odd numbers. No one, however, has yet proved that a pair of mixed parity is impossible. Bratley and McKay conjecture that all odd amicable numbers are multiples of 3, but this too has not yet been proved. All even pairs have a sum that is a multiple of 9, another strange fact no one seems to understand. There is no known formula for generating all amicable pairs, nor has it been established whether the number of such pairs is infinite or finite.

If the chain that leads back to the original number has more than two links, the numbers have been called "sociable." Only two sociable chains are known. In 1918 a French mathematician, P. Poulet, announced a chain of five links (12,496, 14,288, 15,472, 14,536, 14,264) and a truly astounding chain of 28 links (a perfect number) that starts with 14,316 (move 3 to the front and you have pi to four decimals!).

The big unsolved question is whether a "crowd"—a three-link chain—exists. Alanen, Ore and Stemple, in the paper cited above, disclose that if there is a crowd, its lesser number must be more than 1,000,000. At Edinburgh, Bratley and McKay are making extensive computer sweeps for crowds, so far without success. Such searches will surely continue, useless though crowds may be, until such a triple chain is encountered or some clever number theorist proves its impossibility.

Diophantine Analysis and the Problem of Fermat's Legendary "Last Theorem"

11

by Martin Gardner
July 1970

"All right," said Simon. He took a deep breath. "My question is this: Is Fermat's Last Theorem correct?"

The devil gulped. For the first time his air of assurance weakened.

"Whose last what?" he asked in a hollow voice.

—Arthur Porges,
"The Devil and Simon Flagg"

An old chestnut, common in puzzle books of the late 19th century (when prices of farm animals were much lower than today), goes like this. A farmer spent $100 to buy 100 animals of three different kinds. Each cow cost $10, each pig $3 and each sheep 50 cents. Assuming that he bought at least one cow, one pig and one sheep, how many of each animal did the farmer buy?

At first glance this looks like a problem in elementary algebra, but the would-be solver quickly discovers that he has written a pair of simultaneous equations with three unknowns, each of which must have a value that is a positive integer. Finding integral solutions for equations is today called Diophantine analysis. In earlier centuries such analysis allowed integral fractions as values, but now it is usually restricted to whole numbers, including zero and negative integers. Of course in problems such as the one I have cited the values must be positive integers. Diophantine problems abound in puzzle literature, and many have appeared in this department. The well-known problem of the monkey and the coconuts (discussed in *Scientific American* in April, 1958) and the ancient task of finding right-angle triangles with integral sides (October, 1964) are among the classic instances of Diophantine problems.

The term "Diophantine" derives from Diophantus of Alexandria. He was a prominent Greek mathematician of his time, but to this day no one knows in what century he lived. Most authorities place him in the third century A.D. Nothing is known about him except some meager facts contained in a rhymed problem that appeared in a later collection of Greek puzzles. The verse has been quoted so often (see Oystein Ore, *Number Theory and Its History*, page 180) and its algebraic solution is so trivial that I shall not repeat it here. If its facts are correct, we know that Diophantus had a son who died in his middle years and that Diophantus lived to the age of 84. About half of his major work, *Arithmetica*, has survived. Because many of its problems call for a solution in whole numbers, the term Diophantine became the name for such analysis. Diophantus made no attempt at a systematic theory, and almost nothing is known about Diophantine analysis by earlier mathematicians.

Today Diophantine analysis is a vast, complex branch of number theory (not algebra) with an enormous literature. There is a complete theory only for linear equations. No general method is known (it may not even exist) for solving equations with powers of 2 or higher. Even the simplest nonlinear Diophantine equation may be fantastically difficult to analyze. It may have no solution, an infinity of solutions or any finite number. Scores of such equations, so simple a child can understand them, have resisted all attempts either to find a solution or to prove none is possible.

The simplest nontrivial Diophantine equation has the linear form $ax + by = c$, where x and y are two unknowns and a, b and c are given integers. Let us see how such an equation underlies the puzzle in the opening paragraph. Letting x be the number of cows, y the number of pigs and z the number of sheep, we can write two equations:

$$10x + 3y + z/2 = 100$$
$$x + y + z = 100$$

To get rid of the fraction, multiply the first equation by 2. From this result, $20x + 6y + z = 200$, subtract the second equation. This eliminates z, leaving $19x + 5y = 100$. How do we find integral values for x and y? There are many ways, but I shall give only an old algorithm that utilizes continued fractions and that applies to all equations of this form.

Put the term with the smallest coefficient on the left: $5y = 100 - 19x$. Dividing both sides by 5 gives $y = (100 - 19x)/5$. We next divide 100 and $19x$ by 5, putting the remainders (if any) over 5 to form a terminal fraction. In this way the equation is transformed to $y = 20 - 3x - 4x/5$.

It is obvious that if x and y are positive integers (as they must be), x must have a positive value that will make $4x/5$ an integer. Clearly x must be a multiple of 5. The lowest such multiple is 5 itself. This gives y a value of 1 and z (going back to either of the two original equations) a value of 94. We have found a solution: 5 cows, 1 pig, 94 sheep. Are there other solutions? If negative integers are allowed, there are an infinite number, but here we cannot allow negative animals. When x is given a value of 10, or any higher multiple of 5, y becomes negative. The problem therefore has only one solution.

In this easy example the first integral fraction obtained, $4x/5$, does not contain a y term. For equations of the same form but with larger coefficients, the procedure just described must often be repeated many times. The terminal fraction is made equal to a new unknown integer, say a, the term with the lowest coefficient is put on the left, and the procedure is repeated to obtain a new terminal fraction. Eventually you are sure to end with a fraction that has only one unknown and is simple enough so that you can see what values the unknown must have to make the fraction integral. By working backward through whatever series of equations has been necessary, the original problem is solved. (An outline of this standard algorithm can be found in Helen Abbott Merrill's *Mathematical Excursions,* a Dover paperback, Chapter 12.)

For an example of an equation similar to the one just analyzed that has *no* solution, assume that cows cost \$5, pigs \$2 and sheep 50 cents. The two equations are handled exactly as before. The first is doubled to eliminate the fraction and the second is subtracted, producing the Diophantine equation $9x + 3y = 100$. Using the procedure of continued fractions, you end with $y = 33 - 3x - 1/3$, which shows that if x is integral, y cannot be. In this case, however, we can tell at once that $9x + 3y = 100$ has no solution by applying the following old theorem. If the coefficients of x and y have a common factor that is not a factor of the number on the right, the equation is unsolvable in integers. In this case 9 and 3 have 3 as a common divisor, but 3 is not a factor of 100. It is easy to see why the theorem holds. If the two terms on the left are each a multiple of n, so will their sum be; therefore the term on the right also must be a multiple of n. An even simpler instance would be $4x + 8y = 101$. The left side of the equality obviously must be an even integer, so that it cannot equal the odd number on the right. It is also good to remember that if all three given numbers do have a common factor, the equation can immediately be reduced by dividing all terms by the common divisor.

As an example of a variant of the basic problem that has a finite number (more than one) of positive-integer solutions, consider the case in which cows cost \$4, pigs \$2 and sheep a third of a dollar. The solutions will be given at the back of the book.

Many geometric problems are equivalent to finding integral solutions for Diophantine equations. (An example was given last month: finding integral values for the crossed-ladders problem.) A well-known unsolved problem of this type is finding a brick-shaped solid (called a rectangular parallelepiped) whose three edges, three face diagonals and single space diagonal (from one corner to the corner on the opposite side of the solid's center) are all integers [*see illustration below*]. The problem is equivalent to finding integer solutions for the seven unknowns in the following set of equations:

$$a^2 + b^2 = c^2$$
$$a^2 + d^2 = e^2$$
$$b^2 + d^2 = f^2$$
$$b^2 + e^2 = g^2$$

The problem has not been shown to be impossible, nor has it been solved. John Leech, a British mathematician, has been searching for a solution, and I am indebted to him for the following information. The smallest brick with integral edges and face diagonals (only the space diagonal is nonintegral) has edges of 44, 117 and 240. This was known by Leonhard Euler to be the minimum solution. If all values are integral except a face diagonal, the smallest brick has edges of 104, 153 and 672, a result also known to Euler. (The brick's space diagonal is 697.) The third case, where only an edge is nonintegral, has not, as far as Leech knows, been considered before. It too has solutions, but the numbers are, as Leech puts it, "hideous." He suspects that the smallest such brick may be one with edges of 3, 4 and the square root of 136,990,339,200. Of course the brick's volume is also irrational.

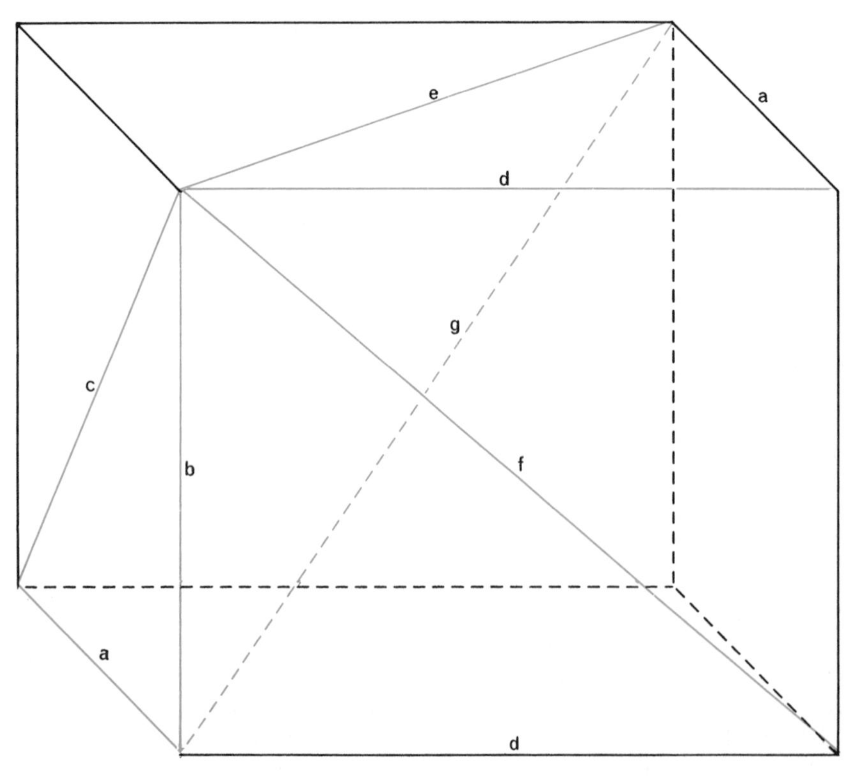

The integral brick, an unsolved Diophantine problem

A much easier geometric problem, which I took from a puzzle book by L. H. Longley-Cook, is illustrated at the top of the opposite page. A rectangle (the term includes the square) is drawn on graph paper as shown and its border cells are shaded. In this case the shaded cells do not equal the unshaded cells of the interior rectangle. Is it possible to draw a rectangle of proportions such that the border—one cell wide—contains the same number of cells as there are within the border? If so, the task is to find *all* such solutions. The Diophantine equation that is involved can be solved easily by a factoring trick, which is explained in the solution section.

In ancient times the most famous Diophantine problem, posed by Archimedes, became known as the "cattle problem." It involves eight unknowns, but the integral solutions are so huge (the smallest value contains more than 200,000 digits) that it was not fully calculated until five years ago, when a computer managed to do it. The interested reader will find a good discussion of the cattle problem in Eric Temple Bell's *The Last Problem* (1961, pages 151–157) and the final solution, by H. C. Williams and others, in the journal *Mathematics of Computation* (October, 1965).

The greatest of all Diophantine problems, which is still far from solved, is the "last theorem" of Pierre de Fermat, the 17th-century French amateur number theorist. (He was a jurist by profession.) Every mathematician knows how Fermat, reading a French translation of Diophantus' *Arithmetica,* added a note in Latin to the eighth problem of the second book, where an integral solution is asked for $x^2 + y^2 = a^2$. Fermat wrote that such an equation had no solution in integers for powers greater than 2. (When the power is 2, the solution is called a "Pythagorean triple" and there are endless numbers of solutions.) In brief, Fermat asserted that $x^n + y^n = a^n$ has no solution in integers if n is a positive integer greater than 2. "I have discovered a truly marvelous demonstration," Fermat concluded his note, "which this margin is too narrow to contain."

To this day no one knows if Fermat really had such a proof. Because the greatest mathematicians since Fermat have failed to find a proof, the consensus is that Fermat was mistaken. Lingering doubts arise from the fact that Fermat always *did* have a proof whenever he said he did. For example, consider the Diophantine equation $y^3 = x^2 + 2$. It is easy to find by trial and error that it has the solutions $3^3 = 5^2 + 2$ and $3^3 = -5^2 + 2$. To prove, however, that there are no other integral solutions, Bell writes in *Men of Mathematics*, "requires more innate intellectual capacity... than it does to grasp the theory of relativity." Fermat said he had such a proof although he did not publish it. "This time he was not guessing," Bell continues. "The problem is hard; he asserted that he had a proof; a proof was later found." Fermat did publish a relatively elementary proof that $x^4 + y^4 = a^4$ has no solution, and later mathematicians proved the impossibility of the more difficult $x^3 + y^3 = a^3$. The cases of $n = 5$ and $n = 7$ were settled early in the 19th century.

It is not hard to show that Fermat's last theorem is true for all exponents n except primes greater than 2. There is a special class of primes for which, in 1941, the theorem was shown to be true if n is no greater than 253,747,889. For all other prime values of n it was proved in 1964 that n must be at least 25,013. (See J. L. Selfridge and B. W. Pollack, *Notices of the American Mathematical Society*, Vol. 11, 1964, page 97.)

Attempts to prove Fermat's last theorem are so frustrating, in view of how easy it looks, that the theorem provides a means by which a mathematician outwits the devil in a fantasy story from which I took the quotation at the beginning of this article. (The story is reprinted in Clifton Fadiman's entertaining anthology, *Fantasia Mathematica.*) It is the deepest unsolved problem in Diophantine theory. Some mathematicians believe it may be true but unprovable, now that Kurt Gödel has shown, in his famous undecidability proof, that arithmetic contains theorems that cannot be established inside the deductive system of arithmetic. (If Fermat's last theorem is Gödelian-undecidable, it would have to be true, because if it were false, it would be decidable by a single counterexample.) I earnestly ask readers not to send me proofs. I am not competent to evaluate them. Ferdinand Lindemann, the first to prove (in 1882) that pi is transcendent, once published a long proof of Fermat's last theorem that turned out, according to Bell, to have its fatal mistake right at the beginning. Dozens of other fallacious proofs have been published by leading mathematicians. When David Hilbert was asked why he never tackled the problem, his reply was: "Before beginning I should put in three years of intensive study, and I haven't that much time to squander on a probable failure."

The mathematics departments of many large universities return all proofs of Fermat's last theorem with a form letter stating that the paper will be evaluated only after an advance payment of a specified fee. Edmund Landau, the German mathematician, used a form letter that read: "Dear Sir/Madam: Your proof of Fermat's last theorem has been received. The first mistake is on page ___, line ___." Landau would then assign the filling in of the blanks to a graduate student.

Donald E. Knuth whimsically asks for a proof of Fermat's last theorem as the last exercise at the end of his preface to the first volume of his series *The Art of Computer Programming* (1968). His answer states that someone who read a preliminary draft of the book reported that he had a truly remarkable proof but

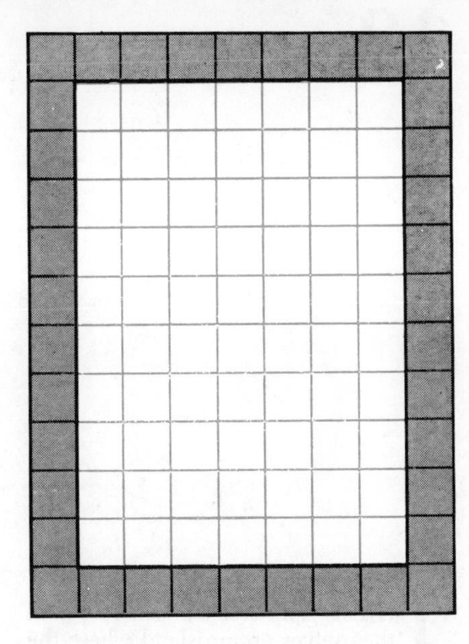

A simple Diophantine problem

that the margin of the page was too small to contain it.

Euler failed to prove Fermat's last theorem, but he made a more general conjecture that, if it is true, would include the truth of Fermat's last theorem as a special case. Euler suggested that no nth power greater than 2 can be the integral sum of fewer than n nth powers. As we have seen, it has long been known that the conjecture holds when n is 3, for this is merely Fermat's last theorem with powers of 3. It is not yet known whether or not $x^4 + y^4 + z^4 = a^4$ has a solution.

I close on a happier note. In 1966 Leon J. Lander and Thomas R. Parkin, using a computer, found the first counterexample to Euler's conjecture. It was first published in *Bulletin of the American Mathematical Society* (Vol. 72, 1966, page 1079) about two centuries after Euler made his guess, in an article of just five lines. The counterexample, the smallest when n is 5, is

$$27^5 + 84^5 + 110^5 + 133^5 = 144^5.$$

12 The Theory of Numbers

by Paul S. Herwitz
July 1951

Concerning shipwrecked sailors, the curious habit of Pierre de Fermat, calculating machines, wine jugs, magic squares and other aspects of the fascinating properties of integers

THREE shipwrecked sailors found themselves on an island where the only food was coconuts. They gathered a large number of coconuts and decided to get some needed sleep before they divided the pile into three equal shares. During the night one of the sailors awakened and, not trusting his companions, decided to take his share of the collection without waiting until morning. He found that after throwing away one of the coconuts he could divide those remaining into three equal shares. He buried his share, left the rest in a pile and went back to sleep. Later one of the other sailors awakened and proceeded to go through the same routine: he threw away one coconut, took one third of the remainder and buried them, then went back to sleep. Still later the third sailor, no less suspicious than his mates, went through exactly the same procedure. What is the least number of coconuts the sailors must have collected originally so that there would be a whole number of coconuts left after all this? (We leave to the psychologists the problem of deciding how the sailors reacted to the shrinkage of the pile in the morning.)

Let x represent the number of coconuts collected by the sailors. The first sailor, after burying his share of them, left $2/3(x-1)$, the second left $2/3[2/3(x-1)-1]$; the final equation is $2/3\{2/3[2/3(x-1)-1]-1\}=y$, the number of coconuts left in the morning. This simplifies to $8x-27y=38$. As a general equation this has an infinite number of possible solutions for varying values of x, but we know from the terms of our particular problem that x and y must be positive whole numbers and that x must be the smallest positive integer that will permit y to be a positive integer. The last condition gives us one, and only one, answer, namely, the least number of coconuts the sailors originally collected was 25, and the next morning there would have been six left in the pile.

Had four sailors landed on the island and suffered the same mutual mistrust, our problem would have reduced to the equation $81x-256y=525$. If we generalize the problem to apply to any number of sailors, n (where n is more than 1), the final equation is $(n-1)^n x-n^n y=(n-1)^n+n(n-1)^{n-1}+n^2(n-1)^{n-2}+\ldots+n^{n-1}(n-1)$. (The series of dots indicates that there may be more terms, or fewer terms, than we have written.) This equation is itself a specific instance of the general equation $ax+by=c$, where all the numbers are integers and c is a constant. Many problems lead to

SUSPICIOUS SAILORS awoke one by one to divide a pile of coconuts into three equal shares. What is the smallest number of coconuts that could have been in the original pile? The drawings do not show all the coconuts.

equations which are special cases of this general equation. Such an equation is called Diophantine, after the Greek mathematician Diophantus, who first discussed it around A.D. 250.

FERMAT wrote a baffling theorem about numbers on the margin of a book. Generations of mathematicians have sought the proof, without success.

THE Diophantine equation is one of the cornerstones of the Theory of Numbers. This branch of study has as its aim the discovery of properties of the integers. The fact that these principles are frequently illustrated by application to the solution of puzzles and brain-teasers, such as that of the sailors and the coconuts, should not mislead anyone into thinking that the Theory of Numbers is nothing but a scientific curiosity. Simple as many properties of the integers may appear, their proofs have often entailed many years of study by numerous fine mathematicians, and number theory has long been an important branch of mathematics.

Pierre de Fermat, the celebrated 17th-century French mathematician, is considered to be the father of the modern theory of numbers. Fermat uncovered many interesting properties which are by no means obvious or superficial. Probably as famous as the ancient problem of trisecting an angle is what is known as Fermat's Last Theorem. Fermat was accustomed to write remarks and theorems, without their proofs, in the margins of his books. In his copy of Diophantus' *Arithmetica* he stated the following "theorem": *The equation* $x^n + y^n = z^n$ *has no nontrivial solution in integers for n greater than* 2. By a trivial solution is meant a solution such as x, y and z all equal to zero. Fermat claimed to have a proof of this "theorem," but to this day no general proof of it has been found for all values of n, although many proofs are known for particular values of n. The best known special case of Fermat's equation is the Pythagorean theorem, in which n = 2. Pythagoras' principle can be stated as follows: If x and y represent the lengths of the sides of a right triangle and z the length of the hypotenuse, then $x^2 + y^2 = z^2$. This is a Diophantine equation, since there are more unknowns than equations, and integral solutions are desired.

OF first importance in the study of the Theory of Numbers is the concept of prime and composite numbers. A prime number of course is one that is evenly divisible only by plus or minus itself and by +1 and −1. All other integral numbers are composite, which means that they can be divided by some number other than themselves or 1 and yield a whole number as the quotient. A composite number can always be written as a product of two or more primes. The first few primes are 2, 3, 5, 7, 11, 13, 17, 19, 23. The composite number 12 can be written as the product of three primes: 2 × 2 × 3. Many unsuccessful attempts have been made to find a simple algebraic formula that will yield only prime numbers. One of the best known was suggested by Fermat: he believed that $2^{2^n} + 1$ was a prime for all positive integral values of n. For n = 1, 2, 3 and 4 we have respectively the primes 5, 17, 257 and 65,537. But the 18th-century Swiss mathematician Leonhard Euler showed that when n = 5, the result is not a prime number; $2^{2^5} + 1 = 4{,}294{,}967{,}297$, which is a composite number that can be factored into 641 × 6,700,417. In fact, it has not been proved that $2^{2^n} + 1$ is a prime for any value of n greater than 4. The formula is laborious and difficult to investigate, because even relatively small values of n produce very large numbers: for instance, if n = 7, Fermat's number is greater than 34×10^{37}, that is, 34 followed by 37 zeros. There are other formulas that yield primes up to a certain point, such as $n^2 - n + 41$, which is prime for integral values of n less than 41, and $n^2 - 79n + 1{,}601$, prime for integral values of n less than 80. But no one has found a formula that produces primes for all values of n.

The distribution of the primes among the integers is highly irregular. The 19th-century German mathematician Peter Gustav Lejeune Dirichlet proved that in every arithmetic progression there are an infinite number of primes. An arithmetic progression is a sequence of numbers each of which is obtained from the preceding one by addition of a certain number; for example, 1, 3, 5, 7 is a progression formed by adding 2 in each case. Dirichlet's proof falls into the branch of mathematics known as analytic number theory—as contrasted to algebraic number theory. The analytic theory applies the calculus and the theory of functions to properties of the integers. The truth of Dirichlet's theorem is readily demonstrable by relatively simple considerations in certain particular cases, but for the general arithmetic progression a, a + d, a + 2d and so on it has been necessary until recently to apply highly technical methods to prove the theorem. In 1949 the young mathematician Atle Selberg, of the Institute for Advanced Study, published a proof which has placed Dirichlet's problem in a new light. Selberg's work has revived interest in this problem and, along with several related proofs, has brought him international recognition.

An unsolved problem concerning the distribution of the primes is the problem of prime pairs. All primes larger than 2 are odd numbers, since any even number is divisible by 2. In the sequence of all odd numbers (1, 3, 5, 7, 9, 11 and so on) it has been noted that certain pairs of consecutive numbers are pairs of primes, as 3 and 5, 5 and 7, 11 and 13, 17 and 19, 29 and 31. Although it is thought that the number of such prime pairs is infinite, no one has yet succeeded in proving this to be so.

Another type of problem concerning the primes was posed in a letter that Christian Goldbach, a German teacher of mathematics, wrote to Euler in 1742. Goldbach said he believed that every even integer could be written as the sum of two primes and asked Euler whether he could prove this or could find an example to disprove it. Though Goldbach's only importance in the history of mathematics lies in this conjecture, his name will be remembered, for the problem has never been solved.

NO DISCUSSION of the Theory of Numbers could be considered complete without mention of the great German mathematician Carl Friedrich Gauss (1777-1855). In his *Disquisitiones arithmeticae* Gauss made monumental contributions to the theory. We shall consider here only one of the many subjects he discussed in this work. This is the notion of congruence of numbers, which Gauss defined somewhat as follows: If the difference of two integers, $a-b$, is divisible by a third integer c, then we say a is congruent to b with respect to the modulus c, or simply a is congruent to b modulo c. For example, 19 is congruent to 7 modulo 3, since $19-7=12$ and 12 is divisible by 3. The meanings of the words "congruence" and "modulus" give a hint as to the significance of this concept. The former word comes from the Latin *congruere,* meaning to coincide or agree, and the latter from the Latin *modulus,* a small measure. The number c acts as a measure of the "sameness" of the numbers a and b, in a certain sense.

Now when one number is divisible by another, the first of course is a multiple of the second; thus the number 6, divisible by 3, is a multiple of 3 obtained by multiplying 3 by 2. The notion of congruence, therefore, may be expressed by the equation $a-b=kc$, in which k represents the multiplier. This in turn can be written $a=b+kc$. So far we have nothing essentially new. The importance of Gauss' work springs from an implication inherent in a slight modification of notation which he introduced. Gauss wrote the second of the above equations as $a \equiv b \pmod{c}$, which reads: a is congruent to b modulo c. From this we may infer that the importance in the relationship between a and b is that they differ by a multiple of c, and the value of that multiplier (k in the original equation) is relatively unimportant.

From a slightly different standpoint, if in the equation $a=b+kc$ the integer b is less than c, but greater than or equal to zero, then b represents the remainder upon division of a by c. For example, 19 is congruent to 1 modulo 3 may be written $19/3=6+1/3$, or $19=1+6\times 3$. Here $a=19$, $b=1$, $c=3$, $k=6$, and the division of 19 by 3 leaves the remainder 1. Again, 18 is congruent to zero modulo 3 may be written $18/3=6$, or $18=0+6\times 3$. In this case $a=18$, $b=0$, $c=3$, $k=6$, and the remainder of the division of 18 by 6 is 0. In the same way 20 is congruent to 2, 21 is congruent to zero, 22 is congruent to 1, all taken modulo 3. In other words, with respect to the modulus 3, all numbers are congruent to one of the numbers 0, 1 or 2. If the modulus is represented by the general term c, then with respect to c every number is congruent to some number from 0 up to $c-2$ and $c-1$. In effect this manner of considering the integers places every integer in one of c *classes,* each class represented by one of the numbers 0, 1, and so on, up to $c-1$. For example, modulo 3 the numbers 18, 21, 24, etc., would be placed in the class represented by the number 0, since zero is the remainder left when each of these numbers is divided by 3; the numbers 17, 20, 23, 26, etc., would be placed in the class represented by the number 2, since 2 is their remainder after division by 3, and so on. The gain from this treatment is that we need not consider an infinite number of integers but only a finite number of classes. In a sense all members of a particular class are essentially the same, since they leave the same remainder upon division by their modulus.

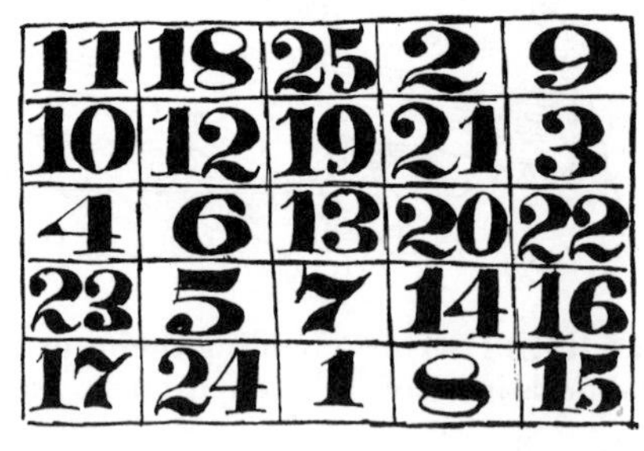

11	18	25	2	9
10	12	19	21	3
4	6	13	20	22
23	5	7	14	16
17	24	1	8	15

MAGIC SQUARE adds up to 65 in each row, column and two diagonals.

Although this way of dealing with numbers may seem somewhat involved, actually we make use of it every day without examining the process. Suppose, for example, we had been telling the time of day from the year zero by giving each hour a new number. At the time of writing of this article it would be some time after 17,093,328 o'clock. That would be a pretty cumbersome way of telling time. But modulo 24, it is shortly after midnight. What we do is to consider the hours of the day as members of 24 classes modulo 24, represented by the names one o'clock, two o'clock and so on up to 24 o'clock. (In practice we break the day into two halves and use a 12-hour clock, distinguishing one half from the other by a.m. and p.m.) Thus there are 24 "hour-classes" in a day, or two sets of 12 "hour-classes" each. Similarly, there are seven "day-classes" in a week and 12 "month-classes" in a year. Consideration of the day and month classes and certain astronomical data permitted Gauss to state congruences and rules for their use which enable us to determine easily the dates of Easter and other holidays, depending upon both the lunar and solar calendars. Similar congruences assist us in finding the day of the week on which any given date falls in any year, so that we can tell in what years January 13 will fall on a Friday, when the 4th of July will next give us a long week-end, and so on.

There is a well-known rule of arithmetic that a number is divisible by 9 only if the sum of its digits is divisible by 9. This rule, and similar rules, may be proved easily by means of the concept of congruence. Let us take the number 234, the sum of whose digits is $2+3+4=9$. The number 234 may be written $2\times 10^2+3\times 10+4$. We represent this expression by N, and the expression $2+3+4$ by n. Then $N-n$ is $(2\times 10^2+3\times 10+4)-(2+3+4)$. This may be written $(2\times 10^2-2)+(3\times 10-3)+(4-4)$, and this simplifies to $2\times 99+3\times 9$. Since 9 and 99 are both divisible by 9, the difference $N-n$ is divisible by 9; in other words, N is congruent to n modulo 9. When the sum of the digits of a number is 9, as in our example, the remainder after dividing the number by 9 is zero. When the sum is a number greater than zero but less than 9, this sum represents the remainder that will be left after division by 9. When the sum of digits is greater than 9, we may apply the process to determine whether or not this new number is divisible by 9. For example, the number 73,506,816 is congruent to 36 (the sum of its digits), which is congruent to 9 modulo 9. Therefore, this number is divisible by 9. On the other hand, the number 73,506,818 is congruent to 38; 38 is congruent to 11; 11 is congruent to 2 modulo 9. Thus 73,506,818 leaves the remainder 2 when divided by 9.

Any number divisible by 9 is also divisible by 3, since 3 divides 9. We may state, then, a new rule: a number is divisible by 3 only if the sum of its digits is divisible by 3. Similar considerations allow us to state other rules: a number is divisible by 2 if its last digit on the right is divisible by 2; a number is divisible by 4 or 25, respectively, if the number formed by its last two digits is divisible by 4 or 25, respectively; a number is divisible by 5 if its last digit is zero or 5.

A check on multiplication called "Casting Out Nines" has as its basis the rule for division by 9. The check proceeds in this way: to find out whether the product c is correct in the multiplication $a\times b=c$, we find the remainder after dividing a by 9 and the remainder after dividing b by 9 and multiply the two remainders together. This product must equal the remainder after dividing c by 9. If it does not, the value of c obtained in the original multiplication was in error. What makes the check easy is that to obtain the remainders we need only add up the digits of the respective numbers. As an example let us check the multiplication $6{,}743\times 826=5{,}569{,}718$. The sum of the first number's digits is $6+7+4+3=20$. Dividing 20 by 9 gives a remainder of 2, as can be shown by adding its digits: $2+0=2$. Similarly the second number yields $8+2+6=16$, which divided by 9 gives a remainder of

7. Multiplying the remainders, $2 \times 7 = 14$. The sum of these digits is $1+4=5$. The sum of the digits in the product, 5,569,718, is 41, and its digits also add up to 5, so the answer is probably correct. As a practical method the device of casting out nines was probably known in India before A.D. 800; it came into use in Europe in the Middle Ages. But the mathematical theory behind the method could not be explained easily until the arrival of modern number theory; imagine the difficulty of explaining the theoretical basis of these manipulations without using the ideas of congruence.

A VERY interesting modern application of the Theory of Numbers is made in the field of electronic computing machinery. Some of these machines use the binary instead of the decimal system of numbers. We have noted that 234 can be written $2 \times 10^2 + 3 \times 10 + 4$; this is a sum of multiples of powers of 10 (4 may be written as 4×10^0). The number 234 can also be written as a sum of multiples of powers of 2, thus: 1×2^7 (128) $+ 1 \times 2^6$ (64) $+ 1 \times 2^5$ (32) $+ 0 \times 2^4$ (0) $+ 1 \times 2^3$ (8) $+ 0 \times 2^2$ (0) $+ 1 \times 2$ (2) $+ 0 \times 2^0$ (0). Here we use the "base" 2 instead of 10 as in the decimal system. Instead of writing down a long string of powers of 2 we can "suppress the base" and write only the multipliers of the powers of 2, which in order from left to right are 11101010. (The reader should notice that he has been doing exactly the same thing in the decimal system whenever he has written a number larger than 10; he has suppressed the base 10.) Then decimal 234 is the same as binary 11101010; similarly, decimal 15 is binary 1111 and decimal 2 is binary 10. In an electronic computing machine the binary system has this great advantage: the only digits used are 0 and 1. In the "language" of the machine each 1 can be represented by an electrical impulse and each 0 by the absence of an impulse. In this way an electronic computing machine can handle operations with numbers very rapidly.

To return to the realm of games and puzzles, we might mention two typical problems that have been investigated by number theorists. The first is the familiar wine-jug problem, one version of which goes: A man has a five-gallon jug and a three-gallon jug. He wishes to purchase four gallons of wine from an innkeeper who has only a full eight-gallon jug. How does the man measure exactly four gallons without spilling a drop? With some thought you may figure out that he measures out three gallons in the three-gallon jug and pours this into the five-gallon jug, pours three more into the three-gallon jug and fills the five-gallon jug from this, leaving one gallon in the three-gallon jug, then empties the five-gallon jug into the eight-gallon jug and pours into the five-gallon jug the one gallon plus three more measured in the three-gallon jug. This problem has been solved by number theory in general terms for jugs holding A, B and C gallons, with D gallons to be measured out. That is, it is known for what values of A, B, C and D the problem has a solution, and a systematic way of measuring can be stated.

The second problem is the so-called magic square. A magic square is an array of n^2 integers arranged in n rows and n columns in such a way that the sum of the numbers in each row, column or principal diagonal (upper left to lower right and upper right to lower left) is magic; that is, each of these sums is equal to $n/2(n^2+1)$. Magic squares have fascinated man for ages, and many mathematicians have studied them. One method of constructing magic squares is based on notions of congruence. A simple method of constructing such squares, applicable to any square of odd n, is illustrated on the opposite page. In one of the squares shown, for example, $n=5$ and the magic sum is 65. Each box in the square (known as a cell) contains a number from 1 up to n^2, in this case $n^2=25$. The numbers are placed in the cells in their consecutive order according to the following rules:

Begin by placing the number 1 in the middle cell of the bottom row; after filling a cell in the bottom row, place the next number in the cell at the top of the next column on the right; as far as possible fill the cells in a downward diagonal line from left to right; upon filling a cell in the last right-hand column of the square, place the next number in the cell at the extreme left of the next lower row; if the next cell on the left-to-right downward diagonal is occupied, place the next number in the cell immediately above the one last filled; after filling the cell in the lower right-hand corner, place the next number in the cell immediately above.

MAGIC squares and wine jugs, calculating machines, prime numbers and congruences, Diophantine equations and shipwrecked sailors—all these furnish reasons that help to explain why the Theory of Numbers has interested so many people. Professionals and amateurs alike have been attracted by the subtle fascination of these problems, and the mark of their work has been felt in nearly every branch of mathematics. Problems in the Theory of Numbers are among the most challenging in mathematics, possibly because some of them are so difficult. Fortunately in mathematics as in all other scientific endeavors the difficulty experienced in solving a problem drives man to a continuing search for its solution. For he believes that the science of mathematics is a logical discipline, and as such is subject to complete understanding.

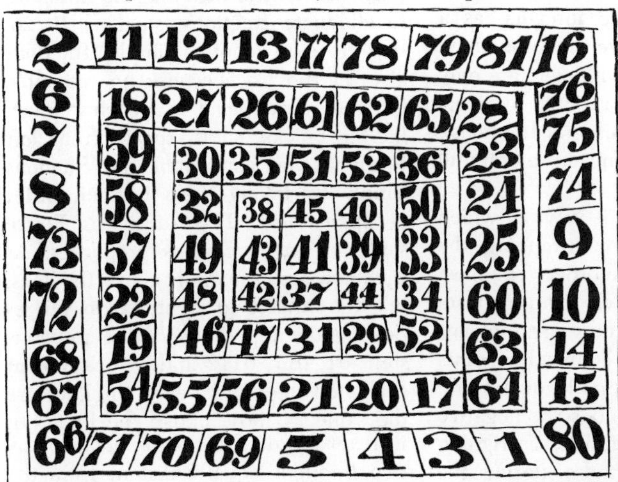

BORDERED SQUARE is magic overall, and each square formed by throwing away a border is also magic.

11	18	13	74	81	76	29	36	31
16	14	12	79	77	75	34	32	30
15	10	17	78	73	80	33	28	35
56	63	58	38	45	40	20	27	22
61	59	57	43	41	39	25	23	21
60	55	62	42	37	44	24	19	26
47	54	49	2	9	4	65	72	67
52	50	48	7	5	3	70	68	66
51	46	53	6	1	8	69	64	71

COMPOSITE SQUARE is similarly magic overall and at the same time magic for each of its smaller squares.

13

The Hierarchy of Infinities and the Problems It Spawns

by Martin Gardner
March 1966

A graduate student at Trinity
Computed the square of infinity.
But it gave him the fidgets
To put down the digits,
So he dropped math and took up divinity.
—ANONYMOUS

In 1963 Paul J. Cohen, a 29-year-old mathematician at Stanford University, found a surprising answer to one of the great problems of modern set theory: Is there an order of infinity higher than the number of integers but lower than the number of points on a line? To make clear exactly what Cohen proved, something must first be said about those two lowest known levels of infinity.

It was Georg Ferdinand Ludwig Philipp Cantor who first discovered that beyond the infinity of the integers—an infinity to which he gave the name aleph-null—there are not only higher infinities but also an infinite number of them. Leading mathematicians were sharply divided in their reactions. Henri Poincaré called Cantorism a disease from which mathematics would have to recover, and Hermann Weyl spoke of Cantor's hierarchy of alephs as "fog on fog."

On the other hand, David Hilbert said, "From the paradise created for us by Cantor, no one will drive us out," and Bertrand Russell once praised Cantor's achievement as "probably the greatest of which the age can boast." Today only mathematicians of the intuitionist school and a few philosophers are still uneasy about the alephs. Most mathematicians long ago lost their fear of them, and the proofs by which Cantor established his "terrible dynasties" (as they have been called by the Argentine writer Jorge Luis Borges) are now universally honored as being among the most brilliant and beautiful in the history of mathematics.

Any infinite set of things that can be counted 1, 2, 3 . . . has the cardinal number $\aleph_0$ (aleph-null), the bottom rung of Cantor's aleph ladder. Of course, it is not possible actually to count such a set; one merely shows how it can be put into one-to-one correspondence with the counting numbers. Consider, for example, the infinite set of primes. It is easily put in one-to-one correspondence with the positive integers:

$$\begin{array}{cccccc} 1 & 2 & 3 & 4 & 5 & 6\ldots \\ \downarrow & \downarrow & \downarrow & \downarrow & \downarrow & \downarrow \\ 2 & 3 & 5 & 7 & 11 & 13\ldots \end{array}$$

The set of primes is therefore an aleph-null set. It is said to be "countable" or "denumerable." Here we encounter a basic paradox of all infinite sets. Unlike finite sets, they can be put in one-to-one correspondence with a *part* of themselves or, more technically, with one of their "proper subsets." Although the primes are only a small portion of the positive integers, as a completed set they have the same aleph number. Similarly, the integers are only a small portion of the rational numbers (the integers plus all integral fractions), but the rationals form an aleph-null set too.

There are all kinds of ways in which this can be proved by arranging the rationals in a countable order. The most familiar way is to attach them, as fractions, to an infinite square array of lattice points and then count the points by following a zigzag path, or a spiral path if the lattice includes the negative rationals. Here is another, intriguing method of ordering and counting the positive rationals that was proposed by the American logician Charles Sanders Peirce. (See *Collected Papers of Charles Sanders Peirce*, Harvard University Press, 1933, pages 578–580.)

Start with the fractions 0/1 and 1/0. (The second fraction is meaningless, but that can be ignored.) Sum the two numerators and then the two denominators to get the new fraction 1/1, and place it between the previous pair: 0/1, 1/1, 1/0. Repeat this procedure with each pair of adjacent fractions to obtain two new fractions that go between them:

$$\frac{0}{1}\ \frac{1}{2}\ \frac{1}{1}\ \frac{2}{1}\ \frac{1}{0}\,.$$

The five fractions grow, by the same procedure, to nine:

$$\frac{0}{1}\ \frac{1}{3}\ \frac{1}{2}\ \frac{2}{3}\ \frac{1}{1}\ \frac{3}{2}\ \frac{2}{1}\ \frac{3}{1}\ \frac{1}{0}\,.$$

In this continued series every rational number will appear once and only once, and always in its simplest fractional form. There is no need, as there is in other methods of ordering the rationals, to eliminate fractions, such as 10/20, that are equivalent to simpler fractions also on the list, because no reducible fraction ever appears. If at each step you fill the cracks, so to speak, from left to right, you can count the fractions simply by taking them in their order of appearance. This series, as Peirce said, has many curious properties. At each new step the digits above the lines, taken from left to right, begin by repeating the top digits of the previous step: 01, 011, 0112 and so on. And at each step the digits below the lines are the same as those above the lines but in reverse order. As a consequence, any two fractions equally distant from the central 1/1 are reciprocals of each other. Note also that for any adjacent pair, a/b, c/d, we can write such equalities as $bc - ad = 1$, and $c/d - a/b = 1/bd$. The series is closely related to what are called

Farey numbers (after the English geologist John Farey, who first analyzed them), about which there is now a considerable literature.

It is easy to show that there is a set with a higher infinite number of elements than aleph-null. To explain one of the best of such proofs a deck of cards is useful. First consider a finite set of three objects, say a key, a watch and a ring. Each subset of this set is symbolized by a row of three cards [*see the illustration on this page*]; a face-up card [*white*] indicates that the object above it is in the subset, a face-down card [*gray*] indicates that it is not. The first subset consists of the original set itself. The next three rows indicate subsets that contain only two of the objects. They are followed by the three subsets of single objects and finally by the empty (or null) subset that contains none of the objects. For any set of n elements the number of subsets is 2^n. Note that this formula applies even to the empty set, since $2^0 = 1$ and the empty set has the empty set as its sole subset.

This procedure is applied to an infinite but countable (aleph-null) set of elements at the left in the illustration on the next page. Can the subsets of this infinite set be put into one-to-one correspondence with the counting integers? Assume that they can. Symbolize each subset with a row of cards, as before, only now each row continues endlessly to the right. Imagine these infinite rows listed in any order whatever and numbered 1, 2, 3... from the top down. If we continue forming such rows, will the list eventually catch all the subsets? No—because there is an infinite number of ways to produce a subset that cannot be on the list. The simplest way is to consider the diagonal set of cards indicated by the arrow and then suppose every card along this diagonal is turned over (that is, every face-down card is turned up, every face-up card is turned down). The new diagonal set cannot be the first subset because its first card differs from the first card of subset 1. It cannot be the second subset because its second card differs from the second card of subset 2. In general it cannot be the nth subset because its nth card differs from the nth card of subset n. Since we have produced a subset that cannot be on the list, even when the list is infinite, we are forced to conclude that the original assumption is false. The set of all subsets of an aleph-null set is a set with the cardinal number 2 raised to the power of aleph-null. This proof shows that such a set cannot be matched one

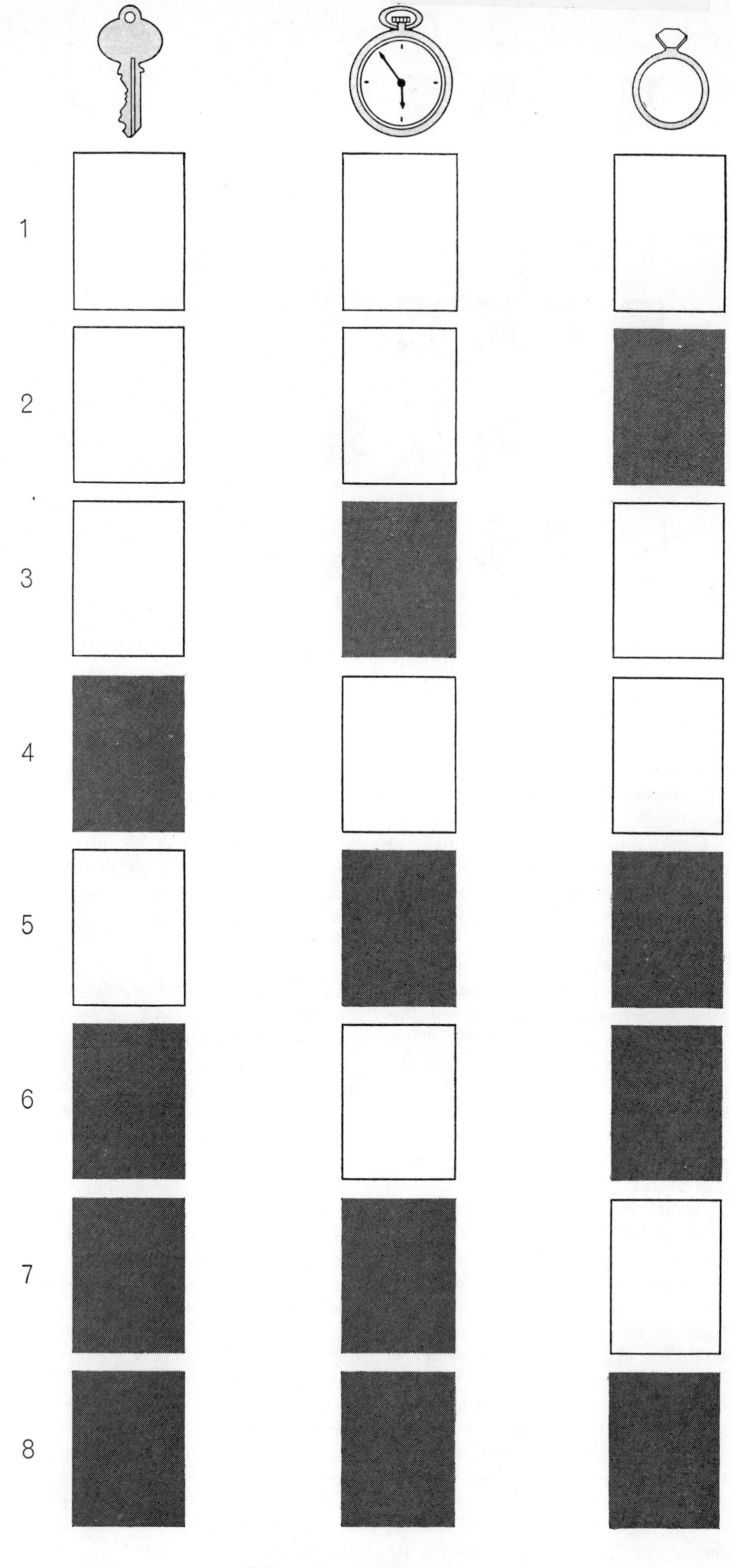

Subsets of a set of three elements

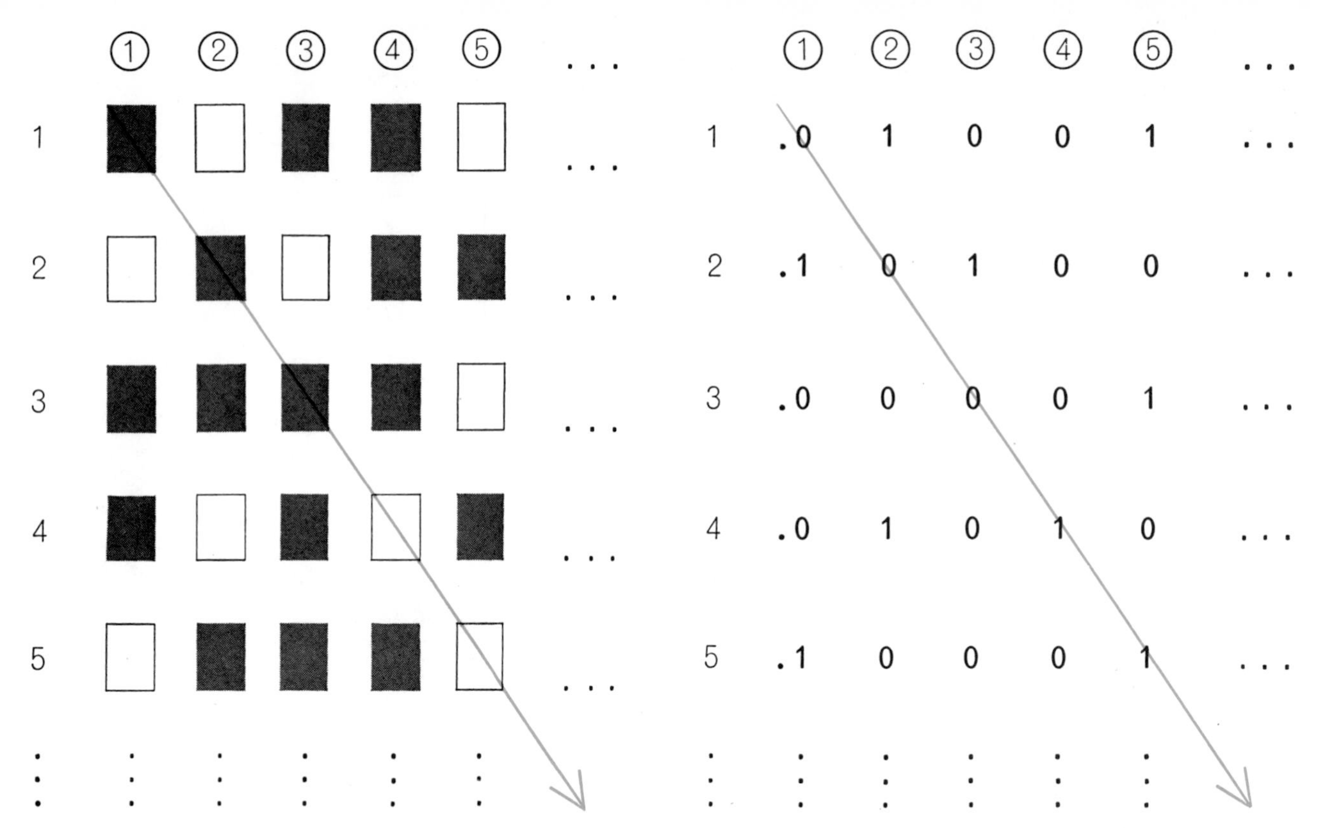

A countable infinity has an uncountable infinity of subsets **(left)** ***that correspond to the real numbers*** **(right)**

to one with the counting integers. It is a higher aleph, an "uncountable" infinity.

Cantor's famous diagonal proof, in the form just given, conceals a startling bonus. It proves that the set of real numbers (the rationals plus the irrationals) is also uncountable. Consider a line segment, its ends numbered 0 and 1. Every rational fraction from 0 to 1 corresponds to a point on this line. Between any two rational points there is an infinity of other rational points; nevertheless, even after all rational points are identified, there remains an infinity of unidentified points—points that correspond to the unrepeating decimal fractions attached to such algebraic irrationals as the square root of 2, and to such transcendental irrationals as pi and *e*. Every point on the line segment, rational or irrational, can be represented by an endless decimal fraction. But these fractions need not be decimal; they can also be written in binary notation. Thus every point on the line segment can be represented by an endless pattern of 1's and 0's, and every possible endless pattern of 1's and 0's corresponds to exactly one point on the line segment.

Now, suppose each face-up card at the left in the illustration above is replaced by 1 and each face-down card by 0, as shown at the right in the illustration. We have only to put a binary point in front of each row and we have an infinite list of different binary fractions between 0 and 1. But the diagonal set of symbols, after each 1 is changed to 0 and each 0 to 1, is a binary fraction that cannot be on the list. From this we see that there is a one-to-one correspondence of three sets: the subsets of aleph-null, the real numbers (here represented by binary fractions) and the totality of points on a line segment. Cantor gave this higher infinity the cardinal number *C*, for the "power of the continuum." He believed it was also $\aleph_1$ (aleph-one), the first infinity greater than aleph-null.

By a variety of simple, elegant proofs Cantor showed that *C* was the number of such infinite sets as the transcendental irrationals (the algebraic irrationals, he proved, form a countable set), the number of points on a line of infinite length, the number of points on any plane figure or on the infinite plane, and the number of points in any solid figure or in all of three-space. Going into higher dimensions does not increase the number of points. The points on a line segment one inch long can be matched one to one with the points in any higher-dimensional solid, or with the points in the entire space of any higher dimension.

The distinction between aleph-null and aleph-one (we accept, for the moment, Cantor's identification of aleph-one with *C*) is important in geometry whenever infinite sets of figures are encountered. Imagine an infinite plane tessellated with hexagons. Is the total number of vertices aleph-one or aleph-null? The answer is aleph-null; they are easily counted along a spiral path [*see top illustration on page 77*]. On the other hand, the number of different circles of one-inch radius that can be placed on a sheet of typewriter paper is aleph-one because inside any small square near the center of the sheet there are aleph-one points, each the center of a different circle with a one-inch radius.

Consider in turn each of the five symbols J. B. Rhine uses in his "ESP" test cards [*see bottom illustration on page 77*]. Can it be drawn an aleph-one number of times on a sheet of paper, assuming that the symbol is drawn with ideal lines of no thickness and that there is no overlap or intersection of any lines? (The drawn symbols need not be the same size, but all must be similar in shape.) It turns out that all except one can be drawn an aleph-one number of times. Can the reader show, before the answer is given next month, which symbol is the exception?

The two alephs are also involved in recent cosmological speculation. Richard Schlegel, a physicist at Michigan State University, has called attention in several papers to a strange contradiction inherent in the "steady state" theory. According to that theory, the number of atoms in the cosmos at the present time is aleph-null. (The cosmos is regarded as infinite even though an "optical horizon" puts a limit on what can be seen.) Moreover, atoms are steadily increasing in number as the universe expands. Infinite space can easily accommodate any finite number of doublings of the quantity of atoms, for whenever aleph-null is multiplied by two, the result is aleph-null again. (If you have an aleph-null number of eggs in aleph-null boxes, one egg per box, you can accommodate another aleph-null set of eggs simply by shifting the egg in box 1 to box 2, the egg in box 2 to box 4, and so on, each egg going to a box whose number is twice the number of the egg's previous box. This empties all the odd-numbered boxes, which can then be filled with another aleph-null set of eggs.) But if the doubling goes on for an aleph-null number of times, we come up against the formula of 2 raised to the power of aleph-null—that is, $2 \times 2 \times 2 \ldots$ repeated aleph-null times. As we have seen, this produces an aleph-one set. Consider only two atoms at an infinitely remote time in the past. By now, after an aleph-null series of doublings, they would have grown to an aleph-one set. But the cosmos, at the moment, cannot contain an aleph-one set of atoms. Any collection of distinct physical entities (as opposed to the ideal entities of mathematics) is countable and therefore, *at the most,* aleph-null.

In his paper, "The Problem of Infinite Matter in Steady-State Cosmology" (*Philosophy of Science,* Vol. 32, January, 1965, pages 21–31), Schlegel finds a clever way out. Instead of regarding the past as a completed aleph-null set of finite time intervals (to be sure, ideal instants in time form an aleph-one continuum, but Schlegel is concerned with those finite time intervals during which doublings of atoms occur), we can view both the past and the future as infinite in the inferior sense of "becoming" rather than completed. Whatever date is suggested for the origin of the universe (remember, we are dealing with the steady-state model, not with a "big bang" or oscillating theory), we can always set an earlier date. In a sense there is a "beginning," but we can push it as far back as we please. There is also an "end," but we can push it as far forward as we please. As we go back in time, continually halving the number of atoms, we never halve them more than a finite number of times, with the result that their number never shrinks to less than aleph-null. As we go forward in time, doubling the number of atoms, we never double more than a finite number of times; therefore the set of atoms never grows larger than aleph-null. In either direction the leap is never made to a completed aleph-null set of time intervals. As a result the set of atoms never leaps to aleph-one and the disturbing contradiction does not arise.

Cantor was convinced that his endless hierarchy of alephs, each obtained by raising 2 to the power of the preceding aleph, represented all the alephs there are. There are none in between. Nor is there an Ultimate Aleph, such as certain Hegelian philosophers of the time identified with the Absolute. The endless hierarchy of infinities itself, Cantor argued, is a better symbol of the Absolute.

All his life Cantor tried to prove that there is no aleph between aleph-null and *C,* the power of the continuum, but he never found a proof. In 1938 Kurt Gödel showed that Cantor's conjecture, which became known as the "continuum hypothesis," could be assumed to be true, and that this could not conflict with the axioms of set theory.

What Cohen proved in 1963 was that the opposite could also be assumed. One can posit that *C* is *not* aleph-one; that there is at least one aleph between aleph-null and *C,* even though no one has the slightest notion of how to specify a set (for example a certain subset of the transcendental numbers) that would have such a cardinal number. This too is consistent with set theory. Cantor's hypothesis is undecidable. Like the parallel postulate of Euclidean geometry, it is an independent axiom that can be affirmed or denied. Just as the two assumptions about Euclid's parallel axiom divided geometry into Euclidean and non-Euclidean, so the two assumptions about Cantor's hypothesis now divide the

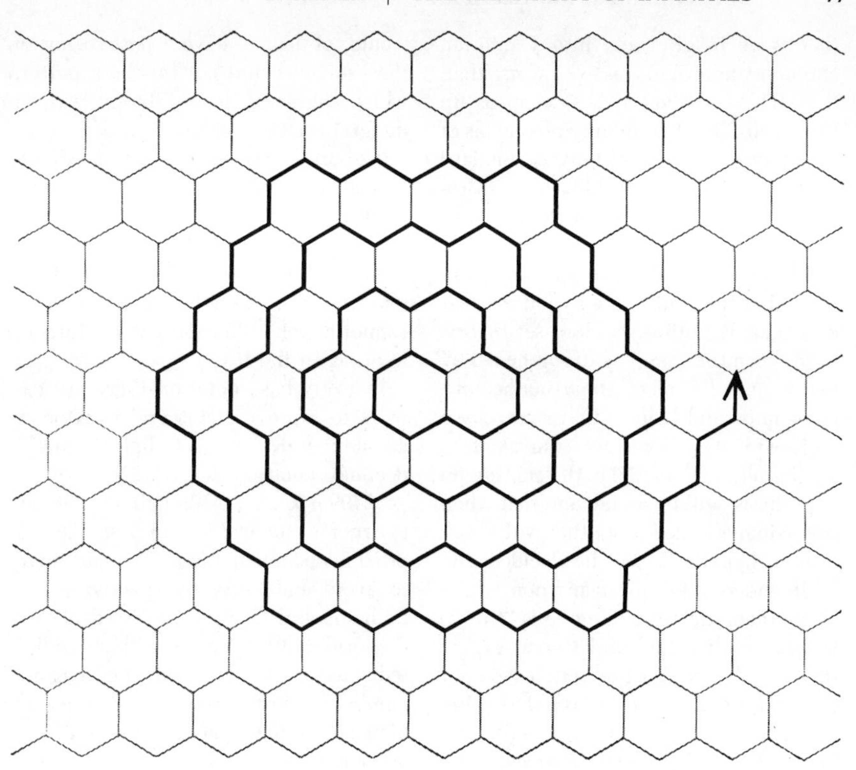

Spiral counts the vertices of a hexagonal tessellation

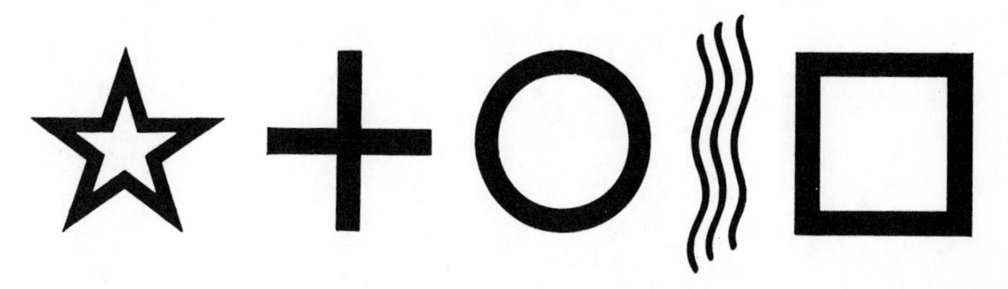

Five "ESP" symbols

theory of infinite sets into Cantorian and non-Cantorian. It is even worse than that. The non-Cantorian side opens up the possibility of an infinity of systems of set theory, all as consistent as standard theory now is, and all differing with respect to assumptions about the power of the continuum.

Of course Cohen did no more than show that the continuum hypothesis was undecidable within standard set theory, even when the theory is strengthened by the axiom of choice. Many mathematicians hope and believe that some day a "self-evident" axiom, not equivalent to an affirmation or denial of the continuum hypothesis, will be found, and that when this axiom is added to set theory the continuum hypothesis will be decided. (By "self-evident" they mean an axiom which all mathematicians will agree is "true.") Indeed, both Gödel and Cohen expect this to happen and are convinced that the continuum hypothesis is in fact false, in contrast to Cantor, who believed and hoped it was true. So far, however, these remain only pious Platonic hopes. What is undeniable is that set theory has been struck a gigantic cleaver blow, and exactly what will come of the pieces no one can say.

Addendum

In giving a binary version of Cantor's famous diagonal proof that the real numbers are uncountable, I deliberately avoided complicating it by considering the fact that every integral fraction between 0 and 1 can be represented as an infinite binary fraction in two ways. For example, ¼ is .01 followed by aleph-null zeroes, and also .001 followed by aleph-null ones. This raises the possibility that the list of real binary fractions might be ordered in such a way that complementing the diagonal would produce a number on the list. The constructed number would, of course, have a *pattern* not on the list, but could not this be a pattern which expressed, in a different way, an integral fraction on the list?

The answer is no. The proof assumes that all possible infinite binary patterns are listed, therefore every integral fraction appears *twice* on the list, once in each of its two binary forms. It follows that the constructed diagonal number cannot match either form of any integral fraction on the list.

In every base notation there are two ways to express an integral fraction by an aleph-null string of digits. Thus in decimal notation ¼ = .2500000 . . . = .2499999. . . . Although it is not necessary for the validity of the diagonal proof in decimal notation, it is customary to avoid ambiguity by specifying that each integral fraction be listed only in the form that terminates with an endless sequence of nines, then the diagonal number is constructed by changing each digit on the diagonal to a different digit other than nine or zero.

Until I discussed Cantor's diagonal proof in *Scientific American*, I had not realized how strongly the opposition to this proof has persisted; not so much among mathematicians as among engineers and scientists. I received many letters attacking the proof. William Dilworth, an electrical engineer, sent me a clipping from the *LaGrange Citizen*, LaGrange, Ill., January 20, 1966, in which he is interviewed at some length about his rejection of Cantorian "numerology." Dilworth first delivered his attack on the diagonal proof at the International Conference on General Semantics, New York, 1963.

One of the most distinguished of modern scientists to reject Cantorian set theory was the physicist P. W. Bridgman. He published a paper about it in 1934, and in his *Reflections of a Physicist* (Philosophical Library, 1955) he devotes pages 99–104 to an uncompromising attack on transfinite numbers and the diagonal proof. "I personally cannot see an iota of appeal in this proof," he writes, "but it appears to me to be a perfect nonsequitur—my mind will not do the things that it is obviously expected to do if it is indeed a proof."

The heart of Bridgman's attack is a point of view widely held by philosophers of the pragmatic and operationalist schools. Infinite numbers, it is argued, do not "exist" apart from human behavior. Indeed, all numbers are merely names for something that a person *does*, rather than names of "things." Because one can count twenty apples, but cannot count an infinity of apples, "it does not make sense to speak of infinite numbers as 'existing' in the Platonic sense, and still less does it make sense to speak of infinite numbers of different orders of infinity, as does Cantor."

"An infinite number," declares Bridgman, "is a certain aspect of what one does when he embarks on carrying out a process . . . an infinite number is an aspect of a *program* of action."

The answer to this is that Cantor *did* specify precisely what one must "do" to define a transfinite number. The fact that one cannot carry out an infinite procedure no more diminishes the reality or usefulness of Cantor's alephs than the fact that one cannot fully compute the value of pi diminishes the reality or usefulness of pi. It is not, as Bridgman maintained, a question of whether one accepts or rejects the Platonic notion of numbers as "things." For an enlightened pragmatist, who wishes to ground all abstractions in human behavior, Cantorian set theory should be no less meaningful or potentially useful than any other precisely defined abstract system such as, say, group theory or a non-Euclidean geometry.

Number

by Philip J. Davis
September 1964

With geometry it is one of the two pillars at the base of mathematics. The concept of number is enlarged by building up number systems and by seeking to break them down into their most primitive elements

By popular definition a mathematician is a fellow who is good at numbers. Most mathematicians demur. They point out that they have as much difficulty as anybody else in reconciling their bank statements, and they like to refer to supporting anecdotes, such as that Isaac Newton, who was Master of the Mint, employed a bookkeeper to do his sums. They observe further that slide rules and electronic computers were developed as crutches to help mathematicians.

All of this is obviously irrelevant. Who, if not the mathematician, is the custodian of the odd numbers and the even numbers, the square numbers and the round numbers? To what other authority shall we look for information and help on Fibonacci numbers, Liouville numbers, hypercomplex numbers and transfinite numbers? Let us make no mistake about it: mathematics is and always has been the numbers game par excellence. The great American mathematician G. D. Birkhoff once remarked that simple conundrums raised about the integers have been a source of revitalization for mathematics over the centuries.

Numbers are an indispensable tool of civilization, serving to whip its activities into some sort of order. In their most primitive application they serve as identification tags: telephone numbers, car licenses, ZIP-code numbers. At this level we merely compare one number with another; the numbers are not subjected to arithmetical operations. (We would not expect to arrive at anything significant by adding the number of Leonard Bernstein's telephone to Elizabeth Taylor's.) At a somewhat higher level we make use of the natural order of the positive integers: in taking a number for our turn at the meat counter or in listing the order of finish in a race. There is still no need to operate on the numbers; all we are interested in is whether one number is greater or less than another. Arithmetic in its full sense does not become relevant until the stage at which we ask the question: How many? It is then that we must face up to the complexities of addition, subtraction, multiplication, division, square roots and the more elaborate dealings with numbers.

The complexity of a civilization is mirrored in the complexity of its numbers. Twenty-five hundred years ago the Babylonians used simple integers to deal with the ownership of a few sheep and simple arithmetic to record the motions of the planets. Today mathematical economists use matrix algebra to describe the interconnections of hundreds of industries, and physicists use transformations in "Hilbert space"—a number concept seven levels of abstraction higher than the positive integers—to predict quantum phenomena.

The number systems employed in mathematics can be divided into five principal stages, going from the simplest to the most complicated. They are: (1) the system consisting of the positive integers only; (2) the next higher stage, comprising the positive and negative integers and zero; (3) the rational numbers, which include fractions as well as the integers; (4) the real numbers, which include the irrational numbers, such as π; (5) the complex numbers, which introduce the "imaginary" number $\sqrt{-1}$.

The positive integers are the numbers a child learns in counting. They are usually written 1, 2, 3, 4..., but they can and have been written in many other ways. The Romans wrote them I, II, III, IV...; the Greeks wrote them α, β, γ, δ...; in the binary number system, containing only the digits 0 and 1, the corresponding numbers are written as 1, 10, 11, 100.... All these variations come to the same thing: they use different symbols for entities whose meaning and order are uniformly understood.

Early man needed only the first few integers, but with the coming of civilization he had to invent higher and higher numbers. This advance did not come readily. As Bernard Shaw remarked in *Man and Superman:* "To the Bushman who cannot count further than his fingers, eleven is an incalculable myriad." As late as the third century B.C. there appears to have been no systematic way of expressing large numbers. Archimedes then suggested a cumbersome method of naming them in his work *The Sand Reckoner.*

Yet while struggling with the names of large numbers the Greek mathematicians took the jump from the finite to

NEW-WORLD NUMBERS, in dot-and-bar notation, record the date of a fragmentary Olmec stela from the state of Vera Cruz in Mexico [*See photo on page 84.*] Each dot equals one unit; each bar equals five. Restored, these numbers show seven periods of 400 "years" (*missing from top*), plus 16 periods of 20 "years" (*the topmost surviving numeral, dot eroded*), plus six "years" of 360 days each, plus 16 "months" of 20 days each, plus 18 days: a total elapsed time of nearly 3,127 "years" since the mythical start of the system. By one method of correlation with the Christian calendar, this is the equivalent of November 4, 291 B.C., and is the second oldest recorded date in the Western Hemisphere.

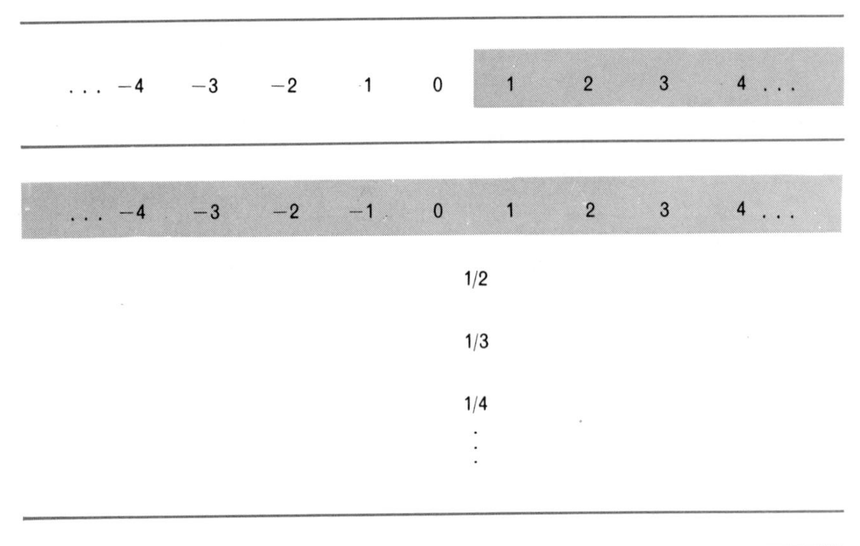

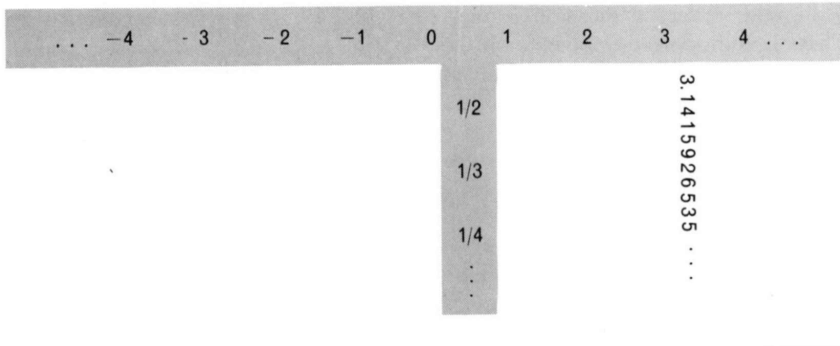

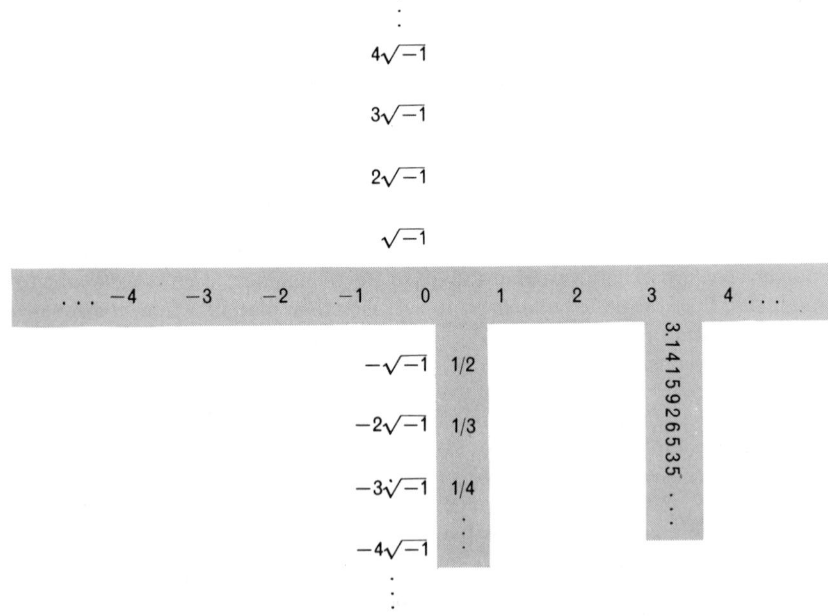

NUMBER CONCEPTS can be arrayed in such a way that each succeeding system embraces all its predecessors. The most primitive concept, consisting of the positive integers alone, is succeeded by one extended to include zero and the negative integers. The next two additions are the rational and the irrational numbers, the latter recognizable by their infinitely nonrepetitive sequence of integers after the decimal. This completes the system of real numbers. The final array represents complex numbers, which began as a Renaissance flight of fancy and have since proved vital to the mathematics of physics and engineering. The complex numbers consist of real numbers combined with the quantity $\sqrt{-1}$, or i.

the infinite. The jump is signified by the three little dots placed after the 4 in the series above. They indicate that there is an integer after 4 and another after the successor to 4 and so on through an unlimited number of integers. For the ancients this concept was a supreme act of the imagination, because it ran counter to all physical experience and to a philosophical belief that the universe must be finite. The bold notion of infinity opened up vast possibilities for mathematics, and it also created paradoxes. Its meaning has not been fully plumbed to this day.

Oddly the step from the positive to the negative integers proved to be a more difficult one to make. Negative numbers seem altogether commonplace in our day, when 10 degrees below zero is a universally understood quantity and the youngest child is familiar with the countdown: "...five, four, three, two, one...." But the Greeks dealt with negative numbers only in terms of algebraic expressions of the areas of squares and rectangles, for example $(a - b)^2 = a^2 - 2ab + b^2$ [*see illustration at top of pages 82 and 83*]. Negative numbers were not fully incorporated into mathematics until the publication of Girolamo Cardano's *Ars Magna* in 1545.

Fractions, or rational numbers (the name they go by in number theory), are more ancient than the negative numbers. They appear in the earliest mathematical writings and were discussed at some length as early as 1550 B.C. in the Rhind Papyrus of Egypt. The present way of writing fractions (for instance 1/4, 1/5, 8/13) and also the present way of doing arithmetic with them date from the 15th and 16th centuries. Today most people probably could not be trusted to add 1/4 and 1/5 correctly. (Indeed, how often do they need to?) The handling of fractions, however, is by no means a dead issue. It recently became a matter of newspaper controversy as a result of the treatment of fractions in some of the new school mathematics courses, with the cancellation school pitted against the anti-cancellation school. The controversy stemmed from a divergence of opinion as to what the practical and aesthetic goals of school mathematics should be; the mystified layman, reading about it over his eggs and coffee, may have been left with the impression that everything he had once been taught about fractions was wrong or immoral.

The irrational numbers also have a long history. In the sixth century B.C. the mathematical school of Pythag-

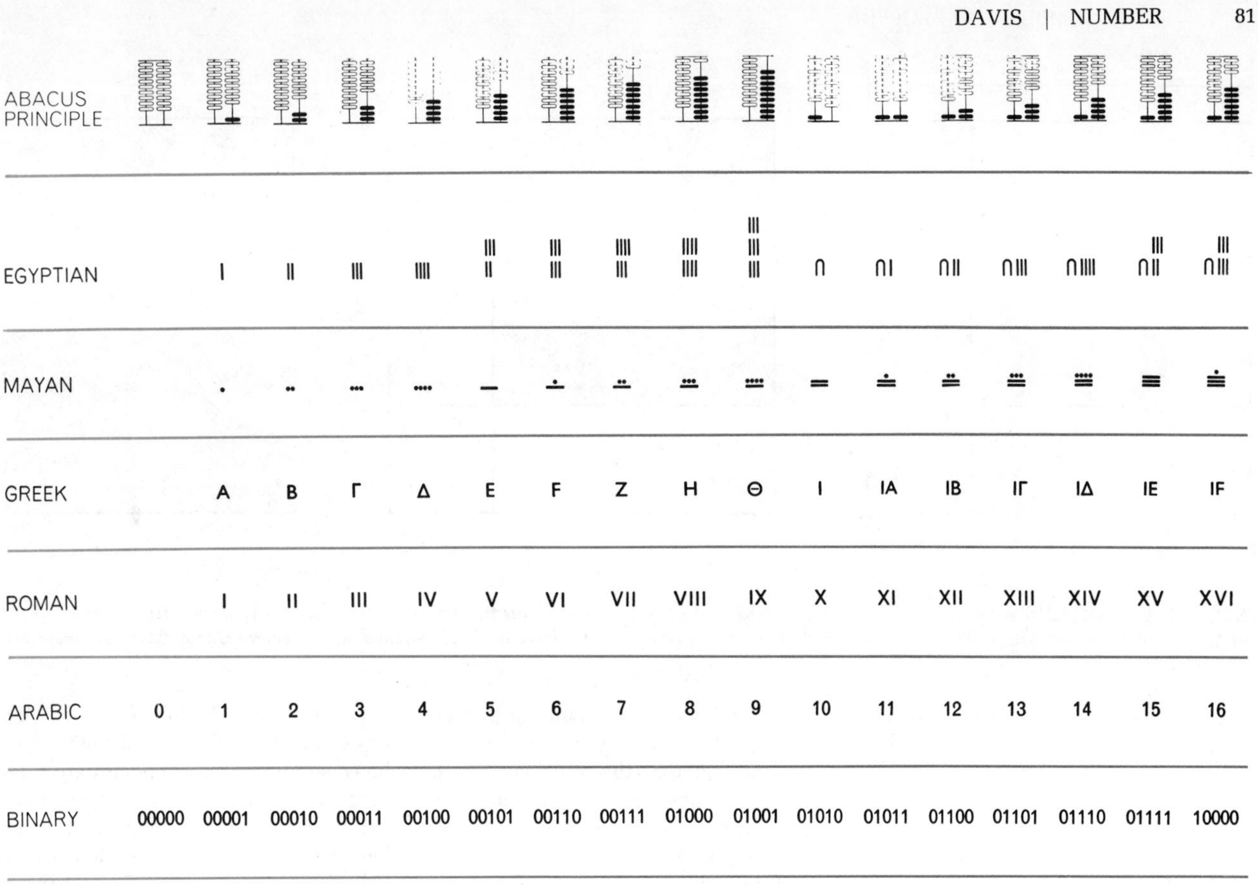

ABACUS PRINCIPLE																	
EGYPTIAN		I	II	III	IIII	III II	III III	IIII III	IIII IIII	III III III	∩	∩I	∩II	∩III	∩IIII	III ∩II	III ∩III
MAYAN		•	••	•••	••••	—	• —	•• —	••• —	•••• —	=	• =	•• =	••• =	•••• =	≡	• ≡
GREEK		A	B	Γ	Δ	E	F	Z	H	Θ	I	IA	IB	IΓ	IΔ	IE	IF
ROMAN		I	II	III	IV	V	VI	VII	VIII	IX	X	XI	XII	XIII	XIV	XV	XVI
ARABIC	0	1	2	3	4	5	6	7	8	9	10	11	12	13	14	15	16
BINARY	00000	00001	00010	00011	00100	00101	00110	00111	01000	01001	01010	01011	01100	01101	01110	01111	10000

ANCIENT AND MODERN NOTATIONS for the numerals from 1 to 16 are arrayed beneath the equivalent values set up on a two-rod abacus. Of the six examples all but two have a base of 10; these are repetitive above that number regardless of whether the symbol is tally-like or unique for each value. The Mayan notation has a base of 20 and is repetitive after the numeral 5. The binary system has a base of 2 and all its numbers are written with only a pair of symbols, 0 and 1. Thus two, or 2^1, is written 10; four, or 2^2, is written 100, and eight, or 2^3, is written 1000. Each additional power of 2 thereafter adds one more digit to the binary notation.

oras encountered a number that could not be fitted into the category of either integers or fractions. This number, arrived at by the Pythagorean theorem, was $\sqrt{2}$: the length of the diagonal of a square (or the hypotenuse of a right triangle) whose sides are one unit long. The Greeks were greatly upset to find that $\sqrt{2}$ could not be expressed in terms of any number a/b in which a and b were integers, that is, any rational number. Since they originally thought the only numbers were rational numbers, this discovery was tantamount to finding that the diagonal of a square did not have a mathematical length! The Greeks resolved this paradox by thinking of numbers as lengths. This led to a program that inhibited the proper development of arithmetic and algebra, and Greek mathematics ran itself into a stone wall.

It took centuries of development and sophistication in mathematics to realize that the square root of two can be represented by putting three dots after the last calculated digit. Today we press the square-root button of a desk calculator and get the answer: $\sqrt{2} = 1.41421\ldots$. Electronic computers have carried the specification of the digits out to thousands of decimal places. Any number that can be written in this form—with one or more integers to the left of a decimal point and an infinite sequence of integers to the right of the point—is a "real" number. We can express in this way the positive integers (for example, $17 = 17.0000\ldots$), the negative integers ($-3 = -3.0000\ldots$) or the rational numbers ($17\frac{1}{5} = 17.20000\ldots$). Some rational numbers do not resolve themselves into a string of zeros at the right; for instance, the decimal expression of one-seventh is $1/7 = 0.142857\ 142857\ 142857\ldots$. What makes these numbers "rational" is the fact that they contain a pattern of digits to the right of the decimal point that repeats itself over and over. The numbers called "irrational" are those that, like the square root of two, have an infinitely nonrepeating sequence of decimal digits. The best-known examples of irrationals are: $\sqrt{2} = 1.4142135623\ldots$ and $\pi = 3.1415926535\ldots$. The irrational numbers are of course included among the real numbers.

It is in the domain of the "complex numbers" that we come to the numbers called "imaginary"—a term that today is a quaint relic of a more naïve, swashbuckling era in arithmetic. Complex numbers feature the "quantity" $\sqrt{-1}$, which, when multiplied by itself, produces -1. Since this defies the basic rule that the multiplication of two positive or negative numbers is positive, $\sqrt{-1}$ (or i, as it is usually written) is indeed an oddity: a number that cannot be called either positive or negative. "The imaginary numbers," wrote Gottfried Wilhelm von Leibniz in 1702, "are a wonderful flight of God's Spirit; they are almost an amphibian between being and not being."

From Renaissance times on, although mathematicians could not say what these fascinating imaginaries were, they used complex numbers (which have the general form $a + b\sqrt{-1}$) to solve equations and uncovered many beautiful identities. Abraham de Moivre discovered the formula $(\cos\theta + \sqrt{-1}\sin\theta)^n$

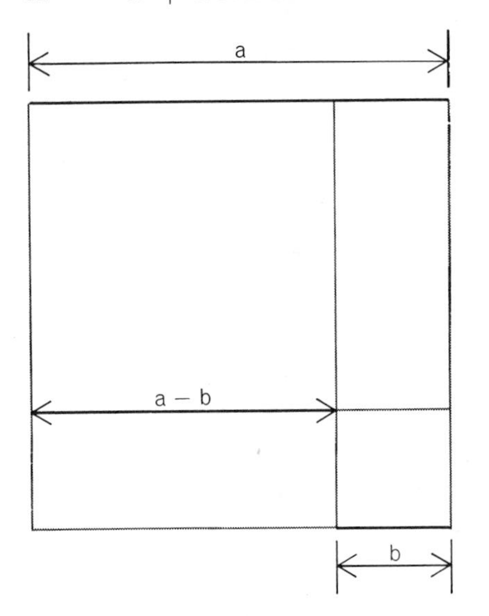

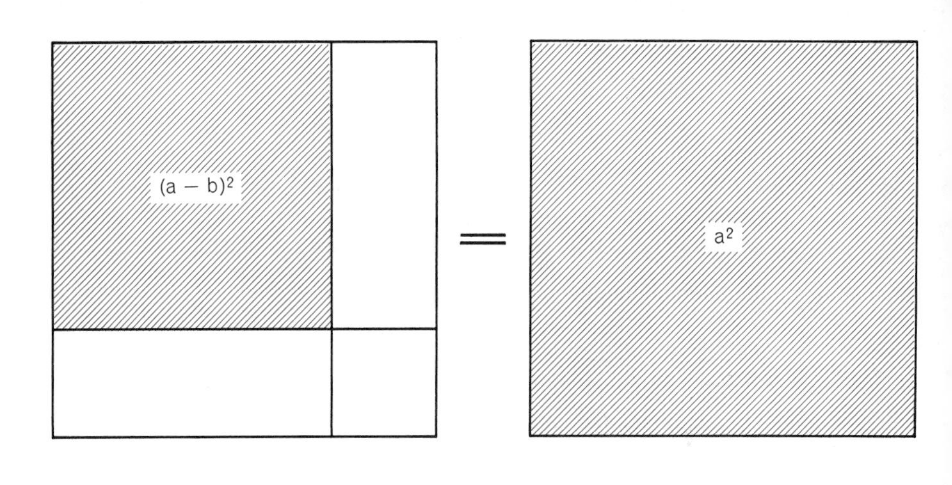

NEGATIVE NUMBERS were visualized by the Greeks in terms of lines and bounded areas. Thus they realized that the square erected on the line $a - b$ was equal in area, after a series of manipulations, to the square on the entire line a. The first manipula-

$= \cos n\theta + \sqrt{-1} \sin n\theta$. Leonhard Euler discovered the related formula

$$e^{\pi\sqrt{-1}} = -1$$

(e being the base of the "natural logarithms," 2.71828...).

The complex numbers remained on the purely manipulative level until the 19th century, when mathematicians began to find concrete meanings for them. Caspar Wessel of Norway discovered a way to represent them geometrically [*see illustration on page 85*], and this became the basis of a structure of great beauty known as the theory of functions of a complex variable. Later the Irish mathematician William Rowan Hamilton developed an algebraic interpretation of complex numbers that represented each complex number by a pair of ordinary numbers. This idea helped to provide a foundation for the development of an axiomatic approach to algebra.

Meanwhile physicists found complex numbers useful in describing various physical phenomena. Such numbers began to enter into equations of electrostatics, hydrodynamics, aerodynamics, alternating-current electricity, diverse other forms of vibrating systems and eventually quantum mechanics. Today many of the productions of theoretical physics and engineering are written in the language of the complex-number system.

In the 19th century mathematicians invented several new number systems. Of these modern systems three are particularly noteworthy: quaternions, matrices and transfinite numbers.

Quaternions were Hamilton's great creation. For many years he brooded over the fact that the multiplication of complex numbers has a simple interpretation as the rotation of a plane. Could this idea be generalized? Would it be possible to invent a new kind of number and to define a new kind of multiplication such that a rotation of three-dimensional space would have a simple interpretation in terms of the multiplication? Hamilton called such a number a triplet; just as Wessel represented complex numbers by a point on a two-dimensional plane, the triplets were to be represented by a point in three-dimensional space.

The problem was a hard nut to crack. It was continually on Hamilton's mind, and his family worried over it with him. As he himself related, when he came down to breakfast one of his sons would ask: "Well, Papa, can you multiply triplets?" And Papa would answer dejectedly: "No, I can only add and subtract them."

One day in 1843, while he was walking with his wife along a canal in Dublin, Hamilton suddenly conceived a way to multiply triplets. He was so elated that he took out a penknife then and there and carved on Brougham Bridge the key to the problem, which certainly must have mystified passersby who read it: "$i^2 = j^2 = k^2 = ijk = -1$."

The letters i, j and k represent hypercomplex numbers Hamilton called quaternions (the general form of a quaternion being $a + bi + cj + dk$, with a, b, c and d denoting real numbers). Just as the square of $\sqrt{-1}$ is -1, so $i^2 = -1$, $j^2 = -1$ and $k^2 = -1$. The key to the multiplication of quaternions, however, is that the commutative law does not hold [*see table on page 95*]. Whereas in the case of ordinary numbers $ab = ba$, when quaternions are reversed, the product may be changed: for example, $ij = k$ but $ji = -k$.

The second modern number concept mentioned above, that of the matrix, was developed more or less simultaneously by Hamilton and the British mathematicians J. J. Sylvester and Arthur Cayley. A matrix can be regarded as a rectangular array of numbers. For example,

$$\begin{pmatrix} 1 & 6 & 7 \\ 2 & 0 & 4 \end{pmatrix}$$

is a matrix. The entire array is thought of as an entity in its own right. Under the proper circumstances it is possible to define operations of addition, subtraction, multiplication and division for such entities. The result is a system of objects whose behavior is somewhat reminiscent of ordinary numbers and which is of great utility in many provinces of pure and applied mathematics.

The third modern concept, that of transfinite numbers, represents a totally different order of idea. It is entertainingly illustrated by a fantasy, attributed to the noted German mathematician David Hilbert and known as "Hilbert's Hotel." It would be appreciated by roomless visitors to the New York World's Fair. A guest comes to Hilbert's Hotel and asks for a room. "Hm," says the manager. "We are all booked up, but that's not an unsolvable problem

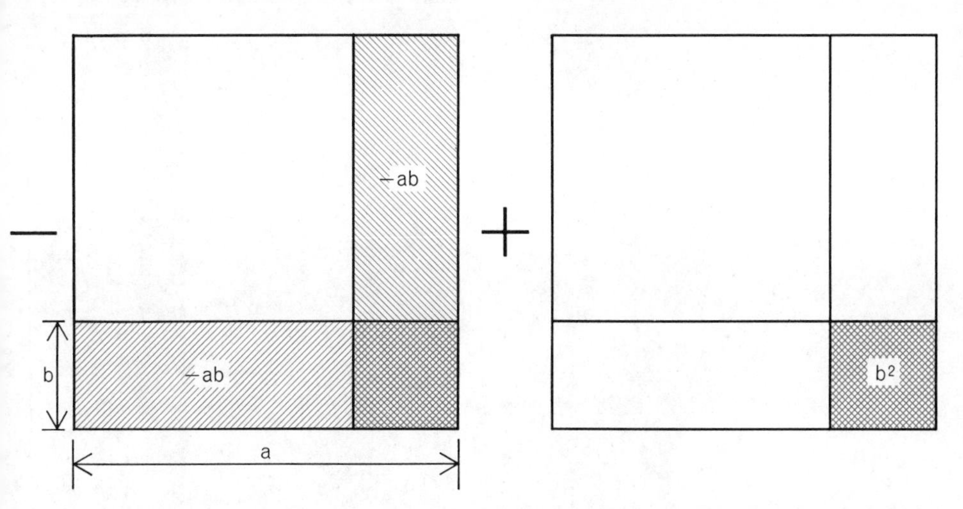

tions require the subtraction of two rectangles of length a and width b from a^2. But these rectangles overlap, and one quantity has been subtracted twice. This is b^2, which is restored.

here; we can make space for you." He puts the new guest in room 1, moves the occupant of room 1 to room 2, the occupant of room 2 to room 3 and so on. The occupant of room N goes into room $N + 1$. The hotel simply has an infinite number of rooms.

How, then, can the manager say that the hotel is "all booked up?" Galileo noted a similar paradox. Every integer can be squared, and from this we might conclude that there are as many squares as there are integers. But how can this be, in view of the known fact that there are integers that are not squares, for instance 2, 3, 5, 6. . . ?

One of the endlessly alluring aspects of mathematics is that its thorniest paradoxes have a way of blooming into beautiful theories. The 19th-century German mathematician Georg Cantor turned this paradox into a new number system and an arithmetic of infinite numbers.

He started by defining an infinite set as one that can be put into a one-to-one correspondence with a part of itself, just as the integers are in a one-to-one correspondence with their squares. He noted that every set that can be put into such correspondence with the set of all the integers must contain an infinite number of elements, and he designated this "number" as $\aleph$ (aleph, the first letter of the Hebrew alphabet). Cantor gave this "first transfinite cardinal" the subscript zero. He then went on to show that there is an infinity of other sets (for example the set of real numbers) that cannot be put into a one-to-one correspondence with the positive integers because they are larger than that set. Their sizes are represented by other transfinite cardinal numbers ($\aleph_1$, $\aleph_2$ and so on). From such raw materials Cantor developed an arithmetic covering both ordinary and transfinite numbers. In this arithmetic some of the ordinary rules are rejected, and we get strange equations such as $\aleph_0 + 1 = \aleph_0$. This expresses, in symbolic form, the hotel paradox.

The transfinite numbers have not yet found application outside mathematics itself. But within mathematics they have had considerable influence and have evoked much logical and philosophical speculation. Cantor's famous "continuum hypothesis" produced a legacy of unsolved problems that still occupy mathematicians. In recent years solutions to some of these problems have been achieved by Alfred Tarski of the University of California at Berkeley and Paul J. Cohen of Stanford University.

We have reviewed the subject matter (or dramatis personae) of the numbers game; it now behooves us to examine the rules of the game. To nonmathematicians this may seem to be an exercise in laboring the obvious. The geometry of Euclid is built on "self-evident" axioms, but rigorous examination of the axioms in the 19th century disclosed loopholes, inconsistencies and weaknesses that had to be repaired in order to place geometry on firmer foundations. But, one may ask, what is there about the simple rules of arithmetic and algebra that needs examination or proof? Shaken by the discoveries of the shortcomings of Euclid's axioms, and spurred by the surprising features of the new number concepts such as the quaternions, many mathematicians of the 19th century subjected the axioms of number theory to systematic study.

Are the laws of arithmetic independent, or can one be derived logically from another? Are they really fundamental, or could they be reduced to a more primitive, simpler and more elegant set of laws? Answers to questions such as these have been sought by the program of axiomatic inquiry, which is

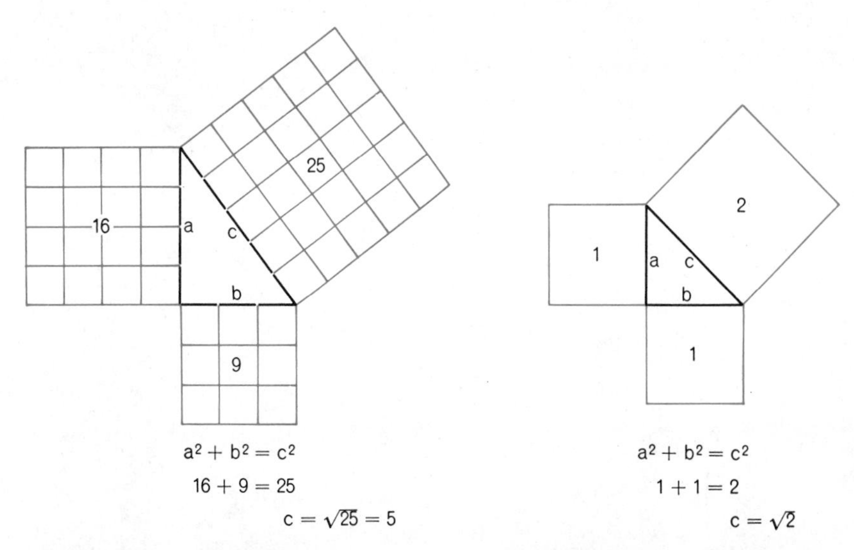

IRRATIONAL NUMBERS seemed paradoxical to the Greeks, who could not imagine numbers that were neither integers nor rational fractions but who could nonetheless express such numbers geometrically. In a right triangle with two sides of unit length 3 and 4 respectively, the hypotenuse is 5 units in length. But no rational fraction is equal to $\sqrt{2}$, the length of the hypotenuse of a right triangle that has sides of length 1. In effect, this says that an easily constructed line of quite tangible length is nonetheless "immeasurable."

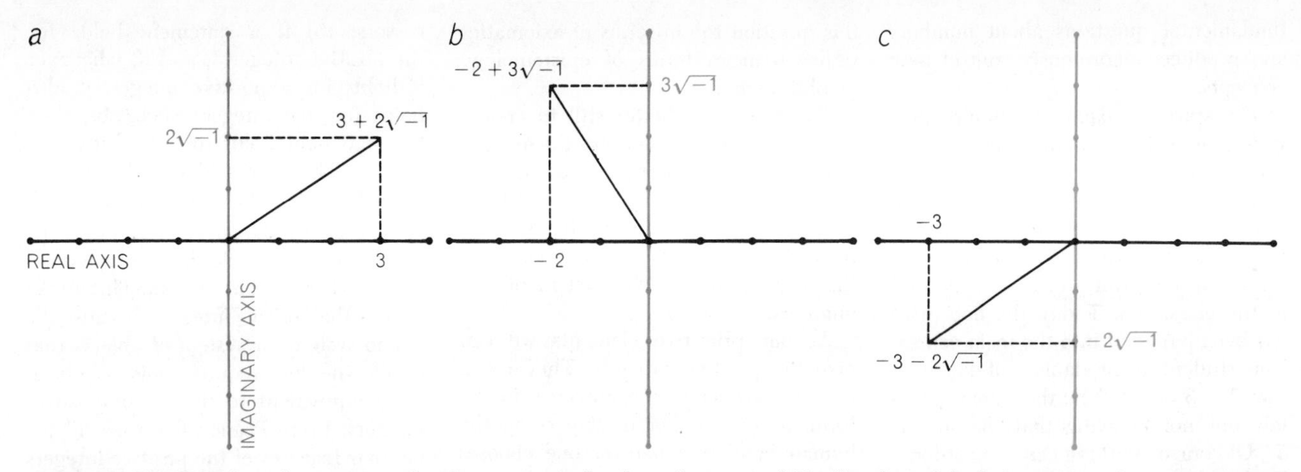

COMPLEX NUMBERS can be represented and even manipulated in a geometric fashion. On the real, or x, axis each unit is 1 or -1. On the imaginary, or y, axis each unit is i, or $\sqrt{-1}$, or else $-i$. Thus all points on the plane can be given complex numbers of the form $x + yi$. If a line through both the origin and any point on the plane (*as shown in "a"*) is rotated through 90 degrees (*as in "b"*), the result is the multiplication of the original complex number by $\sqrt{-1}$. A second rotation, and multiplication by i, appears in *c*.

still going on. It has yielded rigorous and aesthetically appealing answers to some of them, and in the process it has brought forth new concepts such as "rings," "fields," "groups" and "lattices," each with its own set of rules of operation and its own characteristic theory.

One of the major accomplishments, achieved in the 1870's, was the establishment of a set of axioms for the real numbers. It is summed up in the statement that the real-number system is a "complete ordered field." Each of these words represents a group of rules that defines the behavior of the numbers.

First of all, the word "field" means a mathematical system in which addition and multiplication can be carried out in a way that satisfies the familiar rules, namely (1) the commutative law of addition: $x + y = y + x$; (2) the associative law of addition: $x + (y + z) = (x + y) + z$; (3) the commutative law of multiplication: $xy = yx$; (4) the associative law of multiplication: $x(yz) = (xy)z$; (5) the distributive law: $x(y + z) = xy + xz$.

Furthermore, a field must contain a zero element, 0, characterized by the property that $x + 0 = x$ for any element x. It contains a unit element, 1, that has the property that $1 \cdot x = x$. For any given element x of a field there is another element $-x$ such that $-x + x = 0$. This is the foundation on which subtraction is built. Another axiomatic property of a field is the cancellation rule of multiplication, that is, if $xy = xz$, then $y = z$ (provided that x is not equal to zero). Finally, for any element x (other than zero) a field must contain an element $1/x$ such that $x(1/x) = 1$. This is the basis for division. Briefly, then, a field is a system (exemplified by the rational numbers) whose elements can be added, subtracted, multiplied and divided under the familiar rules of arithmetic.

Considering now the second word, a field is "ordered" if the sizes of its elements can be compared. The shorthand symbol used to denote this property is the sign $>$, meaning "greater than." This symbol is required to obey its own set of rules, namely (1) the trichotomy law: for any two elements x and y, exactly one of these three relations is true, $x > y$, $x = y$ or $y > x$; (2) the transitivity law: if $x > y$ and $y > z$, then $x > z$; (3) the law of addition: if $x > y$, then $x + z > y + z$; (4) the law of multiplication: if $x > y$ and $z > 0$, then $xz > yz$.

Finally, what do we mean by the word "complete" in describing the system of real numbers as a "complete ordered field"? This has to do with the problem raised by a number such as $\sqrt{2}$. Practically speaking, $\sqrt{2}$ is given by a sequence of rational numbers such as 1, 1.4, 1.41, 1.414... that provide better and better approximations to it. That is to say, $1^2 = 1$, $(1.4)^2 = 1.96$, $(1.41)^2 = 1.9981$, $(1.414)^2 = 1.999396$.... Squaring these numbers yields a sequence of numbers that are getting closer and closer to 2. Notice, however, that the numbers in the original sequence (1, 1.4, 1.41...) are also getting closer and closer to one another. We would like to think of $\sqrt{2}$ as the "limiting value" of such a sequence of approximations. In order to do so we need a precise notion of what is meant by saying that the numbers of a sequence are getting closer and closer to one another, and we need a guarantee that our system of numbers is rich enough to provide us with a limiting number for such a sequence.

Following the path taken by Cantor, we consider a sequence of numbers in our ordered field. We shall say that the numbers of this sequence are getting closer and closer to one another if the difference of any two numbers sufficiently far out in the sequence is as small as we please. This means, for example, that all terms sufficiently far out differ from one another by at the most 1/10. If one wishes to go out still further, they can be made to differ by at most 1/100, and so forth. Such a sequence of numbers is called a "regular sequence." An ordered field is called a "complete" ordered field if, corresponding to any regular sequence of elements, there is an element of the field that the sequence approaches as a limiting value. This is the "law of completeness": the "gaps" between the rational numbers have been completed, or filled up. It is the final axiomatic requirement for the real-number system.

All these rules may seem so elementary that they hardly need stating, let alone laborious analysis. The program of systematizing them, however, has been vastly rewarding. Years of polishing the axioms have reduced them to a form that is of high simplicity. The rules I have just enumerated have been found to be necessary, and sufficient, to do the job of describing and operating the real-number system; throw any one of them away and the system would not work. And, as I have said, the program of axiomatic inquiry has answered some fundamental questions about numbers and produced enormously fruitful new concepts.

The spirit of axiomatic inquiry pervades all modern mathematics; it has even percolated into the teaching of mathematics in high schools. A high school teacher recently said to me: "In the old days the rules of procedure were buried in fine print and largely ignored in the classroom. Today the fine print has been parlayed into the main course. The student is in danger of knowing that $2 + 3 = 3 + 2$ by the commutative law but not knowing that the sum is 5." Of course anything can be overdone. Exclusive attention to axiomatics would be analogous to the preoccupation of a dance group that met every week and discussed choreography but never danced. What is wanted in mathematics, as in anything else, is a sound sense of proportion.

We have been considering how numbers operate; ultimately we must face the more elementary question: What *are* numbers, after all? Nowadays mathematicians are inclined to answer this question too in terms of axiomatics rather than in terms of epistemology or philosophy.

To explain, or better still to create, numbers it seems wise to try the method of synthesis instead of analysis. Suppose we start with primitive, meaningful elements and see if step by step we can build these elements up into something that corresponds to the system of real numbers.

As our primitive elements we can take the positive integers. They are a concrete aspect of the universe, in the form of the number of fingers on the human hand or whatever one chooses to count. As the 19th-century German mathematician Leopold Kronecker put it, the positive integers are the work of God and all the other types of number are the work of man. In the late 19th century Giuseppe Peano of Italy provided a primitive description of the positive integers in terms of five axioms: (1) 1 is a positive integer; (2) every positive integer has a unique positive integer as its successor; (3) no positive integer has 1 as its successor; (4) distinct positive integers have distinct successors; (5) if a statement holds for the positive integer 1, and if, whenever it holds for a positive integer, it also holds for that integer's successor, then the statement holds for all positive integers. (This last axiom is the famous "principle of mathematical induction.")

Now comes the *fiat lux* ("Let there be light") of the whole business. Axiom: There exists a Peano system. This stroke creates the positive integers, because the Peano system, or system of objects that fulfills the five requirements, is essentially equivalent to the set of positive integers. From Peano's five rules all the familiar features of the positive integers can be deduced.

Once we have the positive integers at our disposal to work with and to mold, we can go merrily on our way, as Kronecker suggested, and construct extensions of the number idea. By operations with the positive integers, for example, we can create the negative integers and zero. A convenient way to do this is by operating with *pairs* of positive integers. Think of a general pair denoted (a,b) from which we shall create an integer by the operation $a - b$. When a is greater than b, the subtraction $a - b$ produces a positive integer; when b is greater than a, the resulting $a - b$ integer is negative; when a is equal to b, then $a - b$ produces zero. Thus pairs of positive integers can represent all the integers—positive, negative and zero. It is true that a certain ambiguity arises from the fact that a given integer can be represented by many different pairs; for instance, the pair (6,2) stands for 4, but so do (7,3), (8,4) and a host of other possible combinations. We can reduce the ambiguity to unimportance, however, simply by agreeing to consider all such pairs as being identical.

Using only positive integers, we can write a rule that will determine when one pair is equal to another. The rule is that $(a,b) = (c,d)$ if, and only if, $a + d = b + c$. (Note that the latter equation is a rephrasing of $a - b = c - d$, but it does not involve any negative integers, whereas the subtraction terms may.) It can easily be shown that this rule for deciding the equality of pairs of integers satisfies the three arithmetical laws governing equality, namely (1) the reflexive law: $(a,b) = (a,b)$; (2) the symmetric law: if $(a,b) = (b,c)$, then $(b,c) = (a,b)$; (3) the transitive law: if $(a,b) = (c,d)$ and $(c,d) = (e,f)$, then $(a,b) = (e,f)$.

We can now proceed to introduce

	1	i	j	k
1	1	i	j	k
i	i	−1	k	−j
j	j	−k	−1	i
k	k	j	−i	−1

MULTIPLICATION TABLE for the quaternions, devised by William Rowan Hamilton, demonstrates the noncommutative nature of these imaginary quantities. For example, the row quantity j, multiplied by the column quantity k, produces i, but row k times column j produces $-i$ instead. Each of the three quantities, multiplied by itself, is equal to -1.

conventions defining the addition and the multiplication of pairs of positive integers, again using only positive terms. For addition we have $(a,b) + (c,d) = (a + c,\ b + d)$. Since (a,b) represents $a - b$ and (c,d) represents $c - d$, the addition here is $(a - b) + (c - d)$. Algebraically this is the same as $(a + c) - (b + d)$, and that is represented by the pair $(a + c,\ b + d)$ on the right side of the equation. Similarly, the multiplication of pairs of positive integers is defined by the formula $(a,b) \cdot (c,d) = (ac + bd,\ ad + bc)$. Here $(a,b)(c,d)$, or $(a - b)(c - d)$, can be expressed algebraically as $(ac + bd) - (ad + bc)$, and this is represented on the right side of the equation by the pair $(ac + bd,\ ad + bc)$.

It can be shown in detail that all the familiar operations with integers (positive, negative and zero), when performed with such pairs of positive integers, will produce the same results.

Having constructed all the integers (as pairs of positive integers), we can go on to create all the other real numbers and even the complex numbers. The rational numbers, or fractions, which are pairs of integers in the ordinary system, can be represented as pairs of pairs of positive integers. For the real numbers, made up of infinite sequences of integers, we must set up infinite sequences of rationals rather than pairs. When we come to the complex numbers, we can again use pairs; indeed, it was for these numbers that the device of number pairs was first employed (by Hamilton). We can think of a complex number, $a + b\sqrt{-1}$, as essentially a pair of real numbers (a,b), with the first number of the pair representing the real element and the second representing the imaginary element of the complex number. Now, pairs will be considered equal only if they contain the same numbers in the same order; that is, $(a,b) = (c,d)$ only if $a = c$ and $b = d$. The rule for addition will be

a

$$\begin{pmatrix} a_1 & a_2 & a_3 \\ a_4 & a_5 & a_6 \\ a_7 & a_8 & a_9 \end{pmatrix} + \begin{pmatrix} b_1 & b_2 & b_3 \\ b_4 & b_5 & b_6 \\ b_7 & b_8 & b_9 \end{pmatrix} = \begin{pmatrix} a_1 + b_1 & a_2 + b_2 & a_3 + b_3 \\ a_4 + b_4 & a_5 + b_5 & a_6 + b_6 \\ a_7 + b_7 & a_8 + b_8 & a_9 + b_9 \end{pmatrix}$$

$$\begin{pmatrix} 7 & 0 & 0 \\ -3 & 1 & -6 \\ 4 & 0 & 7 \end{pmatrix} + \begin{pmatrix} -8 & 0 & 1 \\ 4 & 5 & -1 \\ 0 & 3 & 0 \end{pmatrix} = \begin{pmatrix} 7-8 & 0+0 & 0+1 \\ -3+4 & 1+5 & -6-1 \\ 4+0 & 0+3 & 0+7 \end{pmatrix} = \begin{pmatrix} -1 & 0 & 1 \\ 1 & 6 & -7 \\ 4 & 3 & 7 \end{pmatrix}$$

b

$$\begin{pmatrix} a_1 & a_2 & a_3 \\ a_4 & a_5 & a_6 \\ a_7 & a_8 & a_9 \end{pmatrix} \times \begin{pmatrix} b_1 & b_4 & b_7 \\ b_2 & b_5 & b_8 \\ b_3 & b_6 & b_9 \end{pmatrix} = \begin{pmatrix} a_1b_1 + a_2b_2 + a_3b_3 & a_1b_4 + a_2b_5 + a_3b_6 & a_1b_7 + a_2b_8 + a_3b_9 \\ a_4b_1 + a_5b_2 + a_6b_3 & a_4b_4 + a_5b_5 + a_6b_6 & a_4b_7 + a_5b_8 + a_6b_9 \\ a_7b_1 + a_8b_2 + a_9b_3 & a_7b_4 + a_8b_5 + a_9b_6 & a_7b_7 + a_8b_8 + a_9b_9 \end{pmatrix}$$

$$\begin{pmatrix} 6 & 0 & -1 \\ 1 & -3 & 2 \\ 8 & 5 & 6 \end{pmatrix} \times \begin{pmatrix} 4 & 2 & 3 \\ 0 & 1 & 6 \\ -5 & -1 & 7 \end{pmatrix} = \begin{pmatrix} 24+0+5 & 12+0+1 & 18+0-7 \\ 4+0-10 & 2-3-2 & 3-18+14 \\ 32+0-30 & 16+5-6 & 24+30+42 \end{pmatrix} = \begin{pmatrix} 29 & 13 & 11 \\ -6 & -3 & -1 \\ 2 & 15 & 96 \end{pmatrix}$$

MATRICES are rectangular arrays of numbers, themselves without numerical value, that nonetheless can be treated as entities and thus can be added, subtracted, multiplied or divided in the proper circumstances. Such arrays offer a particularly convenient method for calculating simultaneous changes in a series of related variables. Addition is possible with any pair of matrices having the same number of columns and rows; row by row, each element in each column of the first matrix is added to the corresponding element in the corresponding column of the next, thus forming a new matrix. (The process is shown schematically at the top of the illustration and then repeated, with numerical values, directly below.) Multiplication is a more complex process, in which the two matrices need not be the same size, although they are in the illustration; a 3 × 2 matrix could multiply a 2 × 3 one. Each term in the upper row of the left matrix successively multiplies the corresponding term in the first column of the right matrix; the sum of these three multiplications is the number entered at column 1, row 1 of the product matrix. The upper row of the left matrix is now used in the same way with the second column of the right matrix to find a value for column 2, row 1 of the product matrix, and then multiplies the third column of the right matrix. The entire operation is repeated with each row of the left matrix.

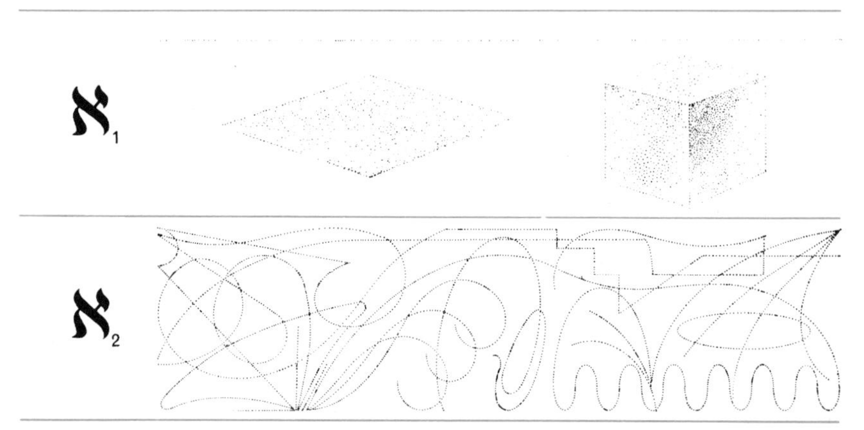

TRANSFINITE CARDINALS exist in infinite number. The most familiar, $\aleph_0$, symbolizes the "number," or the cardinality, of the positive integers or of any set that can be put into a one-to-one correspondence with the positive integers. These are the sets that are countable. The cardinality of the real numbers is larger than the cardinality of the positive integers. It is identical with the cardinality of the points on a line, in a plane or in any portion of a space of higher dimension. These noncountable sets are symbolized by $\aleph_1$. The "number" of all possible point sets is a still larger transfinite cardinal and is symbolized by $\aleph_2$.

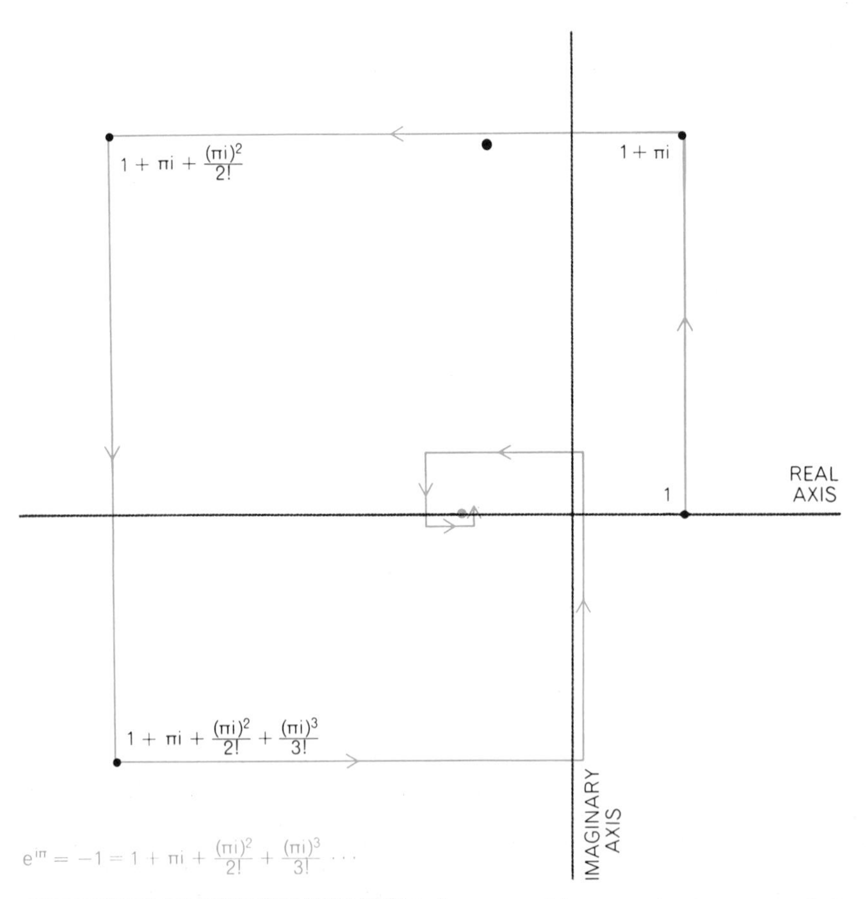

VIRTUOSITY OF COMPLEX NUMBERS is demonstrated by conversion into geometrical form of the formula that relates *e* (the natural base of logarithms), π and $\sqrt{-1}$. The equation (*color*) can be expressed as the sum of a series of vectors. When these are added and plotted on a complex plane, they form a spiral that strangles the point equal to -1.

the same as in the case of the real numbers: $(a,b) + (c,d) = (a + c, b + d)$. This parallels the "ordinary" outcome of the addition of two complex numbers: $(a + b\sqrt{-1}) + (c + d\sqrt{-1}) = (a + c) + (b + d)\sqrt{-1}$. The multiplication formula for complex numbers, $(a,b) \cdot (c,d) = (ac - bd, ad + bc)$, also corresponds to the ordinary multiplication of such numbers: $(a + b\sqrt{-1})(c + d\sqrt{-1}) = (ac - bd) + (ad + bc)\sqrt{-1}$. Pairs of real numbers manipulated according to these rules reproduce all the familiar behavior of the complex numbers. And the mysterious $\sqrt{-1}$, that "amphibian between being and not being," emerges from the sea of axiomatics as the number pair (0,1).

Thus, by four steps of construction and abstraction, we have advanced from the primitive positive integers to the complex numbers. Pairs of positive integers, combined in a certain way, lead to the set of all the integers. Pairs of integers (that is to say now, pairs of pairs of positive integers), combined in a different way, lead to the rational numbers. Infinite sequences of rational numbers lead to real numbers. Finally, pairs of real numbers lead to the complex numbers.

Looking back over the 2,500 years that separate us from Pythagoras, we can make out two streams of thinking about numbers. There is the stream of synthesis, which began with tally marks and went on to build up number concepts of increasing complexity, in much the same way that a complex molecule is built up from atoms. On the other hand, there is a stream of analysis whereby mathematicians have sought to arrive at the essence of numbers by breaking down the complexities to their most primitive elements. Both streams are of enormous importance. Professional mathematicians today tend to play down number as such, favoring the qualitative aspects of their science and emphasizing the logical structure and symbolic potentialities of mathematics. Nevertheless, new ideas about number keep making their way into the mathematics journals, and the modern number theories are just now diffusing rapidly throughout our educational system, even down to the elementary schools. There are programs and committees for teaching advanced number concepts, from set theory to matrices, to students in high school. It seems safe to say that the coming generation will be imbued with an unprecedented interest in the fascinating uses and mysteries of numbers.

III

GEOMETRY

III GEOMETRY

INTRODUCTION

Although number has taken precedence over geometry, especially for scientific work, geometry has not stagnated. In fact it has prospered. Even where geometry must call upon algebra, through the technique of analytic geometry, to secure its results, the geometric concepts and theorems are still the essence. The endurance and continual enrichment of geometry are accounted for by what we have previously noted—that all animate and inanimate objects have shape and size. The more extensive and the deeper the investigation of our world has become, the greater has become the need for the mathematical formulation of what is observed and the need for the invention of geometrical concepts to embrace our multitudinous observations.

That the continuing study of natural and manmade phenomena should necessitate new geometrical creations is readily illustrated by rather simple examples. It so happens that the conic sections—the parabola, ellipse, and hyperbola—were first investigated in Greek times. But it is almost certain that Kepler, pursuing the heliocentric theory of planetary motions first advanced by Copernicus, would have discovered that the ellipse was needed to describe the paths of the planets. If Kepler had not done so, then Newton in deducing the laws of planetary motion from axioms about motion would have found it necessary to introduce the ellipse.

One need not rely upon conjecture. In the seventeenth century, Robert Hooke investigated the motion of a bob attached to a spring and Christian Huygens investigated the pendulum. Both men were interested in designing a clock, and their work led to the creation of the trigonometric functions and the curves representing these functions. Most important among these curves is the now familiar sine curve (see Section VI, "Applications"). Huygens's work on the clock gave prominence to another significant curve, the cycloid. This is the curve that a fixed point on a circle traverses as the circle rolls along a straight line. One of the Gardner articles describes its unusual properties.

In addition to the investigation of nature, sheer intellectual curiosity has led to extensions and advances in geometry. The outstanding example is the two-thousand-year-old attempt to solve the three famous problems of geometry—to trisect any angle, to construct a square equal in area to a given circle, and to construct the side of a cube whose volume is double that of a given cube. These constructions were to be performed only with a straightedge (an unmarked ruler) and a compass. Though we now know, as a consequence of nineteenth-century work, that these constructions cannot be performed under the stated restrictions, many new geometric results were obtained in the efforts to do so. Another of Gardner's articles discusses this subject.

Intellectual curiosity has also led to many theorems of Euclidean geometry that even the Greeks, who produced over a thousand such theorems, did not obtain. Gardner's article on elegant triangle theorems illustrates some of these.

The continuing investigations of Euclidean geometry have led to surprising results. If one were asked what shape the wheels of an automobile must have for the automobile to roll along level ground at a constant height above the ground, the answer would surely be a circle. But it need not be. What other shapes can be used? Gardner's article on curves of constant width answers this question. Actually these curves, other than the circle, could not be used to design smooth-riding automobiles, but one of them is being used in a new type of automobile engine known as the Wankel engine.

Extensions of Euclidean geometry are but one direction in which geometry has developed since Greek times. A vast and important new branch of geometry—projective geometry—arose from a totally unlikely source, painting. The efforts of the Renaissance painters to put on canvas exactly what the eyes see, as opposed to the imaginative, unreal, and even artificial stylistic paintings of their medieval predecessors, necessitated a mathematical investigation of the art of perspective. Just what this investigation produced and how it raised problems leading to a new geometry is presented in Kline's article on projective geometry. This geometry considers questions that had never been raised by the Euclidean geometers, though they do concern ordinary Euclidean figures.

For two thousand years mathematicians, scientists, and philosophers were so firm in their belief that Euclidean geometry presented the one and only possible account of the properties of physical space and of objects in that space that to question this conviction would have been taken as a sign of madness. But genius has often and rightly been described as akin to madness. And it was the genius Gauss who, about 1800, dared to affirm what hindsight tells us should have been apparent at least one or two hundred years earlier: There can be and there are non-Euclidean geometries. Moreover, and more important, such geometries are as useful in describing the properties of objects and space as Euclidean geometry is.

What is non-Euclidean geometry? It is a geometry that starts with axioms different from Euclid's, and, of course, if the axioms are different the theorems are also. Because geometry serves as a description of our world, one who seeks to construct a non-Euclidean geometry would not adopt arbitrary axioms. Changing the Euclidean axiom on parallel lines (which asserts that given a point and line, there is one and only one line through the point and in the plane of the given point and line that does not meet the given line) is the key change that leads to several non-Euclidean geometries that still "fit" the physical world.

The nature of non-Euclidean geometry is fascinating, and some account of it is given in Kline's article "Geometry." The history can be found in the book by Bonola listed in the suggested readings at the end of this section. The theorems of the non-Euclidean geometries will be found in the books by Wolfe, Greenberg, and Meschkowski, and the enormous impact of the creation of non-Euclidean geometry is treated in the books by Kline.

The work on non-Euclidean geometries had another consequence whose import was not as world shaking but which nevertheless taught mathematicians a much needed lesson in humility. Because non-Euclidean geometry was strange and foreign to people brought up on Euclidean geometry, the mathematicians had to be most cautious and critical when trying to prove non-Euclidean theorems. To their surprise and dismay, they found that the presumed rigor of proof accredited to Euclidean geometry was sorely deficient. By using Euclid's "logic," it was possible to "prove" any number of absurd results. Gardner's article on geometric fallacies gives some noteworthy examples. Of course, when these fallacies were recognized, mathematicians made haste to perfect Euclidean proofs.

Our articles include still another extension of geometry—topology. There are problems of mathematics involving geometric figures wherein the precise shape and size are not important but some other features of the structure are. For example, the earth is not a sphere. The overall structure (if one ignores mountains and valleys) is that of a sphere flattened at the top and bottom, or, in mathematical terms, an oblate spheroid. For most research in astronomy, the oblate spheroid can be replaced by the sphere. Even the size of the sphere usually does not matter, though the total amount of matter contained in it may. Again, if one wishes to drive from town A to town B, what matters most is whether they are connected by a road. The precise shape and length of the road may not matter. Topology considers those properties of figures that are independent of size and shape. In other words, if one figure is obtained by expanding or contracting another in a rather arbitrary manner, the two figures may still possess some common properties; topology seeks to ascertain just what these properties are. The article by Euler, one of the greatest mathematicians, several of the articles by Gardner, the article on geometry by Kline, and the article by Tucker and Bailey referred to in the suggested readings illustrate some topological properties.

Geometry, like arithmetic and algebra, is a vast domain, and no mathematician today is a master of all of it. But one can get to know the nature of its regions, just as one can get to know the dominant characteristics of the various regions of one's country. Kline's article on geometry is an attempt to survey the country but does not pretend to explore any region in detail.

SUGGESTED READINGS

Adler, Irving. 1966. *A New Look at Geometry.* John Day, New York.

Arnold, Bradford H. 1962. *Intuitive Concepts in Elementary Topology.* Prentice-Hall, New York.

Barr, Stephen. 1964. *Experiments in Topology.* Crowell, New York.

Blackett, Donald W. 1967. *Elementary Topology.* Academic Press, New York.

Bold, Benjamin. 1969. *Famous Problems of Mathematics.* Van Nostrand-Reinhold, New York.

Bonola, Roberto. 1955. *Non-Euclidean Geometry.* Dover, New York.

Courant, Richard, and Herbert Robbins. 1941. *What Is Mathematics?* Oxford University Press, New York.

Coxeter, H. S. M. 1949. *The Real Projective Plane.* McGraw-Hill, New York.

Coxeter, H. S. M., and S. L. Greitzer. 1967. *Geometry Revisited.* Random House, New York.

Greenberg, Marvin Jay. 1974. *Euclidean and Non-Euclidean Geometries.* W. H. Freeman and Company, San Francisco.

Hilbert, D., and S. Cohn-Vossen. 1952. *Geometry and the Imagination.* Chelsea, New York.

Ivins, William M., Jr. 1964. *Art and Geometry.* Dover, New York.

Kahn, Donald W. 1975. *Topology.* Williams and Wilkins, Baltimore.

Kline, Morris. 1953. *Mathematics in Western Culture.* Oxford University Press, New York.

Kline, Morris. 1967. *Mathematics for Liberal Arts.* Addison-Wesley, Reading, Mass.

Lietzmann, Walther. 1965. *Visual Topology.* American Elsevier, New York.

Lockwood, E. H. 1963. *A Book of Curves.* Cambridge University Press, New York.

Lyusternick, L. A. 1963. *Convex Figures and Polyhedra.* Dover, New York.

Meschkowski, H. 1964. *Noneuclidean Geometry.* Academic Press, New York.

Ore, Oystein. 1963. *Graphs and Their Uses.* Random House, New York.

Tucker, Albert W., and Herbert S. Bailey. 1950. "Topology." *Scientific American,* January.

Wolfe, Harold E. 1945. *Introduction to Non-Euclidean Geometry.* Dryden, New York.

Young, John Wesley. 1930. *Projective Geometry.* Open Court Publishing, Chicago.

Geometric Constructions with a Compass and a Straightedge, and Also with a Compass Alone

by Martin Gardner
September 1969

It is often said that the ancient Greek geometers, following a tradition allegedly started by Plato, constructed all plane figures with a compass and a straightedge (an unmarked ruler). This is not true. The Greeks used many other geometric instruments, including devices that trisected angles. They did believe, however, that compass-and-straightedge constructions were more elegant than those done with other instruments. Their persistent efforts to find compass-and-straightedge ways to trisect the angle, square the circle and duplicate the cube—the three great geometric construction problems of antiquity—were not proved futile for almost 2,000 years.

In later centuries geometers amused themselves by imposing even more severe restrictions on instruments used in construction problems. The first systematic effort of this kind is a work ascribed to the 10th-century Persian mathematician Abul Wefa, in which he described constructions possible with the straightedge and a "fixed compass," later dubbed the "rusty compass." This is a compass that never alters its radius. The familiar methods of bisecting a line segment or an angle are simple examples of fixed-compass-and-straightedge constructions. The illustration below shows how easily a rusty compass can be used for bisecting a line more than twice the length of the compass opening. Many of Abul Wefa's solutions—in particular his method of constructing a regular pentagon, given one of its sides—are extremely ingenious and hard to improve on.

Leonardo da Vinci and numerous Renaissance mathematicians experimented with fixed-compass geometry, but the next important treatise on the subject was *Compendium Euclidis Curiosi*, a 24-page booklet published anonymously in Amsterdam in 1673. It was translated into English four years later by Joseph Moxon, England's royal hydrographer. This work is now known to have been written by a Danish geometer, Georg Mohr, whom we shall meet again in a moment. In 1694 a London surveyor, William Leybourn, in a whimsical book called *Pleasure with Profit*, treated rusty-compass constructions as a form of mathematical play. Above his section on this topic he wrote: "Shewing How (Without Compasses), having only a common Meat-Fork (or such like Instrument, which will neither open wider, nor shut closer), and a Plain Ruler, to perform many pleasant and delightful Geometrical Operations."

In the 19th century the French mathematician Jean Victor Poncelet suggested a proof, later rigorously developed by Jacob Steiner, a Swiss, that *all* compass-and-straightedge constructions are possible with a straightedge and a fixed compass. This conclusion follows at once from their remarkable demonstration that every construction possible with a compass and a straightedge can be done with a straightedge alone, provided that a single circle and its center are given on the plane. Early in the 20th century it was shown that not even the entire "Poncelet-Steiner circle," as it is called, is necessary. All that is needed is one arc of this circle, however small, together with its center! (In such constructions it is assumed that a circle is constructed if its center and a point on its circumference are determined.)

Many well-known mathematicians studied constructions that are possible with such single instruments as a straightedge, a straightedge marked with two points, a ruler with two parallel straightedges, a "ruler" with straightedges meeting perpendicularly or at other angles, and so on. Then in 1797 an Italian geometer, Lorenzo Mascheroni, amazed the mathematical world by publishing *Geometria del compasso*, in which he proved that every compass-and-straightedge construction can be done with a movable compass alone. Since straight lines cannot of course be drawn with a compass alone, it is assumed that two points, obtained by arc intersections, define a straight line.

Compass-only constructions are still called Mascheroni constructions even though it was discovered in 1928 that Mohr had proved the same thing in an obscure little work, *Euclides Danicus*, published in 1672 in Danish and Dutch editions. A Danish student who had found the book in a secondhand book-

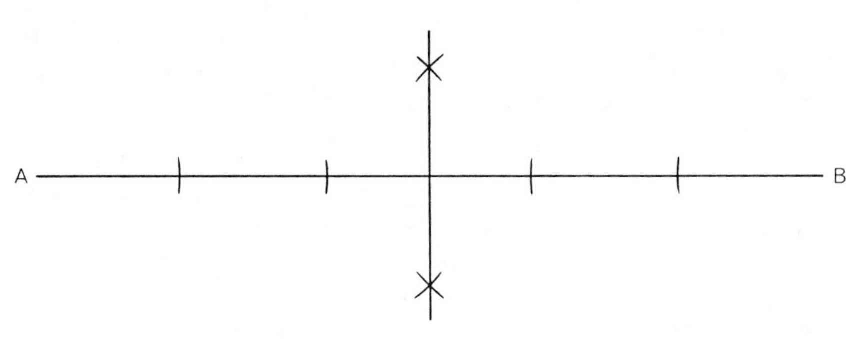

How to bisect a line of any length with a "rusty compass"

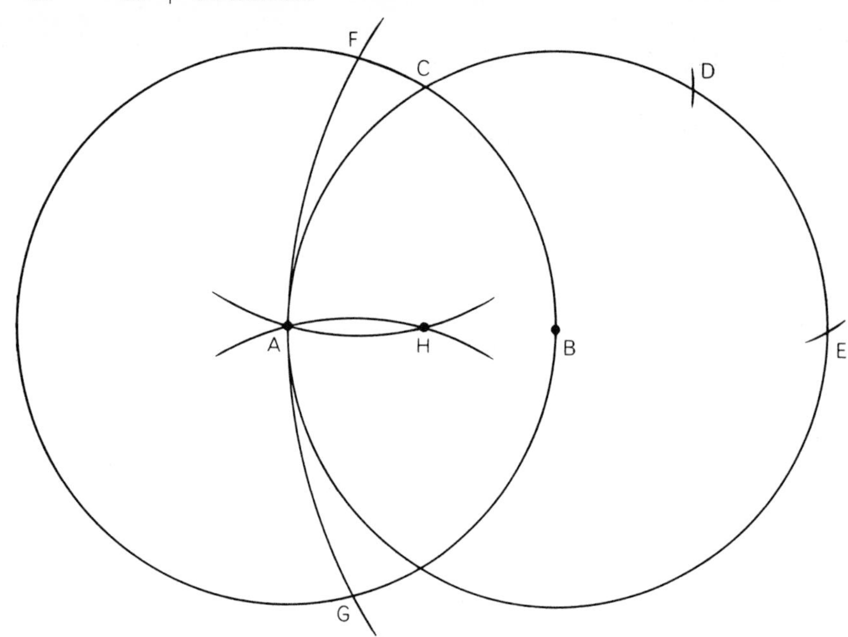

Mascheroni's method of finding point H, *midway between* A *and* B, *using a compass alone*

store in Copenhagen showed it to his mathematics teacher, Johannes Hjelmslev of the University of Copenhagen, who instantly recognized its importance. Hjelmslev published it in facsimile, with a German translation, in Copenhagen in 1928.

Today's geometers have little interest in Mohr-Mascheroni constructions, but because they present so many problems of a recreational nature they have been taken over by puzzle enthusiasts. The challenge is to improve on earlier constructions by finding ways of doing them in fewer steps. Sometimes it is possible to improve on Mohr's or Mascheroni's methods, sometimes not. Consider, for example, the simplest of five solutions by Mascheroni to his problem No. 66, that of finding a point midway between two given points *A* and *B* [*see illustration above*].

Draw two circles of radius *AB*, their centers at *A* and *B*. Keeping the same compass opening, with *C* and *D* as centers, mark points *D* and *E*. (Readers may recall that this is the beginning of the well-known procedure by which a circle is divided into six equal arcs, or three equal arcs if alternate points are taken.) Point *E* will lie on an extension of line *AB* to the right, and *AE* will be twice *AB*. (This procedure obviously can be repeated rightward to double, triple or produce any multiple of length *AB*.) Open the compass to radius *AE* and draw an arc, its center at *E*, that intersects the left circle at *F* and *G*. Close the compass to radius *AB* once more. With centers at *F* and *G* draw the two arcs that intersect at *H*.

H is midway between *A* and *B*. This is easily proved by noting that the two isosceles triangles marked by corners *AFH* and *AFE* share the base angle *FAE* and therefore are similar. *AF* is half of *AE;* consequently *AH* is half of *AB*. For readers acquainted with inversion geometry *H* is the inverse of *E* with respect to the left circle. A simple proof of the construction, by way of inversion geometry, is in *What Is Mathematics?*, by Richard Courant and Herbert Robbins (Oxford, 1941), page 145. Note that if line segment *AB* is drawn at the outset, and the problem is to find its midpoint with compass alone, only one of the last two arcs need be drawn, reducing the number of steps to six. I know of no way to do this with fewer steps.

Another famous problem solved by Mascheroni is locating the center of a given circle. His method is too complicated to reproduce here, but fortunately a simplified approach, of unknown origin, appears in many old books and is given at the top of page 95. *A* is any point on the circle's circumference. With *A* as the center, open the compass to any radius that will draw an arc intersecting the circle at *B* and *C*. With radius *AB* and centers *B* and *C*, draw arcs that intersect at *D*. (*D* may be inside or below the circle, depending on the length of the first compass opening.) With radius *AD* and center *D*, draw the arc giving intersections *E* and *F*. With radius *AE* and centers *E* and *F*, draw arcs intersecting at *G*. *G* is the circle's center. As before, there is an easy proof that starts by observing that the two isosceles triangles marked by *DEA* and *GEA* share the base angle *EAG* and are therefore similar. For the rest of the proof, as well as a proof by inversion geometry, see L. A. Graham's *The Surprise Attack in Mathematical Problems* (Dover, 1968), problem No. 34.

A third well-known problem in Mascheroni's book has become known as "Napoleon's problem" because it is said that Napoleon Bonaparte originally

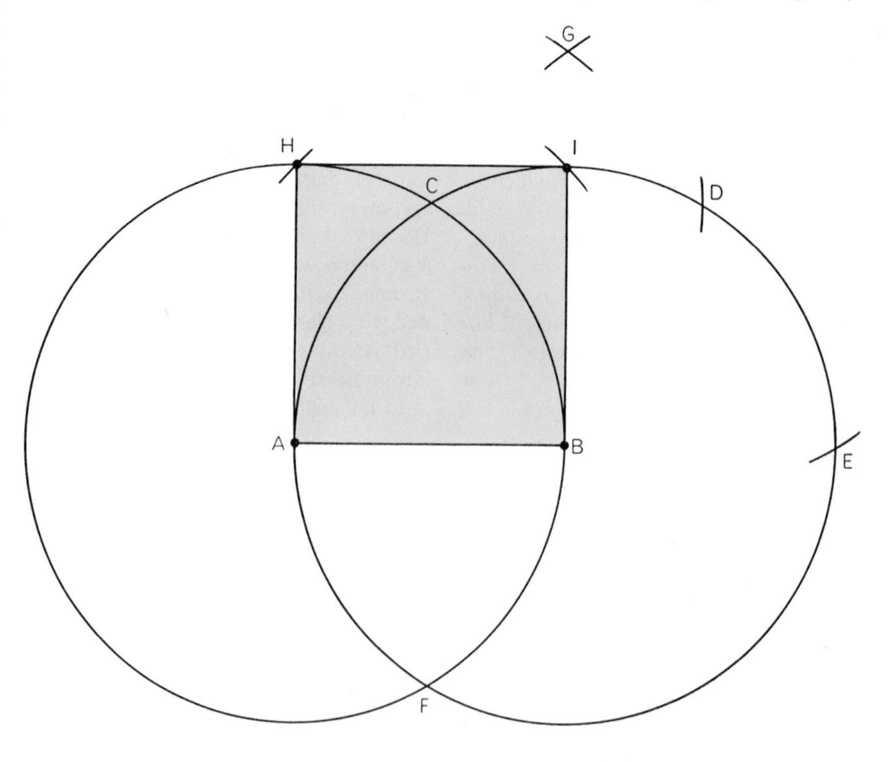

An eight-step way to construct the corners of a square, given adjacent corners A *and* B

proposed it to Mascheroni. It is not generally known that Napoleon was an enthusiastic amateur mathematician, of no great insight but particularly fascinated by geometry, which of course had great military value. He was also a man with unbounded admiration for the creative French mathematicians of his day. Gaspard Monge (known to recreational mathematicians mainly for his youthful analysis of the "Monge shuffle," in which cards are pushed one at a time off the deck by the left thumb to go alternately above and below the cards in the right hand) seems to have been the only man with whom Napoleon had a permanent friendship. "Monge loved me as one loves a mistress," Napoleon once declared. Monge was one of several French mathematicians who were made counts by Napoleon. Whatever Napoleon's ability as a geometer may have been, it is to his credit that he so revolutionized the teaching of French mathematics that, according to several historians of mathematics, his reforms were responsible for the great upsurge of creative mathematics in 19th-century France.

Like Monge, young Mascheroni was an ardent admirer of Napoleon and the French Revolution. In addition to being a professor of mathematics at the University of Pavia, he also wrote poetry that was highly regarded by Italian critics. There are several Italian editions of his collected verse. His book *Problems for Surveyors* (1793) was dedicated in verse to Napoleon. The two men met and became friends in 1796, when Napoleon invaded northern Italy. A year later, when Mascheroni published his book on constructions with the compass alone, he again honored Napoleon with a dedication, this time a lengthy ode.

Napoleon mastered many of Mascheroni's compass constructions. It is said that in 1797, while Napoleon was discussing geometry with Joseph Louis Lagrange and Pierre Simon de Laplace (famous mathematicians whom Napoleon later made a count and a marquis respectively), the little general surprised them by explaining some of Mascheroni's solutions that were completely new to them. "General," Laplace reportedly remarked, "we expected everything of you except lessons in geometry." Whether this anecdote is true or not, Napoleon did introduce Mascheroni's compass work to French mathematicians. A translation of *Geometria del compasso* was published in Paris in 1798, a year after the first Italian edition.

"Napoleon's problem" is to divide a circle, its center given, into four equal

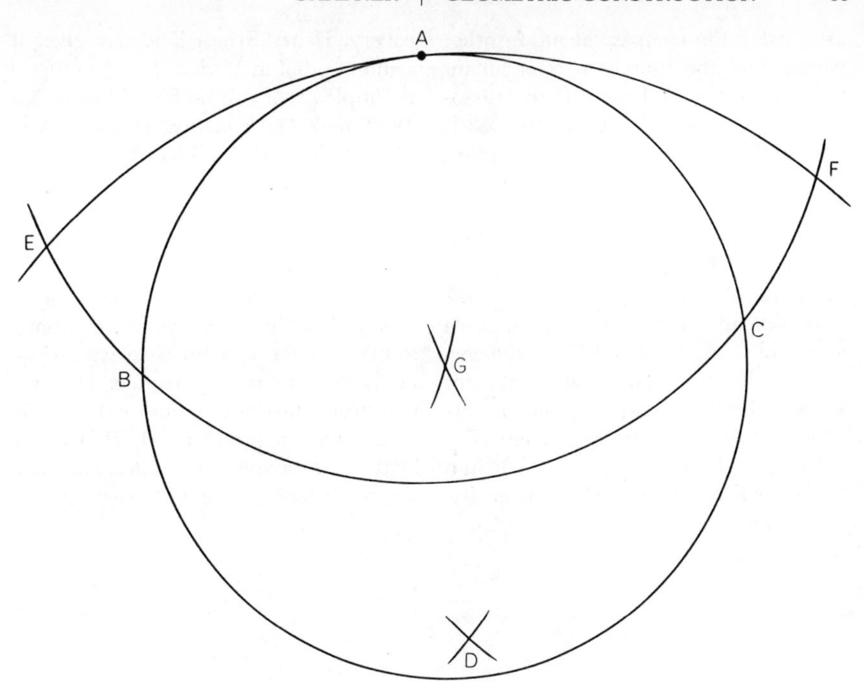

How to find a circle's center, using the compass alone, in six steps

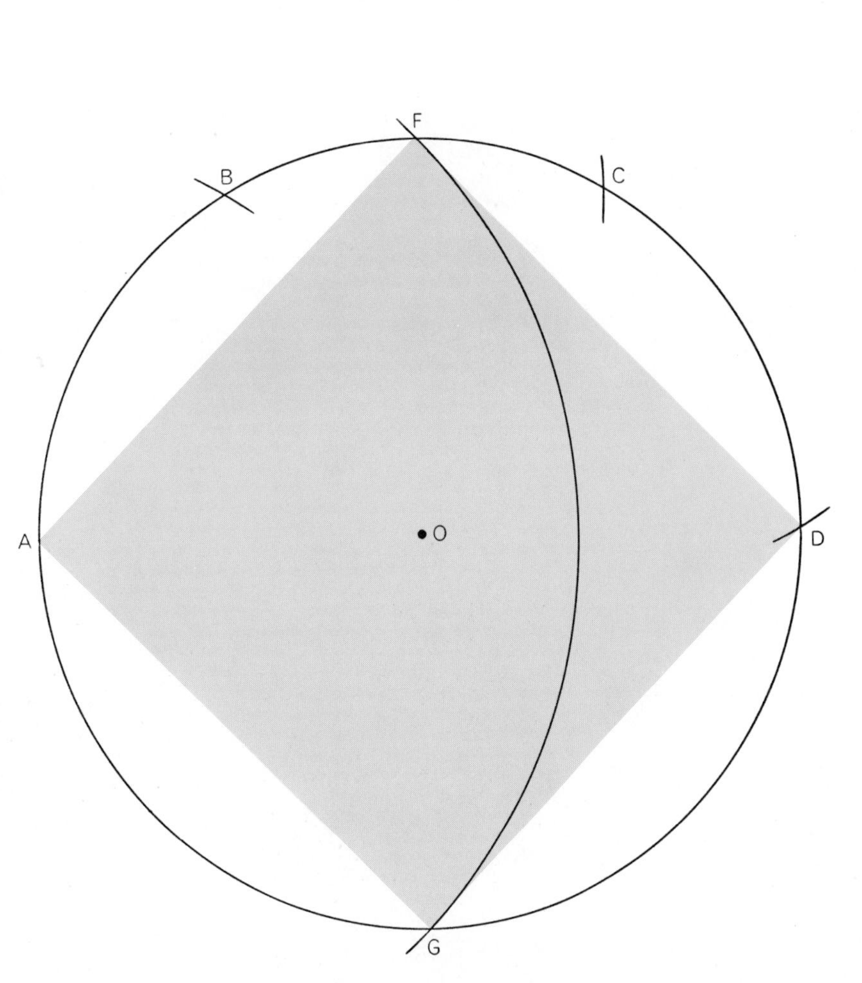

Six-step solution to "Napoleon's problem"

arcs, using the compass alone. In other words, find the four corners of an inscribed square. A beautiful six-arc solution is shown on page 95. With the compass open to the circle's radius, choose any point *A*, then mark spots *B*, *C* and *D*, using *A*, *B* and *C* as centers. Open the compass to radius *AC*. With centers *A* and *D*, draw the arcs intersecting at *E*. With center *A* and radius *OE*, draw the arc that cuts the original circle at *F* and *G*. *A*, *F*, *D* and *G* are the corners of the inscribed square. I do not know if this is Mascheroni's solution (his book has not been translated into English and I have not had access to the Italian or French editions) or a later discovery. Henry Ernest Dudeney gives it without proof in *Modern Puzzles* (1926). A simple proof can be found in Charles W. Trigg's *Mathematical Quickies* (McGraw-Hill, 1967), problem No. 248.

Two related and less well-known Mascheroni problems are: (1) Given two adjacent corner points of a square, find the other two, and (2) Given two diagonally opposite corner points of a square, find the other two. An eight-arc solution to the first problem was separately sent to me by readers Don G. Olmstead and Paul White and can be found, with a proof, in M. H. Greenblatt's *Mathematical Entertainments* (Crowell, 1965), page 139. The bottom illustration on page 94 shows the procedure. *A* and *B* are the two given points. After drawing the two circles, each with radius *AB*, keep the same opening and mark points *D* and *E*, with centers at *C* and *D*. Open the compass to radius *CF*. With *A* and *E* as centers, draw the two arcs that intersect at *G*. With radius *GB* and centers *A* and *B*, draw the arcs that cut the circles at *H* and *I*. *H* and *I* are the two corner points sought.

The best solution I know for the second and more difficult problem requires nine arcs. Readers are invited to search for it, or a better one, or you may turn to the solutions at the end of the book.

16

Elegant Triangle Theorems Not to Be Found in Euclid

by Martin Gardner
June 1970

One might suppose that the humble triangle was so thoroughly investigated by ancient Greek geometers that not much significant knowledge of the polygon with the fewest sides and angles could be added in later centuries. This is far from true. The number of theorems about triangles is infinite, of course, but beyond a certain point they become so complex and sterile that no one can call them elegant. George Polya once defined a geometric theorem's degree of elegance as "directly proportional to the number of ideas you see in it and inversely proportional to the effort it takes to see them." Many elegant triangle discoveries have been made in recent centuries that are both beautiful and important but that the reader is unlikely to have come across in elementary plane geometry courses. This month we shall consider only a minute sample of such theorems, emphasizing those that have suggested puzzle problems.

"Ferst," as James Joyce says in the mathematical section of *Finnegans Wake*, "construct ann aquilittoral dryankle Probe loom!" We begin with a triangle, *ABC*, of any shape [*see illustration below*]. On each side an equilateral triangle is constructed outward [*left*] or inward [*center*]. In both cases, when the centers (the intersections of two altitudes) of the three new triangles are joined by straight lines [*shown in color*], we find we have constructed a fourth equilateral triangle. (The theorem is sometimes given in terms of constructing three isosceles triangles with 30-degree base angles, then joining their apexes, but since these apexes coincide with the centers of equilateral triangles, the two theorems are identical.) If the initial triangle is itself equilateral, the inward triangles give a "degenerate" equilateral triangle, a point. It is a lovely theorem, one that holds even when the original triangle has degenerated into a straight line as shown at the right in the illustration. I do not know who first thought of it—it has been attributed to Napoleon—but many different proofs have been printed in recent decades. An unusual proof using only group theory and symmetry operations is given by the Russian mathematician Isaac Moisevitch Yaglom in *Geometric Transformations* (Random House, New Mathematics Library, 1962), page 93.

Another elegant theorem, in which a circle (like the fourth equilateral triangle of the preceding example) seems to emerge from nowhere, is the famous nine-point-circle theorem. It was discovered by two French mathematicians, who published it in 1821. On any given triangle we locate three triplets of points [*see top illustration on page 98*]:

(1) The midpoints (a, b, c) of the three sides.

(2) The feet (p, q, r) of the three altitudes.

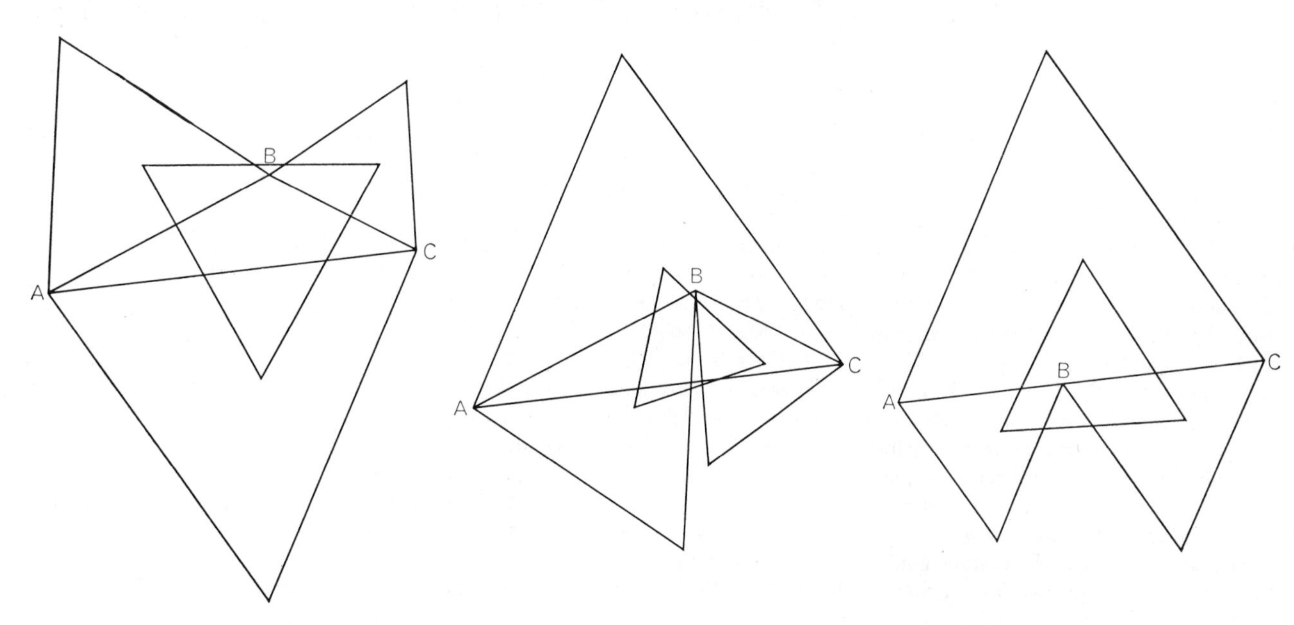

Joining the centers of three equilateral triangles creates a fourth one (color)

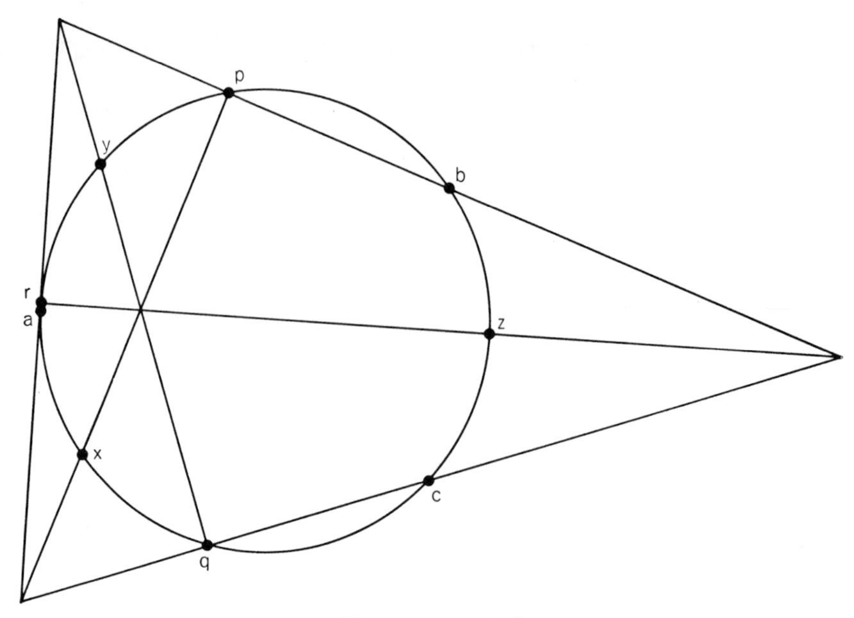

The nine-point circle

(3) The midpoints (x, y, z) of line segments joining each corner to the "orthocenter" (the spot where the three altitudes intersect).

As the illustration shows, those nine points lie on the same circle, a startling theorem that leads to a wealth of other theorems. It is not hard to show, for instance, that the radius of the nine-point circle is exactly half the radius of a circle that circumscribes the original triangle. The fact that the three altitudes of any triangle are concurrent (intersect at the same point) is interesting in itself. It is not in Euclid. Although Archimedes implies it, Proclus, a fifth-century philosopher and geometer, seems to have been the first to state it explicitly.

Three lines joining each midpoint of a side to the opposite vertex are called the triangle's medians [*see the bottom illustration on this page*]. They too are always concurrent, intersecting at what is known as the triangle's centroid. The centroid trisects each median and the three medians carve the triangle into six smaller triangles of equal area. Moreover, the centroid is the triangle's center of gravity, another fact known to Archimedes. Your high school geometry teacher may have demonstrated this by cutting a scalene triangle from cardboard, drawing its medians to find the centroid, then balancing the triangle on a pencil by putting the centroid on the pencil's point.

The median is a special case of a more general line called a "civian" (after a 17th-century Italian mathematician, Giovanni Ceva). A civian is a line from a triangle's vertex to any point on the opposite side. If instead of midpoints we take trisection points, three civians drawn as shown in the top illustration on the next page will cut the triangle into seven regions, each a multiple of 1/21 of the original triangle's area. The central triangle, shown shaded, has an area of 3/21, or 1/7. There are many clever ways to prove this, as well as the results of a more general case where each side of the triangle is divided into n equal parts. If the civians are drawn as before, to the first point from each vertex in a clockwise (or counterclockwise) direction around the triangle, the central triangle (as Howard D. Grossman has shown) has an area of $(n-2)^2/(n^2-n+1)$. A still broader generalization, in which the sides of the original triangle may vary independently in their number of equal parts, is discussed by H. S. M. Coxeter in his *Introduction to Geometry* (Wiley, 1961), Section 13.55. A formula going back to 1896 is given and Coxeter shows how easily it can be obtained by embedding the triangle within a regular lattice of points.

Every triangle has three sides and three angles. Euclid proved three cases in which two triangles are congruent if only three of the six elements are equal (for example two sides and their included angle). Is it possible for two triangles to have five of the six elements identical and yet *not* be congruent? It seems impossible, but there is an infinite set of such "5-con" triangles, as they have been called by Richard G. Pawley, a California mathematics teacher. Two 5-con triangles are congruent if three sides are equal, and therefore the only situation that permits noncongruence is the one in which two sides and three angles are equal. The smallest example of such a pair with integral sides is shown in the bottom illustration (right), next page. Note that the equal sides of 12 and 18 are not corresponding sides. The triangles are necessarily similar, because corresponding angles are equal, but they are not congruent. The problem of finding all such pairs is intimately connected with the golden ratio. A reference on this connection is Chapter 4 of Verner E. Hoggatt, Jr.'s *Fibonacci and Lucas Numbers* (1969), a booklet in the Houghton Mifflin "Mathematics Enrichment Series."

There are many ancient formulas for finding a triangle's sides, angles or area, given certain facts about its altitudes, medians and so on. The expression $\sqrt{s(s-a)(s-b)(s-c)}$, where a, b, c are the sides of any triangle and s is half of the sum of the three sides, gives the triangle's area. This amazingly simple formula was first proved in the *Metrica* of Heron of Alexandria, who is now known to have lived in the first or second century. The formula, Heron's chief claim to mathematical fame, is easily proved by trigonometry. Heron's geometric proof can be found in W. W. Rouse Ball's *A Short Account of the History of Mathematics* (Dover, 1960), Chapter 4. Heron, or Hero as he is sometimes called, is best known today for his delightful treatises on Greek automata and hydraulic toys, such as the perplexing "Hero's fountain," in which a stream of water seems to defy gravity by spouting higher than its source [see "The Amateur Scientist," December, 1966].

A classic puzzle of unknown origin, the solution to which involves similar triangles, has become rather notorious because, as correspondent Dudley F. Church so aptly put it, "its charm lies in the apparent simplicity (at first glance) of its solution, which quickly evolves into an algebraic mess." The problem con-

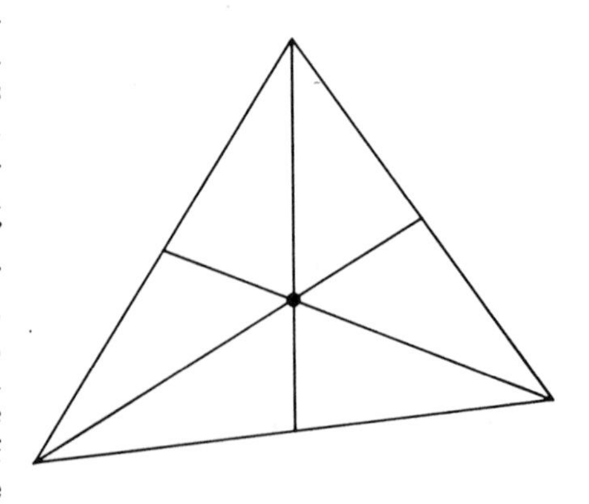

Centroid trisects medians

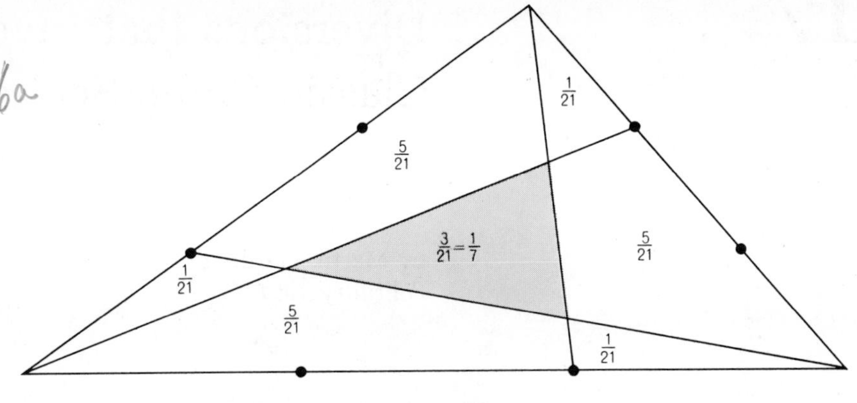

Trisecting civians

cerns two crossed ladders of unequal length. (The problem is trivial if the ladders are equal.) They lean against two buildings as shown in the illustration, below left. Given the lengths of the ladders and the height of their crossing point, what is the width of the space between the buildings? The three given values vary widely in published versions of the puzzle. Here we take a typical instance from William R. Ransom's *One Hundred Mathematical Curiosities* (J. Weston Walch, 1955). The ladders, with lengths of 100 units (a) and 80 units (b), cross 10 units (c) above the ground. By considering similar triangles Ransom arrives at the formula, $k^4 - 2ck^3 + k^2(a^2 - b^2) - 2ck(a^2 - b^2) + c^2(a^2 - b^2) = 0$, which in this case becomes $k^4 - 20k^3 + 3{,}600k^2 - 72{,}000k + 360{,}000 = 0$.

This formidable equation is a quartic, best solved by Horner's method or some other method of successive approximations. The solution gives k a value of about 11.954, from which the width between buildings ($u + v$) is found to be 79.10+. There are many other approaches to the problem. A good trigonometric solution is given in L. A. Graham's *The Surprise Attack in Mathematical Problems* (Dover, 1968), Problem 6. Other ways to solve the problem are in Graham's earlier paperback (Dover, 1959), *Ingenious Mathematical Problems and Methods,* Problem 25.

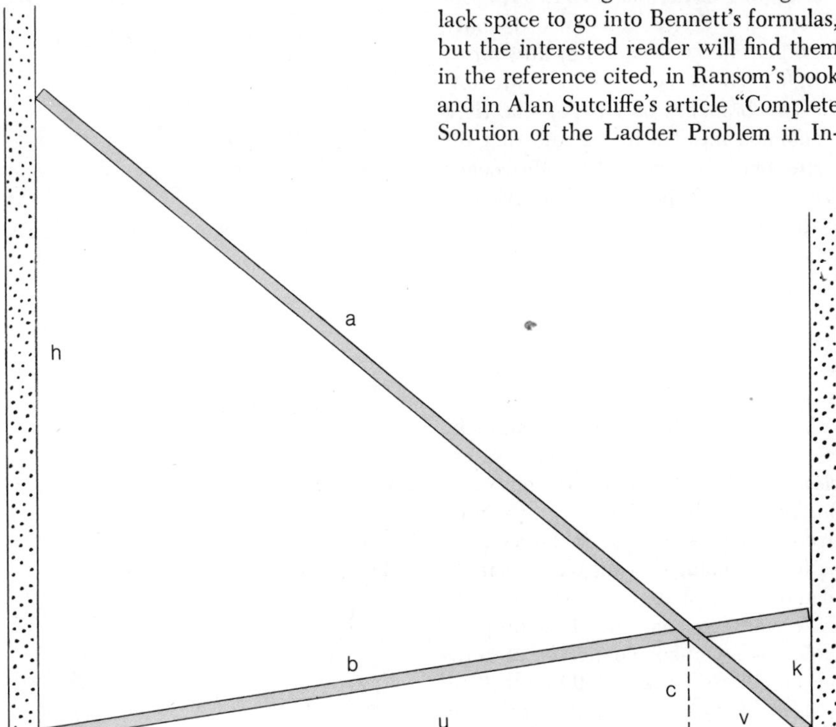

A difficult question arises at once. Are there forms of this problem (assuming unequal ladders) in which all four values are integers? As far as I know this was first answered by Albert A. Bennett of Brown University in *The American Mathematical Monthly,* April, 1941, solution to Problem E433. Bennett's equations have since been rediscovered many times. The smallest integral values, minimizing both the height of the crossing and the width between the buildings, result when the ladders are 119 units and 70 units long, the crossing is 30 units above the ground and the width 56 units. It turns out that if those four values are integral, the lengths of all other line segments in the diagram are also integral. I lack space to go into Bennett's formulas, but the interested reader will find them in the reference cited, in Ransom's book and in Alan Sutcliffe's article "Complete Solution of the Ladder Problem in Integers," *The Mathematical Gazette,* Vol. 47, May, 1963, pages 133–136. In addition there is an infinity of solutions in which the distance between the *tops* of the ladders is also an integer. (See Gerald J. Janusz' answer to Problem 5323, proposed by Sutcliffe, in *The American Mathematical Monthly,* Vol. 73, December, 1966, pages 1125–1127.)

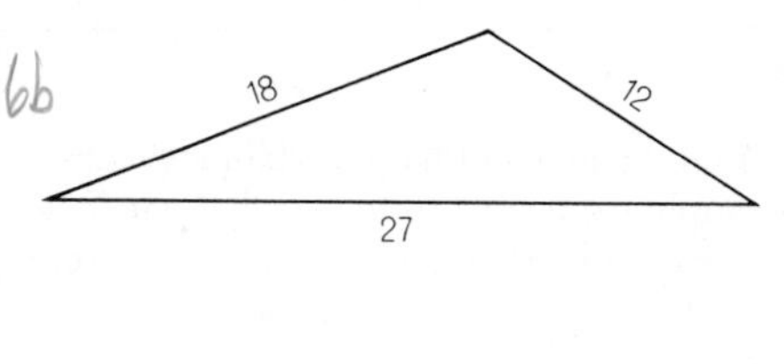

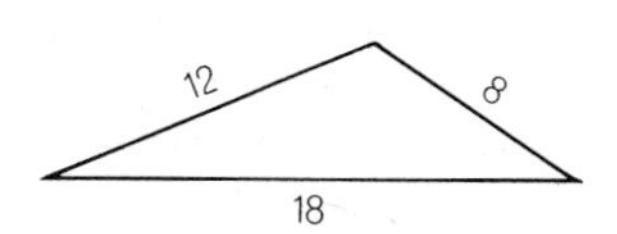

Smallest "5-con" triangle pair

When we are given no more than the distances from a point to the three vertexes of a triangle, there obviously is no way to construct the triangle or to determine the lengths of its sides. If the triangle is known to be equilateral, however, the side *can* be determined. The point may be inside or outside the triangle. A problem of this type, of unknown origin, is frequently sent to me by readers, usually in the following form: A point within an equilateral triangle is three, four and five units from the triangle's corners. How long is the triangle's side?

There are many ways to solve this problem, one of which, using a simple construction and similar triangles, is given on page 238. I know of no version in which all four values are integral.

17

Diversions that Involve One of the Classic Conic Sections: The Ellipse

by Martin Gardner
February 1961

A circle no doubt has a certain appealing simplicity at the first glance, but one look at an ellipse should have convinced even the most mystical of astronomers that the perfect simplicity of the circle is akin to the vacant smile of complete idiocy. Compared to what an ellipse can tell us, a circle has little to say. Possibly our own search for cosmic simplicities in the physical universe is of this same circular kind—*a projection of our uncomplicated mentality on an infinitely intricate external world.*

Eric Temple Bell,
Mathematics: Queen and Servant of Science

Mathematicians have a habit of studying, just for the fun of it, things that seem utterly useless; then centuries later their studies turn out to have enormous scientific value. There is no better example of this than the work done by the ancient Greeks on the noncircular curves of second degree: the ellipse, parabola and hyperbola. They were first studied by one of Plato's pupils. No important scientific applications were found for them until the 17th century, when Kepler discovered that planets move in ellipses and Galileo proved that projectiles travel in parabolas.

Apollonius of Perga, a third century B.C. Greek geometer, wrote the greatest ancient treatise on these curves. His work *Conics* was the first to show how all three curves, along with the circle, could be obtained by slicing the same cone at continuously varying angles. If a plane is passed through a cone so that it is parallel to the base [*see top of illustration at right*], the section is a circle. If the plane is tipped, no matter how slightly, the section becomes elliptical. The more the plane is tipped, the more elongated the ellipse becomes, or, as the mathematician puts it, the more eccentric. One might expect that as the plane became steeper the curve would take on more of a pear shape (since the deeper the slice goes, the wider the cone), but this is not the case. It remains a perfect ellipse until the plane becomes parallel to the side of the cone. At this instant the curve ceases to close on itself; its arms stretch out toward infinity and the curve becomes a parabola. Further tipping of the plane causes it to intersect an inverted cone placed above the other one [*see bottom of illustration at right*]. The two conic sections are now the two branches of a hyperbola. (It is a common mistake to suppose that the plane must be parallel to the cone's axis to cut a hyperbola.) They vary in shape as the cutting plane continues to rotate until finally they degenerate into straight lines. The four curves are called curves of second degree because they are the Cartesian graph forms of all second-degree equations that relate two variables.

The ellipse is the simplest of all plane curves that are not straight lines or circles. It can be defined in numerous ways, but perhaps the easiest to grasp intuitively is this: An ellipse is the locus, or path, of a point moving on a plane so that the sum of its distances from two fixed points is constant. This property underlies a well-known method of drawing an ellipse. Stick two thumbtacks in a sheet of paper, put a loop of string around them and keep the string taut with the point of a pencil as shown in the top illustration on page 101. Moving the pencil around the tacks will trace a perfect ellipse. (The length of the cord cannot vary; therefore the sum of the distances of the pencil point from the two tacks remains constant.) The two fixed points (tacks) are called the foci of the

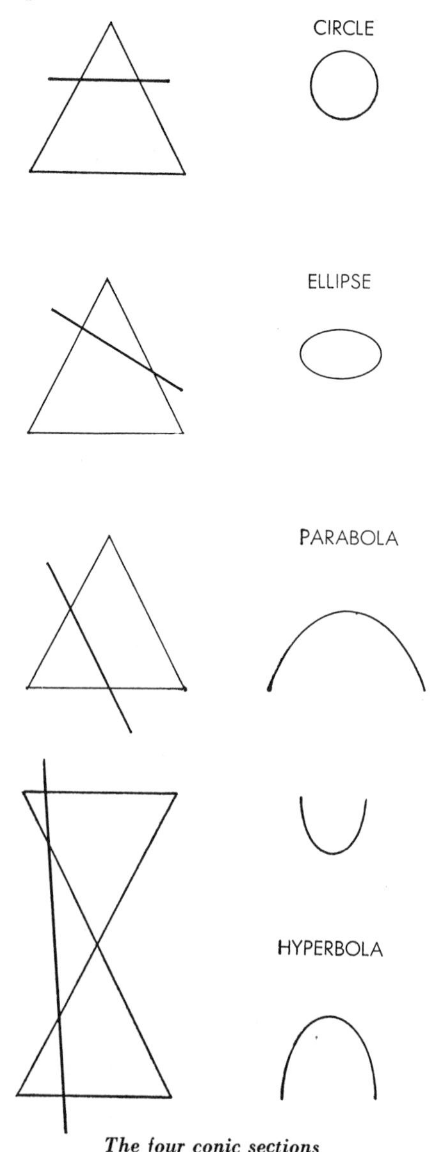

The four conic sections

ellipse. They lie on the major axis. The diameter perpendicular to this axis is the minor axis. If you move the tacks closer together (keeping the same loop of cord), the ellipse becomes less and less eccentric. When the two foci come together, the ellipse becomes a circle. As the foci move farther apart the ellipse becomes more elongated until it finally degenerates into a straight line.

There are many other ways to construct ellipses. One curious method can be demonstrated with a circular cake pan and a cardboard disk having half the diameter of the pan. Put friction tape or masking tape around the inside rim of the pan to keep the disk from slipping when it is rolled around the rim. Anchor a sheet of paper to the bottom of the pan with strips of cellophane tape at the edges. Punch a hole anywhere in the disk with a pencil, place the point of the pencil on the paper and roll the disk around the pan [*see bottom illustration on page 101*]. An ellipse will be drawn on the paper. It is best to hold the pencil lightly with one hand while turning the disk slowly with the other, keeping it pressed firmly against the rim of the pan. If the hole is at the center of the disk, the pencil point will of course trace a circle. The nearer the hole is to the edge of the disk, the greater the eccentricity of the ellipse will be. A point on the circumference of the disk traces an ellipse that has degenerated into a straight line!

Here is another pleasant way to obtain an ellipse. Cut a large circle from a sheet of paper. Make a spot somewhere inside the circle, but not at the center, then fold the circle so that its circumference falls on the spot. Unfold, then fold again, using a different point on the circumference, and keep repeating this until the paper has been creased many times in all directions. The creases form a set of tangents that outline an ellipse [*see top illustration on page 102*].

Though not so simple as the circle, the ellipse is nevertheless the curve most often "seen" in everyday life. The reason is that every circle, viewed obliquely, appears elliptical. In addition, all closed noncircular shadows cast on a plane by circles and spheres are ellipses. Shadows on the sphere itself—the inner curve of a crescent moon, for example—are bordered by great circles, but we see them as elliptical arcs. Tilt a glass of water (it doesn't matter if the glass has cylindrical or conical sides) and the surface of the liquid acquires an elliptical outline.

A ball resting on a table top [*see bottom illustration on page 102*] casts an elliptical shadow that is a cross section of a cone of light in which the ball fits snugly. The ball rests precisely on one focus of the shadow. If we imagine a larger sphere that is tangent to the surface from beneath and fits snugly into the same cone, the larger sphere will touch the shadow at the other focus. These two spheres provide the following famous and magnificent proof (by G. P. Dandelin, a 19th-century Belgian mathematician) that the conic section is indeed an ellipse.

Point A is any point on the ellipse. Draw a line [*shown in color in the illustration*] that passes through A and the

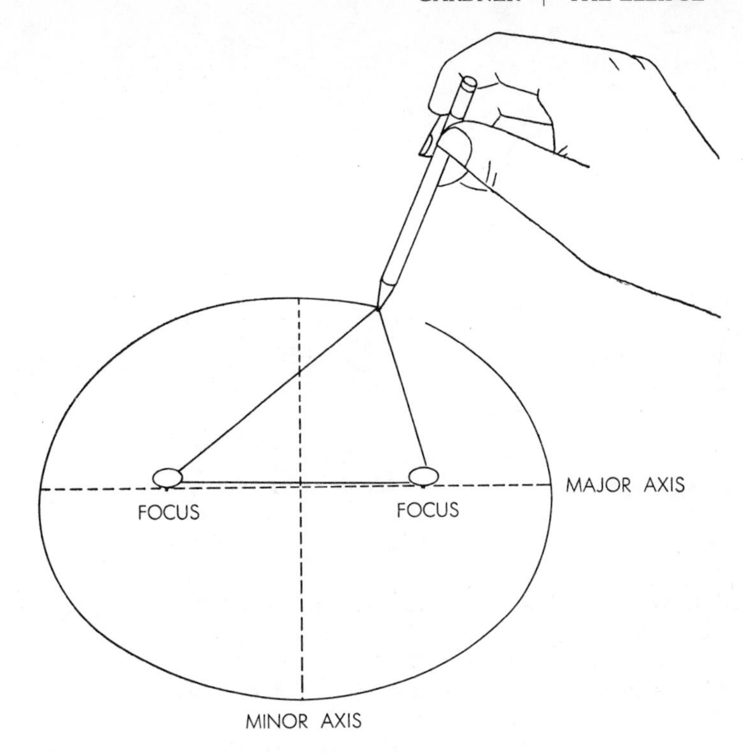

The simplest way to draw an ellipse

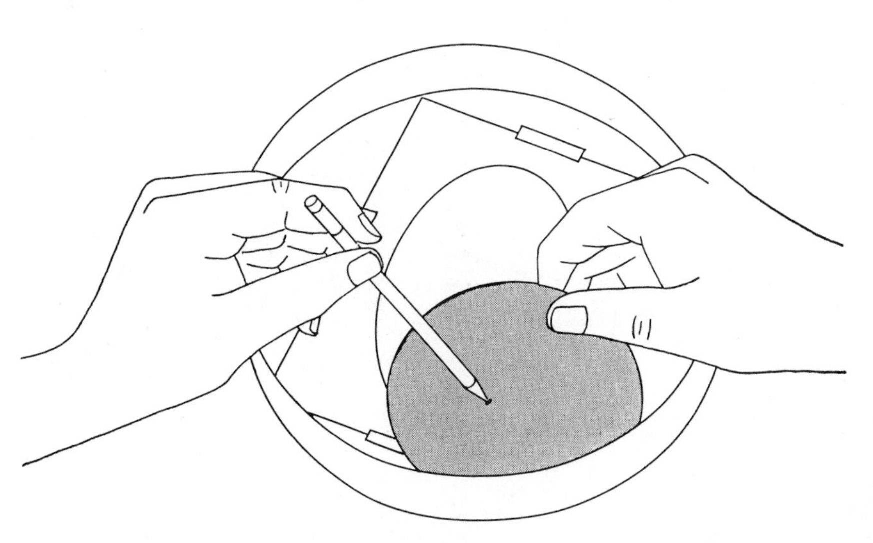

An ellipsograph made with a circular cake pan and a cardboard disk

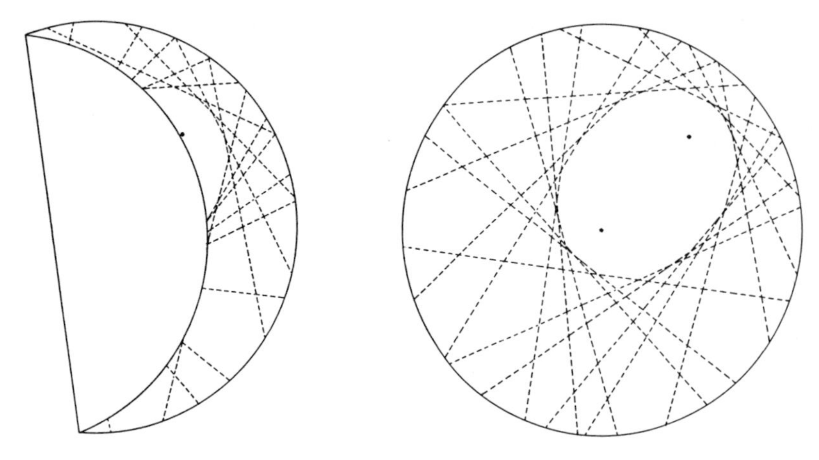

Folding a paper circle so that its edge falls on an off-center spot makes an ellipse

apex of the cone. This line will be tangent to the spheres at points D and E. Draw a line from A to point B, where the small sphere touches the shadow, and a similar line from A to C where the large sphere touches the shadow. AB is equal to AD because both lines are tangents to a sphere from the same fixed point. AE equals AC for the same reason. Adding equals to equals:

$$AD + AE = AB + AC$$

Now AD + AE is the same as the straight line DE. Because of the symmetry of cone and spheres, this line has a constant length regardless of where point A is chosen on the ellipse. If the sum of AD and AE is constant, then the above equation makes the sum of AB and AC a constant also. Since AB and AC are the distances of point A from two fixed points, the locus of A must be an ellipse with B and C as its two foci.

In physics the ellipse turns up most often as the path of an object moving in a closed orbit under the influence of a central force that varies inversely with the square of the distance. Planets and satellites, for example, have elliptical orbits with the center of gravity of the parent body at one of the foci. When Kepler first announced his great discovery that planets move in ellipses, it ran so counter to the general belief that God would not permit the paths of heavenly bodies to be less perfect than circles that Kepler found it necessary to apologize. He spoke of his ellipses as dung that he had been forced to introduce in order to sweep from astronomy the larger amount of dung that had accumulated around attempts to preserve circular orbits. Kepler himself never discovered why the orbits were elliptical; it remained for Newton to deduce this from the nature of gravity. Even the great Galileo to his dying day refused to believe, in the face of mounting evidence, that the orbits were not circular.

An important reflection property of the ellipse is made clear in the top illustration, p. 103. Draw a straight line that is tangent to the ellipse at any point. Lines from that point to the foci make equal angles with the tangent. If we think of the ellipse as a vertical strip of metal on a flat surface, then any body or wave pulse, moving in a straight line from one focus, will strike the boundary and rebound directly toward the other focus. Moreover, if the body or wave is moving toward the boundary at a uniform rate, regardless of the direction it takes when it leaves one focus, it is sure to rebound to the other focus after the same time interval (since the two distances have a constant sum). Imagine a shallow elliptical tank filled with water. We start a circular wave pulse by dipping a finger into the water at one focus of the ellipse. A moment later there is a convergence of circular waves at the other focus.

Lewis Carroll invented and published a pamphlet about a circular billiard table. I know of no serious proposal for an elliptical billiard table, but Hugo Steinhaus (in his book *Mathematical Snapshots*, recently reissued in a revised edition by the Oxford University Press) gives a surprising threefold analysis of

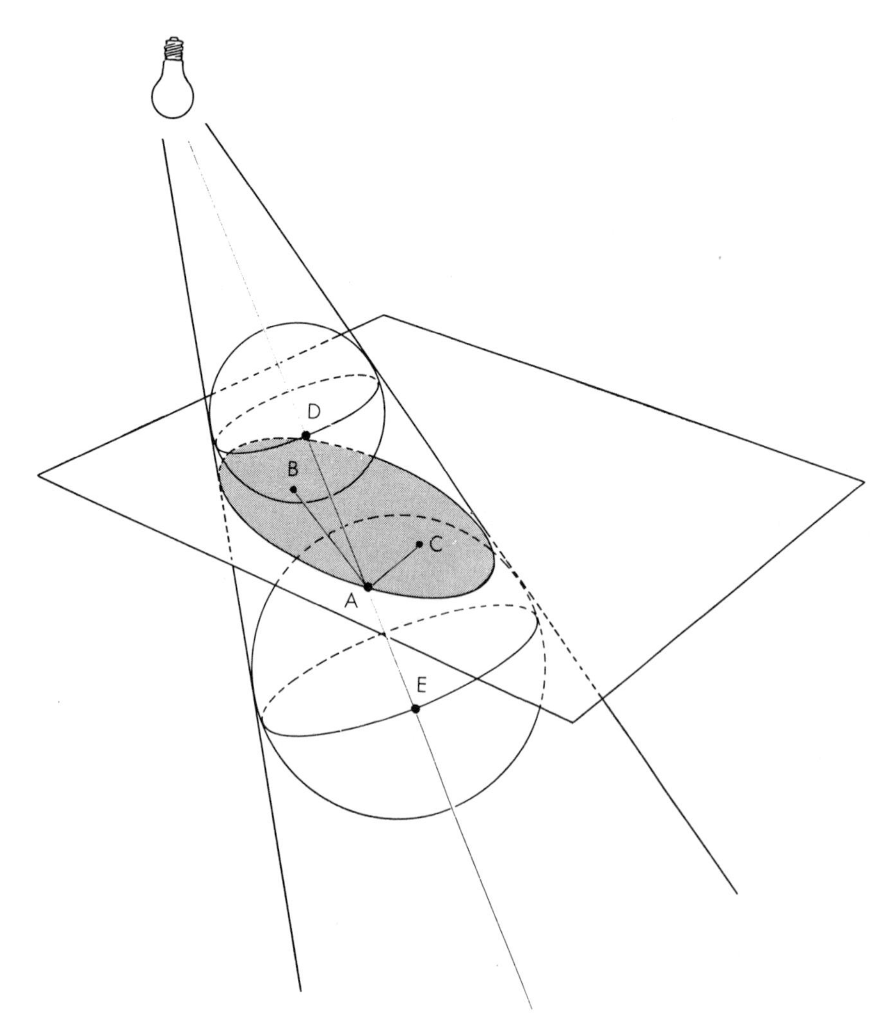

By means of larger sphere it can be shown that shadow of smaller sphere is an ellipse

how a ball on such a table would behave. Placed at one focus and shot (without English) in any direction, the ball will rebound and pass over the other focus. Assuming that there is no friction to retard the motion of the ball, it continues to pass over a focus with each rebound [*see second illustration at right*]. However, after only a few trips the path becomes indistinguishable from the ellipse's major axis. If the ball is not placed on a focus, then driven so that it does not pass between the foci, it continues forever along paths tangent to a smaller ellipse with the same foci [*see third illustration at right*]. If the ball is driven between the foci [*see fourth illustration at right*], it travels endlessly along paths that never get closer to the foci than a hyperbola with the same foci.

In *The Mikado* there are lines about a billiard player forced to play

On a cloth untrue
With a twisted cue,
And elliptical billiard balls!

In *A Portrait of the Artist as a Young Man* James Joyce has a teacher quote these lines, then explain that by "elliptical" W. S. Gilbert really meant "ellipsoidal." What is an ellipsoid? There are three principal types. An ellipsoid of rotation, more properly called a spheroid, is the surface of a solid obtained by rotating an ellipse around either axis. If the rotation is around the minor axis, it generates an oblate spheroid, which is flattened at the poles like the earth. Rotation around the major axis generates the football-shaped prolate spheroid. Imagine a prolate spheroid surface that is a mirror on the inside. If a candle is lighted at one focus, a piece of paper at the other focus will burst into flames.

Whisper chambers are rooms with spheroidal ceilings. Faint sounds originating at one focus can be heard clearly at the other focus. In the U. S. the best-known whispering gallery is in Statuary Hall of the Capitol in Washington, D.C. (No guided tour is complete without a demonstration.) A smaller but excellent whisper chamber is a square area just outside the entrance to the Oyster Bar on the lower level of New York's Grand Central Station. Two people standing in diagonally opposite corners, facing the wall, can hear each other distinctly even when the square area bustles with activity.

Both the oblate and prolate spheroids have circular cross sections if sliced by planes perpendicular to one of the three co-ordinate axes. When all three axes are unequal in length, and sections perpendicular to each are ellipses, the shape is a true ellipsoid [*see bottom illustration on this page*]. This is the shape that pebbles on a beach tend to assume after long periods of being jostled by the waves.

Elliptical "brain teasers" are rare. Here are two easy ones that will be answered in the back of the book.

1. Prove that no regular polygon having more sides than a square can be drawn on a noncircular ellipse so that each corner is on the perimeter of the ellipse.

2. In the paper-folding method of constructing an ellipse, explained earlier, the center of the circle and the spot on the circle are the two foci. Prove that the curve outlined by the creases is really an ellipse.

Addendum

Henry Dudeney, in problem 126 of *Modern Puzzles,* explains the string-and-pins method of drawing an ellipse, then asks how one can use this method for drawing an ellipse with given major and minor axes. The method is simple:

First draw the two axes. The problem now is to find the two foci, A and B, of an ellipse with these axes. Let C be an end of the minor axis. Points A and B are symmetrically located on the major axis at spots such that AC and CB each equals half the length of the major axis. It is easy to prove that a loop of string with a length equal to the perimeter of triangle ABC will now serve to draw the desired ellipse.

Elliptical pool tables actually went on sale in the United States in 1964. A full-page advertisement in *The New York Times* (July 1, 1964) announced that on the following day the game would be introduced at Stern's department store by Broadway stars Joanne Woodward and Paul Newman. Elliptipool, as it is called, is the patented invention of Arthur Frigo, Torrington, Connecticut, then a graduate student at Union College in Schenectady. Because the table's one pocket is at one of the foci, a variety of weird cushion shots can be made with ease.

The eleventh edition of *Encyclopaedia Britannica* in its article on billiards has a footnote that reads: "In 1907 an oval table was introduced in England by way of a change." Neither this table nor Lewis Carroll's circular table had a pocket, however. A design patent (198,571) was issued in July 1964 to Edwin E. Robinson, Pacifica, California, for a circular pool table with four pockets.

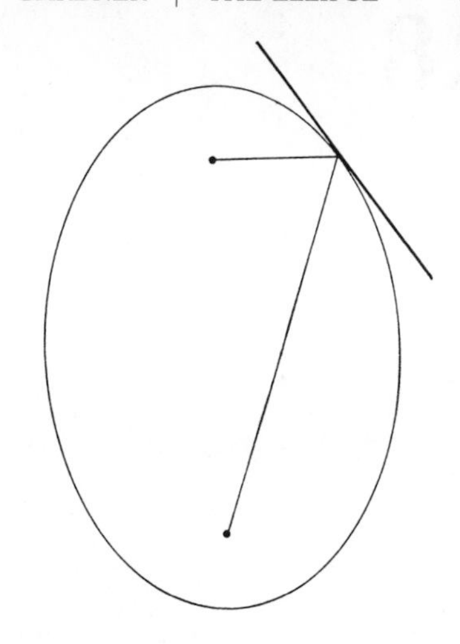

Tangent makes equal angles with two lines

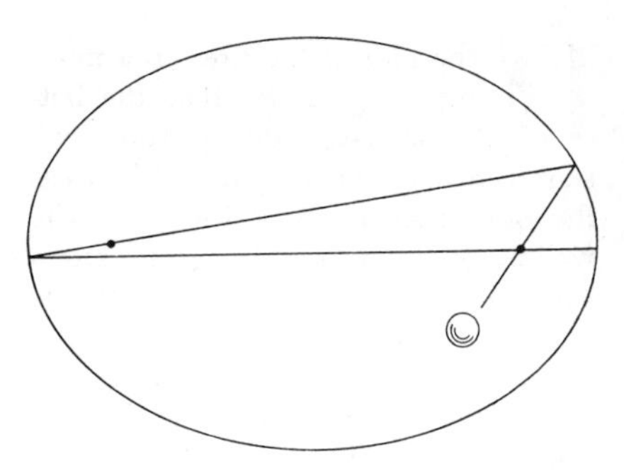

Path of ball driven over focus of ellipse

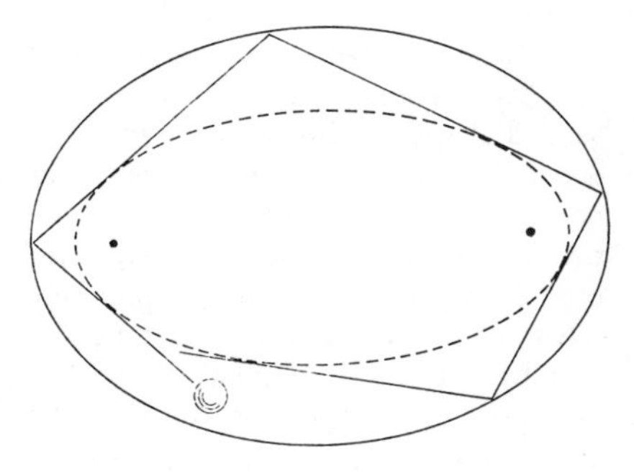

Path of ball that does not go between foci

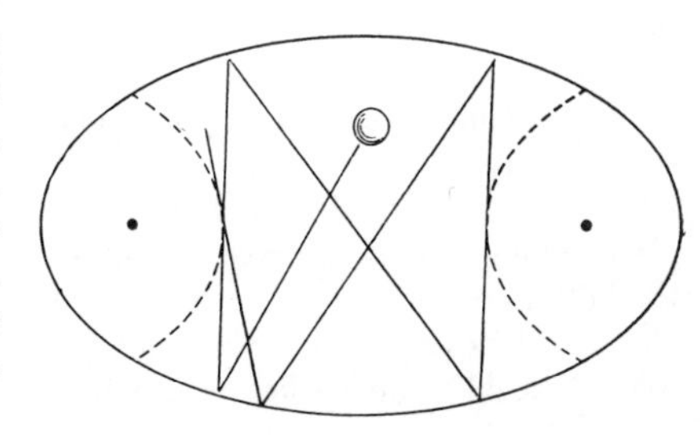

Path of ball that does pass between foci

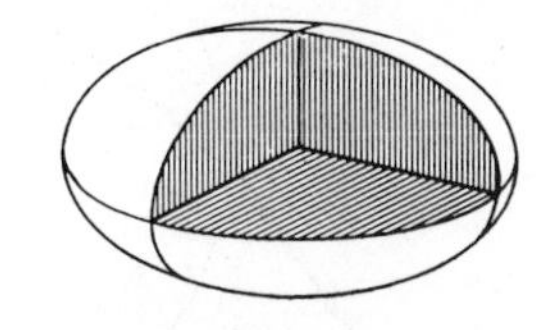

Each section of ellipsoid is elliptical

18

Curious Properties of a Cycloid Curve

by Martin Gardner
July 1964

Do the tops of the tires on a moving car go faster than the bottoms? This odd question will start as many ferocious parlor debates as the old problem about the man who walks around a tree trying to see a squirrel on the opposite side of the trunk. As he walks, the squirrel scurries around the tree, keeping its belly against the trunk so that it always faces the man but with the trunk constantly hiding it from view. When the man has circled the tree, has he also gone around the squirrel?

William James, considering this weighty metaphysical problem in the second chapter of his book *Pragmatism*, concludes that it all depends on what one means by "around." Similarly, the tire question cannot be answered without prior agreement as to precisely what all the words mean. Let us say that by "top" and "bottom" of the tire we mean those points on the tire that are at any given moment close to the top or bottom, and that by "go faster" we refer to the horizontal velocity of those points in relation to the ground. Surprising as it may seem, points near the top do move faster than points near the bottom.

This can be demonstrated by a simple experiment with a coffee can. Cover the bottom of the can with white paper. Using a dark crayon, draw about eight diameters, like the spokes of a wheel, on the circular sheet. Place the can on its side and roll it back and forth past your line of vision. Do *not* follow the can with your eyes; keep your gaze fixed on a distant object so that your eyes do not move as the can rolls by. You will find that the black spokes are visible only in the lower half of the wheel. The upper half is a gray blur. The reason is that the spokes in the upper half are actually moving past your eyes at a much faster rate than the spokes in the lower half. This was such a familiar phenomenon in horse-and-buggy days that artists often indicated the motion of wheels by showing distinct spokes only below the axles.

The illustration below traces the motion of a point on the circumference of a circle as it rolls without slipping along a horizontal line for a distance *AB* that is equal to the circumference of the circle. The position of the circle is shown after each quarter-turn. Assume that the circle rolls with uniform speed. It is easy to see that the point is motionless for an instant on the ground at *A*, gradually increases in speed, reaches its maximum at the highest spot and then accelerates negatively until it touches ground again at *B*. If the wheel continues to roll, the point will trace a series of arches, coming to rest for an instant at the bottom of each cusp. The velocity of the point along the curve conforms to what physicists call a simple harmonic motion. On wheels that have flanges, such as the wheels of a train, points on the flange actually move *backward* while they execute a tiny loop below the level of the track.

The generic name for a curve traced by a point on any type of curve when it rolls without slipping along any other type of curve is "roulette." In this case a circle rolls on a straight line to generate one of the simplest of roulettes, the cycloid. It has been called the "Helen of geometry," not only because of its beautiful properties but also because it has been the object of so many historic quarrels between eminent mathematicians.

No one knows who first recognized the cycloid as a curve worth studying. There is no mention of it before 1500. The first important treatise on the curve was written in 1644 by the Italian physicist Evangelista Torricelli, a student of Galileo's. Fourteen years later Blaise Pascal, who had abandoned mathematics for a life of religious contemplation, found himself suffering from a terrible toothache. To take his mind off the pain he began thinking about the cycloid. The pain stopped. Regarding this as a sign that God was not displeased with his thoughts, Pascal spent the next eight days in furious research on the curve. His remarkable results were issued first as a series of challenges to other mathematicians and then as a treatise on the

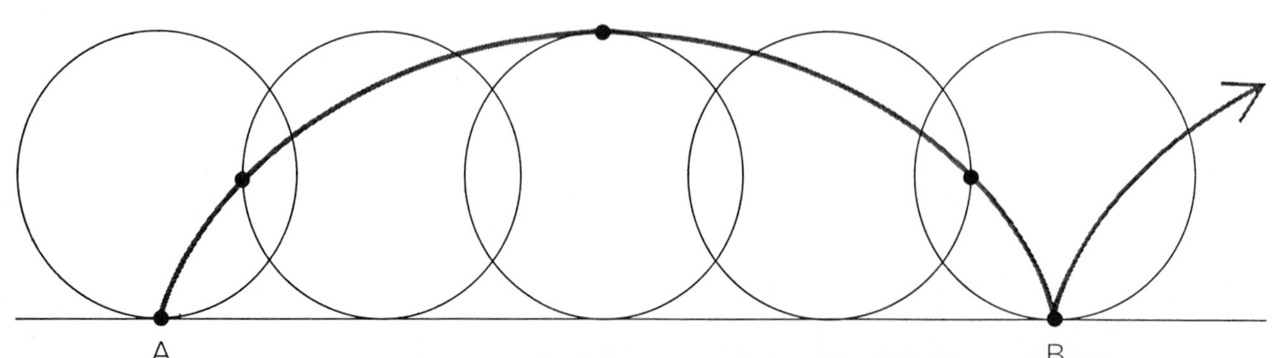

How a cycloid is generated by a point on a rolling circle

cycloid.

One of the simplest questions to ask about the cycloid—although by no means the easiest to answer—is: How long is it? Assume that the generating circle has a diameter of 1. The base line *AB* will, of course, be pi, an irrational number. Everyone expected the length of the curve to be irrational also. Sir Christopher Wren, the distinguished English architect, apparently was the first to show (in 1658) that the length of the cycloidal arch, from cusp to cusp, is precisely four times the diameter of the circle.

The area below the arch had been measured previously and it too had been a surprise. Galileo had guessed the area to be pi times the area of the generating circle, an estimate obtained by the direct method of cutting the arch from thin material and comparing its weight with that of the circle cut from the same material. Torricelli astounded his colleagues in Italy by proving that the area under the arch is exactly three times the area of the circle. Actually this had been shown earlier by the French mathematician Gilles Personne de Roberval, but Torricelli did not know it. In France, René Descartes insisted that the entire problem was trivial. He worked out a simpler way to find the area and challenged Roberval to construct tangents to the cycloid. This led to a long, bitter dispute between the two men. Today all these problems are solved in first-year calculus classes (where the curve is called the "student's curve" because the answers are so simple), but in the 17th century calculus was still primitive.

The mechanical properties of the cycloid are as remarkable as its geometric ones. In high school physics one learns that the time it takes a pendulum to swing back and forth is the same regardless of how wide the swing is, but this is only approximate. When the swings are wide, there are slight deviations. In what path should a pendulum swing so that its period is exactly the same regardless of amplitude? Such a curve, called an isochrone, was first discovered by the Dutch physicist Christian Huygens, who published his discovery in 1673. If we turn two cycloidal arches upside down, as shown in the top illustration at the right, and let a pendulum on a cord swing between them, the pendulum will trace what is called the involute of the cycloid. It turns out that the involute is another cycloid of the same size, and that the cycloidal pendulum is isochronal.

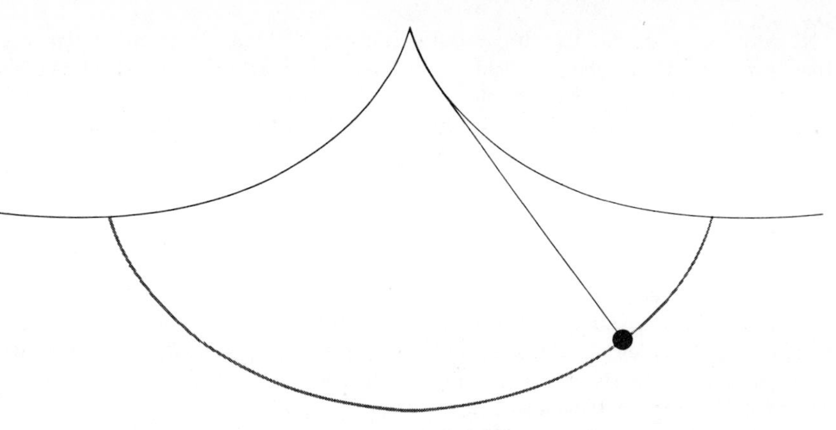

Isochronal pendulum between cycloidal cheeks traces a cycloid

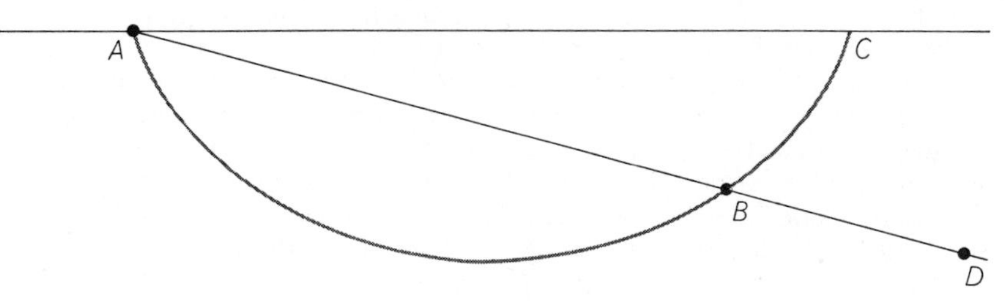

Constructing the curve of quickest descent between A *and* B

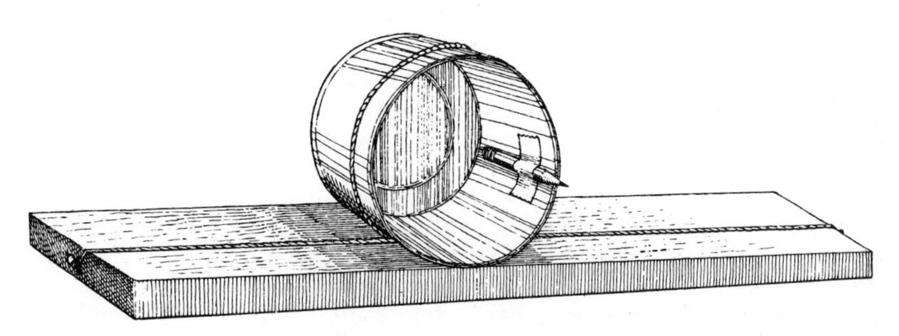

Coffee-can device for drawing a cycloid

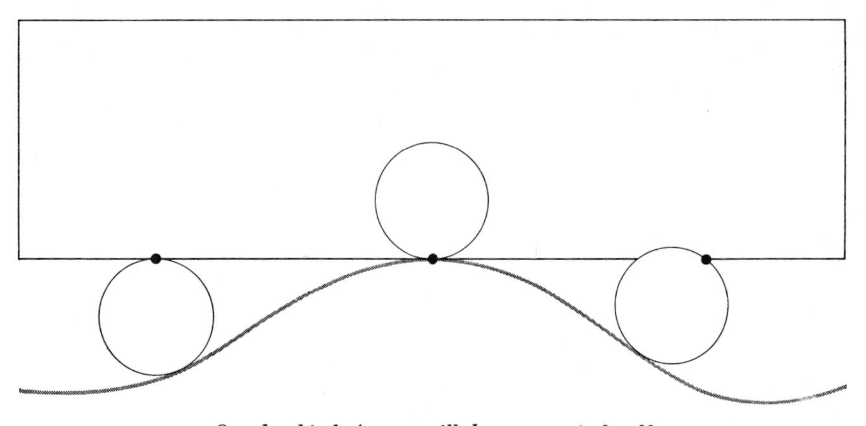

On what kind of curve will the car remain level?

For small swings a circular arc is so nearly the same as the central portion of a cycloid that the circular pendulum is almost isochronal, but if the swings vary even a small amount, the "circular error" is cumulative. For example, if a seconds pendulum has a circular arc of two degrees, an increase to three degrees will cause it to lose about .66 second per day. Huygens constructed a pendulum clock—the first ever made—using a flexible pendulum that swung between two cycloidal cheeks. Unfortunately friction on the cheeks produced a greater error than the cycloidal path corrected; clockmakers found it more practical to arrange things so that a circular pendulum would keep a constant amplitude.

It was Huygens who also discovered that the cycloid is the tautochrone, or curve of equal descent. Imagine a marble rolling without friction down an inverted cycloid. No matter where you start it on the curve, it will reach the bottom in the same length of time. Consider a bowl with sides that curve in such a way that any cross section through the center of the bowl will be a cycloid. Marbles placed at various heights on the sides of the bowl and released simultaneously will reach the center of the bowl at the same instant. Each marble moves with a simple harmonic motion, as does the isochronal pendulum.

The brachistochrone, or curve of *quickest* descent, was not discovered for another score of years. Suppose you are given two points: *A* and *B*. *B* is lower than *A* but not directly below it. The problem is to find a curve connecting *A* and *B* such that a marble, rolling without friction, will travel from *A* to *B* in the shortest possible length of time. This problem was first posed in 1696 by Johann Bernoulli, the Swiss mathematician and physicist, in *Acta Eruditorum,* a famous scientific journal of the day. It was first solved by Johann's brother Jakob (with whom Johann was feuding), but it was also solved by Johann, Leibniz, Newton and others. Newton solved it, along with a related problem, in 12 hours. (The problem reached him at 4:00 P.M.; he had the solution by 4:00 A.M. and sent it off in the morning.) The brachistochrone turned out to be, as the reader has no doubt guessed, the cycloid. Johann Bernoulli's proof has become a classic of nonrigorous, intuitive reasoning. He found the problem equivalent to one concerning the path of a light ray refracted by transparent layers of steadily decreasing density. The interested reader will find his elegant proof clearly explained in *What Is Mathematics?* by Richard Courant and Herbert Robbins (Oxford University Press, 1941), as well as in Ernst Mach's earlier work, *Science of Mechanics* (Open Court Publishing Company, 1893).

Suppose we are given two points, *A* and *B* [*see second illustration from top on preceding page*], and we wish to find the brachistochrone that connects them. What we must find first is the radius of the circle that, when rolled against line *AC*, will generate a cycloid starting at *A* and passing through *B*. To do this we place a circle of any size whatever under *AC* and mark a point on its circumference at *A*. The circle is rolled along *AC* until this point crosses *AB*. Assume that it crosses at *D*. Since all cycloids have similar shapes, we know that *AD* is to *AB* as the radius of the large circle we have just used is to the radius of the smaller circle we seek. This smaller circle, rolled along *AC*, will generate a cycloid from *A* to *B*.

Note that in this case the marble actually rolls *uphill* to reach *B*. Nevertheless, it reaches *B* in a shorter time than it would by rolling along a straight line, the arc of a circle or any other curve. Even when *A* and *B* are on the same horizontal level, a frictionless marble rolls from *A* to *B* in the shortest possible time. (On a straight horizontal line, of course, it would not roll at all.)

An industrious reader should have little difficulty constructing a model for demonstrating the brachistochrone. To draw a large cycloid the coffee can mentioned earlier can be used. A piece of string looped once around it and fastened to the ends of a plank will keep the can from slipping as it rolls along the plank [*see third illustration from top on preceding page*]. A black crayon is taped to the inside of the can so that when the can is rolled along a wall the crayon will trace a cycloid on a sheet of paper fastened to the wall. Using this trace as a pattern, one can bend stiff wire into a cycloid down which a heavy nut will slide or a double cycloidal track down which a marble will roll. The track can also be formed by the cut edges of two rectangular sheets of plywood or heavy cardboard, mounted vertically, with small strips of wood glued between them to keep the edges separated just enough to carry the marble. Similar tracks should be made to carry a second marble down a circular arc and a third marble down a straight line. The three tracks are placed side by side so that the marbles can be released simultaneously by a pencil held horizontally. (Steel balls can be held by electromagnets and released by pushing a button.) If the three tracks lead into one horizontal track, three differently colored marbles will invariably enter the single track in the same order: the cycloid marble will lead, followed by the marble traveling on the circular arc and then by the one on the straight line.

The cycloid has other mechanical properties of interest. It is, as Galileo guessed, the strongest possible arch for a bridge, and for this reason many concrete viaducts have cycloidal arches. Cogwheels are often cut with cycloidal sides to reduce friction by providing a rolling contact as the gears mesh.

We have seen how a circle, rolled on a straight line, generates a cycloid. Stanley C. Ogilvy reverses this situation in one of his books by asking: Along what kind of curve can a circle be rolled so that a point on its circumference traces a straight line? To dramatize this question, imagine a railroad car with each wheel attached at its rim to the axle, as shown in the bottom illustration on the preceding page. How shall we curve a track so that when this curious car is rolled along the track it will remain level at all times and never bob up and down?

Curves of Constant Width, One of Which Makes It Possible to Drill Square Holes

by Martin Gardner
February 1963

If an enormously heavy object has to be moved from one spot to another, it may not be practical to move it on wheels. Axles might buckle or snap under the load. Instead the object is placed on a flat platform that in turn rests on cylindrical rollers. As the platform is pushed forward, the rollers left behind are picked up and put down again in front.

An object moved in this manner over a flat, horizontal surface obviously does not bob up and down as it rolls along. The reason is simply that the cylindrical rollers have a circular cross section, and a circle is a closed curve possessing what mathematicians call "constant width." If a closed convex curve is placed between two parallel lines and the lines are moved together until they touch the curve, the distance between the parallel lines is the curve's "width" in one direction. An ellipse clearly does not have the same width in all directions. A platform riding on elliptical rollers would wobble up and down as it rolled over them. Because a circle has the same width in all directions, it can be rotated between two parallel lines without altering the distance between the lines.

Is the circle the only closed curve of constant width? Most people would say yes, thus providing a sterling example of how far one's mathematical intuition can go astray. Actually there is an infinity of such curves. Any one of them can be the cross section of a roller that will roll a platform as smoothly as a circular cylinder! The failure to recognize such curves can have and has had disastrous consequences in industry. To give one example, it might be thought that the cylindrical hull of a half-built submarine could be tested for circularity by just measuring maximum widths in all directions. As will soon be made clear, such a hull can be monstrously lopsided and still pass such a test. It is precisely for this reason that the circularity of a submarine hull is always tested by applying curved templates.

The simplest noncircular curve of constant width has been named the Reuleaux triangle after Franz Reuleaux (1829–1905), an engineer and mathematician who taught at the Royal Technical High School in Berlin. The curve itself was known to earlier mathematicians, but Reuleaux was the first to demonstrate its constant-width properties. It is easy to construct. First draw an equilateral triangle, *ABC* [*see upper illustration at right*]. With the point of a compass at *A*, draw an arc, *BC*. In a similar manner draw the other two arcs. It is obvious that the "curved triangle" (as Reuleaux called it) must have a constant width equal to the side of the interior triangle.

If a curve of constant width is bounded by two pairs of parallel lines at right angles to each other, the bounding lines necessarily form a square. Like the circle or any other curve of constant width, the Reuleaux triangle will rotate snugly within a square, maintaining contact at all times with all four sides of the square [*see the lower illustration at right*]. If the reader cuts a Reuleaux triangle out of cardboard and rotates it inside a square hole of the proper dimensions cut in another piece of cardboard, he will see that this is indeed the case.

As the Reuleaux triangle turns within a square, each corner traces a path that is almost a square; the only deviation is at the corners, where there is a slight rounding. The Reuleaux triangle has many mechanical uses, but none is so bizarre as the use that derives from this property. In 1914 Harry James Watts, an English engineer then living in Turtle Creek, Pa., invented a rotary drill based on the Reuleaux triangle and capable of drilling square holes! Since 1916 these curious drills have been manufactured by the Watts Brothers Tool Works in Wilmerding, Pa. "We have all heard about left-handed monkey wrenches, fur-lined bathtubs, cast-iron bananas," reads one of their descriptive leaflets. "We have all classed these things with the ridiculous and refused to believe that anything like that could ever happen, and right then along comes a tool that drills square holes."

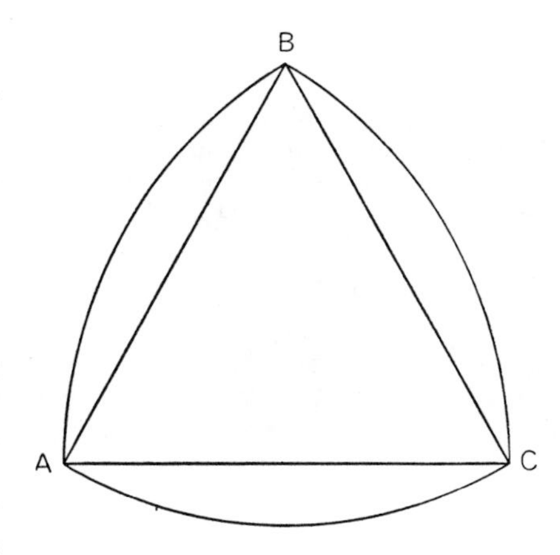

Construction of Reuleaux triangle

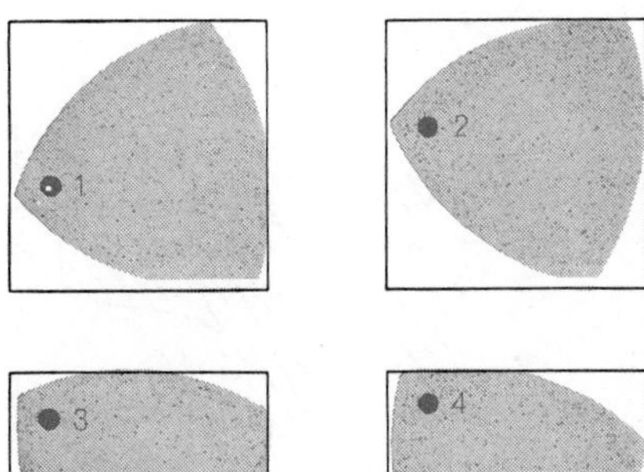

Reuleaux triangle rotating in square

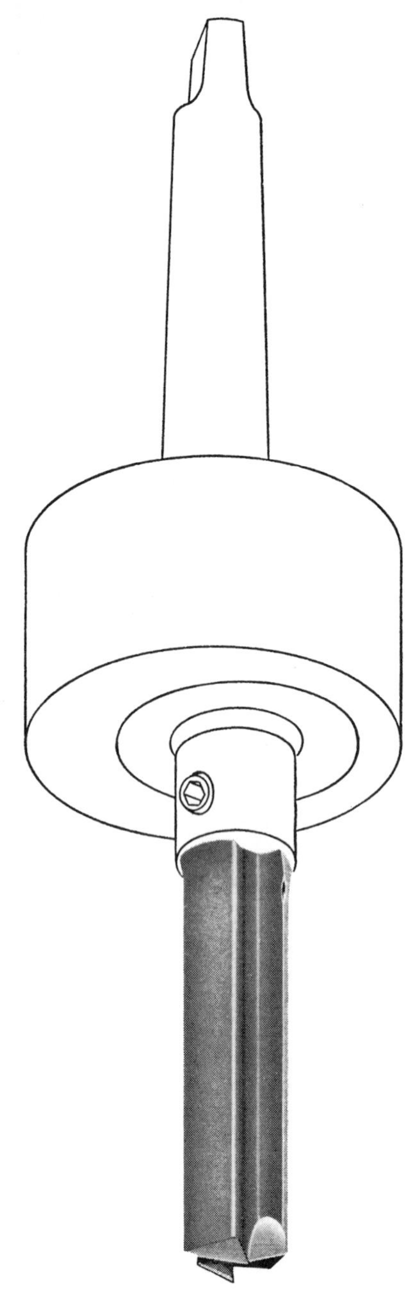

Watts chuck and drill

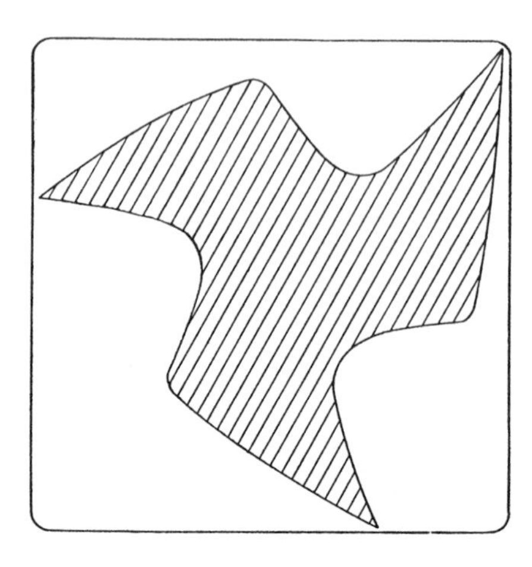

Cross section of drill in hole

The Watts square-hole drill is shown at the left. Below it is a cross section of the drill as it rotates inside the hole it is boring. A metal guide plate with a square opening is first placed over the material to be drilled. As the drill spins within the guide plate, the corners of the drill cut the square hole through the material. As you can see, the drill is simply a Reuleaux triangle made concave in three spots to provide for cutting edges and outlets for shavings. Because the center of the drill wobbles as the drill turns, it is necessary to allow for this eccentric motion in the chuck that holds the drill. A patented "full floating chuck," as the company calls it, does the trick. (Readers who would like more information on the drill and the chuck can check U.S. patents 1,241,175; 1,241,176; and 1,241,177; all dated September 25, 1917.)

The Reuleaux triangle is the curve of constant width that has the smallest area for a given width (the area is $\frac{1}{2}(\pi - \sqrt{3})\,w^2$, where w is the width). The corners are angles of 120 degrees, the sharpest possible on such a curve. These corners can be rounded off by extending each side of an equilateral triangle a uniform distance at each end [*see top illustration on page 109*]. With the point of a compass at *A* draw arc *DI*; then widen the compass and draw arc *FG*. Do the same at the other corners. The resulting curve has a width, in all directions, that is the sum of the same two radii. This of course makes it a curve of constant width. Other symmetrical curves of constant width result if you start with a regular pentagon (or any regular polygon with an odd number of sides) and follow similar procedures.

There are ways to draw unsymmetrical curves of constant width. One method is to start with an irregular star polygon (it will necessarily have an odd number of points) such as the seven-point star shown in black in the bottom illustration on page 109. All of these line segments must be the same length. Place the compass point at each corner of the star and connect the two opposite corners with an arc. Because these arcs all have the same radius, the resulting curve (shown in color) will have constant width. Its corners can be rounded off by the method used before. Extend the sides of the star a uniform distance at all points (shown with broken lines) and then join the ends of the extended sides by arcs drawn with the compass point at each corner of the star. The rounded-corner curve, which is shown in black, will be another curve of constant width.

The illustration at the top left on page 110 demonstrates another method. Draw as many straight lines as you please, all mutually intersecting. Each arc is drawn with the compass point at the intersection of the two lines that bound the arc. Start with any arc, then proceed around the curve, connecting each arc to the preceding one. If you do it carefully, the curve will close and will have constant width. (Proving that the curve must close and have constant width is an interesting and not difficult exercise.) The preceding curves were made up of arcs of no more than two different circles, but curves drawn in this way may have arcs of as many different circles as you wish.

A curve of constant width need not consist of circular arcs. In fact, you can draw a highly arbitrary convex curve from the top to the bottom of a square and touching its left side [*arc* ABC *in the illustration at top right on page 110*], and this curve will be the left side of a uniquely determined curve of constant width. To find the missing part, rule a large number of lines, each parallel to a tangent of arc *ABC* and separated from the tangent by a distance equal to the side of the square. This can be done

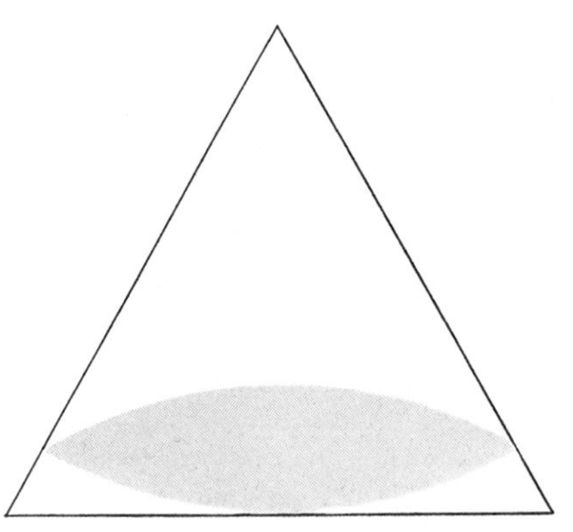

Least-area rotor in equilateral triangle

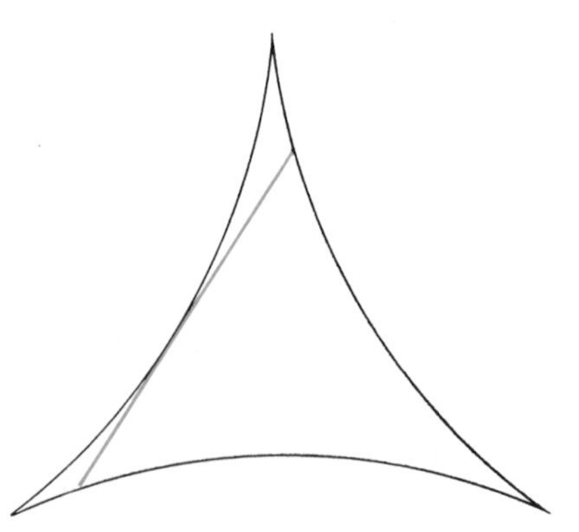

Line rotated in deltoid curve

quickly by using both sides of a ruler. The original square must have a side equal to the ruler's width. Place one edge of the ruler so that it is tangent to arc *ABC* at one of its points, then use the ruler's opposite edge to draw a parallel line. Do this at many points, from one end of arc *ABC* to the other. The missing part of the curve is the envelope of these lines. In this way you can obtain rough outlines of an endless variety of lopsided curves of constant width.

It should be mentioned that the arc *ABC* cannot be completely arbitrary. Roughly speaking, its curvature must not at any point be less than the curvature of a circle with a radius equal to the side of the square. It cannot, for example, include straight line segments. For a more precise statement on this, as well as detailed proofs of many elementary theorems involving curves of constant width, the reader is referred to the excellent chapter on such curves in *The Enjoyment of Mathematics*, by Hans Rademacher and Otto Toeplitz.

If you have the tools and skills for woodworking, you might enjoy making a number of wooden rollers with cross sections that are various curves of the same constant width. Most people are nonplused by the sight of a large book rolling horizontally across such lopsided rollers without bobbing up and down. A simpler way to demonstrate such curves is to cut from cardboard two curves of constant width and nail them to opposite ends of a wooden rod about six inches long. The curves need not be of the same shape, and it does not matter exactly where you put each nail as long as it is fairly close to what you guess to be the curve's "center." Hold a large, light weight empty box by its ends, rest it horizontally on the attached curves and roll the box back and forth. The rod wobbles up and down at both ends, but the box rides as smoothly as it would on circular rollers!

The properties of curves of constant width have been extensively investigated. One startling property, not easy to prove. is that the perimeters of all curves with constant width n have the same length. Since a circle is such a curve, the perimeter of any curve of constant width n must of course be πn, the same as the circumference of a circle with diameter n.

The three-dimensional analogue of a curve of constant width is the solid of constant width. A sphere is not the only such solid that will rotate within a cube, at all times touching all six sides of the cube; this property is shared by all solids of constant width. The simplest example

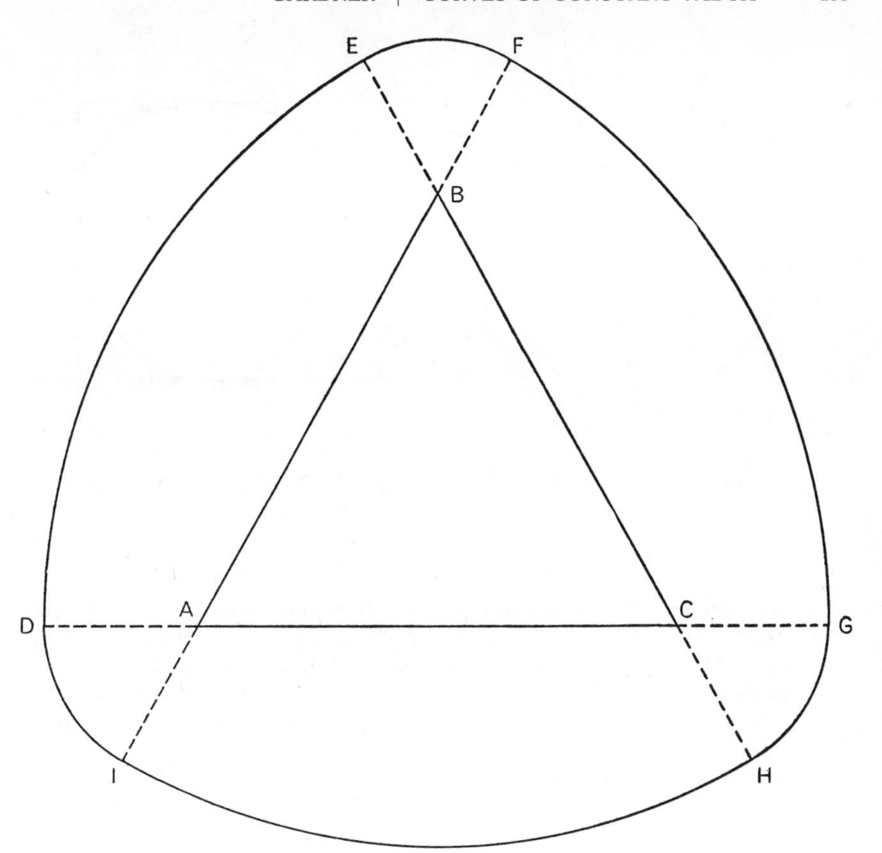

Symmetrical rounded-corner curve of constant width

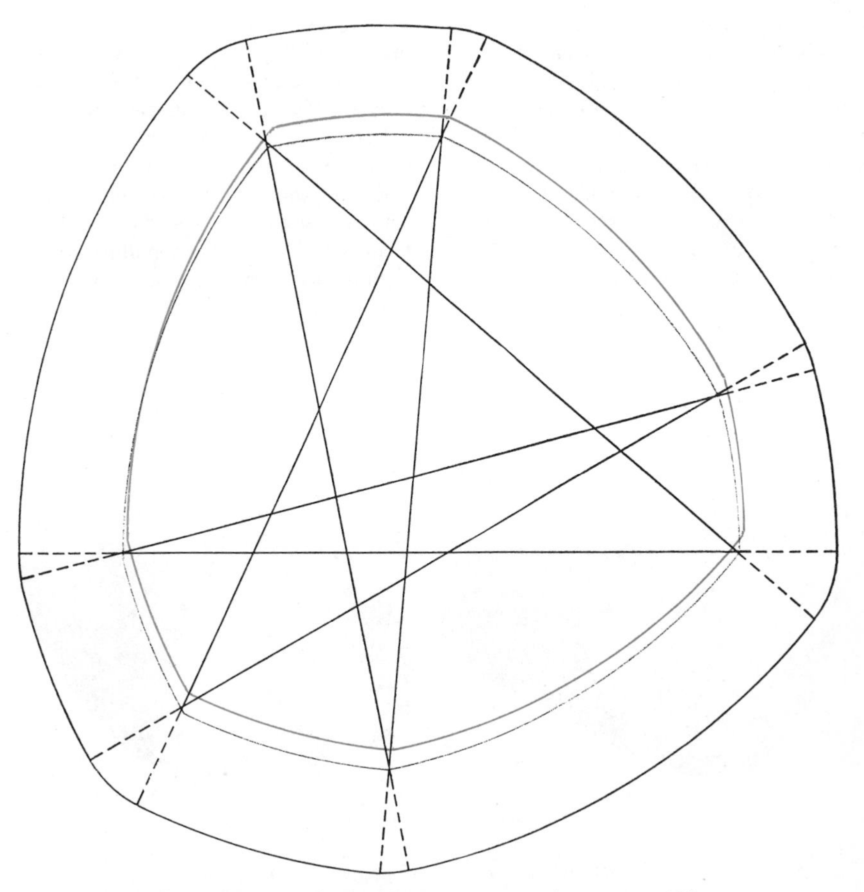
Star-polygon method of drawing a curve of constant width

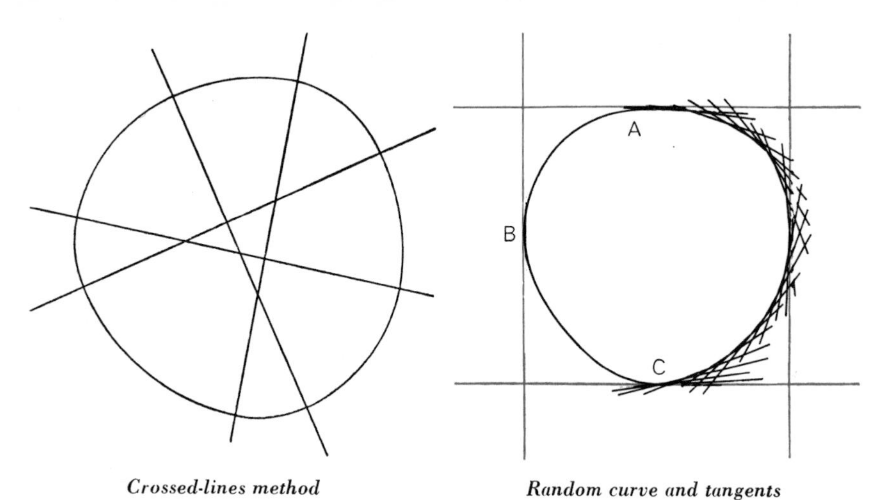

Crossed-lines method

Random curve and tangents

of a nonspherical solid of this type is generated by rotating the Reuleaux triangle around one of its axes of symmetry [*see drawing at left below*]. There is an infinite number of others. The solids of constant width that have the smallest volumes are derived from the regular tetrahedron in somewhat the same way the Reuleaux triangle is derived from the equilateral triangle. Spherical caps are first placed on each face of the tetrahedron, then it is necessary to alter three of the edges slightly. These altered edges may either form a triangle or radiate from one corner. The solid at the right in the illustration below is an example of a curved tetrahedron of constant width.

Since all curves of the same constant width have the same perimeter, it might be supposed that all solids of the same constant width have the same surface area. This is not the case. It was proved, however, by Hermann Minkowski (the Polish mathematician who made such great contributions to relativity theory) that all *shadows* of solids of constant width (when the projecting rays are parallel and the shadow falls on a plane perpendicular to the rays) are curves of the same constant width. All such shadows have equal perimeters (π times the width).

Michael Goldberg, an engineer with the Bureau of Naval Weapons in Washington, has written many papers on curves and solids of constant width, and he is recognized as being this country's leading expert on the subject. He has introduced the term "rotor" for any convex figure that can be rotated inside a polygon while at all times touching every side.

The Reuleaux triangle is, as we have seen, the rotor of least area in a square. The least-area rotor for the equilateral triangle is shown in the top illustration at right on page 108. This lens-shaped figure (it is not, of course, a curve of constant width) is formed with two 60-degree arcs of a circle having a radius equal to the triangle's altitude. Note that as it rotates its corners trace the entire boundary of the triangle, with no rounding of corners. Mechanical reasons make it difficult to rotate a drill based on this figure, but Watts Brothers makes other drills, based on rotors for higher-order regular polygons, that drill sharp-cornered holes in the shape of pentagons, hexagons and even octagons. In three-space, Goldberg has shown, there are nonspherical rotors for the regular tetrahedron and octahedron, as well as the cube, but none for the regular dodecahedron and icosahedron. Almost no work has been done on rotors in dimensions higher than three.

Closely related to the theory of rotors is a famous problem named the Kakeya needle problem after the Japanese mathematician Sôichi Kakeya, who first posed it in 1917. The problem is as follows: What is the plane figure of least area in which a line segment of length 1 can be rotated 360 degrees? The rotation obviously can be made inside a circle of unit diameter, but that is far from the smallest area.

For many years mathematicians believed the answer was the deltoid curve shown in the bottom right illustration on p. 108. (The deltroid is the curve traced by a point on the circumference of a circle as it rolls around the inside of a larger circle, when the diameter of the small circle is either one-third or two-thirds that of the larger one.) If you break a toothpick to the size of the line segment shown, you will find by experiment that it can be rotated inside the deltoid as a kind of one-dimensional rotor. Note how its end points remain at all times on the deltoid's perimeter.

In 1927, 10 years after Kakeya popped his question, the Russian mathematician Abram Samoilovitch Besicovitch (then living in Copenhagen) dropped a bombshell. He proved that the problem had no answer. More accurately, he showed that the answer to Kakeya's question is that there is *no* minimum area. The area can be made as small as one wants. Imagine a line segment that stretches from the earth to the moon. We can rotate it 360 degrees within an area as small as the area of a postage stamp. And if that is too large, we can reduce it to the area of Lincoln's nose on a postage stamp.

Besicovitch' proof is too complicated to give here, but one can get the general idea by studying the illustration on the following page (reproduced from C.

Two solids of constant width

Besicovitch' proof

Stanley Ogilvy's book *Through the Mathescope*). By sliding the line segment back and forth from point to point, it can be turned completely around within the figure, which has an area smaller than the deltoid. By increasing the number of cusps the area can be made as small as desired. Of course, as the area diminishes and the cusps increase we have to work longer in moving the line segment back and forth from cusp to cusp, but we can make the area as small as we please and still do the trick in a finite number of moves.

For readers who would like to work on a much easier problem: What is the smallest *convex* area in which a line segment of length 1 can be rotated 360 degrees? (A convex figure is one in which a straight line, joining any two of its points, lies entirely on the figure. Squares and circles are convex; Greek crosses and crescent moons are not.)

Addendum

Although Watts was the first to acquire patents on the process of drilling square holes with Reuleaux-triangle drills, the procedure was apparently known earlier. Derek Beck, in London, wrote that he had met a man who recalled having used such a drill for boring square holes when he was an apprentice machinist in 1902, and that the practice then seemed to be standard. I have not, however, been able to learn anything about the history of the technique prior to Watt's 1917 patents.

20 Projective Geometry

by Morris Kline
January 1955

Renaissance painters created it to represent three-dimensional reality in two dimensions. Their invention finally transcended Euclidean geometry and today forms an integral part of physics

In the house of mathematics there are many mansions and of these the most elegant is projective geometry. The beauty of its concepts, the logical perfection of its structure and its fundamental role in geometry recommend the subject to every student of mathematics.

Projective geometry had its origins in the work of the Renaissance artists. Medieval painters had been content to express themselves in symbolic terms. They portrayed people and objects in a highly stylized manner, usually on a gold background, as if to emphasize that the subject of the painting, generally religious, had no connection with the real world. An excellent example, regarded by critics as the flower of medieval painting, is Simone Martini's "The Annunciation." With the Renaissance came not only a desire to paint realistically but also a revival of the Greek doctrine that the essence of nature is mathematical law. Renaissance painters struggled for over a hundred years to find a mathematical scheme which would enable them to depict the three-dimensional real world on a two-dimensional canvas. Since many of the Renaissance painters were architects and engineers as well as artists, they eventually succeeded in their objective. To see how well they succeeded one need only compare Leonardo da Vinci's "Last Supper" with Martini's "Annunciation" [*see opposite page*].

The key to three-dimensional representation was found in what is known as the principle of projection and section. The Renaissance painter imagined that a ray of light proceeded from each point in the scene he was painting to one eye. This collection of converging lines he called a projection. He then imagined that his canvas was a glass screen interposed between the scene and the eye. The collection of points where the lines of the projection intersected the glass screen was a "section." To achieve realism the painter had to reproduce on canvas the section that appeared on the glass screen.

Two woodcuts by the German painter Albrecht Dürer illustrate this principle of projection and section [*see below*]. In "The Designer of the Sitting Man" the artist is about to mark on a glass screen a point where one of the light rays from the scene to the artist's eye intersects the screen. The second woodcut, "The Designer of the Lute," shows the section marked out on the glass screen.

Of course the section depends not only

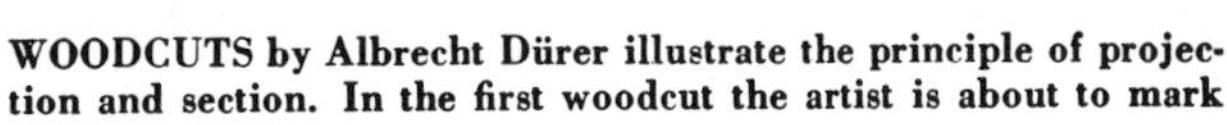

WOODCUTS by Albrecht Dürer illustrate the principle of projection and section. In the first woodcut the artist is about to mark the point at which a light ray from the scene to his eye intersects a glass screen. In the second a scene is marked out on the screen.

upon where the artist stands but also where the glass screen is placed between the eye and the scene. But this just means that there can be many different portrayals of the same scene. What matters is that, when he has chosen his scene, his position and the position of the glass screen, the painter's task is to put on canvas precisely what the section contains. Since the artist's canvas is not transparent and since the scenes he paints sometimes exist only in his imagination, the Renaissance artists had to derive theorems which would specify exactly how a scene would appear on the imaginary glass screen (the location, sizes and shapes of objects) so that it could be put on canvas.

The theorems they deduced raised questions which proved to be momentous for mathematics. Professional mathematicians took over the investigation of these questions and developed a geometry of great generality and power. Let us trace its development.

THE ANNUNCIATION by Simone Martini is an outstanding example of the flat, stylized painting of the medieval artists. The figures were symbolic and framed in a gold background.

Suppose that a square is viewed from a point somewhat to the side [*Figure 1*]. On a glass screen interposed between the eye and the square, a section of its projection is not a square but some other quadrilateral. Thus square floor tiles, for instance, are not drawn square in a painting. A change in the position of the screen changes the shape of the section, but so long as the position of the viewer is kept fixed, the impression created by the section on the eye is the same. Likewise various sections of the projection of a circle viewed from a fixed position differ considerably—they may be more or less flattened ellipses—but the impression created by all these sections on the eye will still be that created by the original circle at that fixed position.

THE LAST SUPPER by Leonardo da Vinci utilized projective geometry to create the illusion of three dimensions. Lines have been drawn on this reproduction to a point at infinity.

To the intellectually curious mathematicians this phenomenon raised a question: Should not the various sections presenting the same impression to the eye have some geometrical properties in common? For that matter, should not sections of an object viewed from different positions also have some properties in common, since they all derive from the same object? In other words, the mathematicians were stimulated to seek geometrical properties common to all sections of the same projection and to sections of two different projections of a given scene. This problem is essentially the one that has been the chief concern of projective geometers in their development of the subject.

DRAWING by da Vinci, made as a study for his painting "The Adoration of the Magi," shows how he painstakingly projected the geometry of the entire scene before he actually painted it.

It is evident that, just as the shape of a square or a circle varies in different sections of the same projection or in different projections of the figure, so also will the length of a line segment, the size of an angle or the size of an area. More than that, lines which are parallel in a physical scene are not parallel in a painting of it but meet in one point; see, for example, the lines of the ceiling beams in da Vinci's "Last Supper." In other words, the study of properties common to the various sections of projections of an object does not seem to lie within the province of ordinary Euclidean geometry.

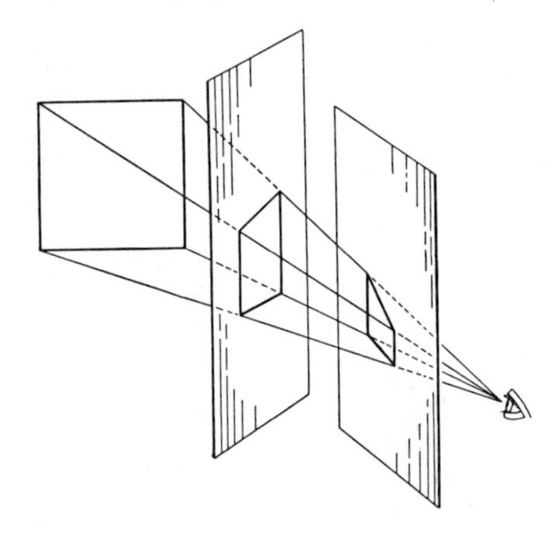

Figure 1 (see text)

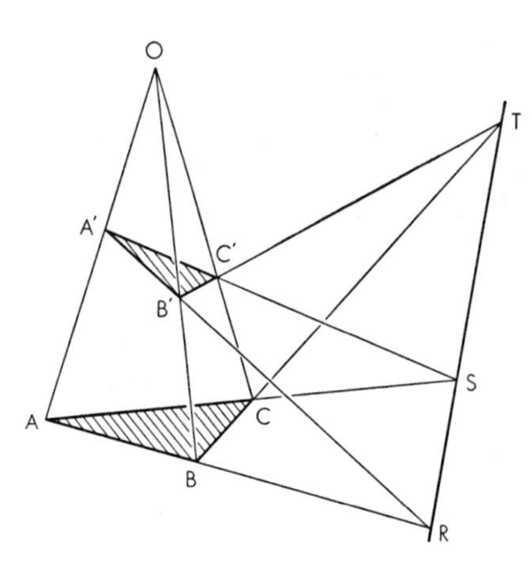

Figure 2

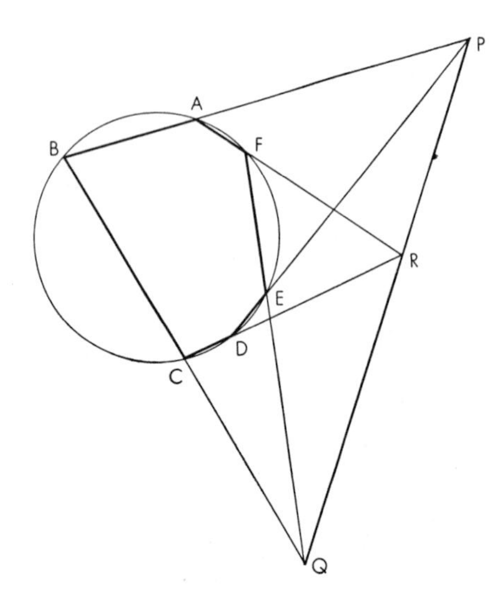

Figure 3

Yet some rather simple properties that do carry over from section to section can at once be discerned. For example, a straight line will remain a line (that is, it will not become a curve) in all sections of all projections of it; a triangle will remain a triangle; a quadrilateral will remain a quadrilateral. This is not only intuitively evident but easily proved by Euclidean geometry. However, the discovery of these few fixed properties hardly elates the finder or adds appreciably to the structure and power of mathematics. Much deeper insight was required to obtain significant properties common to different sections.

The first man to supply such insight was Gérard Desargues, the self-educated architect and engineer who worked during the first half of the 17th century. Desargues's motivation was to help the artists; his interest in art even extended to writing a book on how to teach children to sing well. He sought to combine the many theorems on perspective in a compact form, and he invented a special terminology which he thought would be more comprehensible than the usual language of mathematics.

His chief result, still known as Desargues's theorem and still fundamental in the subject of projective geometry, states a significant property common to two sections of the same projection of a triangle. Desargues considered the situation represented here by two different sections of the projection of a triangle from the point O [*Figure 2*]. The relationship of the two triangles is described by saying that they are perspective from the point O. Desargues then asserted that each pair of corresponding sides of these two triangles will meet in a point, and, most important, these three points will lie on one straight line. With reference to the figure, the assertion is that AB and $A'B'$ meet in the point R; AC and $A'C'$ meet in S; BC and $B'C'$ meet in T; and that R, S and T lie on one straight line. While in the case stated here the two triangle sections are in different planes, Desargues's assertion holds even if triangles ABC and $A'B'C'$ are in the same plane, *e.g.*, the plane of this paper, though the proof of the theorem is different in the latter case.

The reader may be troubled about the assertion in Desargues's theorem that each pair of corresponding sides of the two triangles must meet in a point. He may ask: What about a case in which the sides happen to be parallel? Desargues disposed of such cases by invoking the mathematical convention that any set of parallel lines is to be regarded as having a point in common, which the student is often advised to think of as being at infinity—a bit of advice which essentially amounts to answering a question by not answering it. However, whether or not one can visualize this point at infinity is immaterial. It is logically possible to agree that parallel lines are to be regarded as having a point in common, which point is to be distinct from the usual, finitely located points of the lines considered in Euclidean geometry. In addition, it is agreed in projective geometry that all the intersection points of the different sets of parallel lines in a given plane lie on one line, sometimes called the line at infinity. Hence even if each of the three pairs of corresponding sides of the triangles involved in Desargues's theorem should consist of parallel lines, it would follow from our agreements that the three points of intersection lie on one line, the line at infinity.

These conventions or agreements not only are logically justifiable but also are recommended by the argument that projective geometry is concerned with problems which arise from the phenomenon of vision, and we never actually see parallel lines, as the familiar example of the apparently converging railroad tracks remind us. Indeed, the property of parallelism plays no role in projective geometry.

At the age of 16 the precocious French mathematician and philosopher Blaise Pascal, a contemporary of Desargues, formulated another major theorem in projective geometry. Pascal asserted that if the opposite sides of any hexagon inscribed in a circle are prolonged, the three points at which the extended pairs of lines meet will lie on a straight line [*Figure 3*].

As stated, Pascal's theorem seems to have no bearing on the subject of projection and section. However, let us visualize a projection of the figure involved in Pascal's theorem and then visualize a section of this projection [*Figure 4*]. The projection of the circle is a cone, and in general a section of this cone will not be a circle but an ellipse, a hyperbola, or a parabola—that is, one of the curves usually called a conic section. In any conic section the hexagon in the original circle will give rise to a corresponding hexagon. Now Pascal's theorem asserts that the pairs of opposite sides of the new hexagon will meet on one straight

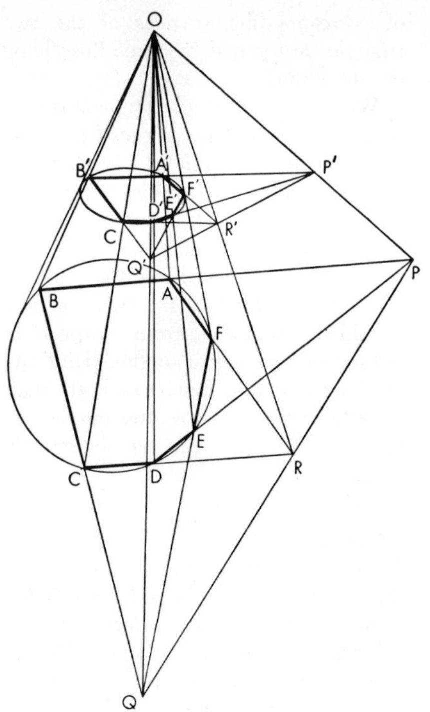

Figure 4

line which corresponds to the line derived from the original figure. Thus the theorem states a property of a circle which continues to hold in any section of any projection of that circle. It is indeed a theorem of projective geometry.

It would be pleasant to relate that the theorems of Desargues and Pascal were immediately appreciated by their fellow mathematicians and that the potentialities in their methods and ideas were eagerly seized upon and further developed. Actually this pleasure is denied us. Perhaps Desargues's novel terminology baffled mathematicians of his day, just as many people today are baffled and repelled by the language of mathematics. At any rate, all of Desargues's colleagues except René Descartes exhibited the usual reaction to radical ideas: they called Desargues crazy and dismissed projective geometry. Desargues himself became discouraged and returned to the practice of architecture and engineering. Every printed copy of Desargues's book, originally published in 1639, was lost. Pascal's work on conics and his other work on projective geometry, published in 1640, also were forgotten. Fortunately a pupil of Desargues, Philippe de la Hire, made a manuscript copy of Desargues's book. In the 19th century this copy was picked up by accident in a bookshop by the geometer Michel Chasles, and thereby the world learned the full extent of Desargues's major work. In the meantime most of Desargues's and Pascal's discoveries had had to be remade independently by 19th-century geometers.

Projective geometry was revived through a series of accidents and events almost as striking as those that had originally given rise to the subject. Gaspard Monge, the inventor of descriptive geometry, which uses projection and section, gathered about him at the Ecole Polytechnique a host of bright pupils, among them Sadi Carnot and Jean Poncelet. These men were greatly impressed by Monge's geometry. Pure geometry had been eclipsed for almost 200 years by the algebraic or analytic geometry of Descartes. They set out to show that purely geometric methods could accomplish more than Descartes's.

It was Poncelet who revived projective geometry. As an officer in Napoleon's army during the invasion of Russia, he was captured and spent the year 1813-14 in a Russian prison. There Poncelet reconstructed, without the aid of any books, all that he had learned from Monge and Carnot, and he then proceeded to create new results in projective geometry. He was perhaps the first mathematician to appreciate fully that this subject was indeed a totally new branch of mathematics. After he had reopened the subject, a whole group of French and, later, German mathematicians went on to develop it intensively.

One of the foundations on which they built was a concept whose importance had not previously been appreciated. Consider a section of the projection of a line divided by four points [*Figure* 5]. Obviously the segments of the line in the section are not equal in length to those of the original line. One might venture that perhaps the ratio of two segments, say $A'C'/B'C'$, would equal the corresponding ratio AC/BC. This conjecture is incorrect. But the surprising fact is that the ratio of the ratios, namely $(A'C'/C'B')/(A'D'/D'B')$, will equal $(AC/CB)/(AD/DB)$. Thus this ratio of ratios, or cross ratio as it is called, is a projective invariant. It is necessary to note only that the lengths involved must be directed lengths; that is, if the direction from A to D is positive, then the length AD is positive but the length DB must be taken as negative.

The fact that any line intersecting the four lines OA, OB, OC and OD contains segments possessing the same cross ratio as the original segments suggests that we assign to the four projection lines meeting in the point O a particular cross ratio, namely the cross ratio of the segments on any section. Moreover, the

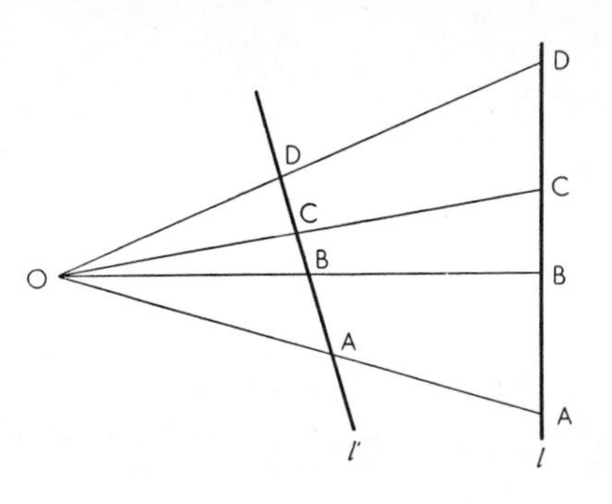

Figure 5

cross ratio of the four lines is a projective invariant, that is, if a projection of these four lines is formed and a section made of this projection, the section will contain four concurrent lines whose cross ratio is the same as that of the original four [*Figure* 6]. Here in the section $O'A'B'C'D'$, formed in the projection of the figure $OABCD$ from the point O'', the four lines $O'A'$, $O'B'$, $O'C'$ and $O'D'$ have the same cross ratio as OA, OB, OC and OD.

The projective invariance of cross ratio was put to extensive use by the 19th-century geometers. We noted earlier in connection with Pascal's theorem that under projection and section a circle may become an ellipse, a hyperbola or a parabola, that is, any one of the conic sections. The geometers sought some common property which would account for the fact that a conic section always gave rise to a conic section, and they found the answer in terms of cross ratio. Given the points O, A, B, C, D, and a sixth point P on a conic section containing the others [*Figure* 7], then a remarkable theorem of projective geometry states that the lines PA, PB, PC and PD have the same cross ratio as OA, OB, OC and OD. Conversely, if P is any point such that PA, PB, PC, and PD have the same

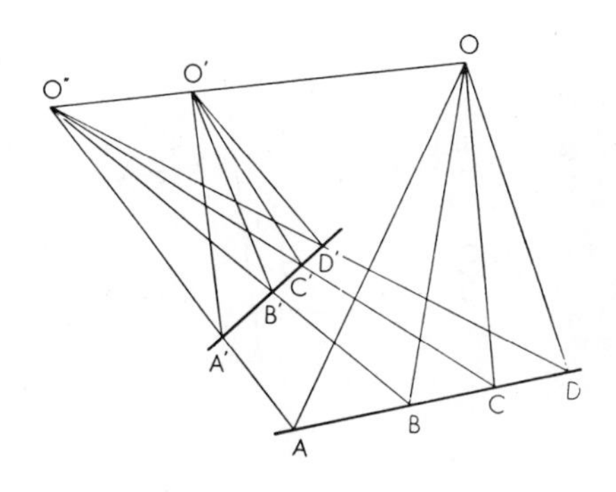

Figure 6

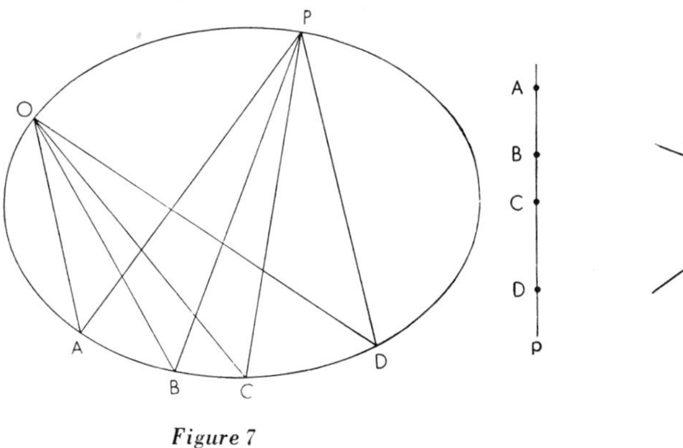

Figure 7

Figure 8

cross ratio as OA, OB, OC and OD, then P must lie on the conic through O, A, B, C and D. The essential point of this theorem and its converse is that a conic section is determined by the property of cross ratio. This new characterization of a conic was most welcome, not only because it utilized a projective property but also because it opened up a whole new line of investigation on the theory of conics.

The satisfying accomplishments of projective geometry were capped by the discovery of one of the most beautiful principles of all mathematics–the principle of duality. It is true in projective geometry, as in Euclidean geometry, that any two points determine one line, or as we prefer to put it, any two points lie on one line. But it is also true in projective geometry that any two lines determine, or lie on, one point. (The reader who has refused to accept the convention that parallel lines in Euclid's sense are also to be regarded as having a point in common will have to forego the next few paragraphs and pay for his stubbornness.) It will be noted that the second statement can be obtained from the first merely by interchanging the words point and line. We say in projective geometry that we have dualized the original statement. Thus we can speak not only of a set of points on a line but also of a set of lines on a point [*Figure 8*]. Likewise the dual of the figure consisting of four points no three of which lie on the same line is a figure of four lines no three of which lie on the same point [*Figure 9*].

Let us attempt this rephrasing for a slightly more complicated figure. A triangle consists of three points not all on the same line and the lines joining these points. The dual statement would read: three lines not all on the same point and the points joining them (that is, the points in which the lines intersect). The figure we get by rephrasing the definition of a triangle is again a triangle, and so the triangle is called self-dual.

Now let us rephrase Desargues's theorem in dual terms, using the fact that the dual of a triangle is a triangle and assuming in this case that the two triangles and the point O lie in one plane. The theorem says:

"If we have two triangles such that lines joining corresponding vertices pass through one point O, then the pairs of corresponding sides of the two triangles join in three points lying on one straight line."

Its dual reads:

"If we have two triangles such that points which are the joins of corresponding sides lie on one line O, then the pairs of corresponding vertices of the two triangles are joined by three lines lying on one point."

We see that the dual statement is really the converse of Desargues's theorem, that is, it is the result of interchanging his hypothesis with his conclusion. Hence by interchanging point and line we have discovered the statement of a new theorem. It would be too much to ask that the proof of the new theorem should be obtainable from the proof of the old one by interchanging point and line. But if it is too much to ask, the gods have been generous beyond our merits, for the new proof can be obtained in precisely this way.

Projective geometry also deals with curves. How should one dualize a statement involving curves? The clue lies in the fact that a curve is after all but a collection of points; we may think of a figure dual to a given curve as a collection of lines. And indeed a collection of lines which satisfies the condition dual to that satisfied by a conic section turns out to be the set of tangents to that curve [*Figure 10*]. If the conic section is a circle, the dual figure is the collection of tangents to the circle [*Figure 11*]. This collection of tangents suggests the circle as well as does the usual collection of points, and we shall call the collection of tangents the line circle.

Let us now dualize Pascal's theorem on the hexagon in a circle. His theorem goes:

"If we take six points, A, B, C, D, E and F, on the point circle, then the lines which join A and B and D and E join in a point P; the lines which join B and C and E and F join in a point Q; the lines which join C and D and F and A join in a point R. The three points P, Q and R lie on one line l."

Its dual reads:

"If we take six lines, a, b, c, d, e and f, on the line circle, then the points

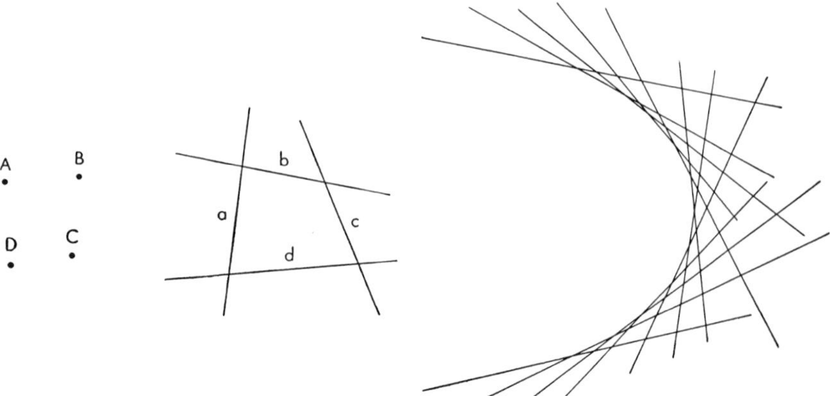

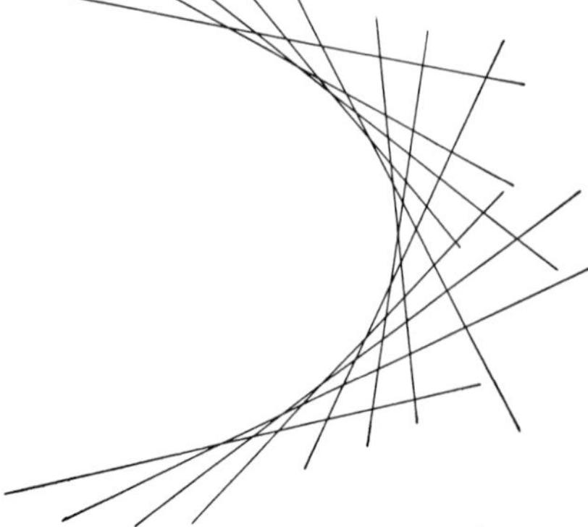

Figure 9

Figure 10

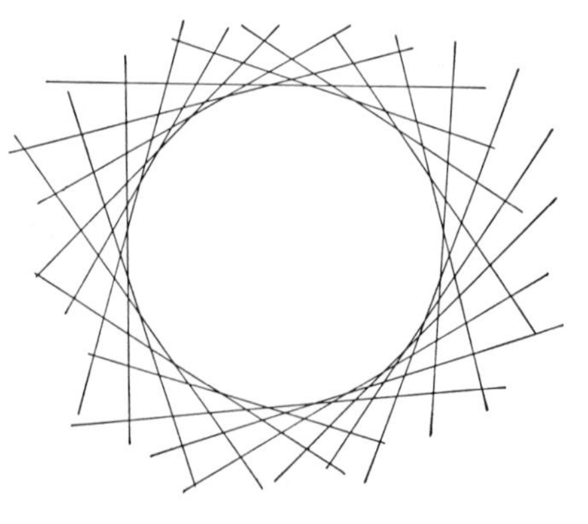

Figure 11

which join a and b and d and e are joined by the line p; the points which join b and c and e and f are joined by the line q; the points which join c and d and f and a are joined by the line r. The three lines p, q and r lie on one point L."

The geometric meaning of the dual statement amounts to this: Since the line circle is the collection of tangents to the point circle, the six lines on the line circle are any six tangents to the point circle, and these six tangents form a hexagon circumscribed about the point circle. Hence the dual statement tells us that if we circumscribe a hexagon about a point circle, the lines joining opposite vertices of the hexagon, lines p, q and r in the dual statement, meet in one point [*Figure 12*]. This dual statement is indeed a theorem of projective geometry. It is called Brianchon's theorem, after Monge's student Charles Brianchon, who discovered it by applying the principle of duality to Pascal's theorem pretty much as we have done.

It is possible to show by a single proof that every rephrasing of a theorem of projective geometry in accordance with the principle of duality must lead to a new theorem. This principle is a remarkable possession of projective geometry. It reveals the symmetry in the roles that point and line play in the structure of that geometry. The principle of duality also gives us insight into the process of creating mathematics. Whereas the discovery of this principle, as well as of theorems such as Desargues's and Pascal's, calls for imagination and genius, the discovery of new theorems by means of the principle is an almost mechanical procedure.

As one might suspect, projective geometry turns out to be more fundamental than Euclidean geometry. The clue to the relationship between the two geometries may be obtained by again considering projection and section. Consider the projection of a rectangle and a section in a plane parallel to the rectangle [*Figure 13*]. The section is a rectangle similar to the original one. If now the point O moves off indefinitely far to the left, the lines of the projection come closer and closer to parallelism with each other. When these lines become parallel and the center of the projection is the "point at infinity," the rectangles become not merely similar but congruent [*Figure 14*]. In other words, from the standpoint of projective geometry the relationships of congruence and similarity, which are so intensively studied in Euclidean geometry, can be studied through projection and section for special projections.

If projective geometry is indeed logically fundamental to Euclidean geometry, then all the concepts of the latter geometry should be defined in terms of projective concepts. However, in projective geometry as described so far there is a logical blemish: our definition of cross ratio, and hence concepts based on cross ratio, rely on the notion of length, which should play no role in projective geometry proper because length is not an invariant under arbitrary projection and section. The 19th-century geometer Felix Klein removed this blemish. He showed how to define length as well as the size of angles entirely in terms of projective concepts. Hence it became possible to affirm that projective geometry was indeed logically prior to Euclidean geometry and that the latter could be built up as a special case. Both Klein and Arthur Cayley even showed that the basic non-Euclidean geometries could be derived as special cases of projective geometry. No wonder that Cayley exclaimed: "Projective geometry is all geometry!"

It remained only to deduce the theorems of Euclidean and non-Euclidean geometry from axioms of projective geometry, and this geometers succeeded in doing in the late 19th and early 20th centuries. What Euclid did to organize the work of three hundred years preceding his time, the projective geometers did recently for the investigations which Desargues and Pascal initiated.

Research in projective geometry is now less active. Geometers are seeking to find simpler axioms and more elegant proofs. Some research is concerned with projective geometry in n-dimensional space. A vast new allied field is projective differential geometry, concerned with local or infinitesimal properties of curves and surfaces.

Projective geometry has had an important bearing on current mathematical research in several other fields. Projection and section amount to what is called in mathematics a transformation, and it seeks invariants under this transformation. Mathematicians asked: Are there other transformations more general than projection and section whose invariants might be studied? In recent times one new geometry has been developed by pursuing this line of thought, namely, topology. It would take us too far afield to consider topological transformations. It must suffice here to state that topology considers transformations more general than projection and section and that it is now clear that topology is logically prior to projective geometry. Cayley was too hasty in affirming that projective geometry is all geometry.

The work of the projective geometers has had an important influence on modern physical science. They prepared the way for the workers in the theory of relativity, who sought laws of the universe that were invariant under transformation from the coordinate system of one observer to that of another. It was the projective geometers and other mathematicians who invented the calculus of tensors, which proved to be the most convenient means for expressing invariant scientific laws.

It is of course true that the algebra of differential equations and some other branches of mathematics have contributed more to the advancement of science than has projective geometry. But no branch of mathematics competes with projective geometry in originality of ideas, coordination of intuition in discovery and rigor in proof, purity of thought, logical finish, elegance of proofs and comprehensiveness of concepts. The science born of art proved to be an art.

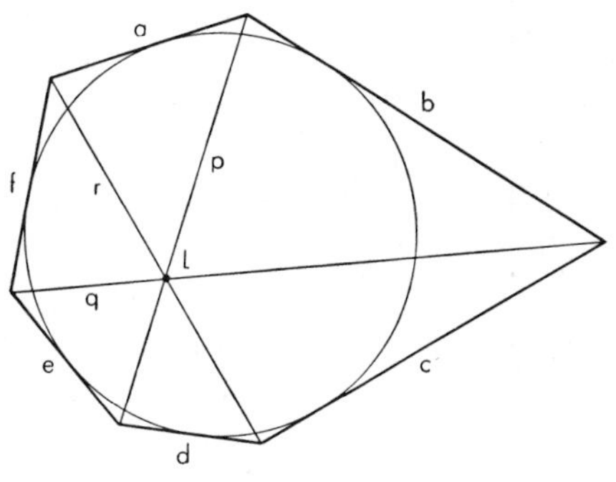

Figure 12

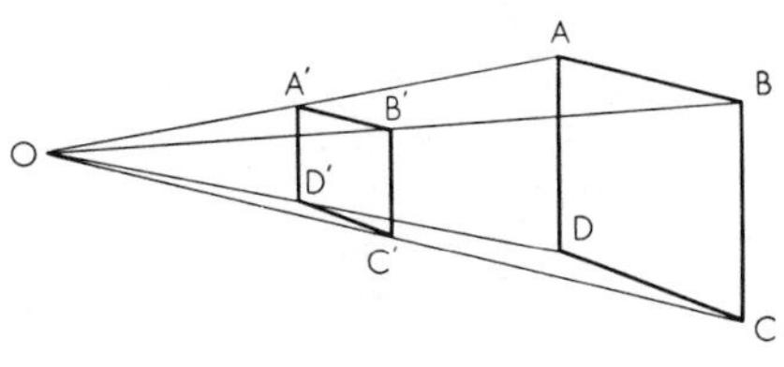

Figure 13

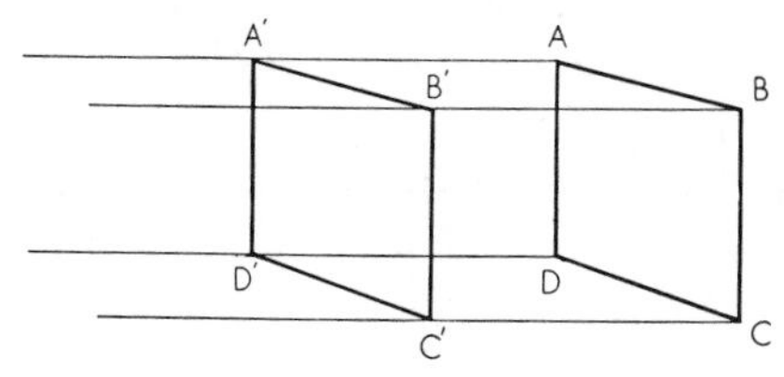

Figure 14

21

Geometric Fallacies: Hidden Errors Pave the Road to Absurd Conclusions

by Martin Gardner
April 1971

"Holmes," I cried, "this is impossible."
"Admirable!" he said. "A most illuminating remark. It is impossible as I state it, and therefore I must in some respect have stated it wrong. . . ."

—SIR ARTHUR CONAN DOYLE,
"The Adventure of the Priory School"

It is commonly supposed that Euclid, the ancient Greek geometer, wrote only one book, his classic *Elements of Geometry*. Actually he wrote at least a dozen, including treatises on music and branches of physics, but only five of his works survived. One of his lost books was a collection of geometric fallacies called *Pseudaria*. Alas, there are no records of what it contained. It probably discussed illicit proofs that led to absurd theorems but in which the errors were not immediately apparent.

Since Euclid's time hundreds of amusing examples of geometric fallacies have been published, some of them genuine mistakes and some deliberately devised. This month we consider five of the best. All are theorems that could have been in Euclid's *Pseudaria*, since none requires more than a knowledge of elementary plane geometry to follow their steps down the garden path to the false conclusion. (Q.E.D.: Quite Entertainingly Deceptive.) The reader is urged to examine each proof carefully, step by step, to see if he can discern exactly where the proof goes wrong before looking at the answers.

Theorem 1: An obtuse angle is sometimes equal to a right angle.

This was one of Lewis Carroll's favorites. The illustration at the left below reproduces Carroll's diagram and labeling. I know of no better way for a high school geometry teacher to convey the importance of deductive rigor than to chalk this diagram on the blackboard and challenge a class to find where the fallacy lies. The construction and proof are described by Carroll as follows (I quote from *The Lewis Carroll Picture Book*, edited by Stuart Dodgson Collingwood, London, 1899; reprinted in the Dover paperback *Diversions and Digressions of Lewis Carroll*, 1961):

Let *ABCD* be a square. Bisect *AB* at *E*, and through *E* draw *EF* at right angles to *AB*, and cutting *DC* at *F*. Then $DF = FC$.

From *C* draw $CG = CB$. Join *AG*, and bisect it at *H*, and from *H* draw *HK* at right angles to *AG*.

Since *AB*, *AG* are not parallel, *EF*, *HK* are not parallel. Therefore they will meet, if produced. Produce *EF*, and let them meet at *K*. Join *KD*, *KA*, *KG*, and *KC*.

The triangles *KAH*, *KGH* are equal, because $AH = HG$, *HK* is common, and the angles at *H* are right. Therefore $KA = KG$.

The triangles *KDF*, *KCF* are equal, because $DF = FC$, *FK* is common, and the angles at *F* are right. Therefore $KD = KC$, and angle $KDC =$ angle KCD.

Also $DA = CB = CG$.

Hence the triangles *KDA*, *KCG* have all their sides equal. Therefore the angles *KDA*, *KCG* are equal. From these equals take the equal angles *KDC*, *KCD*. Therefore the remainders are equal: *i.e.*,

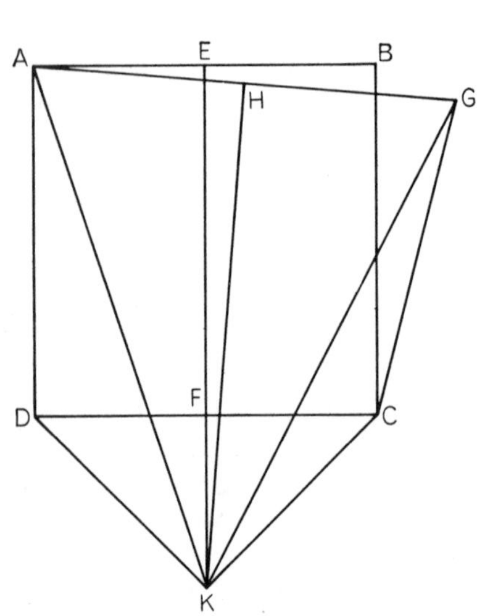

Obtuse angle equals right angle

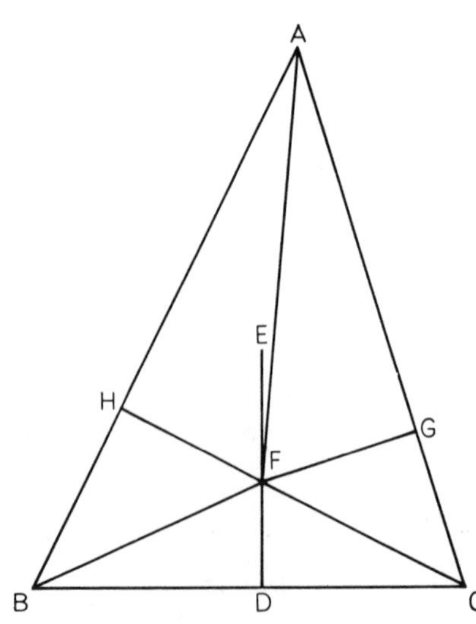

All triangles are isosceles

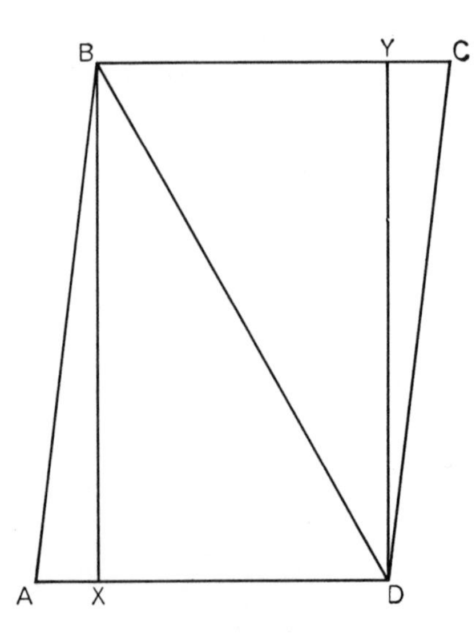

ABCD *is a parallelogram*

the angle GCD = the angle ADC. But GCD is an obtuse angle, and ADC is a right angle.

Therefore an obtuse angle sometimes = a right angle.

Q.E.D.

Theorem 2: Every triangle is isosceles.

This marvelous absurdity is also in *The Lewis Carroll Picture Book.* Carroll probably came on both proofs in the first (1882) edition of W. W. Rouse Ball's *Mathematical Recreations and Essays,* where they appeared for the first time. Carroll has explained it so well that again I give his diagram [*middle illustration, p. 118*] and quote his wording:

Let ABC be any triangle. Bisect BC at D, and from D draw DE at right angles to BC. Bisect the angle BAC.

(1) If the bisector does not meet DE, they are parallel. Therefore the bisector is at right angles to BC. Therefore $AB = AC$, *i.e.*, ABC is isosceles.

(2) If the bisector meets DE, let them meet at F. Join FB, FC, and from F draw FG, FH, at right angles to AC, AB.

Then the triangles AFC, AFH are equal, because they have the side AF common, and the angles FAG, AGF equal to the angles FAH, AHF. Therefore $AH = AG$, and $FH = FG$.

Again, the triangles BDF, CDF are equal, because $BD = DC$, DF is common, and the angles at D are equal. Therefore $FB = FC$.

Again, the triangles FHB, FGC are right-angled. Therefore the square on FB = the squares on FH, HB; and the square on FC = the squares on FG, GC. But $FB = FC$, and $FH = FG$. Therefore the square on HB = the square on GC. Therefore $HB = GC$. Also, AH has been proved = to AG. Therefore $AB = AC$; *i.e.*, ABC is isosceles.

Therefore the triangle ABC is always isosceles.

Q.E.D.

Theorem 3: If a quadrilateral ABCD *has angle* A *equal to angle* C, *and* AB *equals* CD, *the quadrilateral is a parallelogram.*

P. Halsey of London contributed this subtle fallacy to *The Mathematical Gazette,* October, 1959, pages 204–205. On the quadrilateral shown in the illustration at the right on the opposite page draw BX perpendicular to AD, and DY perpendicular to BC. Join BD. Triangles ABX and CYD are congruent, therefore BX equals DY and AX equals CY. It follows that triangles BXD and DYB are congruent, consequently XD equals YB. Since AB equals CD and AD equals BC, the quadrilateral $ABCD$ must be a parallelogram. The proof is strongly convincing, yet the theorem is false. Can the reader provide a counterexample?

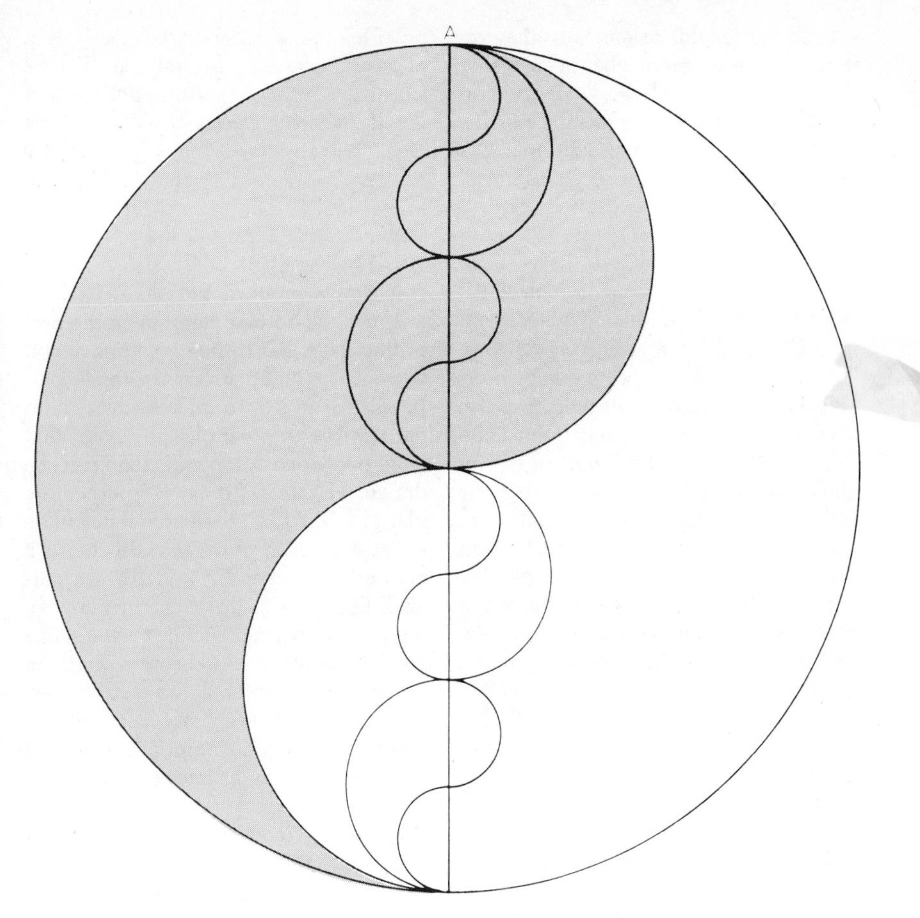

Pi equals 2

Theorem 4: Pi equals 2.

The top illustration on this page is based on the familiar yin-yang symbol of the Orient. Let diameter AB equal 2. Since a circle's circumference is its diameter times pi, the largest semicircle, from A to B, has a length of $2\pi/2 = \pi$. The two next-smallest semicircles, which form the wavy line that divides the yin from the yang, are each equal to $\pi/2$ and so their total length is pi. In similar fashion the sum of the four next-smallest semicircles (each $\pi/4$) also is pi, and the sum of the eight next-smallest semicircles (each $\pi/8$) also is pi. This can be continued endlessly. The semicircles grow smaller and more numerous, but they always add to pi. Clearly the wavy line approaches diameter AB as a limit. Assume that the construction is carried out an infinite number of times. The wavy line must always retain a length of pi, yet when the radii of the semicircles reach their limit of zero, they coincide with diameter AB, which has a length of 2. Consequently pi equals 2.

Theorem 5: Euclid's parallel postulate can be proved by Euclid's other axioms.

First, some historical background. Among Euclid's 10 axioms, his fifth postulate states that if a line A crosses two other lines, making the sum of the interior angles on the same side of A less than 180 degrees, the two lines will intersect on that side of A. A variety of seemingly unrelated theorems can be

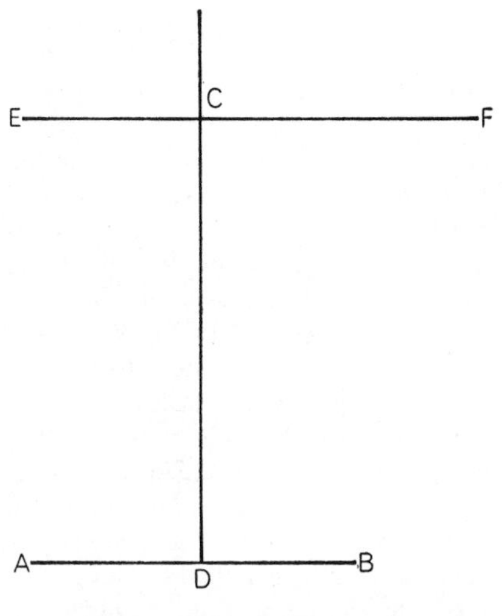

A proof of the parallel postulate

substituted for this axiom since they require it for their proof: the theorem that the interior angles of every triangle add to 180 degrees, or that a rectangle exists, or that similar noncongruent triangles exist, or that through three points not in a straight line only one circle can be drawn, and many others.

Hundreds of attempts have been made since Euclid's time to replace his cumbersome fifth postulate with one that is simpler and more intuitively obvious. The most famous became known as "Playfair's postulate" after the Scottish mathematician and physicist John Playfair. In his popular 1795 edition of Euclid's *Elements* he substituted for the fifth postulate the equivalent but more succinct statement, "Through a given point can be drawn only one line parallel to a given line." Actually this form of the fifth postulate was suggested by Proclus, in a fifth-century Greek commentary on Euclid, as well as by later mathematicians who preceded Playfair, but the parallel postulate still bears Playfair's name.

Whatever form the fifth axiom was given, it always seemed less self-evident than Euclid's other axioms, and some of the greatest mathematicians labored to eliminate it entirely by proving it on the basis of the other nine. (For a good account of this history see W. B. Frankland, *Theories of Parallelism, an Historical Critique*, Cambridge University Press, 1910.) The 18th-century French geometer Joseph Louis Lagrange was convinced that he had produced such a proof by showing (without assuming Euclid's fifth postulate) that the angles of any triangle add to a straight angle. In the middle of the first paragraph of a lecture to the French Academy on his discovery, however, he suddenly said, "Il faut que j'y songe encore" ("I shall have to think it over again"), put his papers in his pocket and abruptly left the hall.

More than a century ago it was established that it is as impossible to prove the fifth postulate as it is to trisect the angle, square the circle or duplicate the cube, yet even in this century "proofs" of the parallel axiom continue to be published. A splendid example is the heart of a 310-page book, *Euclid or Einstein*, privately printed in 1931 by Very Rev. Jeremiah Joseph Callahan, then president of Duquesne University. Since the general theory of relativity assumes the consistency of a non-Euclidean geometry, a simple way to demolish Einstein is to show that non-Euclidean geometry is contradictory. This Father Callahan proceeds to do by a lengthy, ingenious proof of the parallel postulate. It is a pleasant exercise to retrace Father Callahan's reasoning in an effort to find exactly where it goes astray. (For those who give up, the error is exposed by D. R. Ward's "A New Attempt to Prove the Parallel Postulate" in *The Mathematical Gazette*, Vol. 17, pages 101–104, May, 1933.)

A simple proof of the parallel postulate uses the bottom diagram on the preceding page. *AB* is the given line and *C* the outside point. From *C* drop a perpendicular to *AB*. It can be shown, without invoking the parallel postulate, that only one such perpendicular can be drawn. Through *C* draw *EF* perpendicular to *CD*. Again, the parallel postulate is not needed to prove that this too is a unique line. Lines *EF* and *AB* are parallel. Once more, the theorem that two lines, each perpendicular to the same line, are parallel is a theorem that can be established without the parallel postulate, although the proof does require other Euclidean assumptions (such as the one that straight lines are infinite in length) that do not hold in elliptic non-Euclidean geometry. Elliptic geometry does not contain parallel lines, but given Euclid's other assumptions one can assume that parallel lines do exist.

We have apparently now proved the parallel postulate. Or have we?

This and hundreds of other false proofs of Euclid's fifth axiom, or axioms equivalent to it, show how easily intuition can be deceived. It helps one to understand why it took so long for geometers to realize that the parallel postulate was independent of the others, that one may assume either that no parallel line can be drawn through the outside point, or that at least two can. (It turns out that if two can, an infinite number can.) In each case a consistent non-Euclidean geometry is constructible.

Even after non-Euclidean geometries were found to be as free of logical contradiction as Euclidean geometry, many eminent mathematicians and scientists could not believe that non-Euclidean geometry would ever have a useful application to the actual space of the universe. It is well known that Henri Poincaré argued in 1903 that if physicists ever found empirical evidence suggesting that space was non-Euclidean, it would be better to keep Euclidean geometry and change the physical laws. "Euclidean geometry, therefore," he concluded, "has nothing to fear from fresh experiments." Not so well known is the fact that Bertrand Russell and Alfred North Whitehead once voiced the same view. In 1910, in the famous 11th edition of *The Encyclopaedia Britannica*, the article on "Geometry, Non-Euclidean" is by Russell and Whitehead. If scientific observation were ever to conflict with Euclidean geometry, they assert, the simplicity of Euclidean geometry is so overwhelming that it would be preferable "to ascribe this anomaly, not to the falsity of Euclidean geometry [as applied to space], but to the falsity of the laws in question. This applies especially to astronomy."

Six years later Einstein's general theory of relativity made this statement, along with Poincaré's, hopelessly naïve. Not only does non-Euclidean geometry provide a simpler description of the space-time of general relativity; it is even possible that space may close on itself (as it does in Einstein's early model of the universe) to introduce topological properties that are in principle capable of being tested, and that could make the choice of non-Euclidean geometry as the best description of space more than a trivial matter of convention.

Russell was quick to alter the opinion expressed in the *Britannica* article but Whitehead was slow to get the point. In 1922 he wrote an embarrassing book, *The Principle of Relativity*, that attacked Einstein's use of a generalized non-Euclidean geometry (in which curvature varies from spot to spot) by arguing that simplicity demands that the geometry applied to space must be either Euclidean (Whitehead's preference) or, if the evidence warrants it, a non-Euclidean geometry in which the curvature is everywhere constant. Whitehead later had deep misgivings about the book, although his alternative to Einstein still has a few defenders among nonphysicists.

What is the moral of all this? Intuition is a powerful tool in mathematics and science but it cannot always be trusted. The structure of the universe, like pure mathematics itself, has a way of being much stranger than even the greatest mathematicians and physicists suspect.

The Koenigsberg Bridges

22

by Leonhard Euler

Edited by James R. Newman
July 1953

Leonhard Euler, the most eminent of Switzerland's scientists, was a gifted 18th-century mathematician who enriched mathematics in almost every department and whose energy was at least as remarkable as his genius. "Euler calculated without apparent effort, as men breathe, or as eagles sustain themselves in the wind," wrote François Arago, the French astronomer and physicist. It is said that Euler "dashed off memoirs in the half-hour between the first and second calls to dinner." According to the mathematical historian Eric Temple Bell he "would often compose his memoirs with a baby in his lap while the older children played all about him"—the number of Euler's children was 13. At the age of 28 he solved in three days a difficult astronomical problem which astronomers had agreed would take several months of labor; this prodigious feat so overtaxed his eyesight that he lost the sight of one eye and eventually became totally blind. But his handicap in no way diminished either the volume or the quality of his mathematical output. His writings will, it is estimated, fill 60 to 80 large quarto volumes when the edition of his collected works is completed.

The memoir published below is Euler's own account of one of his most famous achievements: his solution of the celebrated problem of the Koenigsberg bridges. The problem is a classic exercise in the branch of mathematics called topology (see "Topology," by Albert W. Tucker and Herbert S. Bailey, Jr.; *Scientific American* January, 1950). Topology is the geometry of distortion; it deals with the properties of an object that survive stretching, twisting, bending or other changes of its size or shape. The Koenigsberg puzzle is a so-called network problem in topology.

In the town of Koenigsberg (where the philosopher Immanuel Kant was born) there were in the 18th century seven bridges which crossed the river Pregel. They connected two islands in the river with each other and with the opposite banks. The townsfolk had long amused themselves with this problem: Is it possible to cross the seven bridges in a continuous walk without recrossing any of them? When the puzzle came to Euler's attention, he recognized that an important scientific principle lay concealed in it. He applied himself to discovering this principle and shortly thereafter presented his simple and ingenious solution. He provided a mathematical demonstration, as some of the townsfolk had already proved to their own satisfaction by repeated trials, that the journey is impossible. He also found a rule which answered the question in general, whatever the number of bridges.

The Koenigsberg puzzle is related to the familiar exercise of trying to trace a given figure on paper without lifting the pencil or retracing a line. In graph form the Koenigsberg pattern is represented by the drawing on the left at the bottom of this page. Inspection shows that this pattern cannot be traced with a single stroke of the pencil. But if there are eight bridges, the pattern is the one at the right, and this one can be traced in a single stroke.

Euler's memoir gives a beautiful explanation of the principles involved and furnishes an admirable example of the deceptive simplicity of topology problems.—James R. Newman

* * *

The branch of geometry that deals with magnitudes has been zealously studied throughout the past, but there is another branch that has been almost unknown up to now; Leibnitz spoke of it first, calling it the "geometry of position" (*geometria situs*). This branch of geometry deals with relations dependent on position alone, and investigates the properties of position; it does not take magnitudes into consideration, nor does it involve calculation with quantities. But as yet no satisfactory definition has been given of the problems that belong to this geometry of position or of the method to be used in solving them. Recently there was announced a problem which, while it certainly seemed to belong to geometry, was nevertheless so designed that it did not call for the determination of a magnitude, nor could it be solved by quantitative calculation; consequently I did not hesitate to assign it to the geometry of position, especially since the solution required only the consideration of position, calculation being of no use. In this paper I shall give an account of the method that I discovered for solving this type of problem, which may serve as an example of the geometry of position.

The problem, which I understand is quite well known, is stated as follows: In the town of Koenigsberg in Prussia there is an island A, called Kneiphof, with the two branches of the river Pregel flowing around it. There are seven bridges—*a*, *b*, *c*, *d*, *e*, *f* and *g*—crossing the two branches [*see illustration at the top of page 123*]. The question is whether a person can plan a walk in such a way that he will cross each

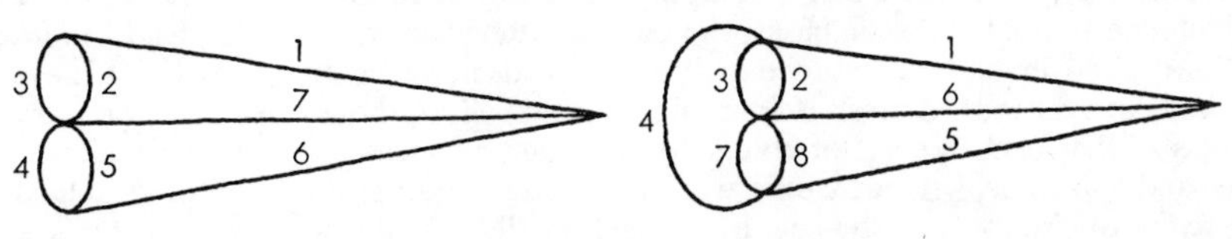

The figure at right can be drawn in one stroke; the one at left cannot

of these bridges once but not more than once. I was told that while some denied the possibility of doing this and others were in doubt, no one maintained that it was actually possible. On the basis of the above I formulated the following very general problem for myself: Given any configuration of the river and the branches into which it may divide, as well as any number of bridges, to determine whether or not it is possible to cross each bridge exactly once.

The particular problem of the seven bridges of Koenigsberg could be solved by carefully tabulating all possible paths, thereby ascertaining by inspection which of them, if any, met the requirement. This method of solution, however, is too tedious and too difficult because of the large number of possible combinations, and in other problems where many more bridges are involved it could not be used at all. . . . Hence I discarded it and searched for another more restricted in its scope; namely, a method which would show only whether a journey satisfying the prescribed condition could in the first instance be discovered; such an approach, I believed, would be simpler.

Leonhard Euler (pronounced oiler); born Basel 1707; died Petrograd 1783

MY ENTIRE method rests on the appropriate and convenient way in which I denote the crossing of bridges, in that I use capital letters, A, B, C, D, to designate the various land areas that are separated from one another by the river. Thus when a person goes from area A to area B across bridge *a* or *b*, I denote this crossing by the letters AB, the first of which designates the area whence he came, the second the area where he arrives after crossing the bridge. If the traveler then crosses from B over bridge *f* into D, this crossing is denoted by the letters BD; the two crossings AB and BD performed in succession I denote simply by the three letters ABD, since the middle letter B designates the area into which the first crossing leads as well as the area out of which the second leads.

Similarly, if the traveler proceeds from D across bridge g into C, I designate the three successive crossings by the four letters ABDC. . . . The crossing of four bridges will be represented by five letters, and if the traveler crosses an arbitrary number of bridges his journey will be described by a number of letters which is one greater than the number of bridges. For example, eight letters are needed to denote the crossing of seven bridges.

With this method I pay no attention to which bridges are used; that is to say, if the crossing from one area to another can be made by way of several bridges it makes no difference which one is used, so long as it leads to the desired area. Thus if a route could be laid out over the seven Koenigsberg bridges so that each bridge were crossed once and only once, we would be able to describe this route by using eight letters, and in this series of letters the combination AB (or BA) would have to occur twice, since there are two bridges, *a* and *b*, connecting the regions A and B. Similarly the combination AC would occur twice, while the combinations AB, BD, and CD would each occur once.

Our question is now reduced to whether from the four letters A, B, C and D a series of eight letters can be formed in which all the combinations just mentioned occur the required number of times. Before making the effort, however, of trying to find such an arrangement we do well to consider whether its existence is even theoretically possible or not. For if it could be shown that such an arrangement is in fact impossible, then the effort expended on finding it would be wasted. Therefore I have sought for a rule that would determine without difficulty, as regards this and all similar questions, whether the required arrangement of letters is feasible.

For the purpose of finding such a rule I take a single region A into which an arbitrary number of bridges, *a*, *b*, *c*, *d*, etc., lead [*middle illustration on the next page*]. Of these bridges I first consider only *a*. If the traveler crosses this bridge, he must either have been in A before crossing or have reached A after crossing, so that according to the above

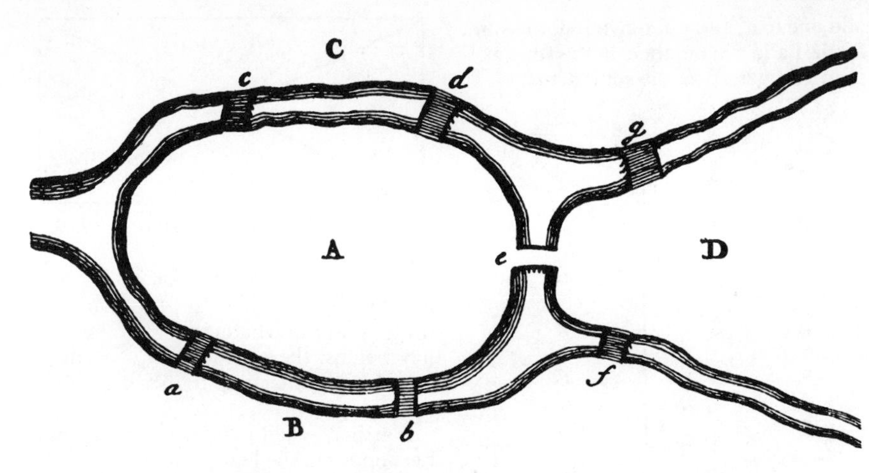

Seven bridges of Koenigsberg crossed the River Pregel

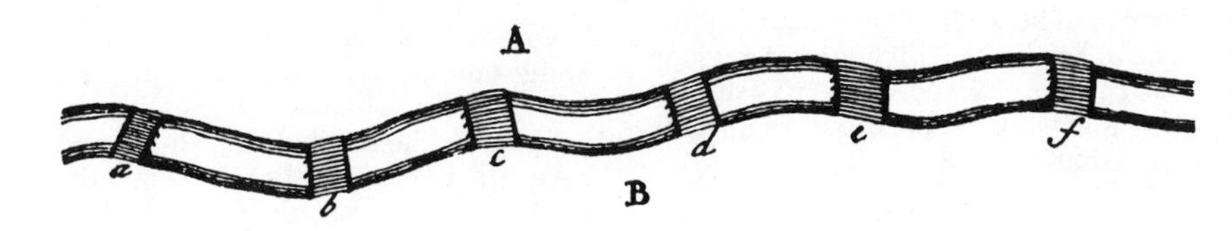

Euler used a simpler case to elucidate his principle

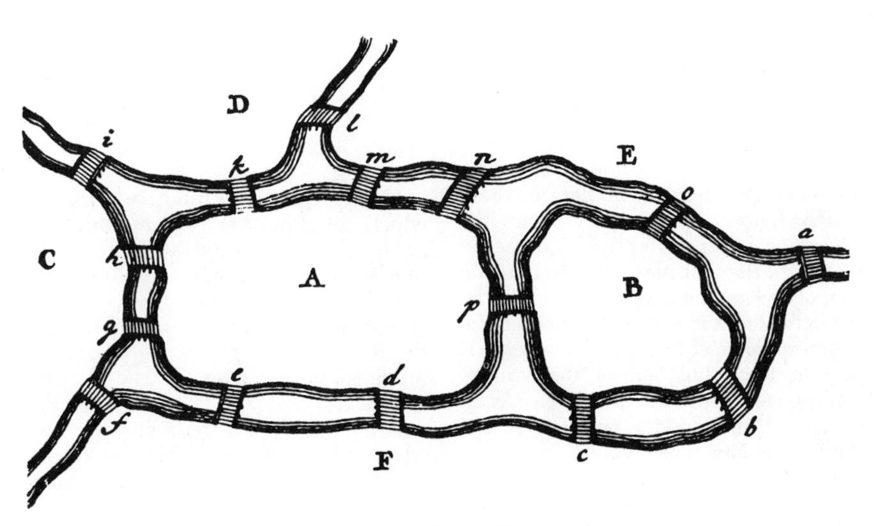

This trip is possible though the Koenigsberg one is not

method of denotation the letter A will appear exactly once. If there are three bridges leading to A and the traveler crosses all three, then the letter A will occur twice in the expression for his journey, whether it begins at A or not. And if there are five bridges leading to A, the expression for a route that crosses them all will contain the letter A three times. If the number of bridges is odd, increase it by one, and take half the sum; the quotient represents the number of times the letter A appears.

LET US now return to the Koenigsberg problem [*top illustration above*]. Since there are five bridges leading to (and from) island A, the letter A must occur three times in the expression describing the route. The letter B must occur twice, since three bridges lead to B; similarly D and C must each occur twice. That is to say, the series of . . . letters that represents the crossing of the seven bridges must contain A three times and B, C and D each twice. But this is quite impossible with a series of eight letters [for the sum of the required letters is nine]. Thus it is apparent that a crossing of the seven bridges of Koenigsberg in the manner required cannot be effected.

Using this method we are always able, whenever the number of bridges leading to a particular region is odd, to determine whether it is possible in a journey to cross each bridge exactly once. Such a route exists if the number of bridges plus one is equal to the sum of the numbers which indicate how often each individual letter must occur. On the other hand, if this sum is greater than the number of bridges plus one, as it is in our example, then the desired route cannot be constructed. The rule that I gave for determining from the number of bridges that lead to A how often the letter A will occur in the route description is independent of whether these bridges all come from a single region B or from several regions, because I was considering only the region A, and attempting to determine how often the letter A must occur.

When the number of bridges leading to A is even, we must take into account whether the route begins in A or not. For example, if there are two bridges that lead to A and the route starts from A, then the letter A will occur twice—once to indicate the departure from A by one of the bridges and a second time to indicate the return to A by the other bridge. However, if the traveler starts his journey in another region, the letter A will occur only once, since by my method of description the single occurrence of A indicates an entrance into as well as a departure from A.

Suppose, as in our case, there are four bridges leading into the region A, and the route is to begin at A. The letter A will then occur three times in the expression for the whole route, while if the journey had started in another region, A would occur only twice. With six bridges leading to A, the letter A will occur four times if A is the starting point, otherwise only three times. In general, if the number of bridges is even, the number of occurrences of the letter A, when the starting region is not A, will be half the number of the bridges; when the route starts from A, one more than half.

Every route must, of course, start in some one region. Thus from the number of bridges that lead to each region I determine the number of times that the corresponding letter will occur in the expression for the entire route as follows: When the number of the bridges is odd, I increase it by one and divide by two; when the number is even, I simply divide it by two. Then if the sum of the resulting numbers is equal to the actual number of bridges plus one, the journey can be accomplished, though it must start in a region approached by an odd number of bridges. But if the sum is one less than the number of bridges plus one, the journey is feasible if its starting point is a region approached by an even number of bridges, for in that case the sum is again increased by one.

MY PROCEDURE for determining whether in any given system of rivers and bridges it is possible to cross each bridge exactly once is as follows: First I designate the individual regions separated from one another by the water as A, B, C, etc. Second, I take the total number of bridges, increase it by one, and write the resulting number at the top of the paper. Third, under this number I write the letters A, B, C, etc., in a column, and opposite each letter I note the number of bridges that lead to that particular region. Fourth, I place an asterisk next to each letter that has an even number opposite it. Fifth, in a third column I write opposite each even number the half of that number, and opposite each odd number I write half of the sum formed by that number plus one. Sixth, I add up the last column of numbers. If the sum is one less than, or equal to, the number written at the top, I conclude that the required journey can be made. But it must be noted that when the sum is one less than the number at the top, the route must start from a region marked with an asterisk, and . . . when these two numbers are equal, it must start from a region that does not have an asterisk.

For the Koenigsberg problem I would set up the tabulation as follows:

Number of bridges 7,
giving 8 (=7+1)

A	5	3
B	3	2
C	3	2
D	3	2

The last column now adds up to more than 8, and hence the required journey cannot be made.

Let us take an example of two islands with four rivers forming the surrounding water [*bottom illustration on the preceding page*]. Fifteen bridges, marked *a, b, c, d,* etc., across the water around the islands and the adjoining rivers. The question is whether a journey can be arranged that will pass over all the bridges, but not over any of them more than once. I begin by marking the regions that are separated from one another by water with the letters A, B, C, D, E, F—there are six of them. Second, I take the number of bridges (15) add one and write this number (16) uppermost. Third, I write the letters A, B, C, etc., in a column and opposite each letter I write the number of bridges connecting with that region, *e.g.*, eight bridges for A, four for B, etc. Fourth, the letters that have even numbers opposite them I mark with an asterisk. Fifth, in a third column I write the half of each corresponding even number, or, if the number is odd, I add one to it, and put down half the sum. Sixth, I add the numbers in the third column and get 16 as the sum. Thus:

		16
A*	8	4
B*	4	2
C*	4	2
D	3	2
E	5	3
F*	6	3
		16

The sum of the third column is the same as the number 16 that appears above, and hence it follows that the journey can be effected if it begins in regions D or E, whose symbols have no asterisk. The following expression represents such a route:

EaFbBcFdAeFfCgAhCiDkAmEnAp-BoElD.

Here I have indicated, by small letters between the capitals, which bridges are crossed.

By this method we can easily determine, even in cases of considerable complexity, whether a single crossing of each of the bridges in sequence is possible. But I should now like to give another and much simpler method, which follows quite easily from the preceding, after a few preliminary remarks. In the first place, I note that the sum of the numbers written down in the second column is necessarily double the actual number of bridges. The reason is that in the tabulation of the bridges leading to the various regions each bridge is counted twice, once for each of the two regions that it connects.

From this observation it follows that the sum of the numbers in the second column must be an even number, since half of it represents the actual number of bridges. Hence . . . if any of the numbers opposite the letters A, B, C, etc., are odd, an even number of them must be odd. In the Koenigsberg problem for instance, all four of the numbers opposite the letters A, B, C, D, were odd, while in the example just given only two of the numbers were odd, namely those opposite D and E.

Since the sum of the numbers opposite A, B, C, etc., is double the number of bridges, it is clear that if this sum is increased by two in the latter example and then divided by two, the result will be the number written at the top. When all the numbers in the second column are even, and the half of each is written down in the third column, the total of this column will be one less than the number at the top. In that case it will always be possible to cross all the bridges. For in whatever region the journey begins, there will be an even number of bridges leading to it, which is the requirement. . . .

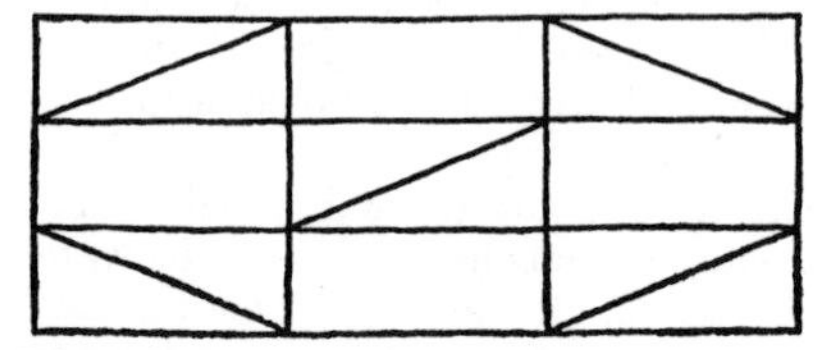
This figure requires only one stroke

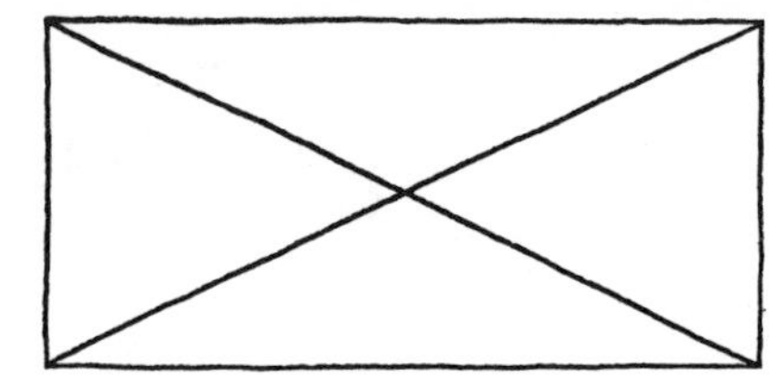
This figure requires two strokes

Further, when only two of the numbers opposite the letters are odd, and the others even, the required route is possible provided it begins in a region approached by an odd number of bridges. We take half of each even number, and likewise half of each odd number after adding one, as our procedure requires; the sum of these halves will then be one greater than the number of bridges, and hence equal to the number written at the top. But [when more than two, and an even number] of the numbers in the second column are odd, it is evident that the sum of the numbers in the third column will be greater than the top number, and hence the desired journey is impossible.

Thus for any configuration that may arise the easiest way of determining whether a single crossing of all the bridges is possible is to apply the following rules:

If there are more than two regions which are approached by an odd number of bridges, no route satisfying the required conditions can be found.

If, however, there are only two regions with an odd number of approach bridges the required journey can be completed provided it originates in one of these regions.

If, finally, there is no region with an odd number of approach bridges, the required journey can be effected, no matter where it begins.

These rules solve completely the problem initially proposed.

After we have determined that a route actually exists we are left with the question how to find it. To this end the following rule will serve: Wherever possible we mentally eliminate any two bridges that connect the same two regions; this usually reduces the number of bridges considerably. Then—and this should not be difficult—we proceed to trace the required route across the remaining bridges. The pattern of this route, once we have found it, will not be substantially affected by the restoration of the bridges which were first eliminated from consideration—as a little thought will show. Therefore I do not think I need say more about finding the routes themselves.

Various Problems Based on Planar Graphs, or Sets of "Vertices" Connected by "Edges"

by Martin Gardner
April 1964

An engineer draws a diagram of an electrical network. A chemist makes a sketch to show how the atoms of a complex molecule are joined by chemical bonds. A genealogist draws an intricate family tree. A military commander plots a network of supply lines on a map. A sociologist traces in an elaborate diagram the power structure of a giant corporation.

What do all these patterns have in common? They are points (representing electrical connections, atoms, people, cities and so on) connected by lines. In the 1930's the German mathematician Dénes König made the first systematic study of all such patterns, giving them the generic name "graphs." (The confusion of this term with the "graphs" of analytic geometry is regrettable, but the term has stuck.) Today graph theory is a flourishing field. It is usually considered a branch of topology (because in most cases only the topological properties of graphs are considered), although it now overlaps large areas of set theory, combinatorial mathematics, algebra, geometry, matrix theory, game theory, logic and many other fields.

König's pioneer book on graphs (published in Leipzig in 1936) has yet to be translated, but an English edition of a later French book, *The Theory of Graphs and its Applications,* by Claude Berge, was published in England in 1962. Last year Oystein Ore's excellent elementary introduction, *Graphs and Their Uses,* was issued as a Random House paperback. Both books are of great recreational interest. Hundreds of familiar puzzles, seemingly unrelated, yield readily to graph theory. This month we center our attention on "planar graphs" and some of their more intriguing puzzle aspects.

A planar graph is a set of points, called vertices, connected by lines, called edges, in such a way that it is possible to draw the graph on a plane without any pair of edges intersecting. Imagine that the edges are elastic strings that can be bent, stretched or shortened as we please. Is the graph shown at the left below planar? (Its four vertices are indicated by spots. The crossing point at the center is not a vertex; think of one line as passing under the other.) Yes, because we can easily remove the intersection by shifting the position of a vertex, as shown in the middle graph, or stretching an edge as shown in the one at the right. All three of these graphs are "isomorphic": they represent three different ways of drawing the *same* planar graph. The edges of any solid polyhedron, such as a cube, are planar graphs because we can always stretch the solid's "skeleton" until it lies on a plane, free of intersections. The skeleton of a tetrahedron is isomorphic with the three graphs shown below.

It is not always easy to decide if a graph is planar. Consider the problem depicted at left, page 126, one of the oldest and most frustrating of all topological teasers. Since the English puzzlist Henry Ernest Dudeney gave it this form in 1917 it has been known as the "utilities problem." Each house must receive gas, water and electricity. Can lines be drawn to connect each house with each utility in such a way that no line intersects another? In other words, is the resulting graph planar?

The answer is no, and it is not difficult to give a rough proof. Assume that only houses A and B are to be connected to the three utilities. To do this without having any line cross another you must divide the plane into three regions as shown in the illustration at right, page 126. Your lines need not be as pictured, but however you draw them your graph will be isomorphic with the one shown. House C must go in one of the three regions. If it goes in X, it is cut off from electricity. If it goes in Y, it is cut off from water. If it goes in Z, it is cut off from gas. The same argument holds when the graph is drawn on a sphere, but not when it is drawn on certain other surfaces. For example, the graph is easily drawn without intersections on the surface of a doughnut.

When every vertex of a graph is connected to each of the other vertices, the graph is said to be "complete." We see in the illustration below that the complete graph for four points is planar. Is the complete graph for five points planar? Again an informal proof (the reader may enjoy working it out for himself) shows that it is not. This proof is equivalent to a proof that it is not possible to draw five regions in such a way that every pair shares a common border

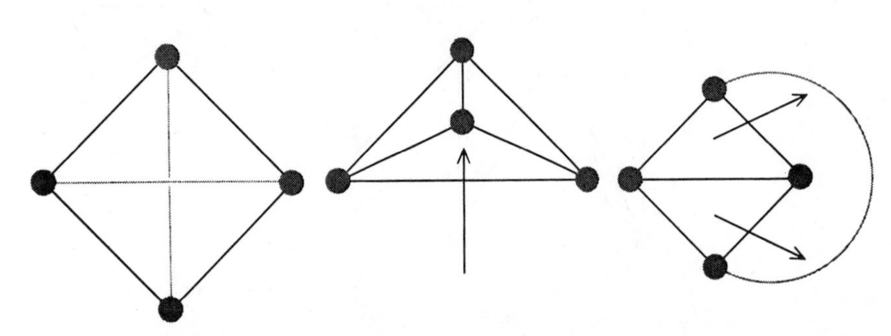

Three ways to draw complete graph for four points

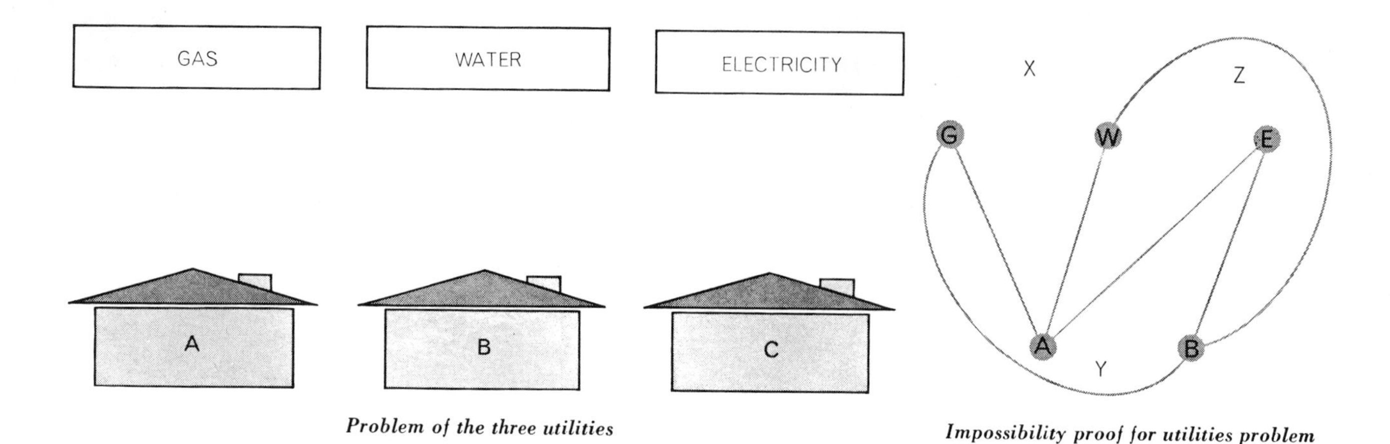

Problem of the three utilities

Impossibility proof for utilities problem

segment, a theorem often confused with the famous four-color map theorem. The two simplest nonplanar graphs are shown below. At the left is the utilities graph (known as a Thomsen graph), at the right is the complete graph for five points.

The fact that a complete graph can be planar only if it has four or fewer points is not without philosophical interest. Many philosophers and mathematicians have tried to answer the question: Why does physical space have three dimensions? In his book *The Structure and Evolution of the Universe* (Harper Torchbooks, 1959) the British cosmologist G. J. Whitrow argues that intelligent life as we know it could not have evolved in a space of *more* than three dimensions because such spaces do not allow stable planetary orbits around a sun. How about spaces of one or two dimensions? Intelligent Linelanders and Flatlanders of the type described in my column in July, 1962, are ruled out, says Whitrow, by graph theory. A brain requires an immense number of nerve cells (points), connected in pairs by nerves (edges) that must not intersect. In three dimensions there is no limit to the number of cells that can be so connected, but in a Flatland the maximum number, as we have seen, would be four.

"Thus," Whitrow writes, "we may conclude that the number of dimensions of physical space is necessarily three, no more and no less, because it is the unique natural concomitant of the evolution of the higher forms of terrestrial life, in particular of Man, *the formulator of the problem.*"

Devising planar graphs is an essential task in many fields of technology. Printed circuits, for instance, will short-circuit if any two paths cross. The reader may wish to test his skill in planar graph construction by considering the two printed-circuit problems shown on page 127. In the upper problem five nonintersecting lines must be drawn within the rectangle, each connecting a pair of spots bearing the same letter (*A* with *A*, *B* with *B* and so on). The two lines *AD* and *BC* are barriers of some sort that may not be crossed. In the lower problem five lines are to be drawn—connecting pairs of spots labeled with the same letter, as before—but in this case all lines must follow the grid. Of course there must be no crossings. The solution is unique.

Another well-known type of graph puzzle is the one that calls for drawing

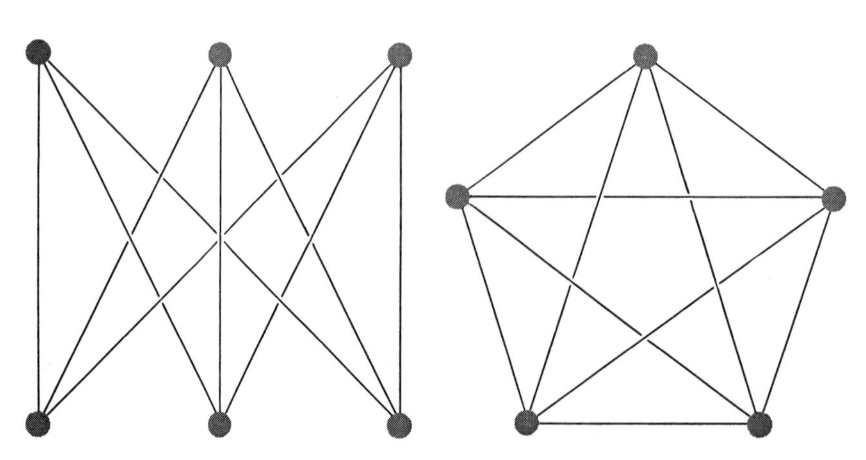

Simplest nonplanar graphs

a given planar graph in one continuous line without taking the pencil from the paper or going over any edge twice. If such a line can be drawn as a closed loop, returning to the vertex from which it started, the graph is said to be an "Euler graph" and the line an "Euler line." In 1736 the Swiss mathematician Leonhard Euler solved a famous problem involving a set of seven bridges in the East Prussian town of Königsberg (now Kaliningrad). Was it possible to walk over each bridge once and only once and return to where one had started? Euler found that the problem was identical with that of tracing a simple graph. He showed, in the first paper ever written on graph theory, that if every vertex of a graph is of "even degree" (has an even number of lines meeting it), it can be traced in one round-trip path. If there are two vertices of odd degree, no round trip is possible, but the graph can be drawn by a line beginning at one odd vertex and ending at the other. If there are $2k$ vertices of odd degree (and the number of odd vertices must always be even), it can be traced by k separate paths, each starting and ending at an odd vertex. The graph for the bridges of Königsberg has four odd vertices, therefore it requires a minimum of two paths (neither of them closed circuits) to traverse all edges.

Any Euler graph can be traversed by an Euler line that makes the entire round trip without intersecting itself. Lewis Carroll, we are told in a biography by his nephew, was fond of asking little girls to draw, with one Euler line, the graph at the left in the top illustration on page 128. It is easily done if lines are allowed to intersect, but it is not so easy if intersections are forbidden. A quick way to solve such puzzles has been proposed by Thomas H. O'Beirne of Edinburgh. One colors alternate regions as shown in the middle drawing, then breaks them apart at certain vertices in any way that will leave the colored areas "simply connected" (connected without enclosing noncolored areas). The perimeter of the colored region is now the Euler line we seek [*drawing at right*]. The reader can try this method on the Euler graph shown in the second illustration from the top on page 128 (proposed by O'Beirne) to see how pleasingly symmetrical an Euler line he can obtain.

An entirely different and, strangely, much more difficult type of graph-traversing puzzle is that of finding a route that passes through each vertex once and only once. Any route that passes through no vertex twice is known in graph theory as an arc. An arc that returns to the starting point is called a circuit. And a circuit that visits every vertex once and only once is called a Hamiltonian line, after Sir William Rowan Hamilton, the 19th-century Irish mathematician, who was the first to study such paths. He showed that a Hamiltonian line could be traced along the edges of each of the five regular solids, and he even sold a toy manufacturer a puzzle based on finding Hamiltonian tours along the edges of the dodecahedron.

It might be supposed that, as in the case of Euler lines, there would be simple rules for determining if a graph is Hamiltonian; the fact is that the two tasks are surprisingly dissimilar. An Euler line must trace every edge once and only once, but it may go through any vertex more than once. A Hamiltonian line must go through each vertex once and only once, but it need not trace every edge. (In fact, it traverses exactly two of the edges that meet at any one vertex.) Hamiltonian paths are important in many fields where one

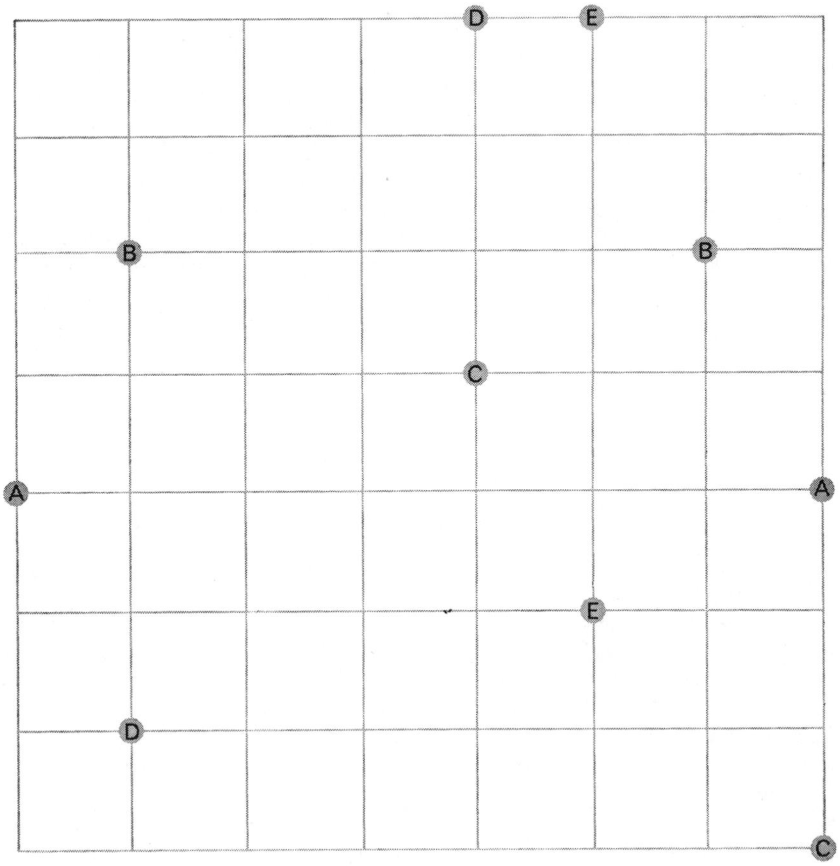

Two printed-circuit problems

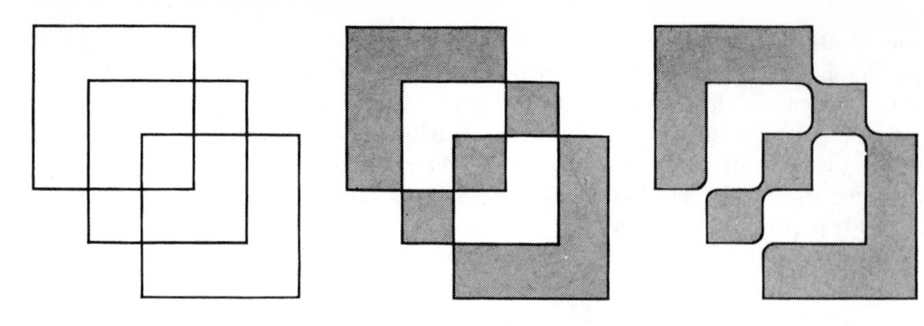

Lewis Carroll's three-square problem

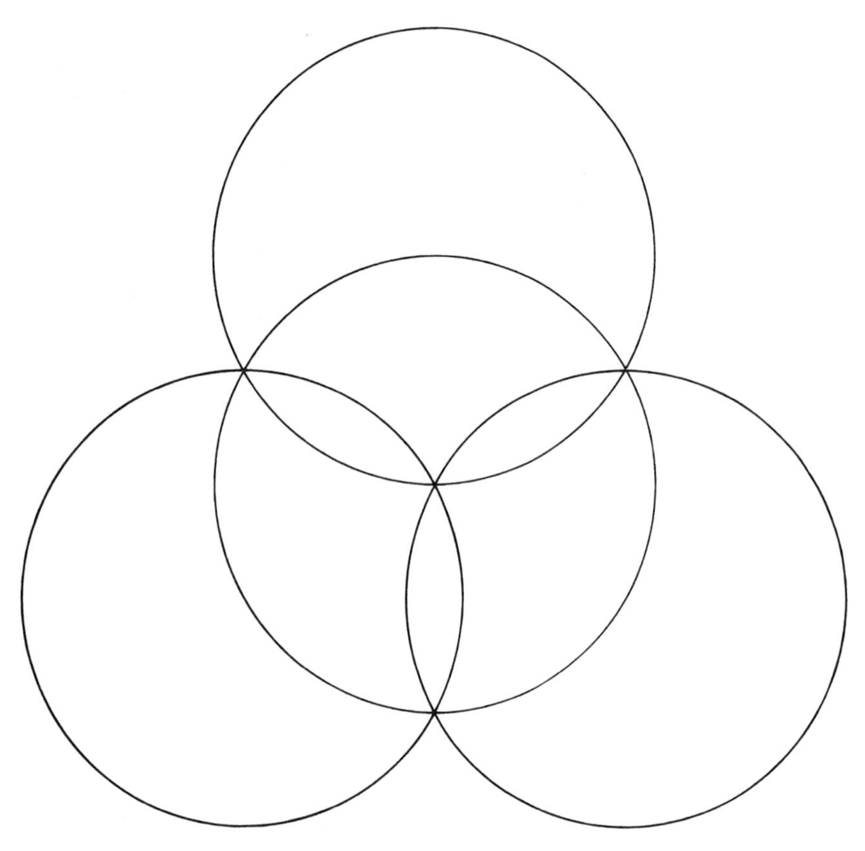

O'Beirne's four-circle problem

would not expect to find them. In operations research, for example, the problem of obtaining the best order in which to carry out a specified series of operations can sometimes be diagramed as a graph on which a Hamiltonian line gives an optimum solution. Unfortunately there is no general method for deciding if a graph is Hamiltonian, or for finding all Hamiltonian lines if it is.

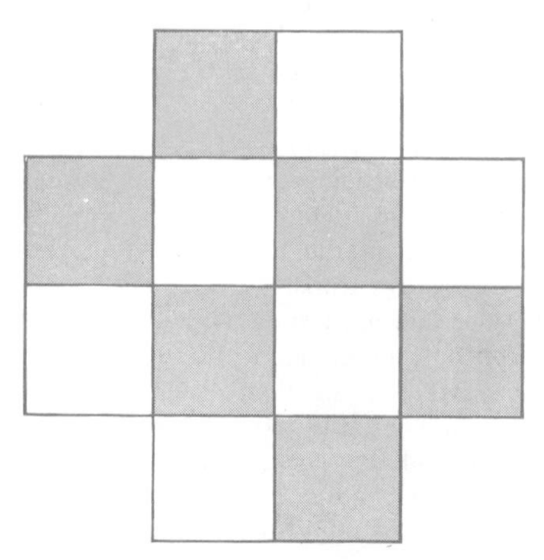

A knight's-tour problem

Many semiregular polyhedrons, but not all, have Hamiltonian skeletons. An exception is the rhombic dodecahedron shown at the right, a form often assumed by crystals of garnet. Even if the path is not required to be closed, there is no way to traverse the skeleton so that each vertex is visited once and only once. The proof, first given by H. S. M. Coxeter, is a clever one. All vertices of degree 4 are shown as black spots, all of degree 3 as colored spots. Note that every black spot is completely surrounded by colored spots and vice versa. Therefore any path through all 14 spots must alternate colored and black. But there are six black spots and eight colored ones! No path of alternating color is possible, either closed or open at the ends.

An ancient chess recreation that at first seems far removed from Hamiltonian paths is the re-entrant knight's tour. It consists of placing the knight on a square of the chessboard, then finding a path of continuous knight's moves that will visit every square once and only once, the knight thereupon returning in one move to the square from which it started. Suppose each cell of the board to be represented by a point and every possible knight's move by a line joining two points. The result is, of course, a graph. Any circuit that visits each vertex once and only once will be a Hamiltonian line, and every such line will trace a re-entrant knight's tour.

Such a tour is impossible on any board with an odd number of cells. (Can the reader see why?) The smallest rectangle on which a closed tour is possible is one with an area of 30 square units (3×10, or 5×6). The six-by-six is the smallest square. No tours, not even open-ended ones, are possible on rectangles with one side less than three. No one knows how many millions of different re-entrant knight's tours can be made on the standard eight-by-eight chessboard. In the enormous literature on the topic the search has usually been confined to paths that exhibit interesting symmetries. Thousands of elegant patterns, such as those shown on the next page, have been discovered. Paths with exact fourfold symmetry (unchanged by any 90-degree rotation) are not possible on the eight-by-eight board, although five such patterns are possible on the six-by-six.

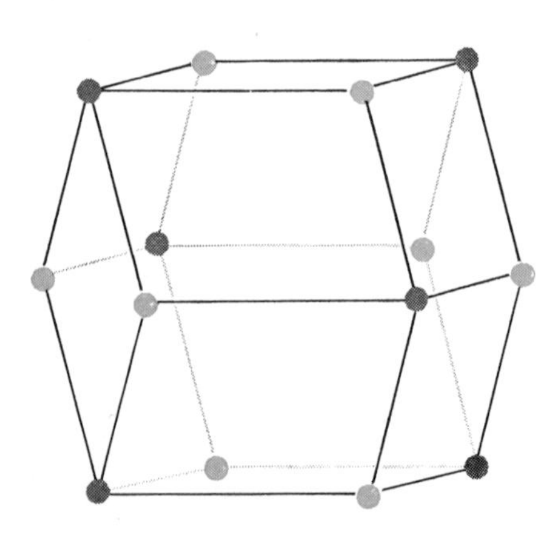

Skeleton of rhombic dodecahedron

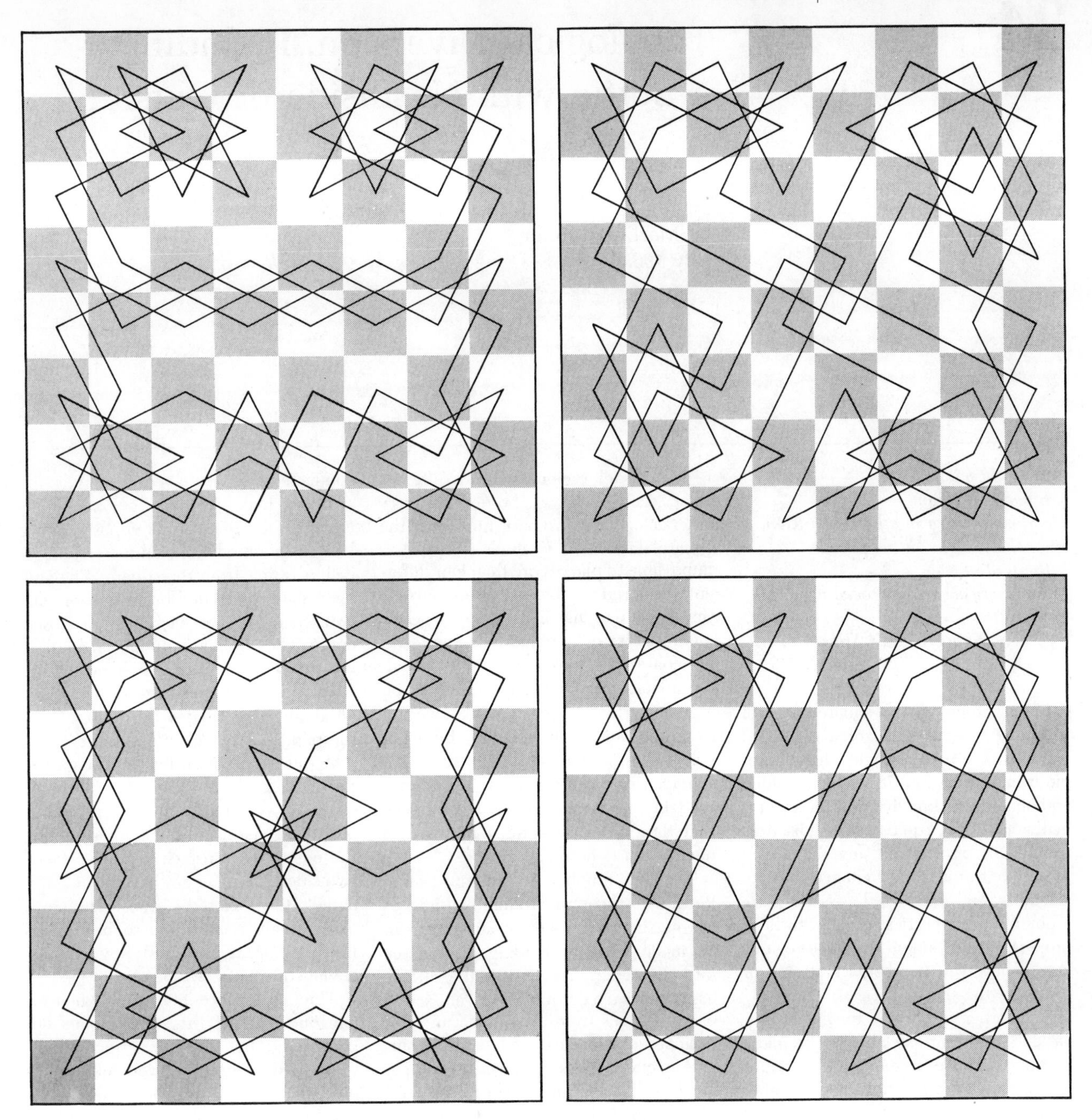

Re-entrant knight's tours

As an introduction to this classic pastime readers are invited to search for a re-entrant knight's tour on a simple 12-cell board [*see illustration on page 128*]. After it has been found, a seemingly more difficult question arises: Is it possible to move the knight over this board in one chain of jumps and make every possible knight's move once and only once? There are 16 different knight's moves. A move is considered "made" whenever a knight connects the two cells by a jump in either direction. Of course, the knight may visit any cell more than once, but it must not make the same move twice. The path need not be re-entrant.

The reader will soon convince himself that such a path is not possible; but what is the smallest number of *separate* paths that will cover all 16 of the possible moves? This can be answered in just a few minutes by applying one of the graph theorems discussed earlier in this article.

24

Topological Diversion, Including a Bottle with No Inside or Outside

by Martin Gardner
July 1963

Three jolly sailors from
Blaydon-on-Tyne
They went to sea in a bottle by Klein.
Since the sea was entirely inside
the hull
The scenery seen was exceedingly dull.

The Space Child's Mother Goose,
by Frederick Winsor

To a topologist a square sheet of paper is a model of a two-sided surface with a single edge. Crumple it into a ball and it is still two-sided and one-edged. Imagine that the sheet is made of rubber. You can stretch it into a triangle or circle, into any shape you please, but you cannot change its two-sidedness and one-edgedness. They are topological properties of the surface, properties that remain the same regardless of how you bend, twist, stretch or compress the sheet.

Two other important topological invariants of a surface are its chromatic number and Betti number. The chromatic number is the maximum number of regions that can be drawn on the surface in such a way that each region has a border in common with every other region. If each region is given a different color, each color will border on every other color. The chromatic number of the square sheet is 4. In other words, it is impossible to place more than four differently colored regions on the square so that each pair has a boundary in common. The term "chromatic number" also designates the minimum number of colors sufficient to color any finite map on a given surface. It is not yet known if 4 is the chromatic number, in this map-coloring sense, for the square, tube, and sphere, but for all other surfaces considered in this article, it has been shown that the chromatic number is the same under both definitions.

The Betti number, named after Enrico Betti, a 19th-century Italian physicist, is the maximum number of cuts that can be made without dividing the surface into two separate pieces. If the surface has edges, each cut must be a "crosscut": one that goes from a point on an edge to another point on an edge. If the surface is closed (has no edges), each cut must be a "loop cut": a cut in the form of a simple closed curve. Clearly the Betti number of the square sheet is 0. A crosscut is certain to produce two disconnected pieces.

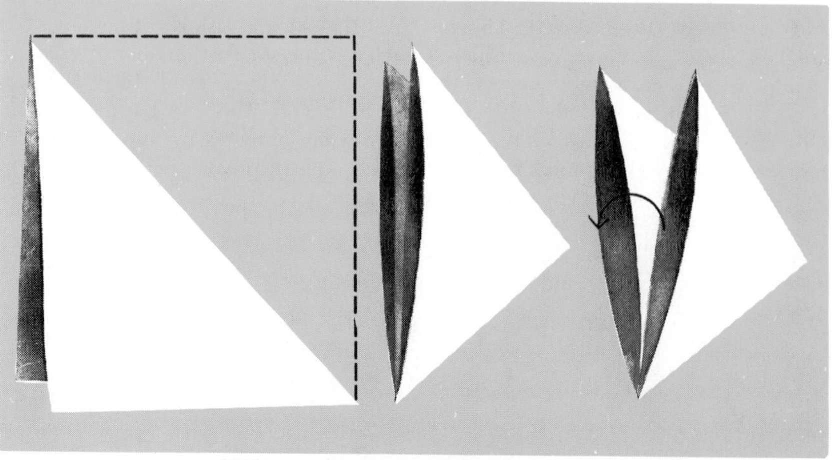

Möbius surface constructed with a square

If we make a tube by joining one edge of the square to its opposite edge, we create a model of a surface topologically distinct from the square. The surface is still two-sided but now there are two separate edges, each a simple closed curve. The chromatic number remains 4 but the Betti number has changed to 1. A crosscut from one edge to the other, although it eliminates the tube, allows the paper to remain in one piece.

A third type of surface, topologically the same as the surface of a sphere or cube, is made by folding the square in half along a diagonal and then joining the edges. The surface continues to be two-sided but all edges have been eliminated. It is a closed surface. The chromatic number continues to be 4. The Betti number is back to 0: any loop cut obviously creates two pieces.

Things get more interesting when we join one edge of the square to its opposite edge but give the surface a half-twist before doing so. You might suppose that this cannot be done with a square piece of paper, but it is easily managed by folding the square twice along its diagonals, as shown in the illustration at left. Tape together the pair of edges indicated by the arrow in the last drawing. The resulting surface is the familiar Möbius strip, first analyzed by A. F. Möbius, the 19th-century German astronomer who was one of the pioneers of topology. The model will not open out, so it is hard to see that it is a Möbius strip, but careful inspection will convince you that it is. The surface is one-sided and one-edged, with a Betti number of 1. Surprisingly, the chromatic number has jumped to 6. Six regions, of six different colors, can be placed on the surface so that each region has a border in common with each of the other five.

Torus surface folded from a square

When both pairs of the square's opposite edges are joined, without twisting, the surface is called a torus. It is topologically equivalent to the surface of a doughnut or a cube with a hole bored through it. The top illustration on this page shows how a flat, square-shaped model of a torus is easily made

Klein bottle: a closed surface with no inside or outside

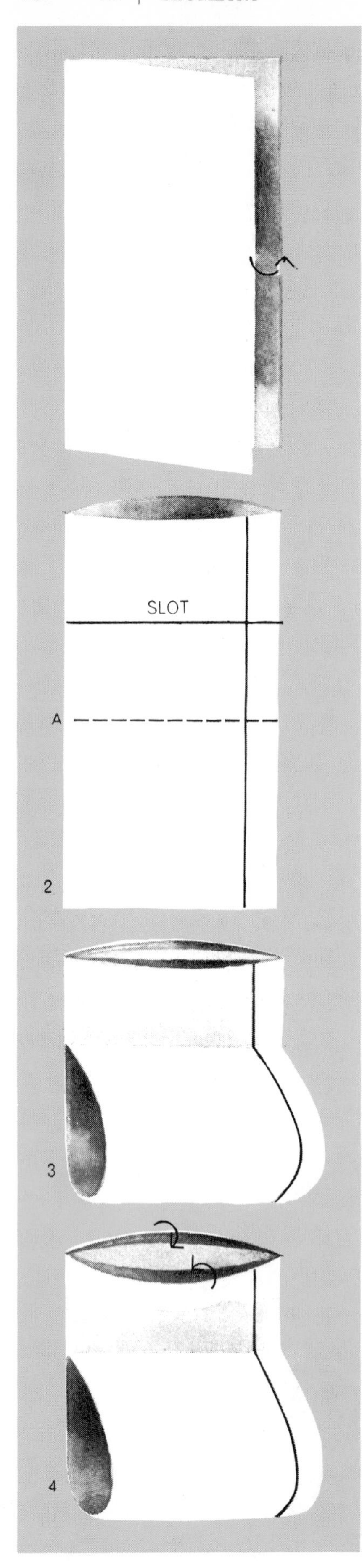

Folding a Klein bottle from a square

by folding the square twice, taping the edges as shown by the solid gray line in the second drawing and the arrows in the last. The torus is two-sided, closed (no-edged) and has a chromatic number of 7 and a Betti number of 2. One way to make the two cuts is first to make a loop cut where you joined the last pair of edges (this reduces the torus to a tube) and then a crosscut where you joined the first pair. Both cuts, strictly speaking, are loop cuts when they are marked on the torus surface. It is only because you make one cut before the other that the second cut becomes a crosscut.

It is hard to anticipate what will happen when the torus model is cut in various ways. If the entire model is bisected by being cut in half either horizontally or vertically, along a center line parallel to a pair of edges, the torus surface receives two loop cuts. In both cases the resulting halves are tubes. If the model is bisected by being cut in half along either diagonal, each half proves to be a square. Can the reader find a way (the answer appears in the solution section in the back of the book) to give the model two loop cuts that will produce two separate bands interlocked like two rings of a chain?

Many different surfaces are closed like the surface of a sphere and a torus, yet one-sided like a Möbius strip. The easiest one to visualize is a surface known as the Klein bottle, discovered in 1882 by Felix Klein, the great German mathematician. An ordinary bottle has an outside and inside in the sense that if a fly were to walk from one side to the other, it would have to cross the edge that forms the mouth of the bottle. The Klein bottle has no edges, no inside or outside. What seems to be its inside is continuous with its outside, like the two apparent "sides" of a Möbius surface.

Unfortunately it is not possible to construct a Klein bottle in three-dimensional space without self-intersection of the surface. The bottom illustration on page 131 shows how the bottle is traditionally depicted. Imagine the lower end of a tube stretched out, bent up and plunged through the tube's side, then joined to the tube's upper mouth. In an actual model made, say, of glass there would be a hole where the tube intersects the side. You must disregard this defect and think of the hole as being covered by a continuation of the bottle's surface. There is no hole, only an intersection of surfaces. This self-intersection is necessary because the model is in three-space. If we conceive of the surface as being embedded in four-space, the self-intersection can be eliminated entirely. The Klein bottle is one-sided, no-edged and has a Betti number of 2 and a chromatic number of 6.

Daniel Pedoe, a mathematician at Purdue University, recently wrote *The Gentle Art of Mathematics.* It is a delightful book, but on page 84 Professor Pedoe slips into a careless bit of dogmatism. He describes the Klein bottle as a surface that is a challenge to the glass blower, but one "which cannot be made with paper." Now, it is true that at the time he wrote this apparently no one had tried to make a paper Klein bottle, but that was before Stephen Barr, a science-fiction writer and an amateur mathematician of Woodstock, N.Y., turned his attention to the problem. Barr quickly discovered dozens of ways to make paper Klein bottles. Here I will describe only one of Barr's Klein bottles; one that enables us to continue working with a square and at the same time follows closely the traditional glass model.

The steps are given in the illustration at the left. First, make a tube by folding the square in half and joining the right edges with a strip of tape as shown in Step 1. Cut a slot about a quarter of the distance from the top of the tube (Step 2), cutting only through the thickness of paper nearest you. This corresponds to the "hole" in the glass model. Fold the model in half along the broken line *A*. Push the lower end of the tube up through the slot (Step 3) and

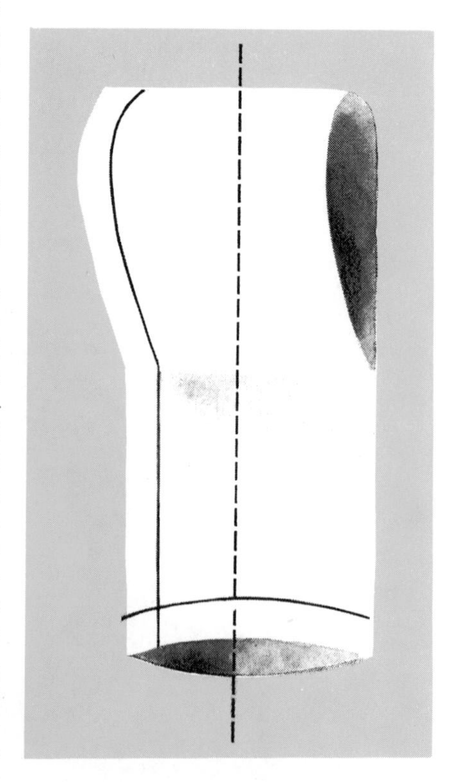

Bisected bottle makes two Möbius strips

join the edges all the way around the top of the model (Step 4) as indicated by the arrows. It is not difficult to see that this flat, square model is topologically identical with the glass bottle shown in the bottom illustration on page 131. In one way it is superior: there is no actual hole. True, you have a slot where the surface self-intersects, but it is easy to imagine that the edges of the slot are joined so that the surface is everywhere edgeless and continuous.

Moreover, it is easy to cut this paper model and demonstrate many of the bottle's astonishing properties. Its Betti number of 2 is demonstrated by cutting the two loops formed by the two pairs of taped edges. If you cut the bottle in half vertically, you get two Möbius bands, one a mirror image of the other. This is best demonstrated by making a tall, thin model [*see illustration p. 132*] from a tall, thin rectangle instead of a square. When you slice it in half along the broken line (actually this is one long loop cut all the way around the surface), you will find that each half opens out into a Möbius strip. Both strips are partially self-intersecting, but you can slide each strip out of its half-slot and close the slot, which is not supposed to be there anyway.

If the bottle can be cut into a pair of Möbius strips, of course the reverse procedure is possible, as described in the following anonymous limerick:

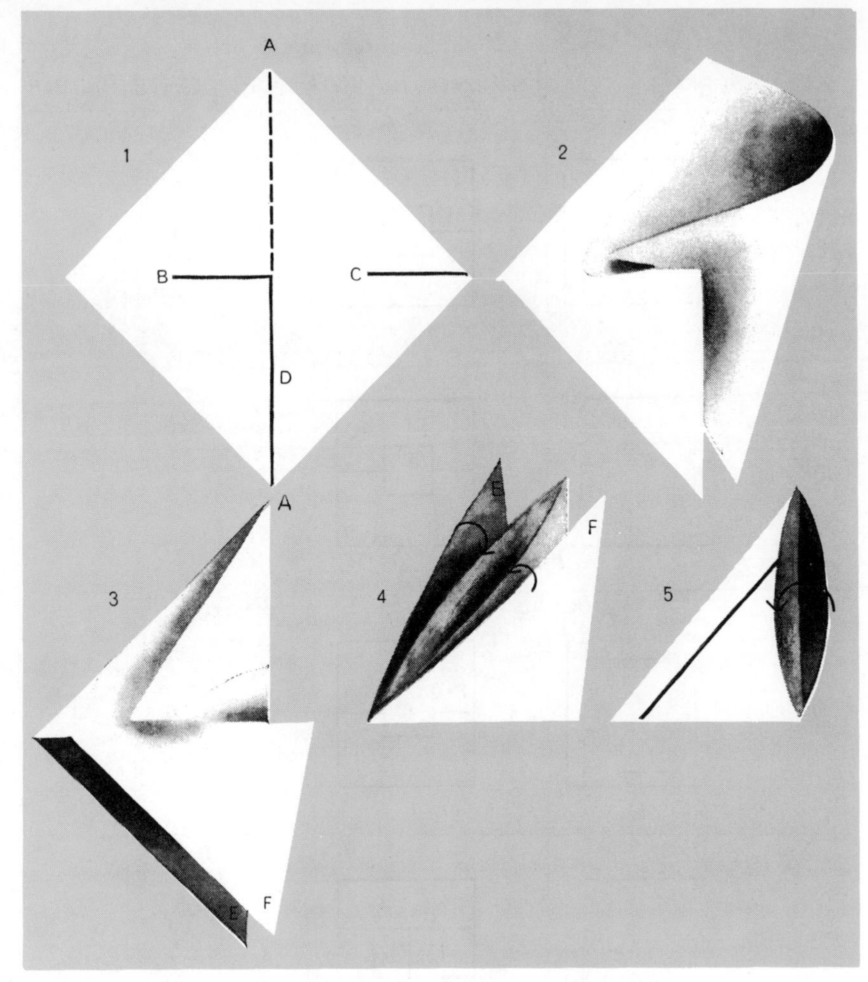

Folding a cross-cap and projective plane from a square

A mathematician named Klein
Thought the Möbius band was divine.
Said he: "If you glue
The edges of two,
You'll get a weird bottle like mine."

Surprisingly, it is possible to make a single loop cut on a Klein bottle and produce not two Möbius strips but only one. A great merit of Barr's paper models is that problems like this can be tackled empirically. Can the reader discover how the cut is made?

The Klein bottle is not the only simple surface that is one-sided and no-edged. A surface called the projective plane (because of its topological equivalence to a plane studied in projective geometry) is similar to the Klein bottle in both respects as well as in having a chromatic number of 6. As in the case of the Klein bottle, a model cannot be made in three-space without self-intersection. A simple Barr method for folding such a model from a square is shown in the illustration above. First cut the square along the solid black lines shown in Step 1. Fold the square along the diagonal *A-A′*, inserting slot *C* into slot *B* as shown in Steps 2 and 3. You must think of the line where the slots interlock as an abstract line of self-intersection. Fold up the two bottom triangular flaps *E* and *F*, one on each side (Step 4), and tape the edges as indicated.

The model is now what topologists call a cross-cap, a self-intersecting Möbius strip with an edge that can be stretched into a circle without further self-intersection. This edge is provided by the edges of cut *D*, originally made along the square's diagonal. Note that unlike the usual model of a Möbius strip, this one is symmetrical: neither right- nor left-handed. When the edge of the cross-cap is closed by taping it as shown in Step 5, the model becomes a projective plane. You might expect it to have a Betti number of 2, like the Klein bottle, but it does not. It has a Betti number of 1. No matter how you loop-cut it, the cut produces either two pieces or a piece topologically equivalent to a square sheet that cannot be cut again without making two pieces. If you remove a disk from anywhere on the surface of the projective plane, the model reverts to a cross-cap.

The chart on page 134 summarizes all that has been said. The square diagrams in the first column show how the edges join in each model. Sides of the same color join each to each, with the direction of their arrows coinciding. Corners labeled with the same letter are corners that come together. Broken lines are sides that remain edges in the finished model. Next to the chromatic number of each model is shown one way in which the surface can be mapped to accommodate the maximum number of colors. It is instructive to color each sheet as shown, coloring the regions on both sides of the paper (as though the paper were cloth through which the colors soaked), because you must think of the sheet as having zero thickness. An inspection of the final model will show that each region does indeed border on every other one.

SURFACE	CHROMATIC NUMBER	SIDES	EDGES	BETTI NUMBER
SQUARE (OR DISK)	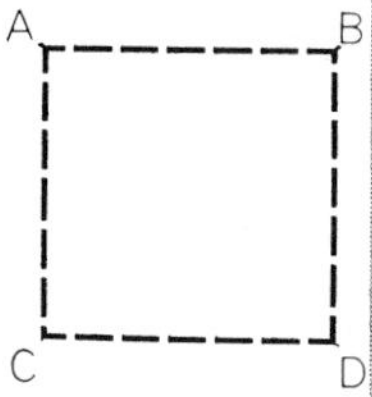 4	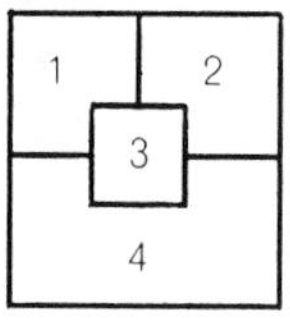 2	1	0
TUBE	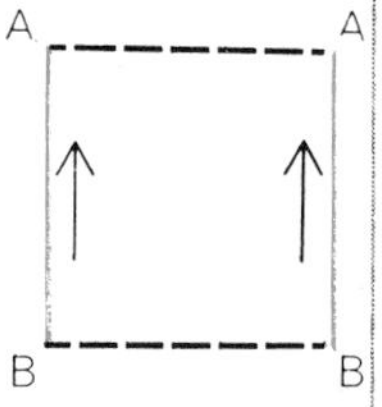 4	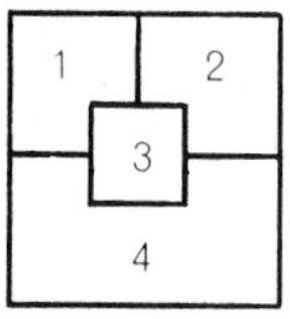 2	2	1
SPHERE	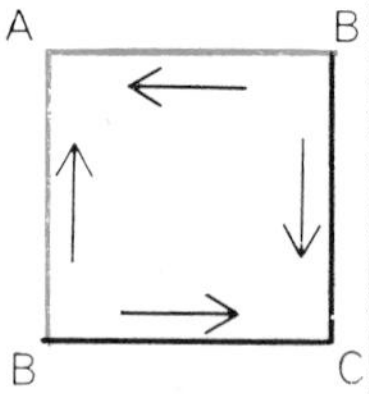 4	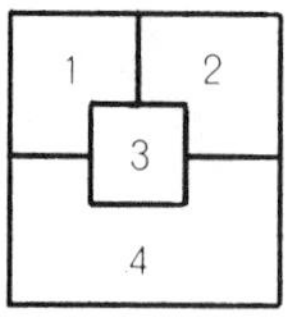 2	0	0
MÖBIUS STRIP	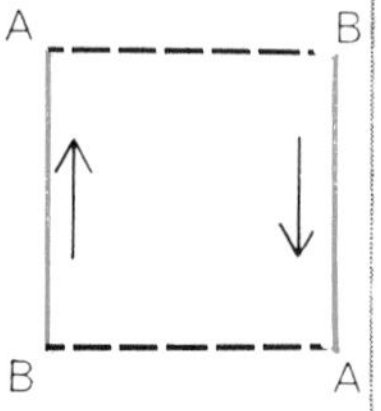 6	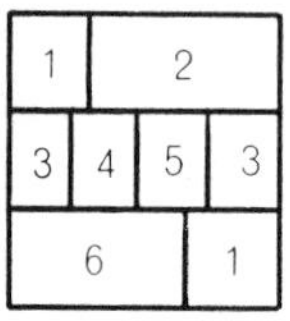 1	1	1
TORUS	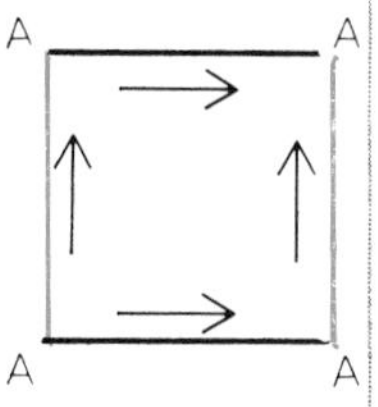 7	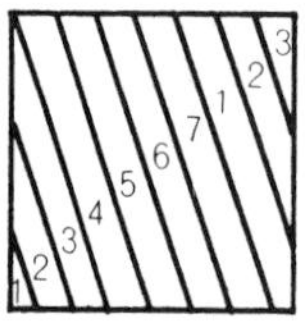 2	0	2
KLEIN BOTTLE	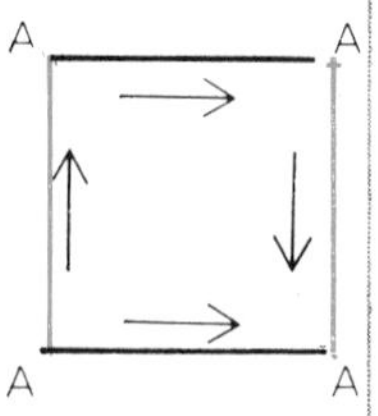 6	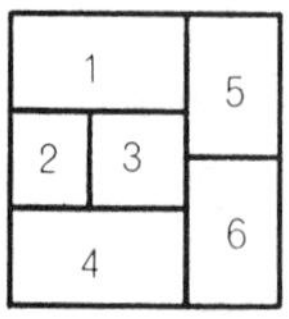 1	0	2
PROJECTIVE PLANE	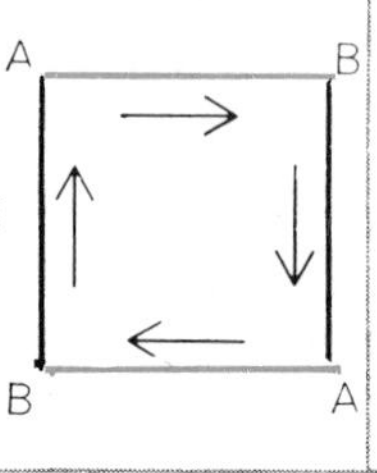 6	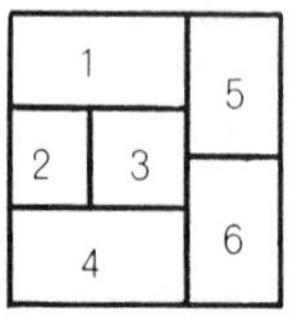 1	0	1

Topological invariants of seven basic surfaces

Geometry

by Morris Kline
September 1964

For 2,000 years geometry meant Euclidean geometry. Then it was found not only that other geometries described physical space equally well but also that geometry was properly the study of all possible spaces

The evolution of mathematics depends on advances in both number and geometry. It cannot be said, however, that these key elements of mathematics have always advanced side by side. Frequently they have competed, and the advance of one has been at the expense of the other. The history of this sometimes strained relation between two disciplines that actually have a common purpose is reminiscent of contrapuntal themes in music.

The first genuine stride of mathematics was taken by geometry. Some primitive mathematics was created by Egyptian and Babylonian carpenters and surveyors in the 4,000 years preceding the Christian era, but it was the classical Greek philosophers who, between 600 B.C. and 300 B.C., gave mathematics its definitive architecture of abstraction and deductive proof, erected the vast structure of Euclidean geometry and dedicated the subject to the understanding of the universe.

Of the several forces that turned the Greeks toward geometry, perhaps the most important was the difficulty Greek scholars had with the concept of the irrational number: a number that is neither a whole number nor a ratio of whole numbers. The difficulty arose in connection with the famous Pythagorean theorem that the length of the hypotenuse of a right triangle is the square root of the sum of the squares of the two sides. In a right triangle with sides of one unit each the hypotenuse must then be $\sqrt{2}$, an irrational number. Such a concept was beyond the Greeks; number to them had always meant whole number or ratio of whole numbers. They resolved the difficulty by banishing it, producing a geometry that affirmed theorems and offered proofs without reference to number. Today this geometry is known as pure geometry or synthetic geometry, the latter an unfortunate term that has only historical justification.

Since the mathematics of the classical Greeks was devoted to deducing truths of nature, it had to be founded on truths. Fortunately there were some seemingly self-evident truths at hand, among them the following: two points determine a line; a straight line extends indefinitely far in either direction; all right angles are equal; equals added to equals yield equals; figures that can be made to coincide are congruent. Some of these axioms make assertions primarily about space itself; others pertain to figures in space.

From these axioms Euclid, in his *Elements,* deduced almost 500 theorems. In other works he and his successors, notably Archimedes and Apollonius, deduced many hundreds more. Because the Greeks chose to work purely in geometry, many of the theorems stated results now regarded as algebraic. For example, the solution of second-degree equations in one unknown ($x^2 - 8x + 7 = 0$ is such an equation) was carried out geometrically and the answer given by Euclid was not a number but a line segment. Thus Euclidean geometry embraced the algebra known at that time.

The welter of theorems might suggest that the Greeks drifted from topic to topic. That would be a false impression. The figures they chose were basic: lines and curves in one category and surfaces in another. In the first category are such figures as the triangle and the conic sections: circle, parabola, ellipse and hyperbola. In the second category are such figures as the cube, sphere, paraboloid, ellipsoid and hyperboloid [*see illustration on page 136*]. Then the Greek geometers tackled basic problems concerning those figures. For instance, what must one know about two figures to assert that they are congruent (identical except for position in space), similar (having the same shape if not the same size) or equivalent (having the same area)? Thus congruence, similarity and equivalence are major themes of Euclidean geometry, and the majority of the theorems deal with these questions.

The classical Greek civilization that gave rise to Euclidean geometry was destroyed by Alexander the Great and rebuilt along new lines in Egypt. Alexander moved the center of his empire from Athens to the city he modestly named Alexandria, and he proclaimed the goal of fusing Greek and Near Eastern civilizations. This objective was ably executed by his successors, the Ptolemys, who ruled Egypt from 323 B.C. until the last member of the family, Cleopatra, was seduced by the Romans. Under the influence of the Near Eastern civilizations, notably the Egyptian and the Persian, the culture of the Alexandrian Greek civilization became more engineering-minded and more practically oriented. The mathematicians responded to the new interests.

Applied science and engineering must in large part be quantitative. What the Alexandrians appended to Euclid's geometry in order to obtain quantitative results was number: arithmetic and algebra. The disturbing fact about these

RAPHAEL'S "SPOSALIZIO," or "Marriage of the Virgin," part of which is reproduced on page 143, indicates how Renaissance painters solved problems of perspective and so contributed to the evolution of projective geometry. The superimposed white lines show how the artist depicted as converging on a "principal vanishing point" lines that in actuality were horizontal, parallel and receding directly from the viewer.

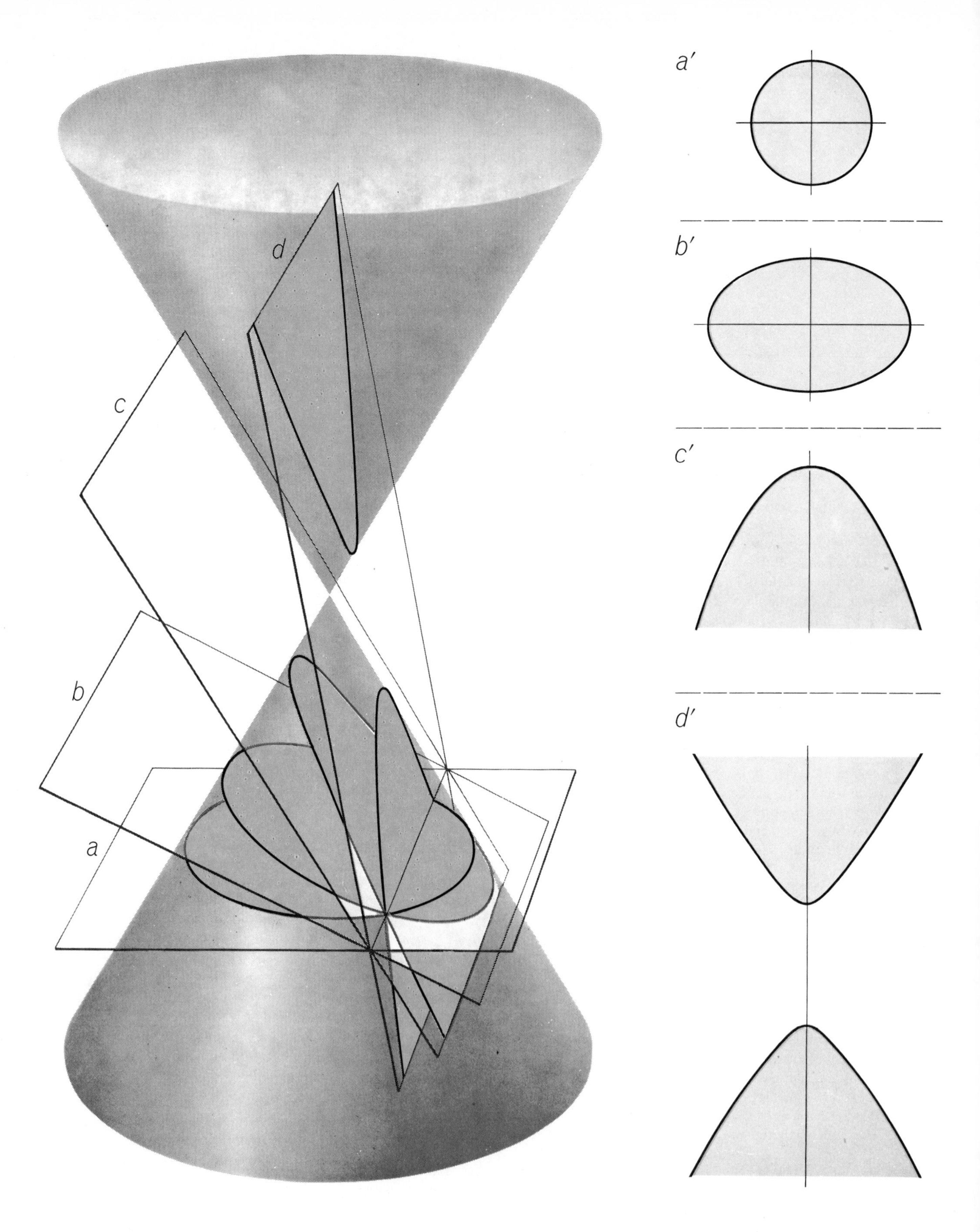

CONIC SECTIONS provide the basic curves with which geometry deals. By following each series of letters, such as *a*, *a′* and *a″*, one can see at left a plane intersecting a cone to produce a curve, at center the resulting curve and at right the corresponding surface. Thus *a′* is a circle and *a″* a sphere, *b′* an ellipse and *b″* an ellipsoid, *c′* a parabola and *c″* a paraboloid, *d′* a hyperbola and *d″*

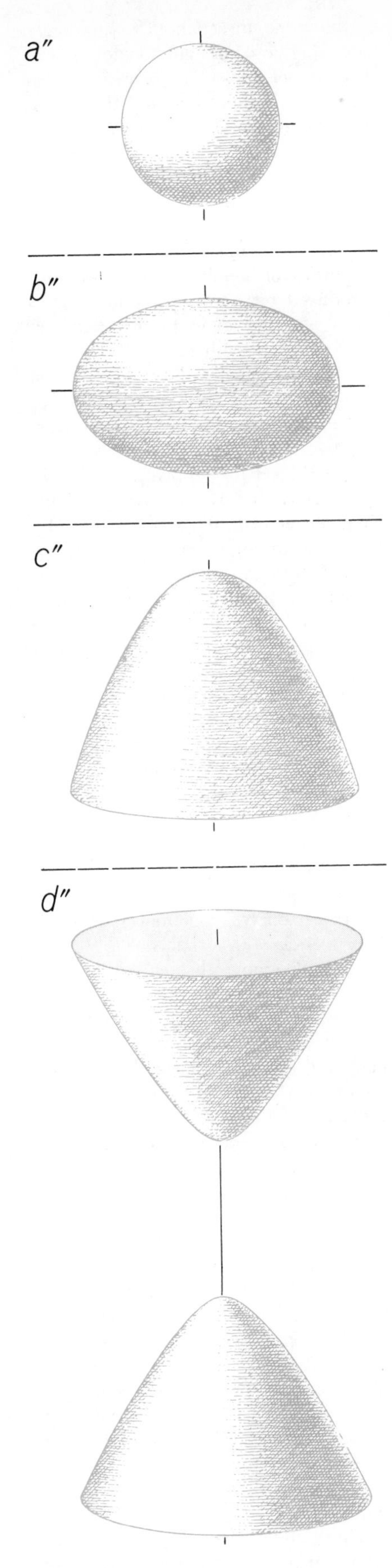

a hyperboloid. Definitions and properties of conic sections were worked out by ancient Greek scholars, notably Apollonius.

subjects was that they did not have a logical foundation; the Alexandrians merely took over the empirically based arithmetical knowledge built up by the Egyptians and Babylonians. Because Euclidean geometry offered the security of proof, it continued for centuries to dominate mathematics. Not until late in the 19th century did mathematicians solve the problem of providing an axiomatic basis for arithmetic and algebra.

Actually geometry consists of several geometries. The first break in the direction of a new geometry was made by Renaissance painters who sought to solve the problem of depicting exactly what the eyes see. Because real scenes are three-dimensional, whereas a painting is flat, it would appear to be impossible to paint realistically. The painters solved their problem by recognizing a fundamental fact about vision. Suppose a man, using one eye, looks through a window at some real scene. He sees the scene because light rays from various points in it travel to his eye. This collection of light rays is called a projection. Since the rays pass through the window, it is possible to mark a point on the window where each light ray pierces it. This collection of points is called a section. What the painters discovered is that the section creates the same impression on the eye as the scene itself does. This is physically understandable [*see top illustration on next page*]. Whether the light rays emanate from particles in the real scene or from points on the window, the same light rays reach the eye. Hence the canvas could contain what appears on the window. Even though this is a one-eye scheme and sight involves two eyes, the painters compensated for the restriction by using diminution of light intensity with distance and by using shadows. How well they succeeded in solving the problems of perspective can be judged by the painting reproduced on page 143.

The use of projection and section raised a basic geometrical question, first voiced by the painters and later taken up by mathematicians. What geometrical properties do an original figure and its section have in common that enable them to create the same impression on the eye? The answer to this question led to new concepts and theorems that ultimately constituted a new branch of geometry called projective geometry [see the article by Morris Kline, "Projective Geometry," page 112 in this volume]. Some of the concepts and theorems are as follows. It is apparent from the top illustration on the next page that the section of the projection of a line is a line and that, if two lines intersect, then a section of the projection of these two intersecting lines will also be two intersecting lines, although the angle between the two lines of the section will generally not be the same as the angle between the two lines in the original figure. It follows that a triangle will give rise to a triangular section and a quadrilateral will give rise to a quadrilateral section.

A more significant example of the properties common to a figure and a section was furnished in the 17th century by the self-educated French architect and engineer Gérard Desargues. In what is now known as Desargues's theorem he showed that for any triangle and any section of any projection of that triangle any pair of corresponding sides will meet in a point and the three points of intersection of the three pairs of corresponding sides lie on one straight line [*see bottom illustration on next page*]. The significance of this and other theorems of projective geometry is that this geometry no longer discusses congruence, similarity, equivalence and other concepts of Euclidean geometry but instead deals with collinearity (points that lie on a line), concurrency (lines that go through a point) and other notions stemming from projection and section.

Projective geometry flourished rather briefly and then was pushed aside temporarily by a rival geometry that appeared on the scene. The rival, which embodied an algebraic approach to geometry, is now called analytic geometry or coordinate geometry. It was motivated by a series of events and discoveries that in the 16th and 17th centuries launched the scientific age in western Europe and brought to the fore the problem of deriving and using the properties of curves and surfaces.

For one thing, the creation by Nicolaus Copernicus and Johannes Kepler of the heliocentric theory of planetary motion made manifest the need for effective methods of working with the conic sections; these curves are the paths of the celestial bodies in such a system. Moreover, by invalidating classical Greek mechanics, which presupposed a stationary earth, the heliocentric theory necessitated a completely new science of motion and therefore the study of curves along which objects move.

Several other forces pushed geometry

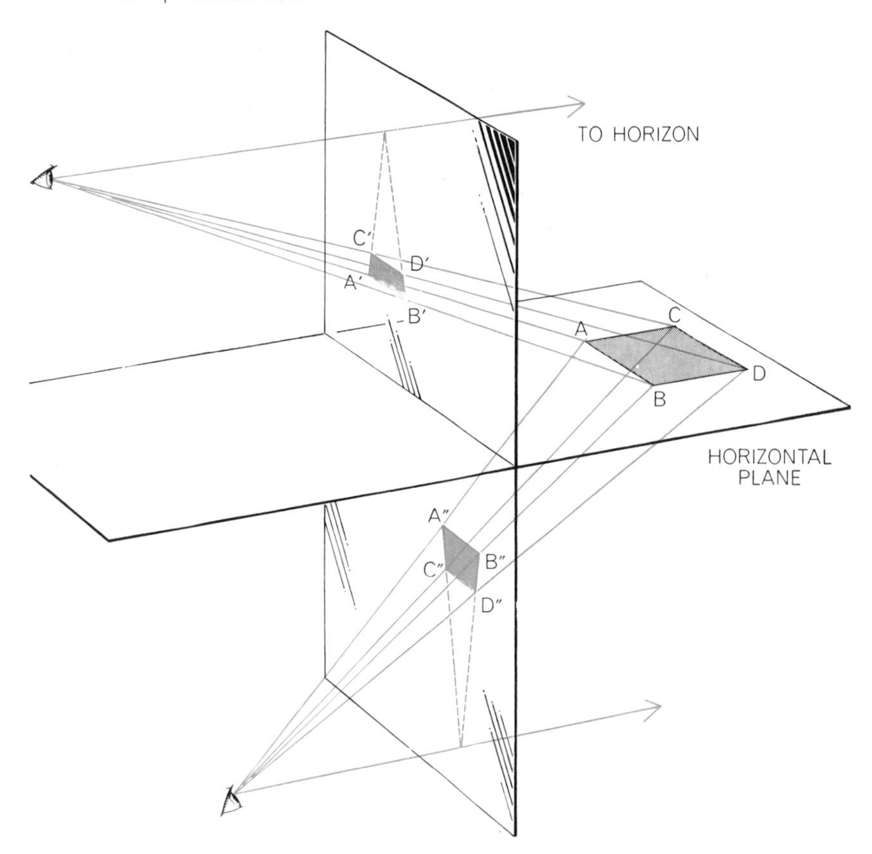

PROJECTION AND SECTION were concepts that arose from the work of artists and helped lead to projective geometry. Projections of a square, such as *ABCD*, to two observers form sections (*color*) on an intersecting plane. In a drawing the square must be represented as a section in order to appear realistic to an observer looking at the drawing.

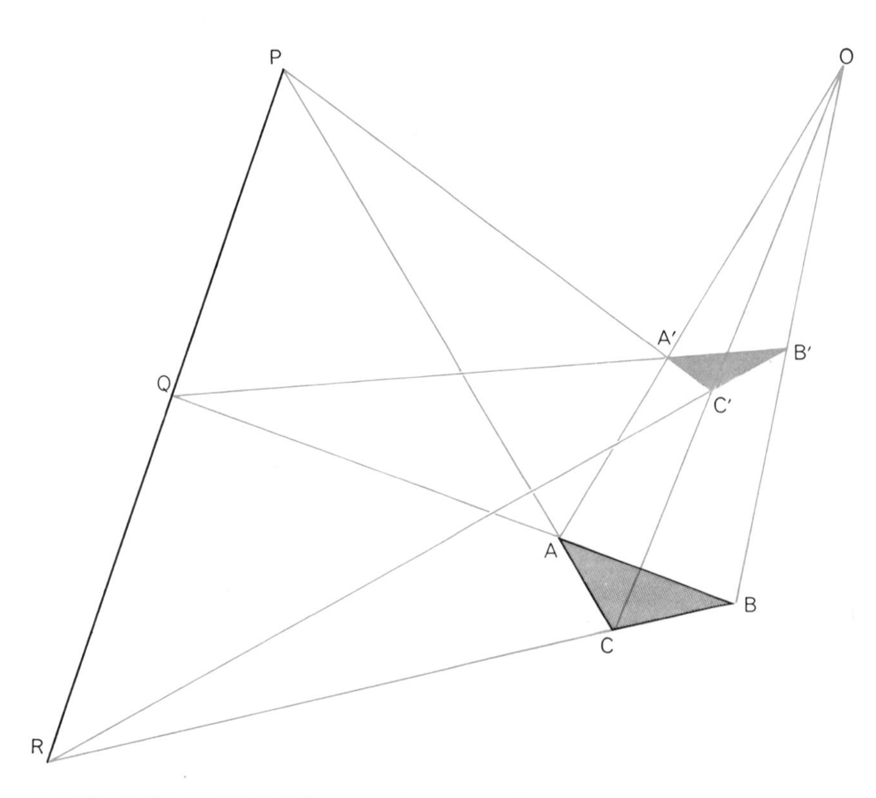

DESARGUES'S THEOREM illustrates the concern of projective geometry with properties common to a figure and its sections. The theorem states that any pair of corresponding sides of a triangle (*ABC*) and a section (*color*) will meet in a point—as, for example, the sides *BC* and *B′C′* meet in point *R*—and that the three points *P*, *Q* and *R* will lie on a line.

in the same direction. The gradually increasing use of gunpowder raised problems of projectile paths. The discovery of the telescope and the microscope motivated the study of lenses. Geographical exploration called for maps and in particular for the correlation of paths on the globe with paths on flat maps. All these problems not only increased the need for knowledge of properties of familiar curves but also introduced new curves. As René Descartes and Pierre de Fermat realized, the Euclidean synthetic methods were too limited to deal with these problems.

Descartes and Fermat, both major contributors to the fast-growing discipline of algebra, saw the potentialities in that subject for supplying methodology to geometry. The analytic geometry they developed replaced curves by equations through the device of a coordinate system. Such a system locates points in a plane or in space by numbers. In a plane the system uses two numbers, an abscissa and an ordinate [*see illustration on opposite page*]. The abscissa expresses the distance of a point from a fixed vertical line, called the *Y* axis; the ordinate expresses the distance of the point from a fixed horizontal line, called the *X* axis. Distances to the right of the *Y* axis or above the *X* axis are positive; distances in the opposite directions are negative.

How does this device enable one to represent curves algebraically? Consider a circle with a radius of five units. A circle, like any other curve, is just a particular collection of points. And if the circle is placed on a coordinate system, then each point on the circle has a pair of coordinates. Since the circle is a particular collection of points, the coordinates of these points are special in some way. The specialized nature is expressed by the equation $x^2 + y^2 = 5^2$. What this equation states is that if one takes the abscissa of any point on the curve and substitutes that for x, and if one takes the ordinate of that same point and substitutes it for y, then the number obtained for $x^2 + y^2$ will be 25. One says that the coordinates of any point on the curve satisfy the equation. Moreover, the coordinates of only those points that do lie on the curve satisfy the equation. In the case of surfaces an equation in three coordinates serves. For example, the equation of a sphere with a five-unit radius is $x^2 + y^2 + z^2 = 25$.

Thus under the Descartes-Fermat scheme points became pairs of numbers, and curves became collections of pairs of numbers subsumed in equations. The properties of curves could be deduced by algebraic processes applied to the equations. With this development the relation between number and geometry had come full circle. The classical Greeks had buried algebra in geometry, but now geometry was eclipsed by algebra. As the mathematicians put it, geometry was arithmetized.

Descartes and Fermat were not entirely correct in expecting that algebraic techniques would supply the effective methodology for working with curves. For instance, those techniques could not cope with slope and curvature, which are fundamental properties of curves. Slope is the rate at which a curve rises or falls per horizontal unit; curvature is the rate at which the direction of the curve changes per unit along the curve. Both rates vary from point to point along all curves except the straight line and the circle. To calculate rates of change that vary from point to point the purely algebraic techniques of Descartes and Fermat are not adequate; the calculus, particularly the differential calculus, must be employed. Indeed, the distinguishing feature of the calculus is its power to yield such rates.

With the aid of the differential calculus the study of curves and surfaces was expedited so much that a new term, differential geometry, was introduced to designate this study. Differential geometry considers a variety of problems beyond the calculation of slope and curvature. It considers in particular the all-important problem of geodesics, or the shortest distance between two points on a surface. Given a surface such as the surface of the earth, what curve joining two given points P and Q on the surface is the shortest distance from P to Q along the surface? If one takes the surface of the earth to be a sphere, the answer is simple. The geodesics are arcs of great circles. (A great circle cuts the sphere in half; the Equator is a great circle but a circle of latitude is not.) If one more accurately takes the surface of the earth to be an ellipsoid, however, the geodesics are more complicated curves and depend on which points P and Q one chooses. The concerns of differential geometry include the curvature of surfaces, map making and surfaces of least area bounded by curves in space, the last of which are so handsomely realized by soap films [*see bottom illustration on page 141*].

From the standpoint of pure geometry the methodologies of analytic geometry and differential geometry were far too successful. Although these subjects treated geometry, the representations of curves were equations and the methods of proof were algebraic or analytic (that is, they involved the use of the calculus). The beautiful geometrical reasoning was abandoned and geometry was submerged in a sea of formulas. The spirit of geometry was banished.

For 150 years the pure geometers remained in the shadows. In the 19th century, however, they found the courage and the vitality to reassert themselves. The revival of geometry was launched by Gaspard Monge (1746–1818), a leading French mathematician and adviser to Napoleon. Monge thought the analysts had sold geometry short and had even handicapped themselves by failing to interpret their analysis geometrically and to use geometrical pictures to help them think. Monge was such an inspiring teacher that he gathered about him a number of very bright pupils, among them Sadi Carnot (1753–1823), Charles J. Brianchon (1796–1832) and Jean Victor Poncelet (1788–1867). These men, imbued by Monge with a fervor for geometry, went beyond the intent of their master and sought to show that geometric methods could accomplish as much and more than the algebraic and analytic methods. To defeat Descartes or, as Carnot put it, "to free geometry from the hieroglyphics of analysis," became the goal.

The geometers, led by Poncelet, turned back to projective geometry, which had been so ruthlessly abandoned in the 17th century. Poncelet, serving as an officer in Napoleon's army, was captured by the Russians and spent the year 1813–1814 in a Russian prison. There he reconstructed without the aid of any books all he had learned from Monge; he then proceeded to create new results in projective geometry.

Projective geometry was actively pursued throughout the 19th century. Curiously an algebraic method, essentially an extension of the method of coordinate geometry, was developed to prove

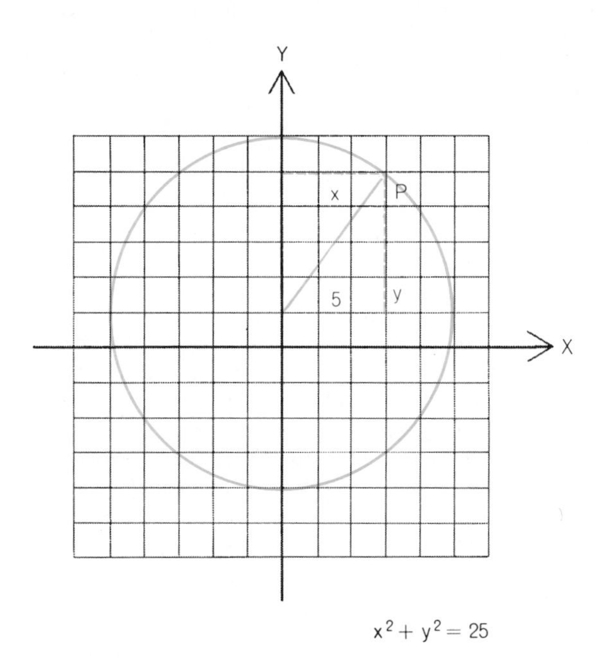

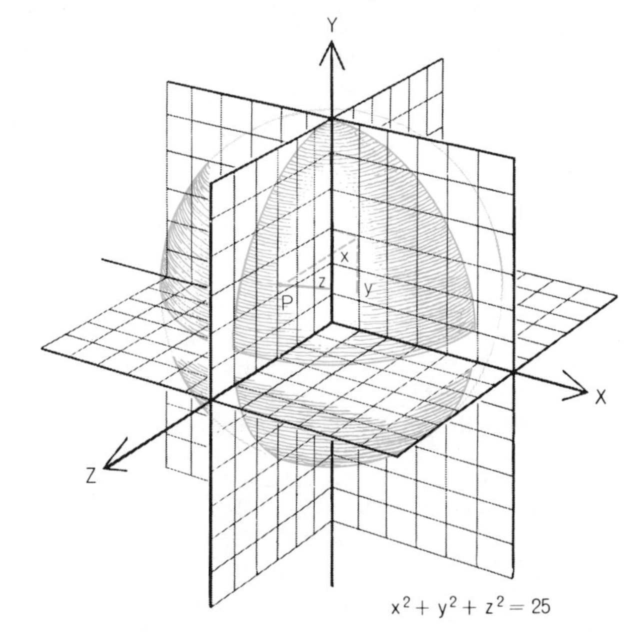

CARTESIAN COORDINATE SYSTEM made it possible to express any shape as an equation. For the circle at left, with a radius of five units, the equation is $x^2 + y^2 = 25$. Any values of x and y that produced 25 in the equation would represent a point on this circle; for the point P, $x = 3$ and $y = 4$. At right is a visualization of the sphere represented by the equation $x^2 + y^2 + z^2 = 25$.

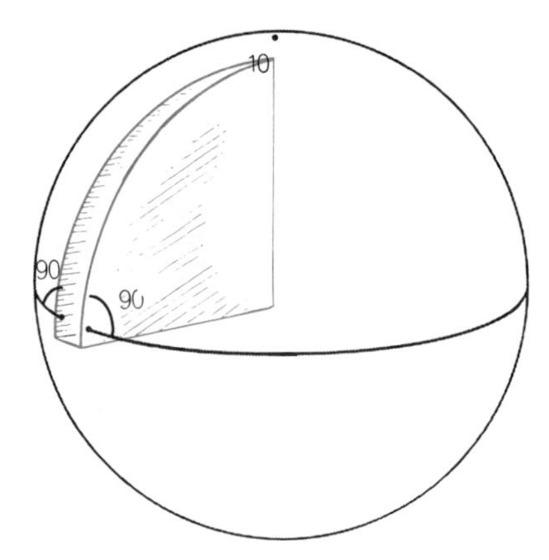

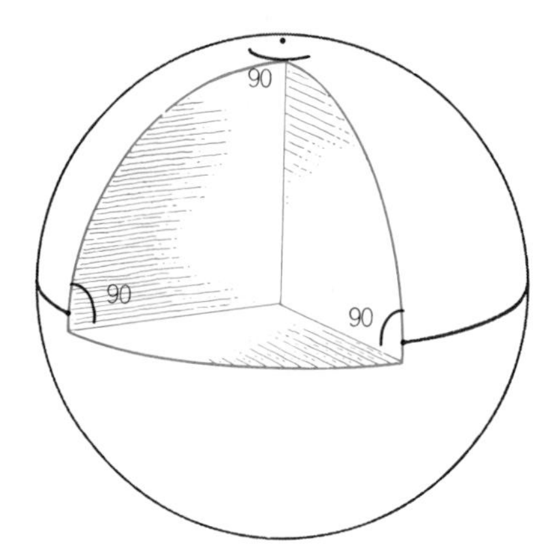

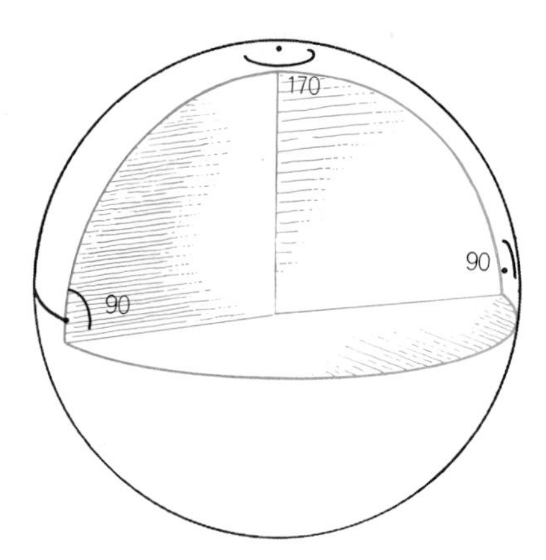

SPHERICAL TRIANGLES can have angles that sum to more than 180 degrees. On the sphere at left the triangle has angles summing to 190 degrees. On the succeeding spheres the angles of the triangles sum respectively to 270 degrees, 350 degrees and 510 de-

its theorems, and to this extent the interests of the pure geometers who launched the revival were subverted. But projective geometry was again put in the shade by another development as dramatic and as weighty as the creation of mathematics by the classical Greeks: the creation of non-Euclidean geometry.

Throughout the long reign of Euclidean geometry many mathematicians were troubled by a slight blemish that seemed to mar the collection of axioms. Apropos of parallel lines, by which is meant two lines in the same plane that do not contain any points in common, Euclid formulated an axiom that reads as follows: If the straight line n cuts the lines l and m so as to make corresponding angles with each line that total less than 180 degrees, then l and m will meet on that side of the line n on which the angles lie. This axiom is essential to the derivation of the most important theorems, among them the theorem that the sum of the angles of a triangle is 180 degrees. The axiom is a bit involved, and there are reasons to believe Euclid himself was not too happy about it. Neither he nor any of the later mathematicians up to about 1800 really doubted the truth of the statement; that is, they had no doubt that it was a correct idealization of the behavior of actual, or physical, lines. What bothered Euclid and his successors was that the axiom was not quite so self-evident as, say, the axiom that any two right angles are equal.

From Greek times on mathematicians sought to replace the axiom on parallels by an equivalent one: an axiom that, together with the other nine axioms of Euclid, would make it possible to deduce the same body of theorems Euclid deduced. Many equivalent axioms were proposed. One of these, which was suggested by the mathematician John Playfair (1748–1819) and is the one usually taught in high schools, states that given a line l and a point P not on l, there is only one line m in the plane of P and l that passes through P and does not meet l.

Playfair's axiom is not only equivalent to Euclid's axiom but it is also simpler and appears to be intuitively convincing; that is, it does seem to state an unquestionable or self-evident property of lines in physical space. Later mathematicians, however, were not satisfied with Playfair's axiom or any of the other proposed equivalents of Euclid's axiom. The reason they were not satisfied was that every proposed substitute directly or indirectly involved an assertion about what happens far out in space. Thus Playfair's axiom asserts that l and m will not meet, no matter how far out these lines are extended. As a matter of fact, Euclid's axiom is superior in this respect because all it asserts is a condition under which lines will meet at some finite distance.

What is objectionable about axioms that assert what happens far out in space? The answer is that they transcend experience. The axioms of Euclidean geometry were supposed to be unquestionable truths about the real world. How can one be sure that two straight lines will extend indefinitely far out into physical space without ever being forced to meet? The problem the mathematicians faced was that Euclid's parallel axiom was not quite self-evident, and that the equivalent axioms, which were seemingly more self-evident, proved on closer examination to be somewhat suspect also.

The problem of the parallel axiom or, as the French mathematician Jean Le Rond d'Alembert put it, "the scandal of geometry," engaged the mathematicians of every period from Greek times up to 1800. The history of these investigations would be worth noting if for no other reason than to see how persistent and critical mathematicians can be. It is necessary here to forgo the history and jump to the results. The truth that destroyed truth was seen clearly by the greatest of all 19th-century mathematicians, Karl Friedrich Gauss (1777–1855). His first point was somewhat technical but essential, namely, that the parallel axiom is independent of the other nine axioms; that is, it is logically possible to choose a contradictory axiom and use it in conjunction with the other nine Euclidean axioms to deduce theorems of a new geometry. Thus one might assume that given a line l and a point P not on l, there is an infinite number of lines through P and in the plane of P and l that do not meet l. Gauss adopted this very axiom and from it and the other nine axioms deduced a number of theorems. Gauss called his new geometry non-Euclidean geometry.

As might be expected, many theorems of the new geometry contradict theorems of Euclidean geometry. The sum of the angles of a triangle in this geometry is always less than 180 degrees. Moreover, the sum varies with the size of the triangle; the closer the area of the triangle is to zero, the closer the angle sum is to 180 degrees.

The existence of a logical alternative to Euclidean geometry was in itself a startling fact. Geometry up to this time had been essentially Euclidean geometry; analytic and differential geometry were merely alternative technical methodologies, and although projective geometry dealt with new concepts and new themes, they were entirely in accord with Euclidean geometry. Non-Euclidean geometry was in conflict with Euclidean geometry.

Gauss's second conclusion was even more disturbing. It was that non-Eu-

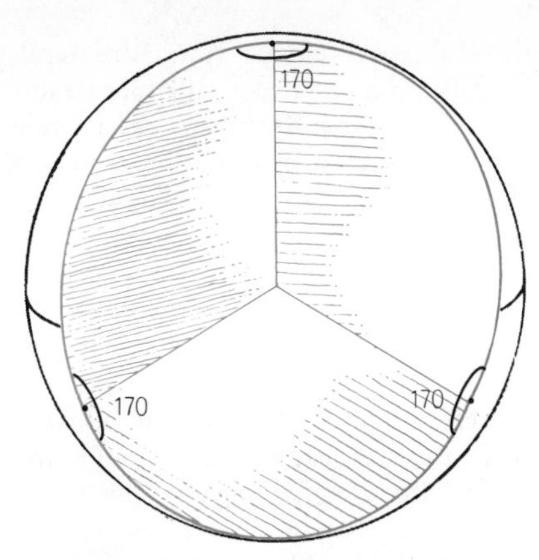

grees. Such triangles typify concepts of Bernhard Riemann's non-Euclidean geometry.

clidean geometry could be used to represent physical space just as well as Euclidean geometry does. This assertion seems at first to be downright nonsense. If the sum of the angles of a triangle is 180 degrees, how could it also be less than 180 degrees? The answer to this seeming impossibility is that the non-Euclidean geometry calls for an angle sum arbitrarily close to 180 degrees when the size of the triangle is small enough. The triangles man usually deals with are small; therefore the angle sums of these triangles might be so close to 180 degrees that measurement of the sum, in view of the inevitable errors of measurement, would not exclude either possibility.

The implications of non-Euclidean geometry are drastic. If both Euclidean and non-Euclidean geometry can represent physical space equally well, which is the truth about space and figures in space? One cannot say. In fact, the choice might not be limited to just these two. This doleful possibility was soon to be realized. The fact gradually forced on the mathematicians is that geometry is not the truth about physical space but the study of possible spaces. Several of these mathematically constructed spaces, differing sharply from one another, could fit physical space equally well as far as experience could decide.

The concept of geometry had then to be revised, but the same was true for the concept of mathematics itself. Since for more than 2,000 years mathematics had been the bastion of truth, non-Euclidean geometry, the triumph of reason, proved to be an intellectual disaster. This new geometry drove home the idea that mathematics, for all its usefulness in organizing thought and advancing the works of man, does not offer truths but is a man-made fable having the semblance of fact.

The new vista opening up in geometry was widened immeasurably by the work of Georg Friedrich Bernhard Riemann (1826–1866). Riemann was one of Gauss's students and undoubtedly acquired from him an interest in the study of the physical world. Riemann's first observation in the field of geometry was that the mathematicians had been deceived into believing the Euclidean parallel axiom was necessarily true. Perhaps they were equally deceived in accepting one or more of the other axioms of Euclid. Riemann fastened at once on the axiom that a straight line is infinite. Experience, he pointed out, does not assure us of the infinitude of the physical straight line. Experience tells us only that in following a straight line we do not come to an end. But neither would one come to an end if one followed the Equator of the earth. In other words, experience tells us only that the straight line is endless or unbounded. If we change the relevant axiom of Euclid accordingly, and if we assume that there are *no* parallel lines, we have another set of axioms from which we can deduce still another non-Euclidean geometry.

In a paper of 1854 entitled "On the Hypotheses Which Underlie Geometry" Riemann launched an even deeper investigation of possible spaces, utilizing only the surest facts about physical space. He constructed a new branch of geometry, now known as Riemannian geometry, that opened up the variety of mathematical spaces a thousandfold [see "The Curvature of Space," by P. Le Corbeiller; *Scientific American* November, 1954.

To appreciate Riemannian geometry one must first perceive that what is chosen as the distance between two points determines the geometry that results. This can be readily seen. Consider three points on the surface of the earth. One can take as the distance between any two the length of the ordinary straight-line segment that joins them through the earth. In this case one obtains a triangle that has all the properties of a Euclidean triangle. In particular, the sum of the angles of this triangle is 180 degrees. One could, however, take as the distance between any two points the distance along the surface of the earth, meaning the distance along the great circle through these points. In this case the three points determine what is called a spherical triangle. Such triangles possess quite different properties. For example, the sum of the angles in them can be any number between 180 degrees and 540 degrees [*see illustration at the top of these two pages*]. This is a fact of spherical geometry.

What Riemann had in mind was a geometry for changing configurations. Suppose one were to try to design a geometry that would fit the surface of a mountainous region. In some places the surface might be flat, in others there might be conical hills and in still others hemispherical hills. The character of the surface changes from place to place,

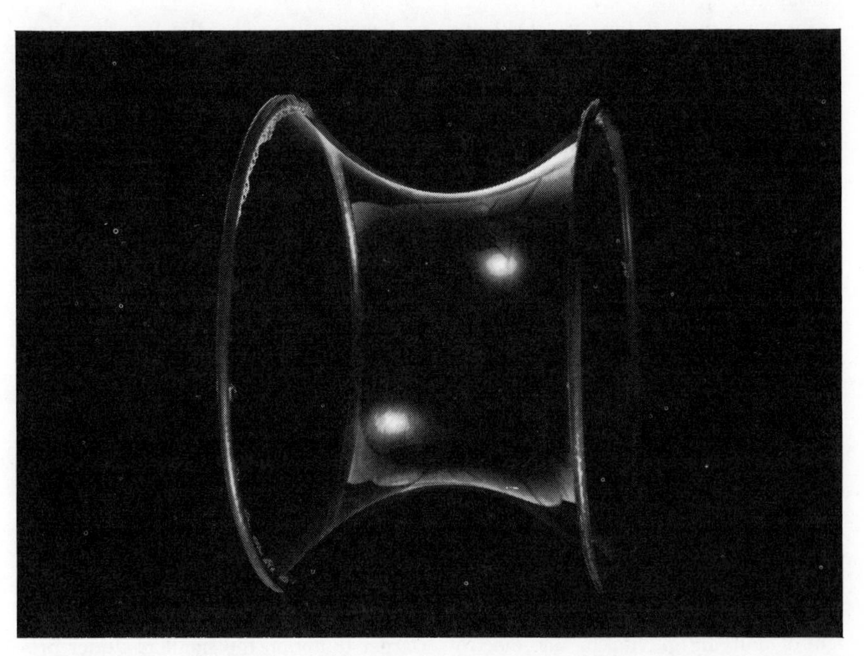

SOAP FILMS, which always assume a shape with the least possible area, illustrate a concern of differential geometry: surfaces of least area bounded by curves in space. Differential geometry is also applicable to problems of map making and curvature of surfaces.

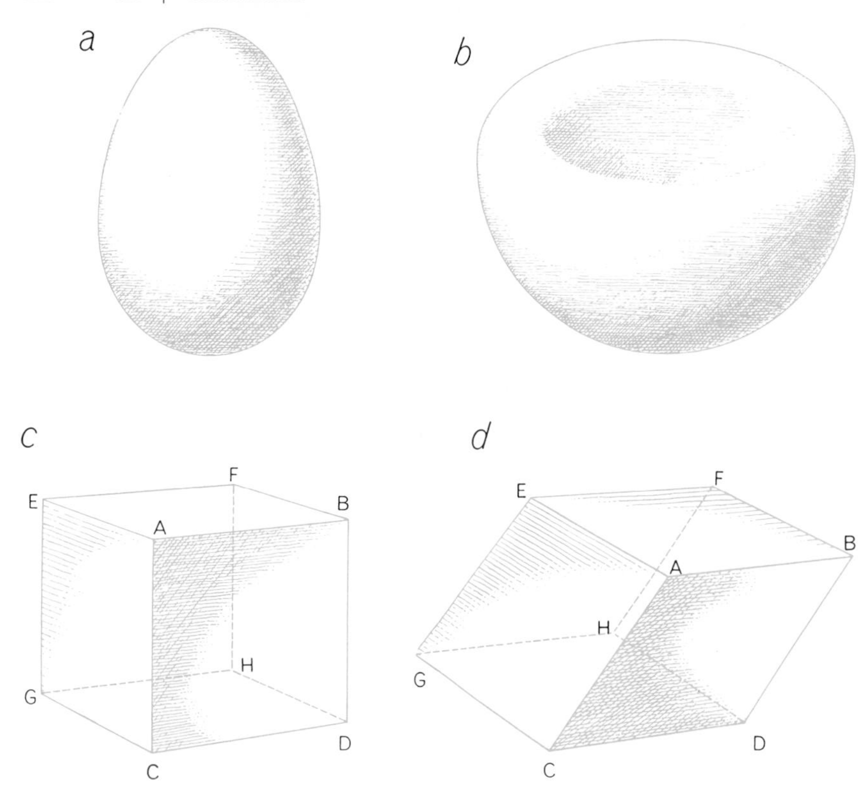

TOPOLOGICAL DEFORMATIONS of familiar shapes are portrayed. A sphere can be deformed into an egg shape (*a*), a squashed-ball shape (*b*), a cube (*c*) and a deformation of the cube (*d*). Each deformation is topologically equivalent to the others and to the sphere.

and so the distance formula that determines the geometry must change from place to place and possibly even from point to point. Riemann proposed, in other words, nonhomogeneous spaces—spaces whose characteristics vary from point to point or spaces with varying curvature.

Riemann died at the age of 40 and was therefore able to do little more than sketch the broad outline of his conception of space. The further development of Riemannian geometry became the task of many men and is still under way. Early in this century the Italian mathematicians Gregorio Ricci and Tullio Levi-Civita made significant contributions. Ricci introduced the tensor calculus, a formalism that enables one to express geometric relations independently of the coordinate system. Levi-Civita brought a concept of parallelism to Riemannian geometry: it provided a way of expressing the Euclidean notion of parallelism for more general spaces.

The creation of the general theory of relativity by Albert Einstein not only stimulated further work in Riemannian geometry but also suggested the problem of unifying gravitation and electromagnetism in one mathematical framework. Toward this end Hermann Weyl in 1918 introduced what he called affinely connected spaces, a concept that uses Levi-Civita's notion of parallelism rather than the notion of distance to relate the points of a space to one another. An expression of distance even more generalized than Riemann's produced the spaces called Finsler spaces.

Riemann was also the founder of topology, another branch of geometry in which research is most active today. During the 1850's he was working with what are now called functions of a complex variable, and he introduced a class of surfaces, called Riemann surfaces, to represent such functions. The properties of the functions proved to be intimately connected with the geometric properties of the surfaces. For any given function, however, the precise shape of the surface was not critical, and so he found it desirable to classify surfaces in accordance with a new principle.

Given two similar figures, for example a large and a small triangle having the same shape, either can be regarded as a deformation, or transformation, of the other, the change being a uniform expansion of the smaller to obtain the larger, or a uniform contraction of the larger to obtain the smaller. Under projection and section the deformation of one figure into another is more radical. Yet even in these deformations a quadrilateral, say, remains a quadrilateral. It is possible to make still more radical deformations. For instance, a circle can be deformed by being bent into an ellipse or into an even more complicated shape, and a sphere can be stretched to assume the shape of an egg. For Riemann's purposes the circle could be replaced by the ellipse and the sphere by the egg shape. On the other hand, a circle, a figure eight and a trefoil were not interchangeable curves, and the sphere, the doughnut-shaped torus and the pretzel-shaped double torus were not interchangeable surfaces.

Hence Riemann was led to consider deformations that permit stretching, bending, contracting and even twisting. Figures that can be obtained from one another by such deformations are said to be homeomorphic, or topologically equivalent. If, however, one tears a figure or contracts it in such a way as to make points coalesce, the new figure is not topologically equivalent to the old one. Thus one can pinch a circle top and bottom and form a figure eight, but the latter is not topologically the same as the original. It is also possible to describe topologically equivalent figures by imagining them to be made of rubber. Then any figure that can be obtained by stretching, bending or contracting but not tearing the rubber would be topologically equivalent to the initial one.

The major problem of topology is to know when two figures are topologically equivalent. This may be difficult to see by looking at the figures, particularly since topology considers three-dimensional and even higher-dimensional figures. For this reason and others one seeks to characterize equivalent figures by some definitive properties so that if two figures possess these properties, they must be topologically equivalent, just as the congruence of two triangles is guaranteed if two sides and the included angle of one triangle are equal to the respective parts of the other. For example, if one draws any closed curve on the surface of a sphere or on an ellipsoid, the curve bounds a region on the surface. This is not true on the torus [*see the illustration on page 144*]. The sphere and the torus are therefore not topologically equivalent. It is possible to characterize closed surfaces in terms of curves that do or do not bound on the surface, but this criterion will not suffice for more complicated surfaces or for higher-dimensional figures.

Although many basic problems of topology remain unsolved, mathematicians make progress where they can,

and in the past 10 years they have turned to the branch called differential topology. In this endeavor they combine the methods of topology and of differential geometry in the hope that two tools will be better than one.

Another enormously active field today is algebraic geometry. Two hundred years ago this subject was an extension of coordinate geometry and was devoted to the study of curves that are more complicated than the conic sections and are represented by equations of degree higher than the second. Since the latter part of the 19th century, however, the proper domain of algebraic geometry has been regarded as the study of the properties of curves, surfaces and higher-dimensional structures defined by algebraic equations and invariant under rational transformations. Such transformations distort a figure more than projective transformations and less than topological transformations do.

Mathematicians, yielding to their propensity to complicate and to algebraicize, have allowed the coordinates in the equations of algebraic geometry to take on complex values and even values in algebraic fields [see "Number," beginning on page 79]. Consequently even the simple equation $x^2 + y^2 = 25$, which when x and y have real values represents the circle discussed previously, can represent a complicated Riemann surface or a structure so unconventional that it can hardly be imagined. The geometry suffers, but the algebra flourishes.

This discussion of geometry as the study of the properties of space and of figures in space may have exhibited the growth, variety and vitality of geometry and the interconnections of the branches with each other and with other divisions of mathematics, but it does not present the full nature of modern geometry. It is often said that algebra is a language. So is geometry.

Today mathematicians pursue the subject of abstract spaces, and one might infer from the term that the pursuit involves some highly idealized, esoteric spaces. This is true, but the major use of the theory of abstract spaces—indeed, historically the motivation for its study—is to expedite the use of classes of functions in analysis. The "points" of an abstract space are usually functions, and the distance between two points is some significant measure of a difference between two functions. Thus one might be interested in studying functions such as x^2, $3x^2$ and $x^3 - 2x$ and be interested in the values of these functions as x varies from 0 to 1. One could define the distance between any two of these functions as the largest numerical difference between the two for all values of x between 0 and 1. Such function spaces prove to be infinite-dimensional. The Hilbert spaces and Banach spaces about which one hears much today are function spaces. On the mathematical side these are important in the subject known as functional analysis, which is now the chief tool in quantum mechanics.

Why talk about spaces when one is really dealing with functions? It is because the geometrical mode of thinking is helpful and even suggestive of theorems about functions. What may be complicated and obscure when formulated analytically may in the geometrical interpretation be intuitively obvious. The study of abstract spaces is, surprisingly, part of topology because the properties of these structures that are important, whether the structures are regarded as actual spaces or as collections of functions, are preserved, or invariant, under topological transformations.

The subject of abstract spaces clearly exhibits the abstractness of modern mathematics. Geometry supplies models not only of physical space but also of any structure whose concepts and properties fit the geometric framework.

In still another vital respect geometry proves to be far more than the receptacle for matter. The present century is witnessing the realization of an assertion by Descartes that physics could be geometrized. In the theory of relativity, one of the two most notable scientific advances of this century (quantum theory is the other), the gravitational effect of gross matter has been reduced to geometry. Just as the geometry of a mountainous region requires a distance formula that varies from place to place to represent the varying shape of the land, so Einstein's geometry has a variable distance formula to represent the different masses in space. Matter determines the geometry, and the geometry as a result accounts for phenomena previously ascribed to gravitation.

Geometry has ingested part of reality and may have to ingest all of it. Today in quantum mechanics physicists are striving to resolve the seemingly contradictory wave and particle properties of subatomic matter, and they may have to generate both from quanta of space. Perhaps matter itself will also dissolve into pure space.

If one assesses today the competition between number and geometry, one must admit that insofar as methodology of proof is concerned, geometry has largely given way to algebra and analysis. The geometric treatment of complicated structures and of course of higher-dimensional spaces can, as Descartes complained of Euclidean geometry, "exercise the understanding only on condition of greatly fatiguing the imagination." Moreover, the quantitative needs of science can be met only by ultimate recourse to number.

Geometry, however, supplies sustenance and meaning to bare formulas. Geometry remains the major source of rich and fruitful intuitions, which in turn supply creative power to mathematics. Most mathematicians think in terms of geometric schemes, even though they leave no trace of that scaffolding when they present the complicated analytical structures. One can still believe Plato's statement that "geometry draws the soul toward truth."

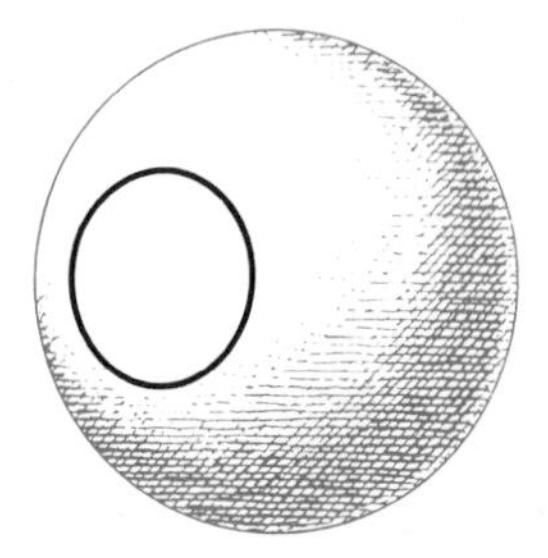

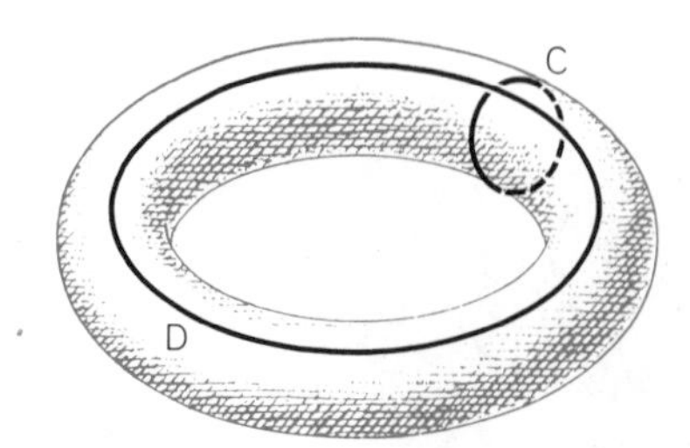

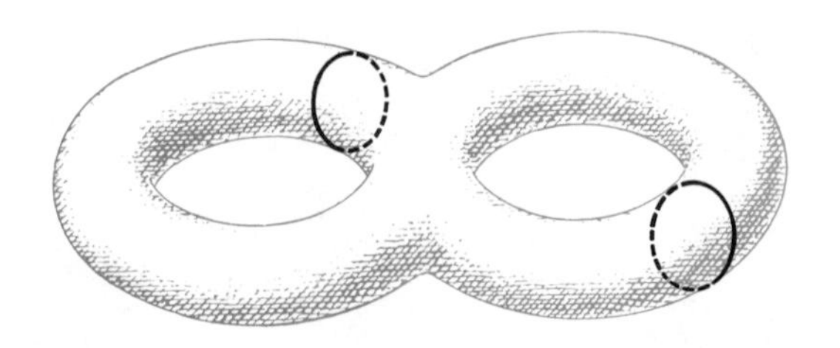

TOPOLOGICAL EQUIVALENCE of surfaces can be determined by drawing closed curves on the figures. If each curve bounds an area on a surface, the surface is topologically equivalent to a sphere. The type of curve drawn on the sphere at left does not bound an area on the torus at center or on the double torus at right; thus the latter figures are not topologically equivalent to a sphere.

IV

STATISTICS AND PROBABILITY

IV STATISTICS AND PROBABILITY

INTRODUCTION

The kind of reasoning most people associate with mathematics is deductive, wherein the conclusions follow inescapably and indubitably from the premises. For example, if all good cars are expensive and if the Hupmobile is a good car, one must accept the fact that the Hupmobile is expensive. This type of reasoning can be used in simple situations of ordinary life and is often employed in involved legal arguments, but it falls far short of what is needed to decide some of the most important problems people face. The career one chooses, the efficacy of a medical treatment, and the measures a government adopts to check inflation or to reduce unemployment are usually not determined by deduction from premises; if they are, the conclusions are most likely to be worthless because reliable premises are not available. Even the most arrogant mathematician, who likes to boast of the power of deductive reasoning, will admit that while the investigation of inanimate matter may yield reliable premises, the study of phenomena involving humanity does not.

Of course sensible mathematicians—and there are such people—have long been aware of this fact, and because most mathematicians seek to be helpful in all human affairs, they have devised methods of arriving at useful conclusions that do not call for deductive reasoning from unquestionable premises. These methods are incorporated in the subjects of statistics and probability. The historical origins of these subjects are intriguing but will not be related here. They can be found in some of the books listed in the suggested readings. (See also the article on Pascal in the first section.) More important is what these subjects involve and accomplish.

The most effective attack on economic, biological, and sociological problems, all of which involve human behavior, is begun by gathering statistics. The value of stocks and bonds on any one day can be found in the daily newspapers. The decennial census of our population, although not entirely accurate, is another example of statistical information. However, even when complete data are available and one wishes to present the substance or the essential information contained in those data, several problems arise. If, with the wages of every employee in some industry available, one wishes to present the average earnings, what kind of average should one use? Should it be the arithmetic mean of all the individual wages or some other kind of average? Actually several kinds of averages—mean, median, and mode—are in common use, and each conveys different information about the earnings of the people involved.

Averages are informative. An average earning of $10,000, if the type of average chosen is suitable, tells us a great deal about the standard of living of the people involved. However, let us take a rather simple situation in which all three averages are the same $10,000, which, let us further suppose, permits a decent living standard. It can still be true that 25% or more of the workers earn far less than a living wage, and 1%, perhaps the executives of large corporations, earn several hundred thousand dollars each. This information is not revealed by the averages. Yet it surely would be important to any government concerned with the welfare of its people or with the formulation of a reasonable tax policy. An additional technique for studying earnings and many other human and societal factors is the frequency distribution. This is a graph that shows the number of people who achieved any one rating, such as income or height from the lowest to the highest.

Still another kind of problem is tackled by using statistics. Does success in academic work ensure financial success in later life? Let us assume that the records of all college graduates for the year 1940 are available. We can then obtain, say, the average grade of each individual and the subsequent average earnings of that individual from graduation to age 60. A mere collation of the data tells us nothing. But statisticians have devised a measure of correlation between two such possibly related factors, and this measure gives us a fairly reliable indication of the extent to which academic success promises financial success.

A deeper problem of statistics arises in connection with sampling. Opinion polls on an issue, for example, the choice of a candidate for public office, are taken by sampling. If one million voters are eligible to vote but only one thousand people are canvassed for their opinions, how reliable are the percentages in favor of each candidate? Clearly these percentages may be in error and yet offer some information on the outcome of the election. The information is reliable only to a certain degree of probability. The technique of sampling then obliges statisticians to consider probability. How can we use sampling and still ensure that the sample does represent the entire population to a very high order of accuracy? Our first article in this section, Warren Weaver's article on statistics, surveys the subject and in particular considers the problems raised by sampling.

We have seen that the attempt to evaluate statistics may involve probability. This use of probability would in itself warrant extensive consideration of the subject. However, problems of probability arise directly. Games of chance are simple examples. What is the probability of throwing a seven on a throw of two dice, or the probability of picking one ace when choosing a single card from the usual deck of 52 or when choosing two cards? Far more serious is the question being debated today of the probability of a nuclear reactor leaking dangerous radiation. Less crucial questions, such as the probability of life on Mars, are also being considered.

To estimate that the probability is high or low does not suffice as a basis for deciding what action to take in a given situation. Clearly the concept of probability must be carefully formulated and mathematical laws or formulas derived that will answer the various questions. It would seem that this is easily done. Consider throwing a die. There are six faces. It is reasonable to argue that all faces are equally likely to turn up. Hence the three should show up one out of six times; that is, the probability is 1/6. But the concept of "equal likelihood" is fraught with difficulties. Suppose two coins are tossed. There appear to be three equally likely possibilities—two heads, two tails, and one head and one tail. Hence the probability of each should be 1/3. In fact, a famous mathematician, Jean Le Rond d'Alembert, argued thus. But the reasoning is incorrect. This is readily seen if we take one coin to be a dime and the other a quarter. The possibilities then are two heads, two tails, a head on the dime and a tail on the quarter, and a tail on the dime and a head on the quarter. Thus there are four equally likely possibilities. We are also prone to use the concept of equal likelihood when the possible outcomes of an event do indeed seem to warrant its use and we have no sound information to gainsay doing so. For example, suppose we have no information before its birth about whether a baby will be a boy or a girl. We might well assume that the probability of either possibility is 1/2. But statistics show that the ratio of boys to girls is 51 to 49; that is, the probability of a boy is 51/100 and of a girl is 49/100.

The concept of equal likelihood is not at all applicable to many important situations in which we would like to know probabilities. Life insurance companies must know the probability that a man aged 50, say, will live 20 more years. In such a situation the probability is calculated by resorting to statistics. The records for a large number of men are examined and if, for example, of 8000 alive at age 50, 5000 are alive at age 70, the probability of a man of age 50 living to be 70 is taken to be 5/8. Here one uses what is called the frequency definition of probability.

The use of statistics to calculate probabilities is also fraught with difficulties. The probability that the sun will rise tomorrow, since it has been observed to rise every day for the past 50,000 days, say, is certainty. But the probability of a person who has lived 50,000 days living one more day is not certainty and in fact decreases as the days pass.

Neither equal likelihood nor statistics serves to calculate the probability that a nuclear reactor might melt down. Were we to build many reactors and rely upon statistics, the need to calculate the probability would be precluded. We see then that the concept of probability must take into account various bases for computation and that even those bases, such as equal likelihood, must be examined with care if they are to yield correct results.

The articles by Carnap and Weaver discuss the concept of probability and point out the uses and the pitfalls. Gardner's article shows us how careful we must be in using either the equal likelihood or the frequency definition. The somewhat more advanced article by Kac listed in the suggested readings gives excellent examples of uses that are certainly not common knowledge and are even surprising. Additional applications will be found in the articles on quality control and operations research in the last section. All these articles bear out what Laplace pointed out 150 years ago, namely, that probability is the basis for deciding the most important questions of life and is therefore one of the most valuable contributions of mathematics.

SUGGESTED READINGS

Cardano, Gerolamo. 1961. *The Book on Games of Chance.* Holt, Rinehart and Winston, New York.
David, F. N. 1962. *Games, Gods and Gambling.* Griffin, London.
Feller, William. 1950. *An Introduction to Probability Theory and Its Uses.* Wiley, New York.
Gnedenko, B. V., and A. Ya. Khinchin. 1961. *An Elementary Introduction to the Theory of Probability.* W. H. Freeman and Company, San Francisco.
Hacking, Ian. 1975. *The Emergence of Probability.* Cambridge University Press, New York.
Huff, Darrell. 1954. *How to Lie with Statistics.* Norton, New York.
Kac, Mark. 1964. "Probability." *Scientific American,* September.
Levinson, Horace C. 1950. *The Science of Chance.* Rinehart, New York.
Moroney, N. J. 1951. *Facts from Figures.* Penguin, Baltimore.
Niven, Ivan. 1965. *Mathematics of Choice.* Random House, New York.
Ormell, C. P. 1968. *An Introduction to Probability and Statistics.* Oliver and Boyd, London.
Reichmann, W. J. 1962. *Use and Abuse of Statistics.* Oxford University Press, New York.
Tanur, Judith M. (ed.). 1972. *Statistics, a Guide to the Unknown.* Holden-Day, San Francisco.

Statistics

26

by Warren Weaver
January 1952

The word usually suggests masses of numerical information, but it also describes that department of mathematics which grapples with the complexity of nature by means of samples

Statistical thinking will one day be as necessary for efficient citizenship as the ability to read and write.
—H. G. Wells

THERE are two main forms of logical thinking—deduction and induction. For the former we are chiefly indebted to the Greeks, who first saw clearly revealed the great power of announcing general axioms or assumptions and deducing from these a useful array of implied propositions Inductive thinking, which has been called "the second great stage of intellectual liberation," did not begin to become a systematic tool of man until late in the 18th century. Induction proceeds in the opposite direction from deduction. Starting from the facts of experience, it leads us to infer general conclusions.

Deductive reasoning is definite and absolute. Its specific inferences follow inescapably from the general assumptions. Inductive reasoning, on the other hand, is uncertain inference. The concrete and special facts of experience, from which inductive reasoning begins, generally do not lead inexorably to categorical general conclusions. Rather they lead to judgments concerning the plausibility of various general conclusions.

Francis Bacon was the first properly to emphasize inductive methods as the basis of scientific procedure, but it was not until 1763 that the English clergyman Thomas Bayes gave the first mathematical basis to this branch of logic. To get an idea of what Bayes did, let us look at an admittedly artificial example. Suppose you have a closed box containing a large number of black and white balls. You do not know the proportion of black to white but have reason to think that the odds are two to one that there are about equal numbers of black and white balls. You reach into this box, take out a sample of balls and find that three fourths of the sample are black. Now before taking this sample you tended strongly to think that the unknown mixture was half white, half black. After taking the sample you clearly should change your thinking and begin to lean toward the view that black balls outnumber the white in the box. Bayes worked out a theorem which indicates exactly how opinions held before the experiment should be modified by the evidence of the sample. Though the usefulness of this theorem itself has proved to be very limited, it was the beginning of the whole modern theory of statistics, and thus of a mathematical theory of inductive reasoning.

WHAT'S this, you will say; is statistics something as general and profound as all that? Isn't statistics merely the name for the numerical information with which propagandists try to convince and sometimes even to confuse us?

The word statistics has two somewhat different meanings. In familiar usage, to be sure, statistics does mean simply numerical information, usually arranged in tables or graphs. It is in this sense that we say *The World Almanac* contains a great deal of useful statistics. But more broadly, and more technically, statistics is the name for that science and art which deals with uncertain inference—which uses numbers to find out something about nature and experience.

The importance of inductive reasoning depends on the basic fact that, apart from trivial exceptions, the events and phenomena of nature are too multiform, too numerous, too extensive or too inaccessible to permit complete observation. As the author of *Ecclesiastes* remarked, "No man can find out the work that God maketh from the beginning to the end." We can't measure cosmic rays everywhere and all the time. We can't try a new drug on everybody. We can't test every shell or bomb we manufacture—for one thing, there would then be none to use. So we have to content ourselves with samples. The measurements involved in every scientific experiment constitute a sample of that unlimited set of measurements which would result if one performed the same experiment over and over indefinitely. This total set of potential measurements is referred to as the population. Almost always one is interested in the sample only insofar as it is capable of revealing something about the population from which it came.

The four principal questions to be asked about samples are these: 1) How can one describe the sample usefully and clearly? 2) From the evidence of this sample how does one best infer conclusions concerning the total population? 3) How reliable are these conclusions? 4) How should samples be taken in order that they may be as illuminating and dependable as possible?

Question 1 pretty well covers the subject matter of elementary statistics. Tables, graphs, bar and pie diagrams and the schematic pictorial representations which can be so useful (and sometimes so deceptive) are all ways of summarizing the evidence of a sample. Averages and other related quantities—arithmetical means, medians, modes, geometric means, harmonic means, quartiles, deciles, and so on—are useful for similar purposes; and these also must be used with discretion if they are to be really illuminating. The arithmetical mean income of a certain Princeton class five years after graduation, for example, is not a very useful figure if the class happens to include one man who has an income of half a million dollars.

Descriptive statistics of this sort is concerned with broad and vague questions like "What's going on here?"; and the answers returned are a not unworthy example of "doing one's damndest with one's mind, no holds barred," to use Percy W. Bridgman's phrase. It is only when we pass to Questions 2, 3 and 4, however, that we get to the heart of modern mathematical statistics.

THESE three questions have to do with different aspects of one common problem: namely, how much can one learn, and how reliably, about a population by taking and analyzing a sample from that population? First of all, what sort of knowledge about a population is possible?

Remember that a population, as one

uses the word in statistics, is a collection—usually a large or even infinite collection—of numbers which are measurements of something. It is not possible in the case of an infinite collection, and usually not feasible in other cases, to describe one at a time all the individual measurements that make up the population. So what one does is to lump similar or nearly similar measurements together, describing the population by telling what fraction of all measurements are of this approximate size, what fraction of that size, and so on. This is done by stating in a table, a graph or a formula just what fraction of the whole population of values fall within any stated interval of values. When this is done graphically, the result is a frequency curve, which describes the distribution of measurements in the population in question. The most widely useful population distribution is the so-called normal or Gaussian probability distribution, which takes the form of the familiar bell-shaped curve. A frequency curve can, of course, be described by stating its mathematical equation.

It is frequently useful to give a condensed description of a distribution. If circumstances make it necessary to be content with only two items of information, then one would usually choose the average (which the statistician calls the arithmetical mean) and the variance. The variance is defined as the average of the squares of the differences between all measurements of the population and the mean of the population. It is a very useful measure of the degree of scatter of the measurements, being relatively small when the distribution clusters closely about the mean and relatively large when the distribution is a widely spread-out one. The square root of the variance is called the standard deviation. A small variance always means, of course, a small standard deviation, and *vice versa*.

The statistician's shorthand usually denotes the arithmetical mean by the Greek letter mu, the variance by sigma squared and the standard deviation by sigma. In the special case of normal distributions, a knowledge of the mean, mu, and of the standard deviation, sigma, is sufficient to pick out of all possible normal distributions the specific one in question. Thus the mathematical formula for the normal distribution curve need involve, in addition to the variable, just the two quantities mu and sigma. More complicated distributions may depend upon more than two.

Using the notions just introduced, we can now restate our last three questions:

2) Using the evidence of a sample, what can one say about the population distribution?

3) How can one characterize the reliability of these estimates?

4) How can one select the sample so as to produce the most reliable estimates?

BEFORE going on to indicate the kind of answer modern statistical theory can give to these three questions, it would be well to stop a moment to consider once more, and somewhat more accurately now, the relation between the descriptive statistician who deals only with our original Question 1, and the mathematical statistician who deals with Questions 2, 3 and 4.

In seeking to summarize and describe a sample, the descriptive statistician is in fact trying to shed some light, however dim and indirect, on the nature of the population. Thus he is often trying to give some sort of informal and loose answer to Question 2; and he frequently succeeds in a really useful way. He differs from the mathematical statistician in that he uses only elementary mathematical tools and is therefore unable to give any really precise answers to Question 2 or any answers at all to Questions 3 and 4.

The problem of drawing inferences concerning a population from a sample is a problem in probability. There is an obvious analogy between this procedure and the artificial case of sampling the box of colored balls. It is important to remember that when you take a sample of colored balls out of an unknown mixture, you cannot make simple probability statements about the mixture unless you start out by having some idea about what is in the box. In technical terms, this means that you have to have knowledge, before the drawing of the sample, of the *a priori* probabilities of all possible mixtures that might be in the box. As we have mentioned earlier, Bayes' theorem furnishes a basis for modifying this prior opinion, but it is powerless to originate an opinion.

Bayes' theorem furnishes a sound and simple procedure. But unfortunately it is very seldom applicable to really serious problems of statistical theory, for the good reason that in such situations one seldom has any positive knowledge of the *a priori* probabilities. Consequently it is necessary for statistics to take recourse to more complicated and more subtle theorems.

It is evident that the statistician can never say for certain what the parent population is merely by sampling it, because the samples will vary. If, for example, you draw from a mixture containing 60 per cent white balls and 40 per cent black, you will by no means get this 60-40 ratio of white to black in every sample you take. However, for a given kind of parent population and with suitable methods of sampling it is possible to work out theoretically the pattern of variation for samples. This knowledge of the pattern of sample variability gives the statistician a toe hold. It permits him

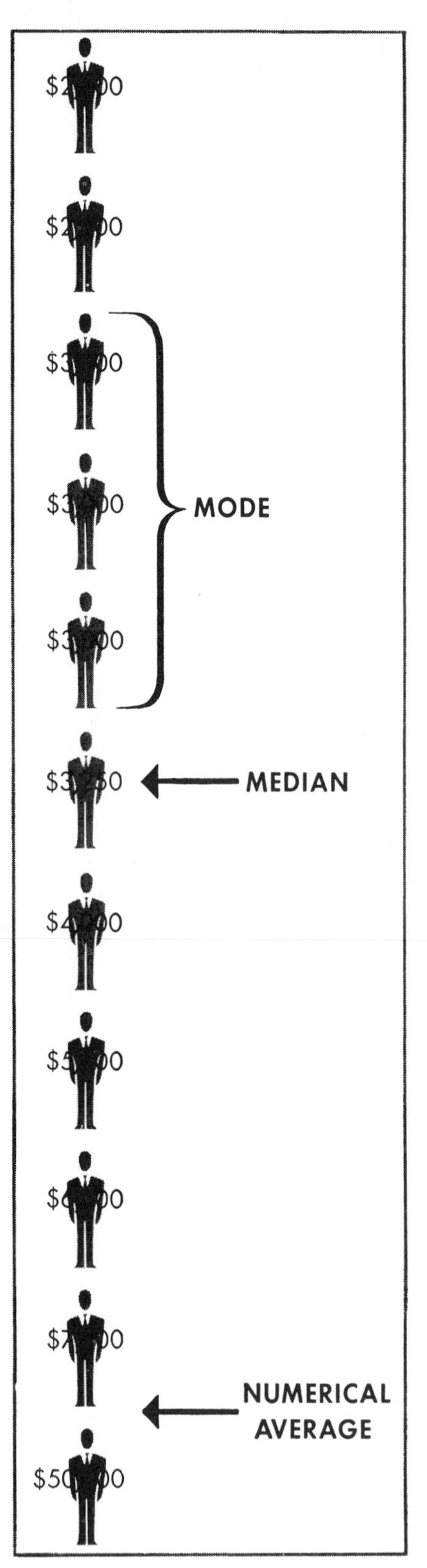

ELEVEN SALARIES illustrate how a single number may be used to characterize a set of numbers. The numerical average of the salaries ($8,159) is heavily influenced by one very large salary ($50,000). Hence the average is not as representative of this set of numbers as the median (the salary in the middle) or the mode (the salary that occurs most often).

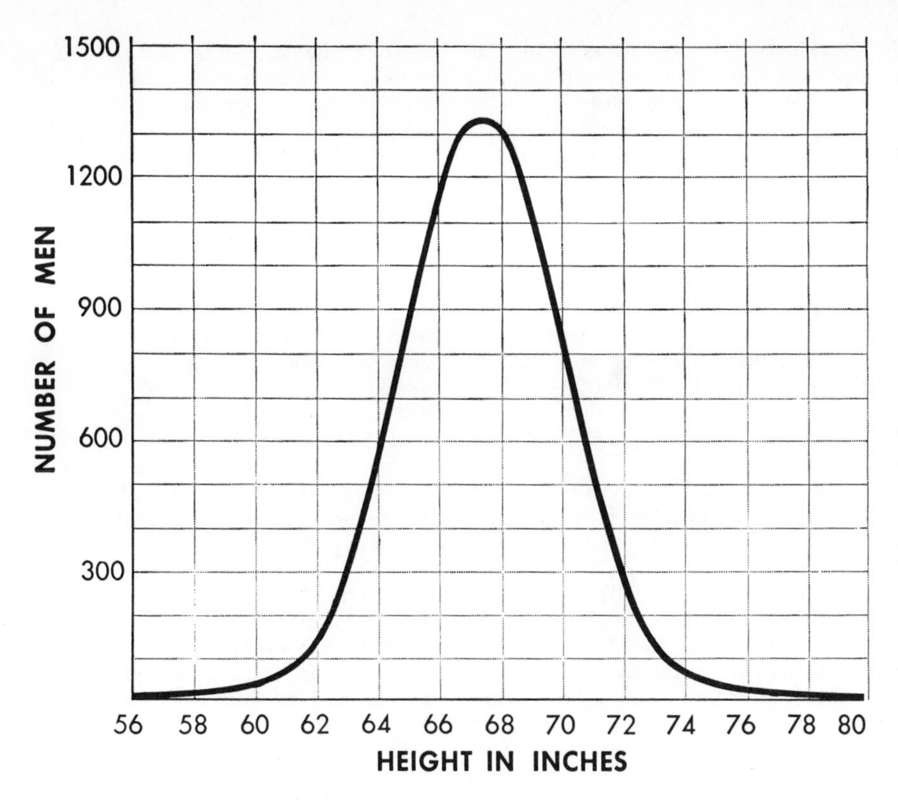

FREQUENCY DISTRIBUTION OF HEIGHT for 8,585 men is expressed as a normal distribution curve. The actual points from which the curve was derived are not shown, but only a theoretical approximation of them.

to look at samples and draw inferences about the parent population.

The pattern of variation depends not only on the population being sampled; it also depends sensitively upon the method of sampling. If you were interested in family sizes in the U. S. and took your evidence solely from houses with eight rooms or more, obviously the sample pattern would be atypical. When samples are deliberately selected in an atypical way, we call that rigging the evidence. But often samples have quite innocently been taken in an atypical way, as Mr. George Gallup will remember.

It turns out that in general the only good method of sampling is a random method. In a random method the sample is picked in accordance with purely probabilistic criteria, personal choice or prejudice being completely excluded. Suppose television tubes are passing an inspector on a moving belt and it is desired to test on the average one of every six tubes in a random way. The inspector could throw two dice each time a tube passed him and take off the tube for test only when he threw a double number. This, of course, would happen on the average once in six throws of the dice. The tubes thus selected would be a random sample of the whole population of tubes.

NOW we must note an important fact about the pattern of variation of random samples. Suppose you have a parent population which is normally distributed, with a certain mean value and a certain standard deviation. Suppose you take from this population random samples consisting of a certain number of items, *n*. Compute the mean for each sample. You will find that this new population of means is normally distributed, just as the parent population was. Because an averaging process has entered in, it is more tightly clustered than was the original population. In fact, its standard deviation is found by dividing the standard deviation of the parent population by the square root of *n*, the number of items in each sample. Thus if the samples each contain 64 items, the standard deviation of the means of these samples will be one eighth of the standard deviation of the parent population.

The fact that samples from a normal population have means which are themselves normally distributed tells us that normally distributed populations have a kind of reproductive character. Their offspring (samples) inherit their most important character (normality). And it is comforting to know further that if large samples are taken from almost any kind of population, their means also have almost normal distributions.

The importance of sample pattern can easily be illustrated by a concrete example in manufacturing. A manufacturer makes large numbers of a part which in one dimension should measure one inch with high accuracy. Random samples of the product are measured. If the samples consistently average more than one inch, he knows there is some systematic error in his manufacturing procedure. But if the mean of the samples is just an inch, and the variations from the mean fall into the pattern of distribution which is theoretically to be expected when the parent population itself is normally distributed about an average of one inch, the manufacturer can conclude that systematic errors have been eliminated, and that his manufacturing process is "under control."

Sampling is not merely convenient; it is often the only possible way to deal with a problem. In the social sciences particularly it opens up fields of inquiry which would otherwise be quite inaccessible. The British Ministry of Labor was able to carry out a most useful study of working-family budgets for the entire nation from detailed figures on the expenditures of only 9,000 families over a period of four weeks. Without sampling such studies would be wholly impracticable.

TO return to our main argument. We are now better prepared to deal with Questions 2 and 3. (We shall omit Question 4, on the design of experiments, because it is a large subject that would make this article far too long.)

Suppose that the mean lifetime of a certain type of electronic tube is known to be 10,000 hours and the standard deviation 800 hours. The engineers now develop a new design of tube. A sample of 64 of the new-type tubes is tested, and the mean life of the 64 tubes in the sample is found to be 10,200 hours—200 hours longer than the mean life of the old population.

Now the new design may actually be no longer-lived than the old. In that case the sample of 64 just happened to be a somewhat better than average sample. Clearly what the engineers want to know is whether the apparent improvement of 200 hours is real or merely due to a chance variation.

The amount of variation one expects from chance can be estimated by comparing the actual deviation with the standard deviation. Since the standard deviation of the means of samples of 64 items is one eighth the standard deviation of the parent population, in this case the standard deviation in the means of such samples would be one eighth of 800, or 100 hours. Hence the apparent improvement of 200 hours in our sample of new-type tubes is twice the standard deviation in the means of such samples.

Probability theory tells the statistician that the odds are 19 to 1 against a difference of this size between the sample and population means occurring merely by chance. He therefore reports: "It seems sensible to conclude that this mildly rare event has not occurred, that on the contrary the sample of 64 in fact came from a new population with a higher mean life. In other words, I conclude that the new design is probably an improvement."

This is one of the common ways of dealing with such a situation. There are various rather complicated and subtle weaknesses in the argument just given, but we need not go into them here. A more satisfactory way of dealing with the same problem would be to apply the modern theory of statistical estimation, which involves the use of so-called confidence intervals and confidence coefficients. Here the statistician proceeds as follows: He says that the sample of 64 tubes comes from a new population which, while assumed normal, has an unknown mean and variance; and he very much wants to know something about the mean of this population, for that information will help him to conclude whether the new design is an improvement.

Now we must remember that statistics deals with uncertain inference. We must not expect the statistician to come to an absolutely firm conclusion. We must expect him always to give a two-part answer to our question. One part of this reply goes: "My best estimate is. . . ." The inescapable other part of his reply is: "The degree of confidence which you are justified in placing in my estimate is. . . ."

Thus we are not surprised that the statistician starts out by choosing a number which he calls a confidence coefficient. He might, for example, choose the confidence coefficient .95. This means that he is about to adopt a course of action which will be right 95 per cent of the time on the average. We therefore know how much confidence we are justified in placing in his results. Having decided on this figure, statistical theory now furnishes him with the width of a so-called confidence interval whose midpoint is the mean of the sample. In our example this interval turns out to be 10,200 plus or minus 195 hours, or from 10,005 to 10,395 hours. The statistician then answers Questions 2 and 3 as follows: "I estimate that the mean of the population of lifetimes of new-design tubes is greater than 10,005 hours and less than 10,395 hours. I can't guarantee that I am correct; but in a long series of such statements I will be right 95 per cent of the time. Since this range is above the mean life of the old tubes, I conclude that the new design is probably an improvement."

If the statistician had originally decided to adopt a procedure that would be correct 99 per cent of the time, his confidence interval would have turned out wider. He could make a less precise statement but make it with greater confidence. Conversely, he could arrange to make a more precise statement with somewhat less confidence.

Finally, let us examine this same question in the still more sophisticated manner that goes under the name "testing of statistical hypotheses." Here one starts by making some sort of guess about the situation, and then goes through a statistical argument to find out whether it is sensible, and how sensible, to discard this guess or retain it.

Thus the statistician might tentatively assume that the new design is equivalent to the old in average tube life. Of course he hopes that this is not true. Although it sounds a little perverse, it is in fact customary to start with a hypothesis that one hopes to disprove.

Here, just as in the previous case, the statistician first picks out a number which is going to tell what confidence we dare have in his statements. Actually he uses something which might be called an "unconfidence coefficient," for it measures the per cent of the time he expects to be wrong, rather than the per cent he expects to be right.

This unconfidence coefficient is technically called the significance level. Let us say that the statistician chooses a significance level of .05, which is exactly equivalent to .95 as a confidence coefficient. Then he calculates the confidence interval for this confidence coefficient. Since we have the same confidence coefficient as before, we already know that this particular confidence interval reaches from 10,005 hours to 10,395 hours. The statistician then reports: "The mean life of the old population (10,000 hours) does not fall within my confidence interval. Therefore theory tells me to discard the assumed hypothesis that the new tube has the same average lifetime as the old. The assumed hypothesis may of course actually be true. But theory further tells me that in cases in which the hypothesis is true, and in which I proceed as I just have done, I will turn out to make mistakes only 5 per cent of the time."

This report, if one thinks it over carefully, is rather incomplete. It says something about the probability of one sort of error—the error of discarding the hypothesis when it is in fact true. But it says nothing about another sort of error—accepting the hypothesis when it is in fact false. In certain situations one of these two mistakes might be very dangerous and costly and the other relatively innocuous. There are available still more refined statistical procedures (called the Neyman-Pearson methods and the theory of decision functions) in which one designs the test so as to make a desirable compromise with respect to the probabilities of the two types of error.

What Is Probability?

by Rudolf Carnap
September 1953

Some mathematicians argue that it is "statistical"; others, that it is "inductive." The author believes that there are two kinds, both essential to the future progress of science

No one reading articles on fundamental questions of science can fail to be impressed with the great importance to science of hypotheses—the daring guesses on slender evidence that go into building new theories. The question I should like to raise in this article is: Can the method of scientific inquiry be made more precise? Can we learn to judge the hypotheses, to weigh the extent to which they are supported by the evidence at hand, as an investigator judges and weighs his data?

The question leads at once into the subject of probability. If you query scientists about the meaning of this term, you will discover a curious situation. Practically everyone will say that probability as used in science has only one meaning, but when you ask what that meaning is, you will get different answers. Most scientists will define it as statistical probability, which means the relative frequency of a given kind of events or phenomena within a class of phenomena, usually called the "population." For instance, when a statistician says the probability that a native of the U. S. has A-type blood is 4/10, he means that four out of 10 people have this type. This meaning of probability has become almost the standard usage in science. But you will also find that there are scientists who define probability in another way. They prefer to use the term in the sense nearer to everyday use, in which it means a measurement, based on the available evidence, of the chances that something is true—as when a jury decides that a defendant is "probably" guilty, or a weather forecaster predicts that it will probably rain tomorrow. This kind of probability amounts to a weighing of the strength of the evidence. Its numerical expression has a meaning quite different from that of statistical probability: if the weather man were to venture to say that the probability of rain tomorrow was 4/10, he would not be describing a statistical fact but would simply mean that, should you bet on it raining tomorrow, you had better ask for odds of 4 to 6.

This concept is called inductive probability. A scientist makes a judgment of the odds consciously or unconsciously, whenever he plans an experiment. Usually the probability ascribed to his hypothesis is stated not in numbers but in comparative terms; that is, the probability is said to be high or low, or one probability is considered higher than another. To some of us it seems that inductive probability could be refined into a more precise tool for science. Given a hypothesis and certain evidence, it is possible to determine, by logical analysis and mathematical calculation, the probability that the hypothesis is correct, or the "degree of confirmation." If we had a system of inductive logic in mathematical form, our inferences about hypotheses in science, business and everyday life, which we usually make by "intuition" or "instinct," might be made more rational and exact. I have made a beginning in the construction of such a system, using the findings of past workers in this field and the exact tools of modern symbolic logic. Before discussing this system, let me review briefly the history of the inductive concept of probability.

The scientific theory of probability began, as a matter of fact, with the inductive concept and not the statistical one. Its study was started in the 16th century by certain mathematicians who were asked by their gambler friends to determine the odds in various games of chance. The first major treatise on probability, written by the Swiss professor Jacob Bernoulli and published posthumously in 1713, was called *Ars Conjectandi*, "The Art of Conjecture"—in other words, the art of judging hypotheses on the basis of evidence. The classical period in the study of probability culminated in the great 1812 work *Théorie analytique des probabilités*, by the French astronomer and mathematician Pierre Laplace. He declared the aim of the theory of probability to be to guide our judgments and to protect us from illusions, and he was concerned primarily not with statistics but with methods for weighing the acceptability of assumptions.

But after the middle of the 19th century the word probability began to acquire a new meaning, and scientists turned more and more to the statistical concept. By the 1920s Robert Aylmer Fisher in England, Richard von Mises and Hans Reichenbach in Germany (both of whom have died within the last few months) and others began to develop new probability theories based on the statistical interpretation. They were able to use many of the mathematical theorems of classical probability, which hold equally well in statistical probability. But they had to reject some. One of the principles they rejected, called the principle of indifference, sharply points up the distinction between inductive and statistical probability.

Suppose you are shown a die and are told merely that it is a regular cube. With no more information than this, you can only assume that when the die is thrown any one of its six faces is as likely to turn up as any other; in other words, that each face has the same probability, 1/6. This illustrates the principle of indifference, which says that if the evidence does not contain anything that would favor one possible event over another, the events have equal probabilities *relative to this evidence*. Now a second observer may have additional evidence: he

	STATISTICAL DISTRIBUTIONS		INDIVIDUAL DISTRIBUTIONS		METHOD I	METHOD II	
	NUMBER OF BLUE	NUMBER OF WHITE			INITIAL PROBABILITY OF INDIVIDUAL DISTRIBUTIONS	INITIAL PROBABILITY OF: STATISTICAL DISTRIBUTIONS	INDIVIDUAL DISTRIBUTIONS
1.	4	0	1.	○ ○ ○ ○	1/16	1/5	1/5 = 12/60
2.	3	1	2.	○ ○ ○ ●	1/16	1/5	1/20 = 3/60
			3.	○ ○ ● ○	1/16		1/20 = 3/60
			4.	○ ● ○ ○	1/16		1/20 = 3/60
			5.	● ○ ○ ○	1/16		1/20 = 3/60
3.	2	2	6.	○ ○ ● ●	1/16	1/5	1/30 = 2/60
			7.	○ ● ○ ●	1/16		1/30 = 2/60
			8.	○ ● ● ○	1/16		1/30 = 2/60
			9.	● ○ ○ ●	1/16		1/30 = 2/60
			10.	● ○ ● ○	1/16		1/30 = 2/60
			11.	● ● ○ ○	1/16		1/30 = 2/60
4.	1	3	12.	○ ● ● ●	1/16	1/5	1/20 = 3/60
			13.	● ○ ● ●	1/16		1/20 = 3/60
			14.	● ● ○ ●	1/16		1/20 = 3/60
			15.	● ● ● ○	1/16		1/20 = 3/60
5.	0	4	16.	● ● ● ●	1/16	1/5	1/5 = 12/60

INDUCTIVE PROBABILITY METHODS are illustrated in an example which is tabulated above. Four balls are to be drawn in succession from an urn. They are identical in every way except that some are blue and some white. Nothing is known, however, about the proportion of blue to white balls in the urn. First we want to decide on the initial probabilities in the experiment—the probabilities before the first ball is drawn. We list (under "Individual Distributions") all the possible ways in which the drawing can turn out. Now we apply the principle of indifference, which says that if the evidence contains nothing that favors one possibility over another, all possibilities must be considered equally probable. There are two ways to apply the principle to this example. The first is illustrated under "Method I." Since there are 16 possible cases, dividing the probability equally among them gives each a probability of 1/16. But there is another way to look at the table. Instead of taking into account the order in which blue and white turn up, we can concentrate only on the total numbers of blue and white in a drawing—all blue, three blue and one white, and so on. This classifies the table into "Statistical Distributions," which are indicated by the brackets on the left. There are five statistical distributions. If the principle of indifference is applied to them, then each has a probability of 1/5, as shown in the first column of "Method II." Now the individual distributions within each statistical distribution are assigned probabilities that are again determined by the principle of indifference. The first statistical distribution (four blue) has only one member, so it gets the full amount of the probability to be distributed, or 1/5, as shown in the second column of "Method II." The second statistical distribution (three blue, one white) has four members, so the probability must be split four ways, 1/20 to each. Similarly, the remaining three statistical distributions are divided into their individual members. At the extreme right hand of the table, all probabilities are converted to a least common denominator of 60 in order to facilitate comparing and combining them. Method II is superior to Method I because it assigns probabilities to future events on the basis of the frequency of their past occurrence.

may know that the die is loaded in favor of one of the faces, without knowing which face it is. The probabilities are still the same for him, because as far as his information goes, each of the six faces has an equal possibility of being loaded. On the other hand, for a third observer who knows that the load favors the face numbered 1 the probabilities change; on the basis of his evidence the probability of the ace is higher than 1/6.

Thus inductive probability depends on the observer and the evidence in his possession; it is not simply a property of the object itself. In statistical probability, which refers to the actual frequency of an event, the principle of indifference is of course absurd. It would be incautious for an observer who knew only that a die had the accurate dimensions of a cube to assert that the six faces would appear with equal frequency. And if he knew that the die was biased in favor of one side, he would contradict his own knowledge. Inductive probability, on the other hand, does not predict frequencies; rather, it is a tool for evaluating evidence in relation to a hypothesis. Both the statistical and inductive concepts of probability are indispensable to science; each has valuable functions to perform. But it is important to recognize the distinctions between the two concepts and to develop the possibilities of both tools.

In the past 30 years the inductive concept of probability, which had been supplanted by the statistical concept, has been revived by a few workers. The first of these was the great English economist John Maynard Keynes. In his *Treatise on Probability* in 1921 he showed how the inductive concept is implicitly used in all our thinking about unknown events, in science as well as in everyday life. Yet Keynes' attempt to develop this concept was too restricted: he believed it was impossible to calculate numerical probabilities except in well-defined situations such as the throw of dice, the possible distributions of cards, and so on. Moreover, he rejected the statistical concept of probability and argued that all probability statements could be formulated in terms of inductive probability.

I believe that he was mistaken in this point of view. Today an increasing number of those who study both sides of the controversy, which has been going on for 30 years, are coming to the conclusion that here, as often before in the history of scientific thinking, both sides are right in their positive theses, wrong in their polemical remarks. The statistical concept, for which a very elaborate mathematical theory exists, and which has been applied fruitfully in many fields in science and industry, need not be abandoned in order to make room for the inductive concept. Statistical probability characterizes an objective situation, *e.g.*, a state of a physical, biological or social system. On the other hand, inductive probability, as I see it, does not occur in scientific statements but only in judgments *about* such statements. Thus it is applied in the methodology of science—the analysis of concepts, statements and theories.

In 1939 the British geophysicist Harold Jeffreys put forward a much more comprehensive theory of inductive probability than Keynes'. He agreed with the classical view that probability can be expressed numerically in all cases. Furthermore, he wished to apply probability to quantitative hypotheses of science, and he set up an axiom system for probability much stronger than that of Keynes. He revived the principle of indifference in a form which seems to me much too strong: "If there is no reason to believe one hypothesis rather than another, the probabilities are equal." It can easily be shown that this statement leads to contradictions. Suppose, for example, that we have an urn known to be filled with blue, red and yellow balls but do not know the proportion of each color. Let us consider as a starting hypothesis that the first ball we draw from the urn will be blue. According to Jeffreys' (and Laplace's) statement of the principle of indifference, if the question is whether the first ball will be blue or not blue, we must assign equal probabilities to both these hypotheses; that is, each probability is 1/2. If the first ball is not blue, it may be either red or yellow, and again, in the absence of knowledge about the actual proportions in the urn, these two have equal probabilities, so that the probability of each is 1/4. But if we were to start with the hypothesis that the first ball drawn would be, say, red, we would get a probability of 1/2 for red. Thus Jeffreys' system as it stands is inconsistent.

In addition, Jeffreys joined Keynes in rejecting the statistical concept of probability. Nevertheless his book *Theory of Probability* remains valuable for the new light it throws on many statistical problems by discussing them for the first time in terms of inductive probability.

I have drawn upon the work of Keynes and Jeffreys in constructing my mathematical theory of inductive probability, set forth in the book *Logical Foundations of Probability*, which was published in 1950. It is not possible to outline here the mathematical system itself. But I shall explain some of the general problems that had to be solved and some of the basic conceptions underlying the construction.

One of the fundamental questions to be decided is whether to accept a principle of indifference, and if so, in what form. It should be strong enough to allow the derivation of the desired theorems, but at the same time sufficiently restricted to avoid the contradictions resulting from the classical form.

The problem can be made clear by an example illustrating a few elementary concepts of inductive logic. We have an urn filled with blue and white balls in unknown proportions. We are going to draw four balls in succession. Taking the order into account, there are 16 possible drawings (all four blue, the first three blue and the fourth white, the first white and the next three blue, and so on). We list these possibilities in a table (*see table on page 154*).

Now what is the initial probability, before we have drawn at all, that we shall draw any one of these 16 distributions? We might assign any probability to the individual distributions, so long as they all added up to 1. Suppose we apply the principle of indifference and say that all the distributions have equal probabilities; that is, each has a probability of 1/16.

Let us state a specific hypothesis and calculate its probability. The hypothesis is, for example, that among the first three balls we draw, just one will be white. Looking at the table, we can see that six out of the 16 possible drawings will give us this result. The probability of our hypothesis, therefore, is the sum of these initial probabilities, or 6/16.

Suppose now that we are given some evidence, *i.e.*, have drawn some balls, and are asked to calculate the probability of a given hypothesis on the basis of this evidence. For instance, we have drawn first a blue ball, then a white ball, then a blue ball. The hypothesis is that the fourth ball will be blue; what is its probability? Here we run into a question as to how we should apply the principle of indifference. Let us try two different methods.

In Method I we start by assigning equal probabilities to the individual distributions. Referring to the table, we see that two of these distributions (Nos. 4 and 7) will give us the sequence blue, white, blue for the first three balls. Its probability is therefore 2/16. In only one of these distributions is the fourth ball blue; its probability is 1/16. The probability of our hypothesis on the basis of the evidence is obtained by dividing one into the other: *i.e.*, 1/16 divided by 2/16, which equals 1/2. In other words, the chances that our hypothesis is correct are 50-50: the fourth ball is just as likely to be white as blue.

But as a guide to judging a hypothesis, this result contradicts the principle of learning from experience. Other things being equal, we should consider one event more probable than another if it has happened more frequently in the past. We would regard a man as unreasonable if his expectation of a future event were the higher the less often he had seen it before. We must be guided by our knowledge of observed events, and in this example the fact that two out of three balls drawn from an unknown urn were blue should lead us to expect the probabilities to favor the fourth's also being blue. Yet a number of philosophers, including Keynes, have proposed Method I in spite of its logical flaw.

There is a second method which gives us a more reasonable result. We first apply the principle of indifference not to individual distributions but to statistical distributions. That is, we consider only the number of blue balls and of white balls obtained in a drawing, irrespective of order. The table shows that there are five possible statistical distributions (four blue, four white, three blue and one white, three white and one blue, two blue and two white). By the principle of indifference we assign equal probabilities to these, so that the probability of each is 1/5. We distribute this value (expressed for arithmetical convenience as 12/60) in equal parts among the corresponding individual distributions (*see last column of table*). Now the probabilities of distributions No. 4 and No. 7 are 3/60 and 2/60, respectively, and the probability of the hypothesis on the basis of the evidence is 3/60 divided by 5/60, or 3/5. In short, the chances that the fourth ball will be blue are not even but 3 to 2, which is more consistent with what experience, meaning the evidence we have acquired, should lead us to expect.

Method II, as well as Method I, leads to contradictions if it is applied in an unrestricted way. If it is used in cases characterized by more than one property difference (such as the difference between blue and white balls in our example) then all the relevant differences must be specified. Thus restricted, this system, which I proposed in 1945, is the first consistent inductive method, so far as I am aware, that succeeded in satisfying the principle of learning from experience. Since then I have found that there are many others. None of them seems as simple to define as Method II, but some of them have other advantages.

Having found a consistent and suitable inductive method, we can proceed to develop a general procedure for calculating, on the basis of given evidence, an estimate of an unknown value of any quantity. Suppose that the evidence indicates a certain number of possible values for a quantity at a given time, *e.g.*, the amount of rain tomorrow, the number of persons coming to a meeting, the price of wheat after the next harvest. Let the possible values be x_1, x_2, x_3, etc., and their inductive probabilities be p_1, p_2, p_3, etc. Then p_1x_1 is the "expectation value" of the first case at the present moment, p_2x_2 of the second case, and so on. The total expectation value of the quantity on the given evidence is the sum of the expectation values for all the possible cases. To take a specific example, suppose there are four prizes in a lottery, a first prize of $200 and three prizes of $50 each. It is known that the probability of a ticket winning the first prize is 1/100, and of a second prize, 3/100; the probability that the ticket will win nothing is therefore 96/100. Applying the method I have described above, a ticket holder can estimate that the ticket is worth to him 1/100 times $200 plus 3/100 times $50 plus 96/100 times 0, or $3.50. It would be irrational to pay more for it.

The same method may be used to make a rational decision in a situation where one among various possible actions is to be chosen. For example, a man considers several possible ways of investing a certain amount of money. He can—in principle, at least—calculate the estimate of his gain for each possible way. To act rationally, he should then choose that way for which the estimated gain is highest.

Bernoulli, Laplace and their followers envisaged a theory of inductive probability which, when fully developed, would supply the means for evaluating the acceptability of hypothetical assumptions in any field of theoretical research and for making rational decisions in the affairs of practical life. They were a great deal farther from this audacious objective than they realized. In the more sober cultural atmosphere of the late 19th and early 20th centuries their idea was dismissed as Utopian. But today a few men dare to think that these pioneers were not mere dreamers.

Probability

28

by Warren Weaver
October 1950

Three centuries ago some sensible questions asked by gamblers founded a branch of mathematics. Today it powerfully assists our understanding of nature

Probability is the very guide of life.
—Cicero, *De Natura*

OVER three centuries ago some gamblers asked the great Italian scientist Galileo why a throw of three dice turns up a sum of 10 more often than a sum of nine. In 1654 the Chevalier de Mere—another gambler—asked the French mathematician and philosopher Pascal why it was unprofitable to bet even money that at least one double six would come up in 24 throws of two dice. This problem of de Mere really started off the mathematical theory of probability, and the end is not yet in sight.

Probability theory has now outgrown its disreputable origin in the gaming rooms, but its basic notions can still be most easily stated in terms of some familiar game.

When you toss a die—one carefully made, so that it is reasonable to believe that it is as likely to land on one of its six faces as on any other—a gambler would say that the odds against any specified number are five to one. A mathematician defines the probability to be one-sixth. Suppose we ask now: What is the probability of getting a three and a four in one roll of two dice? For convenience we make one die white and one red. Since any one of six results on the white die can be paired with any one of six results on the red die, there is now a total of 36 ways in which the two can land—all equally likely. The difference in color makes it clear that a red three and a white four is a different throw from a white three and a red four. The probability of throwing a three and a four is the ratio of 2—the number of favorable cases—to 36, the total number of equally likely cases; that is, the probability is 2/36, or 1/18.

What is the probability of throwing a sum of seven with two dice? An experienced crapshooter knows that seven is a "six-way point," which is his way of saying that there are six favorable cases (six and one, one and six, three and four, four and three, five and two, two and five). So the probability of throwing a sum of seven with two dice is 6/36, or 1/6.

In general, the probability of any event is defined to be the fraction obtained by dividing the number of cases favorable to the event by the total number of equally likely cases. The probability of an impossible event (no favorable cases) obviously is 0, and the probability of an inevitable or certain event (all cases favorable) is 1. In all other cases the probability will be a number somewhere between 0 and 1.

Logically cautious readers may have noticed a disturbing aspect of this definition of probability. Since it speaks of "equally likely," *i.e.*, equally probable, events, the definition sits on its own tail, so to speak, defining probability in terms of probability. This difficulty, which has caused a vast amount of technical discussion, is handled in one of two ways. · · ·

WHEN one deals with purely *mathematical probability*, "equally likely cases" is an admittedly undefined concept, similar to the theoretical "points" and "lines" of geometry. And there are cases, such as birth statistics for males and females, where the ordinary concept of "equally likely cases" is artificial, so that the notion must be generalized. But a logically consistent theory can be erected on the undefined concept of equally likely cases, just as Euclidean geometry is developed from theoretical points and lines. Only through experience can one decide whether any actual events conform to the theory. The answer of experience is, of course, that the theory does in fact have useful application.

The other way of avoiding the dilemma is illustrated by defining the probability of throwing a four with a particular die as the actual fraction of fours obtained in a long series of throws under essentially uniform conditions. This, the "frequency definition," leads to what is called a *statistical probability*.

On the basis of the mathematical definition of probability, a large and fascinating body of theory has been developed. We can only hint here at the range and interest of the problems that can be solved. Two rival candidates in an election are eventually going to receive m and n votes respectively, with m greater than n. They are sitting by their radios listening to the count of the returns. What is the probability that as the votes come in the eventual winner is always ahead? The answer is $m-n/m+n$. A storekeeper sells, on the average, 10 of a certain item per week. How many should he stock each Monday to reduce to one in 20 the chance that he will disappoint a customer by being sold out? The answer is 15. Throw a toothpick onto a floor whose narrow boards are just as wide as the toothpick is long. In what fraction of the cases will the toothpick land so as to cross a crack? The answer is $2/\pi$, where π is the familiar constant we all met when we studied high-school geometry. A tavern is 10 blocks east and seven blocks north of a customer's home. If he is so drunk that at each corner it is a matter of pure chance whether he continues straight or turns right or left, what is the probability that he will eventually arrive home? This is a trivial case of a very general "random walk" problem which has serious applications in physics; it applies, for example, to the so-called Brownian movement of very small particles suspended in a liquid, caused by accidental bumps from the liquid's moving molecules. This latter problem, incidentally, was first solved by Einstein when he was 26 years old.

THERE are laws of chance. We must avoid the philosophically intriguing question as to why chance, which seems to be the antithesis of all order and regularity, can be described at all in terms of laws. Let us consider the Law of Large Numbers, which plays a central role in the whole theory of probability.

The Law of Large Numbers has been established with great rigor and for very general circumstances. The essence of the matter can be illustrated with a simple case. Suppose someone makes a great many tosses of a symmetrical coin, and records the number of times heads and tails appear. One aspect—the more familiar aspect—of the Law of Large Numbers states that by throwing enough times we can make it as probable as desired that the ratio of heads to total throws differ by as little as one pleases from the predicted value 1/2. If you want the ratio to differ from 1/2 by as little as 1/100,000, for example, and if you want to be 99 per cent sure (*i.e.*, the probability = .99) of accomplishing this purpose, then there is a perfectly definite but admittedly large number of throws which will meet your demand. Note that there is no number of throws, however large, that will really *guarantee* that the fraction of heads be within 1/100,000 of 1/2. The law simply states, in a very precise way, that as the number of experiments gets larger and larger, there is a stronger and stronger tendency for the results to conform, *in a ratio sense*, to the probability prediction.

This is the part of probability theory that is vaguely but not always properly understood by those who talk of the "law of averages," and who say that the probabilities "work out" in the long run. There are two points which such persons sometimes misunderstand.

The first of these relates to the less familiar aspect of the Law of Large Numbers. For the same law that tells us that the *ratio* of successes tends to match the probability of success better and better as the trials increase also tells us that as we increase the number of trials the *absolute number* of successes tends to deviate more and more from the expected number. Suppose, for example, that in 100 throws of a coin 40 heads are obtained, and that as one goes on further and throws 1,000 times, 450 heads are obtained. The *ratio* of heads to total throws has changed from 40 per cent to 45 per cent, and has therefore come closer to the probability expectation of 50 per cent, or 1/2. But in 100 throws the absolute number of heads (40) differs by only 10 from 50, the theoretically expected number, whereas in 1,000 throws, the absolute number of heads (450) differs by 50, or five times as much as before, from the expected number (500). Thus the ratio has improved, but the absolute number has deteriorated.

The second point which is often misunderstood has to do with the independence of any throw relative to the results obtained on previous throws. If heads have come up several times in a row, many persons are inclined to think that the "law of averages" makes a toss of tails now rather more likely than heads. Granting a fair, symmetrical coin, this is simply and positively not so. Even after a very long uninterrupted run of heads, a fair coin is, on the next throw, precisely as likely to come up heads as tails. Actually the less familiar aspect of the Law of Large Numbers already mentioned makes it likely that longer and longer uninterrupted sequences of either heads or tails will occur as we go on throwing, although the familiar aspect of the same law assures us that, in spite of these large absolute deviations, the ratio of heads to tails is likely to come closer and closer to one.

All of these remarks, of course, apply to a series of *independent* trials. Probability theory has also been most fruitfully applied to series of dependent trials—that is, to cases, such as arise in medicine, genetics, and so on, where past events do influence present probabilities. This study is called the probability of causes.

Suppose we have a covered box about which we know only that it contains a large number of small colored balls. Suppose that without looking into the box we scoop out a handful and find that one third of the balls we have taken are white and two thirds red. What probability statements can we make about the mixture in the box?

ROULETTE WHEEL makes possible bets against several probabilities. At Monte Carlo red once came up 32 times in a row. This probability is: $1/(2)^{32}$, or about one in 4 billion.

This schematic problem, which sounds so formal and trivial, is closely related to the very essence of the procedure of obtaining knowledge about nature through experimentation. Nature is, so to speak, a large closed box whose contents are initially unknown. We take samples out of the box—*i.e.*, we do experiments. What conclusions can be drawn, how are they to be drawn and how secure are they?

This is a subject which has caused considerable controversy in probability, and in the related field of statistics as well. The problem of the balls as stated above is, as a matter of fact, not a proper problem. The theorem of probability theory which applies here (it is known as Bayes' theorem, and it was first developed by a clergyman) makes clear just how the experimental evidence of the sample justifies one in changing a previously held opinion about the contents of the box; but the application of the theorem requires you to have an opinion prior to the experiment. You cannot construct a conclusion concerning the probability of various mixtures out of the experiment alone. If many repeated experiments continue to give the same indication of one third white and two thirds red, then of course the evidence becomes more and more able to outweigh a previously held contrary opinion, whatever its nature.

Recently there have been developed powerful new methods of dealing with situations of this general sort, in which one wishes to draw all the justified inferences out of experimental evidence. Although Bayes' theorem cannot be applied unless one possesses or assumes prior opinions, it has been found that other procedures, associated with statistical theory rather than pure probability, are capable of drawing most useful conclusions.

What does a probability mean? What does it mean, for instance, to tell a patient: "If you decide to submit to this surgical operation, the probability that you will survive and be cured is .72"? Obviously this patient is going to make only one experiment, and it will either succeed or fail. What useful sense does the number .72 have for him?

The answer to this—and essentially to any question whatsoever that involves the interpretation of a probability—is: "If a large number of individuals just like you and just in your present circumstances were to submit to this operation, about 72 out of every 100 of them would survive and get well. The larger the number of individuals, the more likely it is that the ratio would be very close to 72 in each 100."

This answer may at first seem a little artificial and disappointing. It admittedly involves some entirely unrealizable conditions. A complicated intuitive process is required to translate the statement into a useful aid to the making of decisions. But experience does nevertheless show that it is a useful aid.

A theory may be called right or wrong according as it is or is not confirmed by actual experience. In this sense, can probability theory ever be proved right or wrong?

In a strict sense the answer is no. If you toss a coin you expect to get about half heads. But if you toss 100 times and get 75 heads instead of the expected 50, you have not disproved probability: probability theory can easily reckon the chance of getting 75 heads in 100 tosses. If that probability be written as 1/N, you would then expect that if you tossed 100 coins N times, in about one of those N times you would actually get 75 heads. So suppose you now toss 100 coins N times, and suppose that you get 75 heads

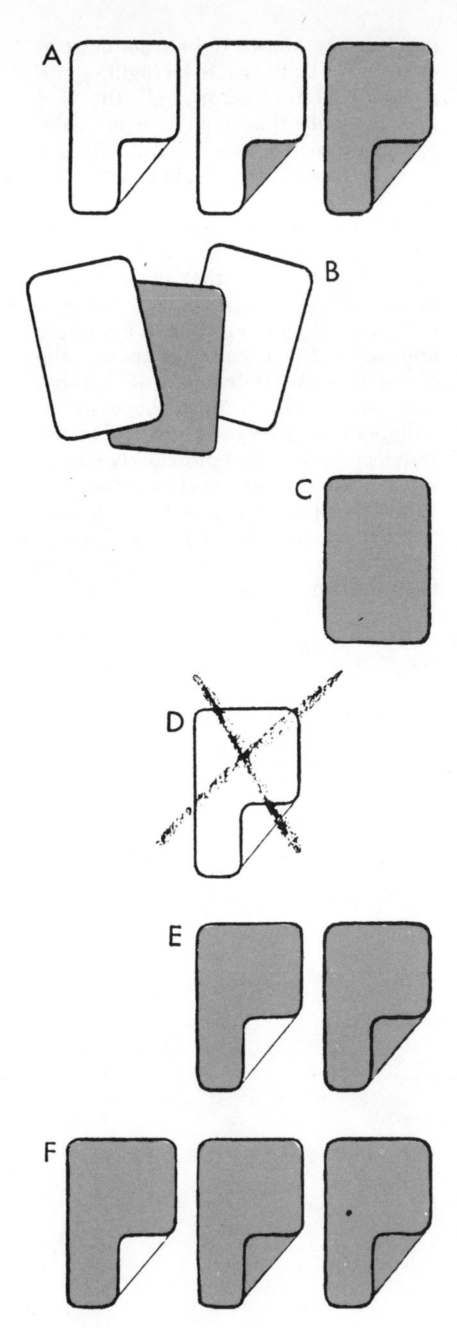

THREE-CARD GAME, classically known as The Problem of Three Chests, illustrates deceptiveness of probability. One card is white on both sides; the second is white on one side and red on the other; the third is red on both sides (A). The dealer shuffles the cards in a hat (B), takes one out and places it flat on the table. The side showing is red (C). The dealer now says: "Obviously this is not the white-white card (D). It must be either red-white or red-red (E). I will bet even money that the other side is red." It is a poor bet for anyone else. Actually there are three possible cases (F). One is that the other side is white. The other two are that it is one or the other side of the red-red card. Thus the chance that the underside is red is 2 to 1.

not just one or two times, as you expect, but say 25 times! Is probability now disproved?

Again no. For the event that has now occurred, although amazingly rare, is still an event whose probability can be calculated, and while its probability is exceedingly small, it is not zero. Thus one goes on, again making a new experiment which consists of many repetitions of the previous experiment. And even if miracles persist in occurring, these would be, from the point of view of probability, not impossible miracles.

Thus in a strict sense probability cannot be proved either right or wrong. But this is, as a matter of fact, a purely illusory difficulty. Although probability cannot be strictly proved either right or wrong, it can be proved useful. The facts of experience show that it works.

THERE are two different—or at least apparently different—types of problems to which probability theory applies. For the first type of problem probability theory is used not so much because we are convinced that we have to use it but because it is so very convenient. For the second type, probability theory seems to be even theoretically unavoidable. We shall see, however, that the distinction between the two cases, while of practical value, is really something of an illusion.

The first type has to do with situations which may be considered deterministic but which are so complex that the outcome is for all practical purposes unpredictable. In this kind of situation we realize that the final result has depended, often in a very sensitive way, on the interaction of a large number of causes. Many of these causes may be somewhat obscure in character, or otherwise impractical of detailed study, but it is at least thinkable that science could, if it were worth-while, analyze every cause in turn and thus arrive at a theory which could predict and explain what happens. When, in such circumstances, we say that the main final result "depends upon chance," we merely mean that, conveniently for us, the very complexity that makes a detailed analysis practically impossible assures an over-all behavior which is describable through the laws of probability.

Perhaps tossing a coin is again the simplest and most familiar illustration of this kind of case. There seems to be no essential mystery about why a coin lands heads or tails. The exact position of the coin above the table, the velocities of movement and spin given by the fingers, the resistance of the air, and so on—one can state what he needs to know in order to compute, by well-known dynamical laws, whether the coin will land heads or tails. But such a study would be very complicated, and would require very precise and extensive quantitative information.

There are many situations of this sort in serious everyday life, where we use probability theory not because it is clear that "chance" plays some obscure and mysterious role but primarily because the situation is so complicated, so intricately affected by so many small causes, that it is prohibitively inconvenient to attempt a detailed analysis. The experience of insurance companies, the occurrence of telephone calls and the resulting demands on telephone traffic and switching equipment, the sampling techniques used when one wishes to estimate the quality of many objects or the opinions of many individuals, the ordinary theory of errors of measurement, problems in epidemiology, the kinetic theory of gases—all these are practical instances in which the causes are too numerous, too complicated, and/or too poorly understood to permit a complete deterministic theory. We therefore deal with these subjects through probability. But in all these cases we would say, with Poincaré, that chance "is only the measure of our ignorance."

THE second type of probability problem at first sight seems very different. Most scientists now believe that some of the most elementary occurrences in nature are essentially and inescapably probabilistic. Thus in modern quantum theory, which forms the basis of our working knowledge of the atom, it seems to be not only impossible but essentially meaningless to attempt to compute just where a certain electron will be at a certain instant. All that one can do is reckon, as through the Schrödinger wave equation, the values of a probability position function. One cannot predict where the electron will be—one can only compute the probability that it will or will not be at a given place or places. And any attempt to frame an experiment that would resolve this probability vagueness, by showing just where the electron is, turns out to be a self-defeating experiment which destroys the conditions under which the original question can be asked.

It is only fair to remark that there remain scientists who do not accept the inevitable role of probability in atomic phenomena. The great example, of course, is Einstein, who has remarked in a characteristically appealing way that "I shall never believe that God plays dice with the world." But it is also fair to remark that Einstein, for all his great genius, is in a small minority on this point.

The problems that involve probability in this inescapable way are of the most fundamental kind. Quantum theory and statistical mechanics, which combine to furnish a large part of the basic theory of the physical universe, are essentially built upon probability. The gene-shuffling which controls inheritance in the living world is subject to probability laws. The inner character of the process of communication, which plays so great and so obvious a role in human life, has recently been found to be probabilistic in nature. The concept of the ever forward flow of time has been shown to depend upon entropy change, and thus to rest upon probability ideas. The whole theory of inference and evidence, and in fact of knowledge in general, goes back to probability.

We are now in a position to see that the two types of probability problems are, if we wish to be logically precise, not so different as we first supposed. Is it correct to think of the fall of a coin as being complicated but determinate, and the position of an electron as being essentially indeterminate? Obviously not. From a large-scale and practical point of view, one could doubtless deal quite successfully with coin-tossing on the basis of very careful actual measurements, plus all the analytical resources of dynamical theory. It remains true, however, that the coin is made of elementary particles whose positions and motions can be known, as science now views the matter, only in a probability sense. Thus we refine our original distinction between the two types of cases by saying that the second type, whatever the scale involved, is essentially indeterminate, whereas the first type involves large-scale phenomena which may usefully be considered determinate, even though these large-scale phenomena depend ultimately on small-scale phenomena which are probabilistic.

SCIENCE deals (as in mathematics) with statements about theory which are logically accurate, but to which the concept of "truth" does not apply; it also deals (as in physics) with statements about nature which can never in a rigorous sense be known to be true, but can at best only be known to be highly probable. It is rather surprisingly the case that the only time man is ever really sure is not when he is dealing with science, but when he is dealing with matters of faith.

There is, moreover, some justification for saying that in science probability almost plays the role that faith does in other fields of human activity. For in a vast range of cases in which it is entirely impossible for science to answer the question "Is this statement true?" probability theory does furnish the basis for judgment as to how likely it is that the statement is true. It is probability which, in an important fraction of cases, enables man to resolve the paradoxical dilemma pointed out by Samuel Butler: "Life is the art of drawing sufficient conclusions from insufficient premises."

29

On the Fabric of Inductive Logic, and Some Probability Paradoxes

by Martin Gardner
March 1976

"The universe, so far as known to us, is so constituted, that whatever is true in any one case, is true in all cases of a certain description; the only difficulty is, to find what description."

—JOHN STUART MILL,
A System of Logic

Imagine that we are living on an intricately patterned carpet. It may or may not extend to infinity in all directions. Some parts of the pattern appear to be random, like an abstract expressionist painting; other parts are rigidly geometrical. A portion of the carpet may seem totally irregular, but when the same portion is viewed in a larger context, it becomes part of a subtle symmetry.

The task of describing the pattern is made difficult by the fact that the carpet is protected by a thick plastic sheet with a translucence that varies from place to place. In certain places we can see through the sheet and perceive the pattern; in others the sheet is opaque. The plastic sheet also varies in hardness. Here and there we can scrape it down so that the pattern is more clearly visible. In other places the sheet resists all efforts to make it less opaque. Light passing through the sheet is often refracted in bizarre ways, so that as more of the sheet is removed the pattern is radically transformed. Everywhere there is a mysterious mixing of order and disorder. Faint lattices with beautiful symmetries appear to cover the entire rug, but how far they extend is anyone's guess. No one knows how thick the plastic sheet is. At no place has anyone scraped deep enough to reach the carpet's surface, if there is one.

Already the metaphor has been pushed too far. For one thing, the patterns of the real world, as distinct from this imaginary one, are constantly changing, like a carpet that is rolling up at one end while it is unrolling at the other end. Nevertheless, in a crude way the carpet can introduce some of the difficulties philosophers of science encounter in trying to understand why science works.

Induction is the procedure by which carpetologists, after examining parts of the carpet, try to guess what the unexamined parts look like. Suppose the carpet is covered with billions of tiny triangles. Whenever a blue triangle is found, it has a small red dot in one corner. After finding thousands of blue triangles, all with red dots, the carpetologists conjecture that all blue triangles have red dots. Each new blue triangle with a red dot is a confirming instance of the law. Provided that no counterexample is found, the more confirming instances there are, the stronger is the carpetologists' belief that the law is true.

The leap from "some" blue triangles to "all" is, of course, a logical fallacy. There is no way to be absolutely certain, as one can be in working inside a deductive system, what any unexamined portion of the carpet looks like. On the other hand, induction obviously works, and philosophers justify it in other ways. John Stuart Mill did so by positing in effect that the carpet's pattern has regularities. He knew this reasoning was circular, since it is only by induction that carpetologists have learned that the carpet is patterned. Mill did not regard the circle as vicious, however, and many contemporary philosophers (R. B. Braithwaite and Max Black, to name two) agree. Bertrand Russell, in his last major work, tried to replace Mill's vague "nature is uniform" with something more precise. He proposed a set of five posits about the structure of the world that he believed were sufficient to justify induction.

Hans Reichenbach advanced the most familiar of several pragmatic justifications. If there is any way to guess what unexamined parts of the carpet look like, Reichenbach argued, it has to be by induction. If induction does not work, nothing else will, and so science might as well use the only tool it has. "This answer is not fallacious," wrote Russell, "but I cannot say that I find it very satisfying."

Rudolf Carnap agreed. His opinion was that all these ways of justifying induction are correct but trivial. If "justify" is meant in the sense that a mathematical theorem is justified, then David Hume was right: there

A cartoon comment on inductive reasoning

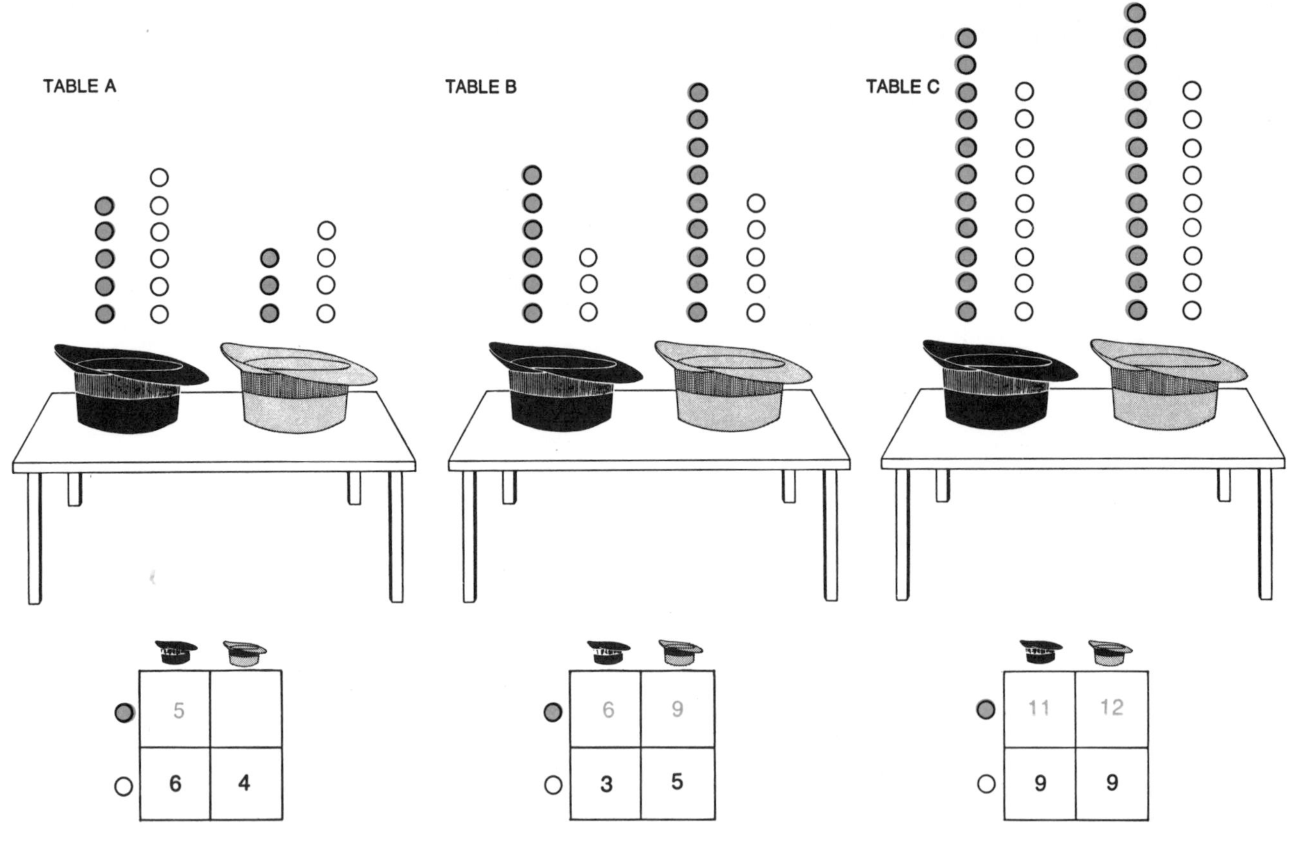

E. H. Simpson's reversal paradox

is no justification. But if "justify" is taken in any of several weaker senses, then, of course, induction can be defended. A more interesting task, Carnap insisted, is to see whether it is possible to construct an inductive logic.

It was Carnap's great hope that such a logic could be constructed. He foresaw a future in which a scientist could express in a formalized language a certain hypothesis together with all the relevant evidence. Then by applying inductive logic he could assign a probability value, called the degree of confirmation, to the hypothesis. There would be nothing final about that value. It would go up or down or stay the same as new evidence accumulated. Scientists already think in terms of such a logic, Carnap maintained, but only in a vague, informal way. As the tools of science become more powerful, however, and as our knowledge of probability becomes more precise, perhaps eventually we can create a calculus of induction that will be of practical value in the endless search for scientific laws.

In Carnap's *Logical Foundations of Probability* and also in his later writings he tried to establish a base for such a logic. How successful he was is a matter of dispute. Some philosophers share his vision (John G. Kemeny for one) and have taken up the task where Carnap left off. Others, notably Karl Popper and Thomas S. Kuhn, regard the entire project as having been misconceived.

Carl G. Hempel, one of Carnap's admirers, has argued sensibly that before we try to assign quantitative values to confirmations we should first make sure we know in a qualitative way what is meant by "confirming instance." It is here that we run into the worst kinds of difficulty [see "Confirmation," by Wesley C. Salmon; SCIENTIFIC AMERICAN, May, 1973].

Consider Hempel's notorious paradox of the raven. Let us approach it by way of 100 playing cards. Some of them have a picture of a raven drawn on the back. The hypothe-

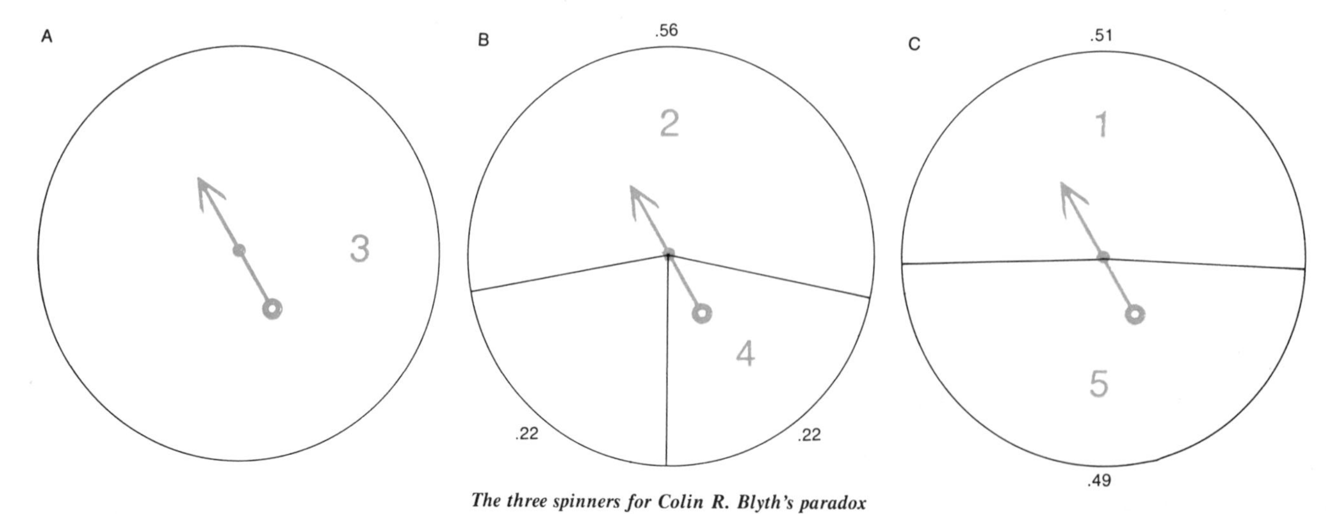

The three spinners for Colin R. Blyth's paradox

sis is "All raven cards are black." You shuffle the deck and deal the cards face up. After turning 50 cards without finding a counterinstance the hypothesis certainly becomes plausible. As more and more raven cards prove to be black, the degree of confirmation approaches certainty and may finally reach it.

Now consider another way of stating the same hypothesis: "All nonblack cards are not ravens." This statement is logically equivalent to the original one. If you test the new statement on another shuffled deck of 100 cards, holding them face up and turning them as you deal, clearly each time you deal a nonblack card and it proves to have no raven on the back, you confirm the guess that all nonblack cards are not ravens. Since this is logically equivalent to "All raven cards are black," you confirm that also. Indeed, if you deal all the cards without finding a red card with a raven, you will have completely confirmed the hypothesis that all raven cards are black.

Unfortunately when this procedure is applied to the real world, it seems not to work. "All ravens are black" is logically the same as "All nonblack objects are not ravens." We look around and see a yellow object. Is it a raven? No, it is a buttercup. The flower surely confirms (albeit weakly) that all nonblack objects are not ravens, but it is hard to see how it has any relevance at all to "All ravens are black." If it does, it equally confirms that all ravens are white or any color except yellow. To make things worse, "All ravens are black" is logically equivalent to "Any object is either black or not a raven." And that is confirmed by any black object whatever (raven or not) as well as by any nonraven (black or not). All of which seems absurd.

Nelson Goodman's "grue" paradox is equally notorious. An object is "grue" if it is green until, say, January 1, 2000, and blue thereafter. Is the law "All emeralds are grue" confirmed by observations of green emeralds? A prophet announces that the world will exist until January 1, 2000, when it will disappear with a bang. Every day the world lasts seems to confirm the prediction, yet no one supposes that it becomes more probable.

To make matters still worse, there are situations in which confirmations make a hypothesis less likely. Suppose you turn the cards of a shuffled deck looking for confirmations of the guess that no card has green pips. The first 10 cards are ordinary playing cards, then suddenly you find a card with blue pips. It is the 11th confirming instance, but now your confidence in the guess is severely shaken. Paul Berent has pointed out a similar example. A man 99 feet tall is discovered. He is a confirming instance of "All men are less than 100 feet tall," yet his discovery greatly weakens the hypothesis.

Confirmations may even falsify a hypothesis. Ten cards with all values from the ace through the 10 are shuffled and dealt face down in a row. The guess is that no card with value n is in the nth position from the left. You turn the first nine cards. Each card confirms the hypothesis. But if none of the turned cards is the 10, the nine cards taken together refute the hypothesis.

Carnap was aware of such difficulties. He distinguished sharply between "degree of confirmation," a probability value based on the total relevant evidence, and what he called "relevance confirmation," which has to do with how new observations alter a confirmation estimate. Relevance confirmation cannot be given simple probability values. It is enormously complex, swarming with counterintuitive arguments. In Chapter 6 of Carnap's *Logical Foundations* he analyzes a group of closely related paradoxes of confirmation relevance that are easily modeled with cards.

For example, it is possible that data will confirm each of two hypotheses but disconfirm the two taken together. Consider a set of 10 cards, half with blue backs and half with green ones. The green-backed cards (with the hearts and spades designated *H* and *S*) are *QH, 10H, 9H, KS, QS*. The blue-backed cards are *KH, JH, 10S, 9S, 8S*. The 10 cards are shuffled and dealt face down in a row.

Hypothesis *A* is that the property of being a face card (a king, a queen or a jack) is more strongly associated with green backs than with blue. An investigation shows that this is true. Of the five cards with green backs, three are face cards as against only two face cards with blue backs. Hypothesis *B* is that the property of being a red card (hearts or diamonds) is also more strongly associated with green backs than with blue. A second investigation confirms this. Three green-backed cards are red but there are only two red cards with blue backs. Intuitively one assumes that the property of being both red and a face card is more strongly associated with green backs than blue, but that is not the case. Only one red face card has a green back, whereas two red face cards have blue backs!

It is easy to think of ways, fanciful or realistic, in which similar situations can arise. A woman wants to marry a man who is both rich and kind. Some of the bachelors she knows have hair and some are bald. Being a statistician, she does some sampling. Project *A* establishes that 3/5 of the men with hair are rich but only 2/5 of the bald men are rich. Project *B* discloses that 3/5 of the men with hair are kind but only 2/5 of the bald men are kind. The woman might hastily conclude that she should marry a man with hair, but if the distribution of the attributes corresponds to that of the face cards and red cards mentioned in the preceding example, her chances of getting a rich, kind man are twice as great if she sets her cap for a bald man.

Another research project shows that 3/5 of a group of patients taking a certain pill are immune to colds for five years, compared with only 2/5 in the control group who were given a placebo. A second project shows that 3/5 of a group receiving the pill were immune to tooth cavities for five years, compared with 2/5 who got the placebo. The combined statistics could show that twice as many among those who got the placebo are free for five years from both colds and cavities compared with those who got the pill.

A striking instance of how a hypothesis can be confirmed by two independent studies, yet disconfirmed by the total results, is provided by the following game. It can be modeled with cards, but to vary the equipment let us do it with 41 poker chips and four hats [*see top illustration on page 162*]. On table *A* is a black hat containing five colored chips and six white chips. Beside it is a gray hat containing three colored chips and four white chips. On table *B* is another pair of black and gray hats. In the black hat there are six colored chips and three white chips. In the gray hat there are nine colored chips and five white chips. The contents of the four hats are shown by the charts in the illustration.

You approach table *A* with the desire to draw a colored chip. Should you take a chip from the black hat or from the gray one? In the black hat five of the 11 chips are colored, so that the probability of getting a colored chip is 5/11. This is greater than 3/7, which is the probability of getting a colored chip if you take a chip from the gray hat. Clearly your best bet is to take a chip from the black hat.

The black hat is also your best choice on table *B*. Six of its nine chips are colored, giving a probability of 6/9, or 2/3, that you will get a colored chip. This exceeds the probability of 9/14 that you will get a colored chip if you choose to take a chip from the gray hat.

Now suppose that the chips from both black hats are combined in one black hat and that the same is done for the chips in the two gray hats [*see table C in top illustration on page 162*]. If you want to get a colored chip, surely you should take a chip from the black hat. The astonishing fact is that this is not true! Of the 20 chips now in the black hat, 11 are colored, giving a probability of 11/20 that you will get a colored chip. This is exceeded by a probability of 12/21 that you will get a colored chip if you take a chip from the gray hat.

The situation has been called Simpson's paradox by Colin R. Blyth, who found it in a 1951 paper by E. H. Simpson. The paradox has turned out to be older, but the name has persisted. Again, it is easy to see how the paradox could arise in actual research. Two independent investigations of a drug, for example, might suggest that it is more effective on men than it is on women, whereas the combined data would indicate the reverse.

One might imagine that such situations are too artificial to arise in statistical research. In a recent investigation to see if there was sex bias in the admissions of men and women to graduate studies at the University of California at Berkeley, however, Simpson's paradox actually turned up. Independent studies of admissions of men and of women in the fall of 1973 showed a positive sex bias against female applicants. Then when the data for men and women were combined, there was a small but statistically significant bias *in favor* of women (see "Sex Bias in Graduate Admissions: Data from Berkeley," by P. J. Bickel, E. A. Hammel and J. W. O'Connell in *Science,* Vol. 187, February 7, 1975, pages 398–404).

Blyth has invented another paradox that is even harder to believe than Simpson's. It can be modeled with three sets of cards or three unfair dice that are weighted to give the required probability distributions to their faces. We shall model it with the three spinners shown in the bottom illustration on page 162, because they are easy to construct by anyone who wants to verify the paradox empirically.

Spinner *A,* with an undivided dial, is the simplest. No matter where the arrow stops, it gives a value of 3. Spinner *B* gives values of 2, 4 or 6 with the respective probability distributions of .56, .22 and .22. Spinner *C* gives values of 1 or 5 with the probabilities of .51 and .49.

You pick a spinner; a friend picks another. Each of you flicks his arrow, and the highest number wins. If you can later change spinners on the basis of experience, which spinner should you choose? When the spins are compared in pairs, we find that *A* beats *B* with a probability of $1 \times .56 = .56$. *A* beats *C* with a probability of $1 \times .51 = .51$. *B* beats *C* with a probability of $(1 \times .22) + (.22 \times .51) + (.56 \times .51) = .6178$. Clearly *A,* which beats both of the others with a probability of more than 1/2, is the best choice. *C* is the worst because it is beaten with a probability of more than 1/2 by both of the others.

Now for the crunch. Suppose you play the game with two others and you have the first choice. The three spinners are flicked and the high number wins. Calculating the probabilities reveals an extraordinary fact. *A* is the worst choice; *C* is the best! *A* wins with a probability of $.56 \times .51 = .2856$, or less than 1/3. *B* wins with a probability of $(.44 \times .51) + (.22 \times .49) = .3322$, or almost 1/3. *C* wins with a probability of $.49 \times .78 = .3822$, or more than 1/3.

Consider the havoc this can wreak in statistical testing. Assume that drugs for a certain illness are rated in effectiveness with numbers 1 through 6. Drug *A* is uniformly effective at a value of 3 (spinner *A*). Studies show that drug *C* varies in effectiveness. Fifty-one percent of the time it has value 1 and 49 percent of the time it has value 5 (spinner *C*). If drugs *A* and *C* are the only two on the market and a doctor wants to maximize a patient's chance of recovery, he clearly chooses drug *A*.

What happens when drug *B,* with values and a probability distribution corresponding to spinner *B,* becomes available? The bewildered doctor, if he considers all three drugs, finds *C* preferable to *A*.

Blyth has an even more mind-blowing way of dramatizing the paradox. Every night a statistician eats at a restaurant that offers apple pie and cherry pie. He rates his satisfaction with each kind of pie in values 1 through 6. The apple pie is uniformly 3 (spinner *A*); the cherry varies in the manner of spinner *C*. Naturally the statistician always takes apple.

Occasionally the restaurant has blueberry pie. Its satisfaction varies in the manner of spinner *B*.

Waitress: "Shall I bring your apple pie?"

Statistician: "No. Seeing that today you also have blueberry, I'll take the cherry."

The waitress would consider that a joke. Actually the statistician is rationally maximizing his expectation of satisfaction. Is there any paradox that points up more spectacularly the kinds of difficulty Carnap's followers must overcome in their efforts to advance his program?

V

SYMBOLIC LOGIC AND COMPUTERS

V SYMBOLIC LOGIC AND COMPUTERS

INTRODUCTION

Both logic and computer science are now vast areas. Whether they should be regarded as subjects in their own right or as subdivisions of mathematics is a matter of opinion. Theoretical physics is entirely mathematical in essence, yet in modern times it has been classed as physics. Certainly both logic and computer science are highly mathematical today and even owe their existence to mathematics. Our articles do not range over the full extent and significance of these two fields; rather, they deal primarily with a relationship between the two. However, it may be helpful to see what the articles do treat in a somewhat larger context.

Though peoples of many civilizations have reasoned for thousands of years, it was the Greeks of the classical period (600–300 B.C.) who decided that mathematics must be built up by deductive reasoning from what appeared to them to be self-evident truths. Valid or correct deductive reasoning takes many forms. The commonest form used in mathematics is the simple syllogism, which says that if all pigs are animals and Pokey is a pig, then Pokey is an animal. In a mathematical context this form is exemplified by the argument: All fractions are rational numbers; 3/4 is a fraction; hence 3/4 is a rational number. A variant on this form might be: All fractions are rational numbers; π is not rational; hence π is not a fraction. In these instances we can readily decide whether the reasoning is valid. However, let us consider the argument: All bears are red; all animals are red; hence all bears are animals. Does the conclusion follow from the two premises? In this case and in many others it is not so easy to answer the question. Through the work of mathematicians the principles of reasoning were sharpened, enlarged, and acquired greater importance. Consequently Aristotle in his *Organon* catalogued in verbal form a number of principles or laws of valid reasoning. A few more principles were added during medieval times, mainly by theologians. To facilitate the use of the laws of logic and possibly to obtain new ones, a number of mathematicians from Leibniz onward conceived the idea that the principles of logic and the concepts used in various areas should be expressed in symbols. Any argument involving the concepts could then be represented in purely symbolic form, and one could readily detect whether the argument was correct. Using symbols, one could also more easily deduce further consequences from given premises by purely "algebraic" means. The suggestion for this approach to logic was derived from ordinary algebra, wherein by Leibniz's time symbolism had clearly demonstrated its value. (See the article by Kreiling in the section on "History.")

Though Leibniz in the seventeenth century and many other men in the eighteenth advanced the idea of an algebra of logic, the first great step in this direction was made by George Boole in the 1840s. Boole, like many other thinkers before him, was impressed with the efficacy of mathematical symbolism. Hence he proceeded to algebraicize logic. Given the logical principles in symbolic form, one can apply them to specific mathematical premises (axioms and theorems) to deduce new theorems. Nor need the algebra of logic be applied only to mathematics; it can be applied to all situations where deductive reasoning is appropriate. Boole's work was the first major step in the algebra of logic, and the beginning of what is called today symbolic logic or mathematical logic.

It so happens that the main concern of Aristotelian logic is syllogistic reasoning, of which the pig and Pokey argument is an example. Syllogistic reasoning deals with classes of objects. Thus the statement "All pigs are animals" says that the class of pigs is contained in the class of animals. By using letters such as a, b, and c for classes of objects and by using special symbols for other concepts, one can express this type of statement in purely symbolic form. Thus the statement "All a is b" was expressed by Boole in the form $a(1 - b) = 0$, wherein $1 - b$ is the class of objects that are not b. The product of a and $1 - b$ represents the objects common to the two classes. Since all a is b, there are no a's in $1 - b$, and so the product contains no objects. This is represented by 0. From a series of such symbolic statements about a and b, one can obtain other correct statements about a and b by algebraic manipulation. Boolean algebra was improved by many successors of Boole, notably William Stanley Jevons and Charles Sanders Peirce.

To help us understand the meaning of the basic and derived principles of logic, John Venn invented a diagrammatic scheme to represent pictorially the relationships among classes that were expressed algebraically by Boole. The diagrams at least confirm the correctness of the logical principles. Gardner's article in this section explains the nature and use of Venn diagrams.

The application of symbolism to logical principles and the deduction of new principles by applying some of the basic principles to other principles initiated a development that has proved to be far weightier than what Boole originally intended. Whereas Boole was content to symbolize and extend Aristotelian logic, logicians have now raised the question that mathematicians raised about Euclidean geometry in Greek times. What principles of reasoning are so surely warranted by the consensus of all human minds that we can adopt them as axioms, and what new theorems of logic can we deduce by means of these axiomatic principles of reasoning? This project may seem to amount to lifting oneself up by one's own bootstraps, but it is not so. Among the basic principles or axioms of logic are principles of inference that permit the deduction of new principles. Thus from the basic principles we can deduce that an assertion that implies a contradiction must be false. This principle is used in the indirect method of proof in mathematics. For example, if the assumption that there is a finite number of prime numbers leads to a contradiction, the assumption must be false and the number of primes must be infinite. By the choice of axiomatic logical principles and the deduction of others, logic becomes organized in exactly the same manner as any branch of mathematics.

Gardner's and Pfeiffer's articles explain more fully the nature of symbolic or mathematical logic. They also point out that there are applications, notably to the design of electrical circuits, that were recognized as recently as 1938 by Claude Shannon. Shannon's discovery came just in time to aid immensely in another development, the improvement of electronic computers.

While symbolic logic was being explored and advanced, another theme was being pursued by mathematicians, namely, methods of speeding up arithmetic calculations. For example, the invention of logarithms in the sixteenth century was so motivated. But the idea of using mechanical devices also occurred, and the first machine was invented by Blaise Pascal and improved by Leibniz.

A detailed history of successive improvements will be found in the book by Goldstine listed in the suggested readings.

Since the 1940s electronic devices have replaced mechanical components. These devices have been improved almost daily to increase the memory or storage capacity (that is, the quantity of information the computer can store and bring back into use while new data are being fed into it) and to increase operating speed so that a million calculations per second is now a reality. It is in the improvement of electronic circuitry that symbolic logic and Shannon's work in particular play a vital role. Evans's article explains this role.

While calculation is still the chief use of computers, it was inevitable that the marvelous nature of the computer should stimulate a search for other uses, and many important ones have been found. Broadly stated, computers can now be used to perform any kind of information processing that can be formulated in the language that a computer can "understand." Hence beyond its enormously valuable service as a calculator, the computer is now used to store immense masses of data, to retrieve any bits of these data, to direct machinery in factories, to perform many of the operations of large businesses (such as the calculation of taxes to be deducted from salaries), to play games (such as chess), to test mathematical conjectures, and even to devise proofs of theorems.

For the purposes just described—arithmetical computations and the storage and retrieval of data—and for many other functions, the computers used are called digital. As the name implies, information fed into the computer and the information derived from it must either be numerical or translatable by some coding scheme into numerical data. There are special languages, called programming languages, in which the information to be put into a computer and the instructions to the computer are expressed. The particular language one uses is suited to the capacities and function of the computer. (See the reference to Strachey in the suggested readings.)

Beyond digital computers there is a large variety of what are called analogue computers. Perhaps the simplest example of an analogue computer is the speedometer of an automobile. As the speed of the automobile varies, the drive shaft communicates a varying centrifugal force to a mechanism that causes the needle in the speedometer to move. The nature of the centrifugal force can be appreciated if one whirls an object tied to the end of a string. The object seeks to pull away from the hand, and the pull is stronger as the object is whirled at higher speeds. Analogue computers record the behavior of quantities that vary continuously, whereas digital computers record discrete changes. One can also say that whereas the digital computer counts, the analogue computer measures. The thermostat in the modern home is another rather trivial example of an analogue computer.

Of course, modern analogue computers are far more complicated than speedometers and thermostats and rely heavily on electrical circuits. Suppose, to take a very simple example, one wished to design an analogue computer to indicate how a mass attached to a spring behaves when the upper end of the spring is fixed, the lower end is attached to the mass, and the mass is set into vibration. This mechanical system is simulated by an electrical system, which is the analogue computer. In the motion of the actual mass, the resistance of the air and the stiffness of the spring act as factors controlling the motion. These factors can be represented in the analogue simulator by variable electrical devices, such as resistors. By setting these variable devices to values corresponding to actual physical values, we can study the effects of air resistance and spring stiffness.

The input to an analogue computer is generally one or more continuously varying electrical signals; the output may be a continuous graph drawn by a pen on paper or a display on an oscilloscope (which is in essence a very simple television set).

The distinction between digital and analogue computers is made clear in Ridenour's article. To the uses of analogue computers that Ridenour describes, one can add many others, such as the use of radar sets by traffic controllers in airports to track airplanes and the use of radar to direct antiaircraft guns.

Other uses of computers are now being realized or are under investigation. Computers linked to oscilloscopes can be programmed to draw complicated geometrical figures, animated cartoons, and series of curves that picture the effects of changing input data. The possibility of getting computers to respond to vocal directions, to translate foreign language materials, and even to reproduce themselves are also receiving serious consideration.

The digital computer is no longer an instrument that laypersons admire from a distance. It has already entered the classroom, where its value in teaching mathematics, physics, and chemistry is certainly positive, although its full potential in aiding teaching remains to be ascertained. Many efforts to determine the best uses of computer-assisted instruction (C.A.I.) are now being made. The computer, in the form of the hand calculator, is now entering the home, retail store, and almost every kind of business. These small computers are of course much more limited in the number of digits they can display and in the operations they can perform. But miniaturization is proceeding apace, and perhaps the day is not far off when all of us will be doing complex calculations on computers.

SUGGESTED READINGS

Bartee, Thomas C. 1975. *Introduction to Computer Science.* McGraw-Hill, New York.

Boole, George. 1952. *Collected Logical Works.* Open Court Publishing, Chicago.

DeLong, Howard. 1970. *A Profile of Mathematical Logic.* Addison-Wesley, Reading, Mass.

Dorn, William S., and Herbert Greenberg. 1967. *Mathematics and Computing.* Wiley, New York.

Eves, Howard, and C. V. Newsom. 1965. *An Introduction to the Foundations and Fundamental Concepts of Mathematics,* second edition. Holt, Rinehart and Winston, New York.

Goldstine, Herman. 1972. *The Computer from Pascal to von Neumann.* Princeton University Press, Princeton, N.J.

Goodstein, R. L. 1971. *Development of Mathematical Logic.* Springer-Verlag, New York.

Higman, Bryan. 1975. *Foundation Course in Computer Science.* American Elsevier, New York.

Katzan, Harry J. 1976. *Computer Systems, Organization and Programming.* Science Research Associates, Chicago.

Langer, Suzanne. 1953. *An Introduction to Symbolic Logic,* second edition. Dover, New York.

Lewis, C. I. 1960. *A Survey of Symbolic Logic.* Dover, New York.

McCracken, Daniel D. 1961. *A Guide to Fortran Programming.* Wiley, New York.

Mendelson, Elliott. 1964. *Introduction to Mathematical Logic.* D. Van Nostrand, New York.

Raphael, Bertram. 1976. *The Thinking Computer, Mind inside Matter.* W. H. Freeman and Company, San Francisco.

Scott, P. E. 1975. *Programming in Basic, a Beginner's Course.* English Universities Press, London.

Singh, Jagit. 1966. *Great Ideas in Information Theory, Language and Cybernetics.* Dover, New York.

Stoll, Robert R. 1974. *Sets, Logic, and Axiomatic Theories,* second edition. W. H. Freeman and Company, San Francisco.

Strachey, Christopher. 1966. "System Analysis and Programming." *Scientific American,* September.

Styazhkin, N. I. 1969. *History of Mathematical Logic from Leibniz to Peano.* M.I.T. Press, Cambridge, Mass.

Tarski, Alfred. 1946. *Introduction to Logic,* second edition. Oxford University Press, New York.

Taube, Mortimer. 1961. *Computers and Common Sense.* Columbia University Press, New York.

Weizenbaum, Joseph. 1976. *Computer Power and Human Reason.* W. H. Freeman and Company, San Francisco.

Boolean Algebra, Venn Diagrams and the Propositional Calculus

by Martin Gardner
February 1969

Aristotle deserves full credit as the founder of formal logic even though he restricted his attention almost entirely to the syllogism. Today, when the syllogism has become a trivial part of logic, it is hard to believe that for 2,000 years it was the principal topic of logical studies, and that as late as 1797 Immanuel Kant could write that logic was "a closed and completed body of doctrine."

"In syllogistic inference," Bertrand Russell once explained, "you are supposed to know already that all men are mortal and that Socrates is a man; hence you deduce what you never suspected before, that Socrates is mortal. This form of inference does actually occur, though very rarely." Russell goes on to say that the only instance he ever heard of was prompted by a comic issue of *Mind*, a British philosophical journal, that the editors concocted as a special Christmas number in 1901. A German philosopher, puzzled by the magazine's advertisements, eventually reasoned: Everything in this magazine is a joke, the advertisements are in this magazine, therefore the advertisements are jokes. "If you wish to become a logician," Russell wrote elsewhere, "there is one piece of sound advice which I cannot urge too strongly, and that is: Do *not* learn the traditional logic. In Aristotle's day it was a creditable effort, but so was the Ptolemaic astronomy."

The big turning point came in 1847 when George Boole (1815–1864), a modest, self-taught son of a poor English shoemaker, published *The Mathematical Analysis of Logic*. This and other papers led to his appointment (although he had no university degree) as professor of mathematics at Queens College (now University College) at Cork in Ireland, where he wrote his treatise *An Investigation of the Laws of Thought, on Which are Founded the Mathematical Theories of Logic and Probabilities* (London, 1854). The basic idea—substituting symbols for all the words used in formal logic—had occurred to others before, but Boole was the first to produce a workable system. By and large neither philosophers nor mathematicians of his century showed much interest in this remarkable achievement. Perhaps that was one reason for Boole's tolerant attitude toward mathematical eccentrics. He wrote an article about a Cork crank named John Walsh (*Philosophical Magazine,* November, 1851) that Augustus De Morgan, in his *Budget of Paradoxes,* calls "the best biography of a single hero of the kind that I know."

The few who appreciated Boole's genius (notably the German mathematician Ernst Schröder) rapidly improved on Boole's notation, which was clumsy mainly because of Boole's attempt to make his system resemble traditional algebra. Today Boolean algebra refers to an "uninterpreted" abstract structure that can be axiomized in all kinds of ways but that is essentially a streamlined, simplified version of Boole's system. "Uninterpreted" means that no meanings whatever—in logic, mathematics or the physical world—are assigned to the structure's symbols.

As in the case of all purely abstract algebras, many different interpretations can be given to Boolean symbols. Boole himself interpreted his system in the Aristotelian way as an algebra of classes and their properties, but he greatly extended the old class logic beyond the syllogism's narrow confines. Since Boole's notation has been discarded, modern Boolean algebra is now written in the symbols of set theory, a set being the same as what Boole meant by a class: any collection of individual "elements." A set can be finite, such as the numbers 1, 2, 3, the residents of Omaha who have green eyes, the corners of a cube, the planets of the solar system or any other specified collection of things. A set also can be infinite, such as the set of even integers or possibly the set of all stars. If we specify a set, finite or infinite, and then consider all its "proper subsets" (they include the set itself as well as the empty set of no members) as being related to one another by inclusion (that

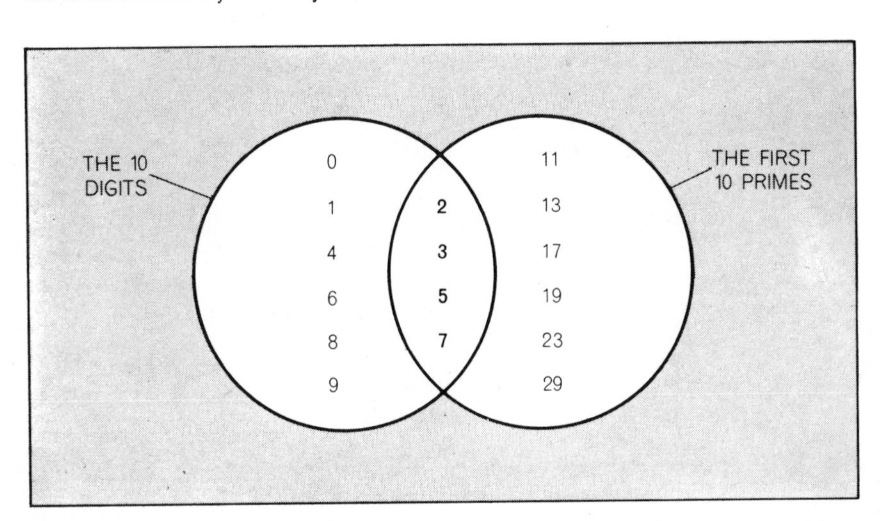

Venn diagram for set intersection

is, the set 1, 2, 3 is included in the set 1, 2, 3, 4, 5), we can construct a Boolean set algebra.

A modern notation for such an algebra uses letters for sets, subsets or elements. The "universal set," the largest set being considered, is symbolized by **U**. The empty, or null, set is $\varnothing$. The "union" of sets a and b (everything in a and b) is symbolized by $\cup$, sometimes called a cup. (The union of 1, 2 and 3, 4, 5 is 1, 2, 3, 4, 5.) The "intersection" of sets a and b (everything common to a and b) is symbolized by $\cap$, sometimes called a cap. (The intersection of 1, 2, 3 and 3, 4, 5 is 3.) If two sets are identical (for example, the set of odd numbers is the same as the set of all integers with a remainder of 1 when divided by 2), this is symbolized by $=$. The "complement" of set a—all elements of the universal set that are not in a—is indicated by a'. (The complement of 1, 2, with respect to the universal set 1, 2, 3, 4, 5, is 3, 4, 5.) Finally, the basic binary relation of set inclusion is symbolized by ϵ; $a \epsilon b$ means that a is a member of b.

As a matter of historical interest, Boole's symbols included letters for elements, classes and subclasses, 1 for the universal class, 0 for the null class, + for class union (which he took in an "exclusive" sense to mean those elements of two classes that are *not* held in common; the switch to the "inclusive" sense, first made by the British logician and economist William Stanley Jevons, had so many advantages that later logicians adopted it), $\times$ for class intersection, $=$ for identity, and the minus sign, $-$, for the removal of one set from another. To show the complement of x, Boole wrote $1 - x$. He had no symbol for class inclusion but could express it in various ways such as $a \times b = a$, meaning that the intersection of a and b is identical with all of a.

The Boolean algebra of sets can be elegantly diagrammed with Venn circles (after the English logician John Venn), which are now being introduced in many elementary school classes. Venn circles are diagrams of an interpretation of Boolean algebra in the point-set topology of the plane. Let two overlapping circles symbolize the union of two sets [*see illustration on preceding page*], which we here take to be the set of the 10 digits and the set of the first 10 primes. The outer rectangle contains the universal set. This includes the area outside both circles, which is shaded to indicate that it is the null set; it is empty because we are concerned solely with the elements inside the two circles. These 16 elements are the union of the

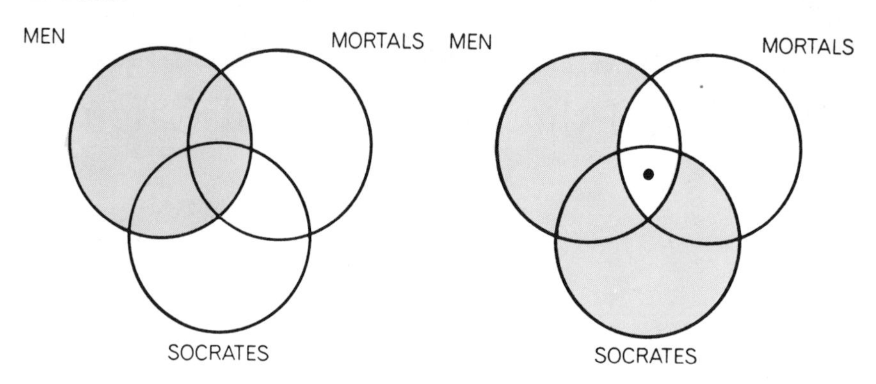

Premise: "All men are mortal"

Premise: "Socrates is a man"

two sets. The overlapping area contains the intersection. It consists of the set 2, 3, 5, 7: digits that are also among the first 10 primes.

Adopting the convention of shading any area known to represent an empty set, we can see how a three-circle Venn diagram proves the ancient syllogism Russell so scornfully cited. The circles are labeled to indicate sets of men, mortal things and Socrates (a set with only one member). The first premise, "All men are mortal," is diagrammed by shading the men circle to show that the class of nonmortal men is empty [*see illustration at top left on this page*]. The second premise, "Socrates is a man," is similarly diagrammed by shading the Socrates circle to show that all of Socrates, namely himself, is inside the men circle [*see illustration at top right on this page*]. Now we inspect the diagram to see if the conclusion, "Socrates is mortal," is valid. It is. All of Socrates (the unshaded part of his circle marked by a dot) is inside the circle of mortal things. By exploiting the topological properties of simple closed curves we have a method of diagramming that is isomorphic with Boolean set algebra.

The first important new interpretation of Boolean algebra was suggested by Boole himself. He pointed out that if his 1 were taken as truth and his 0 as falsehood, the calculus could be applied to statements that are either true or false. Boole did not carry out this program but his successors did. It is now called the propositional calculus. This is the calculus concerned with true or false statements connected by such binary relations as "If p then q," "Either p or q but

BOOLEAN SET ALGEBRA	PROPOSITIONAL CALCULUS
U (UNIVERSAL SET)	T (TRUE)
ϕ (NULL SET)	F (FALSE)
$a, b, c, \ldots$ (SETS, SUBSETS, ELEMENTS)	$p, q, r, \cdots$ (PROPOSITIONS)
$a \cup b$ (UNION: ALL OF a AND b)	$p \vee q$ (DISJUNCTION: EITHER p ALONE OR q ALONE, OR BOTH, ARE TRUE.)
$a \cap b$ (INTERSECTION: WHAT a AND b HAVE IN COMMON)	$p \bullet q$ (CONJUNCTION: BOTH p AND q ARE TRUE.)
$a = b$ (IDENTITY: a AND b ARE THE SAME SET.)	$p = q$ (EQUIVALENCE: IF AND ONLY IF p IS TRUE, THEN q IS TRUE.)
a' (COMPLEMENT: ALL OF **U** THAT IS NOT a)	$\sim p$ (NEGATION: p IS FALSE.)
$a \in b$ (INCLUSION: a IS A MEMBER OF b.)	$p \supset q$ (IMPLICATION: IF p IS TRUE, q IS TRUE.)

Corresponding symbols in two versions of Boolean algebra

not both," "Either p or q or both," "If and only if p then q," "Not both p and q" and so on. The chart below shows the symbols of the propositional calculus that correspond to symbols for the Boolean set algebra.

It is easy to understand the isomorphism of the two interpretations by considering the syllogism about Socrates. Instead of saying "All men are mortal," which puts it in terms of class properties or set inclusion, we rephrase it as, "If x is a man then x is a mortal." Now we are stating two propositions and joining them by the "connective" called "implication." This is diagrammed on Venn circles in exactly the same way we diagrammed "All men are mortal." Indeed, all the binary relations in the propositional calculus can be diagrammed with Venn circles and the circles can be used for solving simple problems in the calculus. It is shameful that writers of most introductory textbooks on formal logic have not yet caught on to this. They continue to use Venn circles to illustrate the old class-inclusion logic but fail to apply them to the propositional calculus, where they are just as efficient. Indeed, they are even more efficient, since in the propositional calculus one is unconcerned with the "existential quantifier," which asserts that a class is not empty because it has at least one member. This was expressed in the traditional logic by the word "some" (as in "Some apples are green"). To take care of such statements Boole had to tie his algebra into all sorts of horribly complicated knots.

To see how easily the Venn circles solve certain types of logic puzzles, consider the following premises about three businessmen, Abner, Bill and Charley, who lunch together every working day:

1. If Abner orders a martini, so does Bill.
2. Either Bill or Charley always orders a martini, but never both at the same lunch.
3. Either Abner or Charley or both always order a martini.
4. If Charley orders a martini, so does Abner.

To diagram these statements with Venn circles we identify having a martini with truth and not having one with falsehood. The eight areas of the overlapping circles shown in the top illustration at the left are labeled to show all possible combinations of truth values for a, b, c, which stand for Abner, Bill and Charley. Thus the area marked a, $\sim b$, c represents Abner's and Charley's having martinis while Bill does not. See if you can shade the areas declared empty by the four premises and then examine the result to determine who will order martinis if you lunch with the three men. The answer is in the back of the book.

There are many other ways to interpret Boolean algebra. It can be taken as a special case of an abstract structure called a ring, or as a special case of another type of abstract structure called a lattice. It can be interpreted in combinatorial theory, information theory, graph theory, matrix theory and metamathematical theories of deductive systems in general. In recent years the most useful interpretation has been in switching theory, which is important in the design of electronic computers but is not limited to electrical networks. It applies to any kind of energy transmission along channels with connecting devices that turn the energy on and off, or switch it from one channel to another.

The energy can be a flowing gas or liquid, as in modern fluid control systems [see "Fluid Control Devices," by Stanley W. Angrist; SCIENTIFIC AMERICAN, December, 1964]. It can be light beams. It can be mechanical energy as in the logic machine Jevons invented for solving four-term problems in Boolean algebra. It can be rolling marbles, as in several computer-like toys now on the market: Dr. Nim, Think-a-Dot and Digi-Comp II. And if inhabitants of another planet have a highly developed sense of smell, their computers could use odors transmitted through tubes to sniffing outlets. As long as the energy either moves or does not move along a channel there is an isomorphism between the two states and the two truth values of the propositional calculus. For every binary connective in the calculus there is a corresponding switching circuit. Three simple examples are shown in the bottom illustration at the left. The bottom circuit is used whenever two widely separated electric light switches are used to control one light. It is easy to see that if the light is off, changing the state of either switch will turn it on, and if the light is on, either switch will turn it off.

This electrical-circuit interpretation of Boolean algebra had been suggested in a Russian journal by Paul S. Ehrenfest as early as 1910 and independently in Japan in 1936, but the first major paper, the one that introduced the interpretation to computer designers, was Claude E. Shannon's "A Symbolic Analysis of Relay and Switching Circuits" in the *Transactions of the American Institute of Electrical Engineers*, Volume 57, December, 1938. It was based on Shannon's 1937 master's thesis at the Massachusetts Institute of Technology, where he is now professor of mathematics.

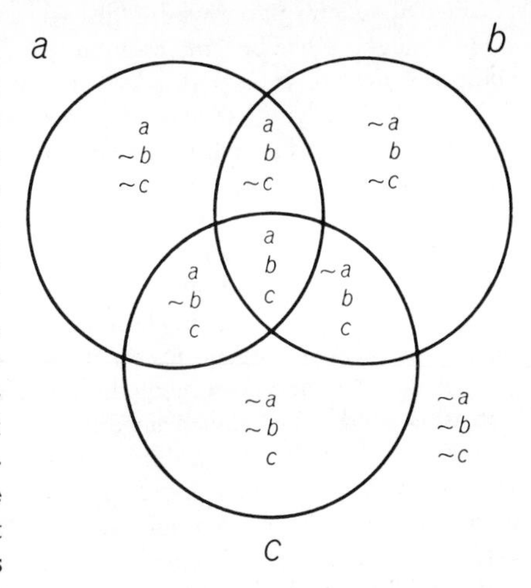

Venn diagram for martini puzzle

"AND" CIRCUIT: BULB LIGHTS ONLY IF BOTH *a* AND *b* ARE CLOSED.

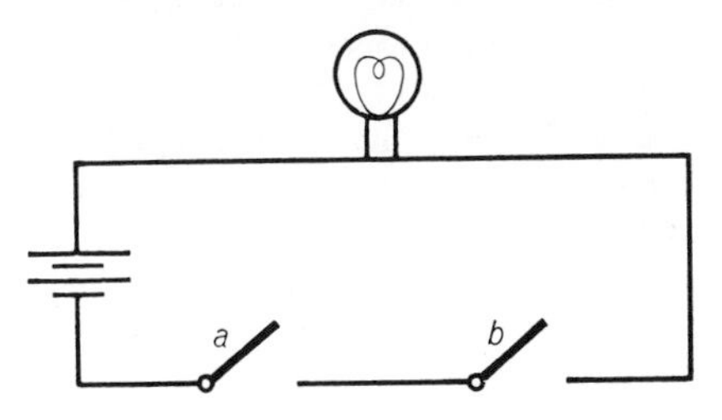

INCLUSIVE "OR" CIRCUIT: BULB LIGHTS ONLY IF *a* OR *b* OR BOTH ARE CLOSED.

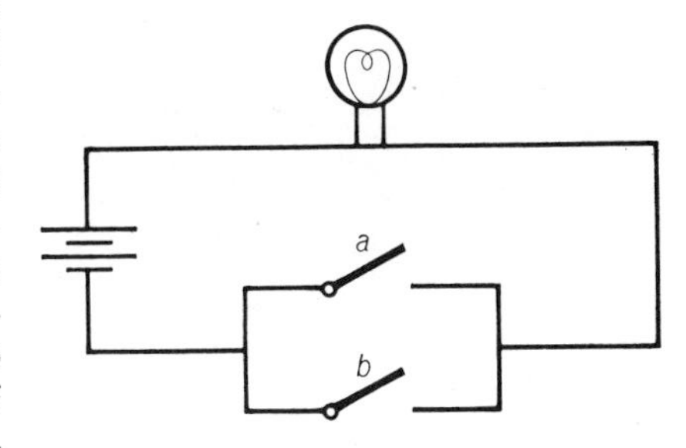

EXCLUSIVE "OR" CIRCUIT: BULB LIGHTS ONLY IF *a* OR *b*, BUT *NOT* BOTH, IS LOWERED.

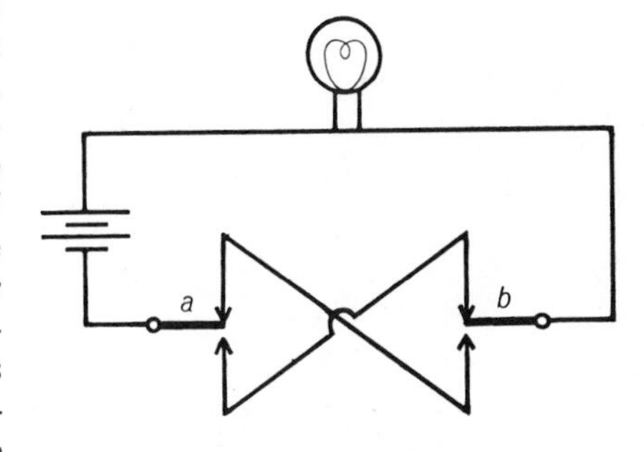

Circuits for three binary relations

Since Shannon's paper was published, Boolean algebra has become essential to computer design. It is particularly valuable in simplifying circuits to save hardware. A circuit is first translated into a statement in symbolic logic, the statement is "minimized" by various clever methods and the simpler statement is translated back to the design of a simpler circuit. Of course in modern computers the switches are no longer magnetic devices or vacuum-tube diodes but transistors and other tiny semiconductors. For a while after the publication of Shannon's historic paper much of the work on the logic of computer design was done with almost no communication between the experts of various countries. Gerard Piel, in his book *Science in the Cause of Man* (Knopf, 1961), reports that American mathematicians employed by several big corporations worked for five years, at a cost of about $200,000, to duplicate work that had already been published in Russia before they started.

Now for one final interpretation of Boolean algebra that is a genuine curiosity. Consider the following set of eight numbers: 1, 2, 3, 5, 6, 10, 15, 30. They are the factors of 30, including 1 and 30 as factors. We interpret "union" as the least common multiple of any pair of those numbers. "Intersection" of a pair is taken to be their greatest common divisor. Set inclusion becomes the relation "is a factor of." The universal set is 30, the null set 1. The complement of a number a is $30/a$. With these novel interpretations of the Boolean relations it turns out that we have a consistent Boolean structure! All the theorems of Boolean algebra have their counterparts in this curious system based on the factors of 30. For example, in Boolean algebra the complement of the complement of a is simply a, or in the propositional-calculus interpretation the negation of a negation is the same as no negation. More generally, only an odd series of negations equals a negation. (In *The New York Times* in 1965 I saw the headline "Albany Kills Bill to Repeal Law Against Birth Control." It took me a while to realize that the three negatives made this a decision *against* birth control.) Let us apply this Boolean law to the number 3. Its complement is $30/3 = 10$. The complement of 10 is $30/10 = 3$, which brings us back to 3 again.

Consider two famous Boolean laws called De Morgan's laws. In the algebra of sets they are

$$(a \cup b)' = a' \cap b'$$
$$(a \cap b)' = a' \cup b' .$$

In the propositional calculus they look like this:

$$\sim(a \vee b) \equiv \sim a \cdot \sim b$$
$$\sim(a \cdot b) \equiv \sim a \vee \sim b .$$

If the reader will substitute any two factors of 30 for a and b, and interpret the symbols as explained, he will find that De Morgan's laws hold. The fact that De Morgan's laws form a pair illustrates the famous duality principle of Boolean algebra. If in any statement you interchange (if and wherever they appear) union and intersection and interchange the universal and the null sets, and also reverse the direction of set inclusion, the result is another valid law. Moreover, these changes can be made all along the steps of the proof of one law to provide a valid proof of the other! (An equally beautiful duality principle holds in projective geometry with respect to interchanges of lines and points.)

The numbers 1, 2, 3, 5, 6, 7, 10, 14, 15, 21, 30, 35, 42, 70, 105, 210—the 16 factors of 210—also form a Boolean algebra when interpreted in the same way, although of course 210 is now the universal set and the complement of a is $210/a$. Can the reader discover, before looking at the solution, a simple way to generate sets of 2^n numbers, where n is any positive integer, that will form Boolean systems of this peculiar kind?

Symbolic Logic

by John E. Pfeiffer
December 1950

It is a language that manipulates ideas as algebra manipulates numbers. It has been applied with notable success to practical problems requiring unusually precise and economical reasoning

WHAT NUMBER added to one fifth of itself equals 21? This problem was too difficult for most of the scholars of ancient Egypt. According to papyrus records, many arithmeticians struggled with it in vain before a patient Egyptian finally arrived at the correct answer about 1600 B.C. Today a ninth-grade algebra student can find the answer in a moment: $x+x/5=21$; therefore $x=17\frac{1}{2}$. What made the problem hard for the Egyptians was that they lacked our handy symbols, *i.e.*, digits for numbers and x for the unknown. Since they had to use words to represent numbers, their operations in arithmetic and algebra were cumbersome and slow.

The substitution of symbols for words is one of the things that has been largely responsible for man's progress in science. Yet in the process of logic—the basic tool with which we must test all ideas and also solve most of our everyday problems—we are still laboring under the Egyptians' handicap. We are at the mercy of the inadequacies and clumsiness of words.

Consider this simple exercise in logic, taken from a textbook on the subject by Lewis Carroll, mathematician and author of *Alice's Adventures in Wonderland*:

No kitten that loves fish is unteachable.

No kitten without a tail will play with a gorilla.

Kittens with whiskers always love fish.

No teachable kitten has green eyes.

No kittens have tails unless they have whiskers.

One, and only one, deduction can be drawn from this set of statements. After considerable trial and error you may find the answer by rewording and rearranging the statements:

Green-eyed kittens cannot be taught.

Kittens that cannot be taught do not love fish.

Kittens that do not love fish have no whiskers.

Kittens that have no whiskers have no tails.

Kittens that have no tails will not play with a gorilla.

The one valid deduction, then, is that green-eyed kittens will not play with a gorilla.

But now take a problem that is somewhat more complicated. The following is adapted from an examination in logic prepared recently by the mathematician Walter Pitts of the Massachusetts Institute of Technology:

If a mathematician does not have to wait 20 minutes for a bus, then he either likes Mozart in the morning or whisky at night, but not both.

If a man likes whisky at night, then he either likes Mozart in the morning and does not have to wait 20 minutes for a bus or he does not like Mozart in the morning and has to wait 20 minutes for a bus or else he is no mathematician.

If a man likes Mozart in the morning and does not have to wait 20 minutes for a bus, then he likes whisky at night.

If a mathematician likes Mozart in the morning, he either likes whisky at night or has to wait 20 minutes for a bus; conversely, if he likes whisky at night and has to wait 20 minutes for a bus, he is a mathematician—if he likes Mozart in the morning.

When must a mathematician wait 20 minutes for a bus?

The reader is not advised to try to work out the solution, for this problem is practically impossible to handle verbally.

ALTHOUGH these particular brain-teasers are artificial and trivial, in form they are quite typical of problems that arise every day in modern engineering and business operations. Many of the problems are so complex that they cannot be solved by the conventional processes of verbal logic. The necessary facts may all be known, but their interrelationships are so complex that no expert can organize them logically. In other words, the bigness of modern machines, business and government is creating more and more problems in reasoning which are too intricate for the human brain to analyze with words alone.

As a result a number of corporations and technicians have recently begun to take an active interest in the discipline known as symbolic logic. This invention, devised by mathematicians, is simply an attempt to use symbols to represent ideas and methods of handling them, just as symbols are employed to solve problems in mathematics. With the shorthand of symbolic logic it becomes possible to deal with such complex problems as the Pitts conundrum about the mathematician waiting for the bus.

Formal logic, as every schoolboy knows, began with the syllogisms of Aristotle, the most famous of which is: "All men are mortal; all heroes are men; therefore all heroes are mortal." The Greek philosopher set forth 14 such syllogisms and believed that they summed up most of the operations of reasoning. Medieval theologians added 5 syllogisms to Aristotle's 14. For hundreds of years these 19 syllogisms were the foundation of the teaching of logic.

Not until the 19th century did anyone successfully apply symbols and algebra to logic, in place of the verbalisms of Aristotle and his followers. In 1847 an English schoolteacher and mathematician named George Boole published a pamphlet called *The Mathematical Analysis of Logic—Being an Essay Towards a Calculus of Deductive Reasoning*. In it he stated a set of axioms from which more complex statements could be deduced. The statements were in algebraic terms, with symbols such as x and y representing classes of objects or ideas, and the deductions were arrived at by algebraic operations. Thus Boole became the inventor of symbolic logic. His work was followed up by mathematicians in many countries. Their chief aim was to use symbolic logic to solve logical paradoxes and other fundamental problems of mathematical thinking. By 1913 Alfred North Whitehead and Bertrand Russell, using a system of symbols invented by the Italian mathematician Giuseppe Peano, had developed a formal "mathematical logic," which they presented in their *Principia Mathematica* (see "Mathematics," by Sir Edmund Whittaker; SCIENTIFIC AMERICAN, September, 1950).

Today symbolic logic is an important

branch of mathematics, occupying the full time of about 200 mathematicians in the U. S. alone. But the main subject of this article is its practical applications in engineering and business.

LET US first take a few simple illustrations to indicate some of the basic symbols and operations employed in symbolic logic. Any single proposition, however simple or complex, is represented by a letter of the alphabet. For example, the letter *a* can stand for the statement "The sun is shining," or for something more involved, like "The three-power commission has been directed to look into the question of whether or not a West German federal police force should be created." Then certain special symbols are used to show relations between propositions. A dot, for example, stands for the word "and." Thus the two-proposition statement "The sun is shining and it is Thursday" can be represented by the expression $a \cdot b$.

The symbol $\supset$ stands for the logical relationship "if . . . then." Thus the assertion "If you love cats, then you are a true American" can be written $a \supset b$. Now by the use of other symbols and by operations similar to those in ordinary algebra, this statement can be transformed into a fully equivalent expression in another form. For example, using the symbol *v*, which stands for the word "or," and a superposed bar, representing the negative, the expression becomes $\bar{a}$ *v* *b*, meaning "You do not love cats or you are a true American." The statement can also be transformed into one containing the symbol for "and." Thus $\overline{\bar{a} \cdot \bar{b}}$ means "It is not the case both that you do not love cats and that you are a true American," or in ordinary English: "You cannot be indifferent or hostile to cats and also be a true American."

It is important to bear in mind that the symbols have nothing to do with the truth or falsity of the propositions themselves, just as algebra is not concerned with whether its symbols stand for apples or hours. The operations of symbolic logic can only show that, given certain premises, certain conclusions are valid and others are invalid. In this case, assuming that only cat-lovers are true Americans, if you are not a cat-lover the only logically valid conclusion is that you are not a true American, however debatable the proposition may be as a moral principle. The establishment of factually accurate premises is outside the province of logic; its concern is with the validity of the conclusions drawn from a given set of facts or assumptions.

By means of simple signs such as those here illustrated, symbolic logic reduces complex logical problems to manageable proportions. The symbols, like the schoolboy's algebra signs, do much of the logician's thinking for him. Large numbers of propositions can be related to one another in easy algebraic terms; equations can be arranged and rearranged, simplified and expanded, and the results, upon retranslation into English, can reveal new forms of statements that are equivalent to the original or can disclose inconsistencies.

THE FIRST application of symbolic logic to a business problem was made in 1936 by the mathematician Edmund C. Berkeley, who is also the designer of the small mechanical brain known as Simple Simon (SCIENTIFIC AMERICAN, November, 1950). Berkeley, then with the Prudential Life Insurance Company, applied symbolic logic to a difficult problem having to do with the rearrangement of premium payments by policyholders. Every year hundreds of thousands of persons request changes in the schedule of payments on their policies, and there is a bewildering array of factors that must be taken into account in making such changes. The company had devised two sets of rules, intended to take care of all possible cases. Were the two rules equivalent? Berkeley suspected that they were not; that there might be cases in which one rule would call for one method of rearranging the payments and the other for a different method.

His problem was to prove that such cases existed. It was hopeless to try to analyze the possibilities by ordinary verbal logic. One part of one of the rules, for example, stipulated that if a policyholder was making premium payments several times a year, with one of the payments falling due on the policy anniversary, and if he requested that the schedule be changed to one annual payment on the policy anniversary, and if he was paid up to a date which was not an anniversary, and if he made this request more than two months after the issue date, and if his request also came within two months after a policy anniversary—then a certain action should be taken. These five ifs alone can occur in 32 combinations, and there were many other factors involved.

Berkeley decided to reduce the many clauses and possible combinations and actions to the algebraic shorthand of symbolic logic. The stipulation detailed above, for example, could be written $a \cdot b \cdot \bar{c} \cdot d \cdot e \supset C$, meaning that if the conditions *a*, *b*, $\bar{c}$, *d* and *e* existed, then the action C was called for. By an algebraic analysis Berkeley was able to show that there were four types of cases in which the two rules would indeed conflict, and an examination of the company's files revealed that such cases actually existed. The upshot of Berkeley's work was that the two rules were combined into one simpler and consistent rule.

Symbolic logic has since been used in many other insurance problems. Mathematicians at Equitable, Metropolitan, Aetna and other companies have applied it to the analysis of war clauses and employee eligibility under group contracts. And other corporations have found symbolic logic very helpful in analyzing their contracts. Contracts between large corporations may run into many pages of fine print packed with stipulations, contingencies and a maze of ifs, ands and buts. Are the clauses worded as simply as they might be? Are there loop-

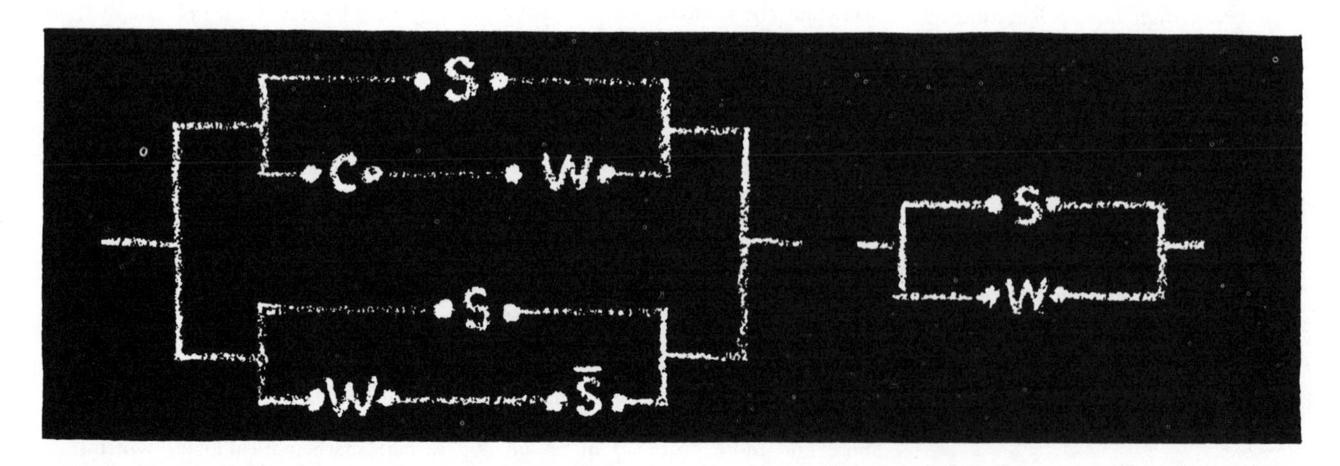

SWITCHING CIRCUITS may be analyzed and simplified by symbolic logic (*see text*). The switches are represented by symbols. Each of these two circuits has the same functions. The one at the left is "redundant."

holes or inconsistencies? A symbolic analysis can readily answer such questions, and lawyers have begun to call on mathematicians to go over their contracts.

Another interesting use of the technique is in checking the accuracy of censuses and of polling reports. If a public opinion poll-taker reports that he has interviewed 100 persons, of whom 70 were white, 10 were women and 5 were Negro men, it is easy enough to see that something is wrong with his figures. But take an actual case such as this: A census of 1,000 cotton-mill employees listed 525 Negroes, 312 males, 470 married persons, 42 Negro males, 147 married Negroes, 86 married males, 25 married Negro males. Are these numbers consistent? Symbolic logic can give the answer quickly.

IN ENGINEERING symbolic logic is particularly useful for the analysis of electric circuits. A circuit can be likened to a contract—it has alternatives, contingencies and possible loopholes, the chief difference being that it uses patterns of switches instead of words and clauses.

More than a dozen years ago Claude E. Shannon, then still a student at the Massachusetts Institute of Technology, began to explore the application of symbolic logic to such problems. At the Bell Telephone Laboratories he has recently completed an elaborate analysis of switching circuits by "engineering logic."

Suppose, to take a simple example, the problem is to simplify the six-switch circuit schematized in the left-hand drawing on the preceding page. The switches are given various symbols. The one labeled C is independent of all the others. The two W switches are connected so that they open and close together. The two S switches also operate together. The sixth switch is designated $\bar{S}$ (not-S, or the opposite of S), because it is open when the S switches are closed and *vice versa.*

There are four possible paths across this circuit from one side to the other. Current will flow across it when the upper S switch is closed, when the C and the upper W switches are closed, when the lower S switch is closed and when the lower W and $\bar{S}$ switches are closed. In the language of symbolic logic this sentence becomes $S\ v\ W \cdot C\ v\ S\ v\ \bar{S}\ W$, with the symbol v, as we have seen, meaning "or." It is at once evident that we can drop one S, since S is equivalent to the expression $S\ v\ S$. The statement now becomes $W \cdot C\ v\ S\ v\ \bar{S} \cdot W$. Next, we can simplify further by dropping the $\bar{S}$, for $S\ v\ W$ is the logical equivalent of $S\ v\ \bar{S} \cdot W$—just as the statement "Williams struck out or Williams did not strike out and walked" is the same as "Williams struck out or walked." This reduces the circuit to $S\ v\ W\ v\ W \cdot C$. A further analysis shows that $W\ v\ W \cdot C$ is equivalent to W. Logically speaking, the statement "Williams walked or Williams walked and was left at first base" provides only one unequivocal piece of information, namely that Williams walked. So the entire circuit boils down to $S\ v\ W$. It can be redesigned in a simple form, illustrated in the right-hand drawing on the preceding page, which eliminates four "redundant" switches and is fully equivalent to the original.

To use symbolic logic on a problem as simple as this would be like killing a mouse with an elephant gun. But in designing more complex circuits the method may save considerable time and money. At the Bell Laboratories, for example, a group of engineers some time ago undertook to design a special coding instrument. Applying conventional methods of analysis, they produced a 65-contact circuit for the job after several days of work. Then an engineer trained in symbolic logic, starting from scratch without seeing their design, designed an equally successful circuit, with 18 fewer contacts, in only three hours. Today more than 50 Bell engineers use symbolic logic in their work. The method has been applied successfully to a wide variety of problems, but it is not the final answer to all circuit difficulties. Its use is limited mainly to telephone equipment with about nine two-contact relays, which may be in 512 possible positions. In its present infant state even this powerful method of analysis cannot handle the breath-taking complexity of large central exchange stations where a single telephone call may cause the opening and closing of 10,000 contacts.

Perhaps the chief use of symbolic logic is in the design of large-scale electronic calculating machines. Eniac, the first of these machines, contains about 20,000 tubes and 500,000 soldered connections. One of the most important problems in the attempts to build more efficient and more elaborate computers is to reduce the number of tubes, and symbolic logic has been helpful in simplifying the circuits. For example, in building the Mark III all-electronic computer at the Naval Proving Ground in Dahlgren, Va., the engineers decided that a nine-tube circuit was about the minimum that would serve for its adding units. But Theodore Kalin and William Burkhart of the Harvard Computation Laboratory, applying symbolic logic, reduced it to six tubes.

THESE applications merely suggest the fruitful future that lies ahead for symbolic logic, not only in business and engineering but in science. Wherever complex problems in logical analysis arise, the new shorthand may help to find solutions. One such field is biology, which is beset with a host of complex logical problems. Already Walter Pitts and Warren McCulloch of the University of Illinois Medical School have begun to employ the symbolic logic of the *Principia Mathematica* in an effort to analyze some of the relationships among the 10 billion nerve cells in the human brain. Norbert Wiener of M.I.T. emphasizes that the new study of cybernetics, which analyzes similarities between the brain and computing machines, leans heavily on modern logic.

Although its applications are steadily widening, the major part of the work being done in symbolic logic is still in the field of mathematics. In mathematics this new tool has had so powerful an influence during the past four decades that today some consider mathematics to be only a branch of logic. Mathematicians are applying symbolic logic to examine some of the basic assumptions upon which mathematical theories have been built—assumptions that have long been taken for granted as "obvious" but have never been subjected to rigorous analysis. They are using it to try to resolve verbal paradoxes, which have always baffled logicians: *e.g.*, "All rules have exceptions," a rule which denies itself, since by its own assertion this statement must also have exceptions and therefore cannot be true. Many other basic problems in logic and mathematics are being explored by the new analysis.

Indeed, modern logicians, assisted by the powerful new technique, have punched the classical Aristotelian system of logic full of holes. Of the 19 syllogisms stated by Aristotle and his medieval followers, four are now rejected, and the rest can be reduced to five theorems. Modern logic has abandoned one of Aristotle's most basic principles: the law of the excluded middle, meaning that a statement must be either true or false. In the new system a statement may have three values: true, false or indeterminate. A close analogy to this system in the legal field is the Scottish trial law, which allows three verdicts—guilty, not guilty or "not proven."

BECAUSE the use of symbols sometimes makes it possible to determine by purely routine operations whether or not a particular statement follows from given assumptions, symbolic logicians have experimented in designing logical machines. Kalin and Burkhart have, for example, built one that can check Aristotelian syllogisms or solve certain insurance problems, and workers at the University of Manchester in England are developing a more elaborate machine.

Not even symbolic logic will ever produce a machine that can do all man's thinking for him. But some logicians believe that symbolic logic may lead to the construction of synthetic languages that will help to free scientific thinking from the murky tyranny of words.

Computer Logic and Memory

by David C. Evans
September 1966

A large modern computer can contain nearly half a million switching elements and 10 million high-speed memory elements. They operate with the simplest of all logics: the binary logic based on 0 and 1

Electronic digital computers are made of two basic kinds of components: logic elements (often called switching elements) and memory elements. In virtually all modern computers these elements are binary, that is, the logic elements have two alternative pathways and the memory elements have two states. Accordingly all the information handled by such computers is coded in binary form. In short, the information is represented by binary symbols, stored in sets of binary memory elements and processed by binary switching elements.

To make a digital computer it is necessary to have memory elements and a set of logic elements that is functionally complete. A set of logic elements is functionally complete if a logic circuit capable of performing any arbitrary logical function can be synthesized from elements of the set. Let us examine one such functionally complete set that contains three distinct types of circuit designated *and, or* and *not*. Such circuits can be depicted with input signals at the left and output signals at the right [*see middle illustration on next page*]. Since the logic elements are binary, each input and output is a binary variable that can have the value 0 or 1. In an electrical circuit the logical value 0 corresponds to a particular voltage or current and the logical value 1 to another voltage or current. For each symbolic circuit one can construct a "truth table," in which are listed all possible input states and the corresponding output states. Each truth table, in turn, can be represented by a Boolean statement (named for the 19th-century logician George Boole) that expresses the output of the circuit as a function of the input. Truth tables and Boolean statements are shown in the illustrations on the next page. In the case of the *and* circuit the output variable C has the value 1 if, and only if, the input variables A and B both have the value 1. In the Boolean statement the operation *and* is designated by the dot; it reads "C is equal to A *and* B." In the *or* circuit C has the value 1 if at least one of the input variables has the value 1. The Boolean statement is read "C is equal to A *or* B." The *not* circuit has for its output the logical complement of the input. Its Boolean statement is read "B is equal to *not* A." The *and* and *or* circuits described have only two input variables. Circuits that have a larger number of input variables are normally used.

There are a number of other functionally complete sets of logic elements. Two sets are particularly interesting because each contains only one element, in one case called *nand* (meaning "not and") and in the other case called *nor* (meaning "not or"). The bottom illustration at the left on the next page shows a symbolic representation of a two-input *nand* circuit with its truth table. Although a practical *nand* circuit is designed as an entity, it is evident that it can be realized by an *and* and a *not* circuit. The reader can easily devise *and, or* and *not* circuits from *nand* circuits to demonstrate to himself that the *nand* circuit is also functionally complete.

With *and* and *not* circuits it is not difficult to construct a decoding circuit that will translate binary digits into decimal digits. The top illustration on page 352 shows such a circuit and its truth table. The decimal digits are each represented by a four-digit binary code (A_0, A_1, A_2, A_3). In the decoding circuit, which yields the first four decimal digits, the input signals A_0, A_1, A_2, A_3 are applied. The signal at each of the numbered outputs is 0 unless the input code is the code for one of the numbered outputs, in which case the signal at that output is 1.

The circuits that store information in a computer can be divided into two classes: registers and memory circuits. Registers are combined with logic circuits to build up the arithmetic, control and other information-processing parts of the computer. The information stored in registers represents the instantaneous state of the processing part of the computing system. The term "memory" is commonly reserved for those parts of a computer that make possible the general storage of information, such as the instructions of a program, the information fed into the program and the results of computations. Memory devices for such storage purposes will be discussed later in this article.

THIN-FILM MEMORY (*opposite page*) consists of an array of rectangular storage elements, only four millionths of an inch thick, deposited on a thin glass sheet. The rectangles are oriented in one of two magnetic states, corresponding to 0 or 1, when electric currents are passed through conductors (*vertical stripes*) printed on the back of the glass. The films can be switched in a few billionths of a second. The states can be made visible if the thin-film surface is illuminated with plane-polarized light and photographed through a suitably adjusted polarizing filter. The magnetic film causes a slight rotation in the plane of polarization of the reflected light. Here the predominantly dark rectangles are in the 1 state; the light rectangles are in the 0 state. The photograph is a 100-diameter enlargement of a thin-film memory developed by the Burroughs Corporation for use in its newest computers.

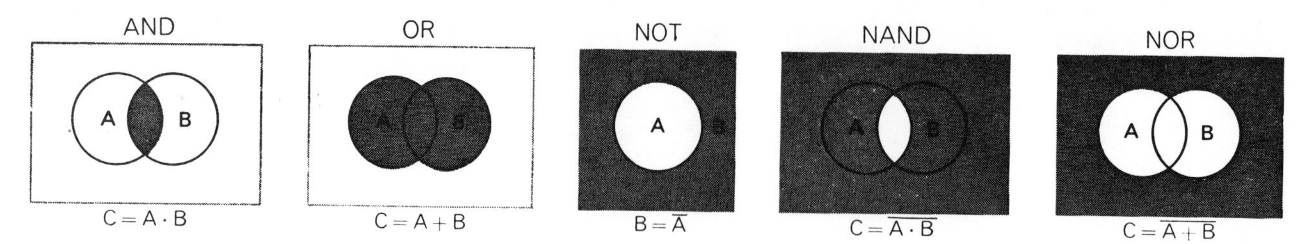

VENN DIAGRAMS use circles to symbolize various logic concepts and relations. Circles represent statements that can be either true or false; they are placed in a universe, or field, that represents all other statements. The logical relation *and* is represented by the shaded area where two circles overlap. This area, *C*, is "true" only if both circles, *A* and *B*, are true; it is "false" if either *A* or *B* or both are false. The logical relation *or* (the "inclusive or") is represented by shading the entire area within both circles. This area, *C*, is true when either *A* or *B* or both are true. *Not* is represented by a circle, *A*, surrounded by a universe, *B*, which is not *A*. The equations below the Venn diagrams are Boolean statements. The dot in the *and* statement stands for "and." The plus sign in the *or* statement stands for "or." The $\overline{A}$ in the *not* statement signifies "not *A*." *Nand* and *nor* stand respectively for "not and" and "not or," as is made clear in the shading of their Venn diagrams. Such diagrams are named for John Venn, a 19th-century English logician.

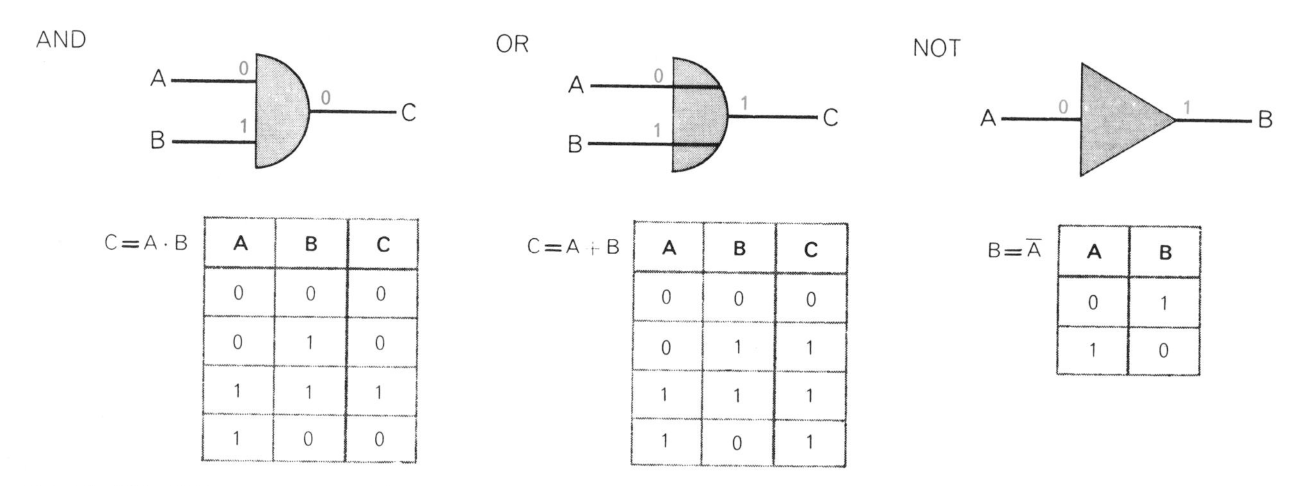

$C = A \cdot B$

A	B	C
0	0	0
0	1	0
1	1	1
1	0	0

$C = A + B$

A	B	C
0	0	0
0	1	1
1	1	1
1	0	1

$B = \overline{A}$

A	B
0	1
1	0

AND, OR AND *NOT* constitute a set of binary logic elements that is functionally complete. The three symbols represent circuits that can carry out each of these logic functions. Input signals, either 0 or 1, enter the circuits at the left; outputs leave at the right (*colored digits are examples*). Below each circuit is a "truth table" that lists all possible input states and corresponding output states.

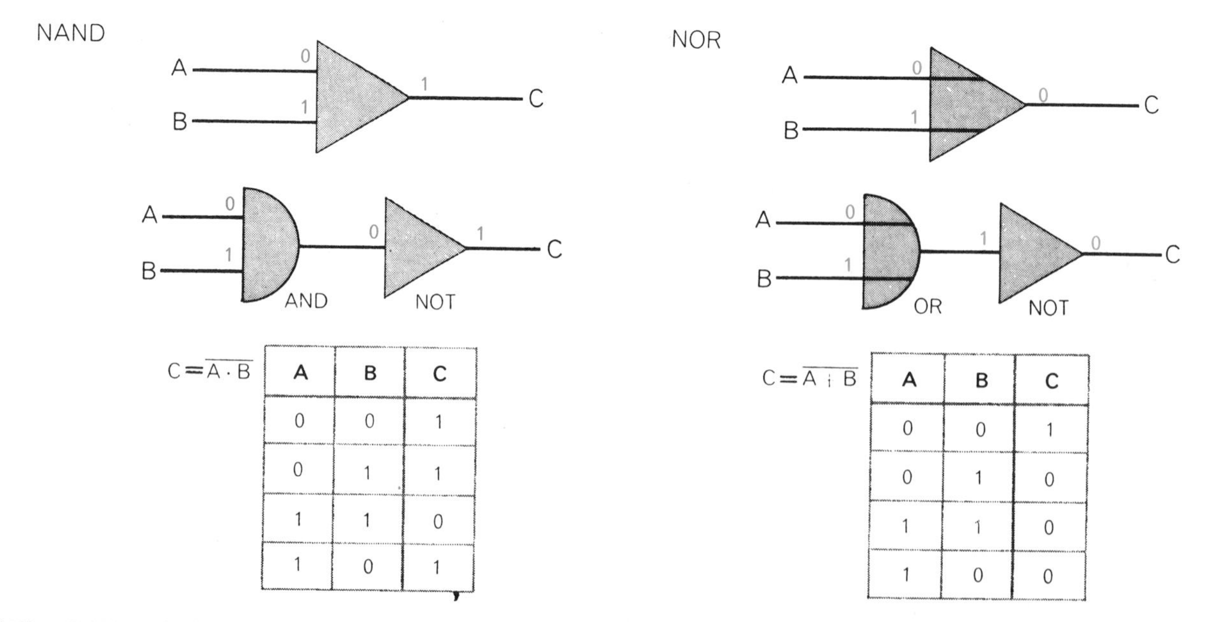

$C = \overline{A \cdot B}$

A	B	C
0	0	1
0	1	1
1	1	0
1	0	1

$C = \overline{A + B}$

A	B	C
0	0	1
0	1	0
1	1	0
1	0	0

NAND CIRCUIT, which contains only one logic element, is functionally complete; it can do everything that *and*, *or* and *not* circuits can perform collectively. The two-input *nand* circuit symbolized at top is equivalent to the combined *and* and *not* circuit. Outputs of the *nand* truth table are opposite to those of the *and* table.

NOR CIRCUIT is also functionally complete. The two-input *nor* circuit symbolized at top is equivalent to the combined *or* and *not* circuit shown immediately below. The *nor* truth table is the converse of the *or* table. Electronic embodiment of a circuit that can serve as either *nand* or *nor* appears on the opposite page.

Registers are usually made up of one-bit storage circuits called flip-flops. A typical flip-flop circuit, called a set-reset flip-flop, has four terminals [*see bottom illustration on next page*]. It is convenient to refer to such a flip-flop by giving it the name of the variable it happens to store; thus a flip-flop for storing the variable A will be named A. If the inputs to the terminals S and R are 0, the flip-flop will be in one of two states. If A has the value 1, it is in the set state; if it has the value 0, it is in the reset state. It can be switched to the set state by applying a 1 signal to the S terminal and switched to the reset state by applying a 1 to the R terminal. The application of 1's to the S and R terminals at the same time will not yield a predictable result. The flip-flop can therefore be regarded as remembering the most recent input state.

Memories for general storage could be made up of logic circuits and flip-flops, but for practical reasons this is not done. A memory so constructed would be large and expensive and would require much power; moreover, the stored information would be lost if the power were turned off.

We are now ready to consider how logic circuits and registers can be combined to perform elementary arithmetical operations. The upper illustration on page 353 includes a truth table describing one-digit binary addition. The inputs to the adder are the binary digits X and Y, together with the "input carry" C_{i-1}. The outputs are the sum digit S and the "carry out" C_i. Also illustrated is an implementation of the binary adder using *and, or* and *not* logic elements. A logic circuit such as this binary adder, which contains only switching elements and no storage circuits, is called a combinatorial circuit.

In a computer employing binary arithmetic the arithmetic unit may have to process numbers consisting of 60 or more digits in order to produce results with the desired precision. (A computer able to handle 60-digit numbers is said to have 60 bits of precision.) Numbers of such length can be added in two general ways. One way is to use an adder for each digit; the other is to use a single "serial" adder and process the digits sequentially. When an adder is used for each digit, the assembly is called a parallel adder. The lower illustration on page 353 shows a four-digit parallel adder. The inputs for this adder are two four-digit binary numbers: $X_3\ X_2\ X_1\ X_0$ and $Y_3\ Y_2\ Y_1\ Y_0$. The adder produces the five binary-digit sum $S_4\ S_3\ S_2\ S_1\ S_0$. This four-digit adder is also a combinatorial circuit. The X and Y inputs to the parallel adder can be provided by two four-bit registers of four flip-flops each. The inputs are all provided at the same time. The sum can be stored in a five-bit register that has previously had all its stages reset to 0.

For the serial adder we need a means of delivering the digits of the inputs to the adder in sequence and of storing the sum digits in sequence. To implement these requirements special registers that have the ability to shift information from one stage to the next are employed; such a register is called a shift register. Each of the three shift registers of a serial binary adder has an input from the terminal called SHIFT [*see bottom figure on page 184*]. Normal-

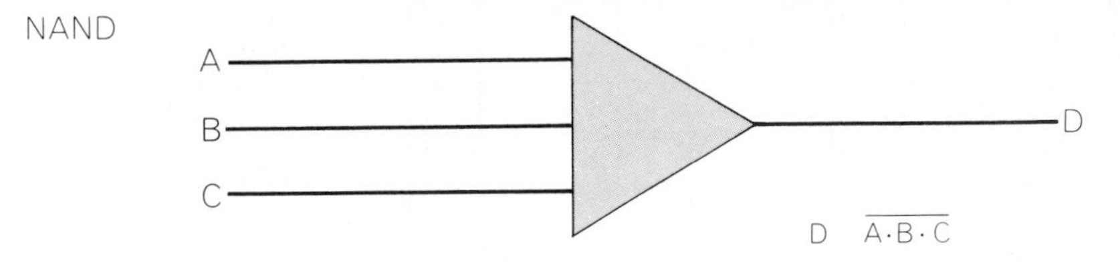

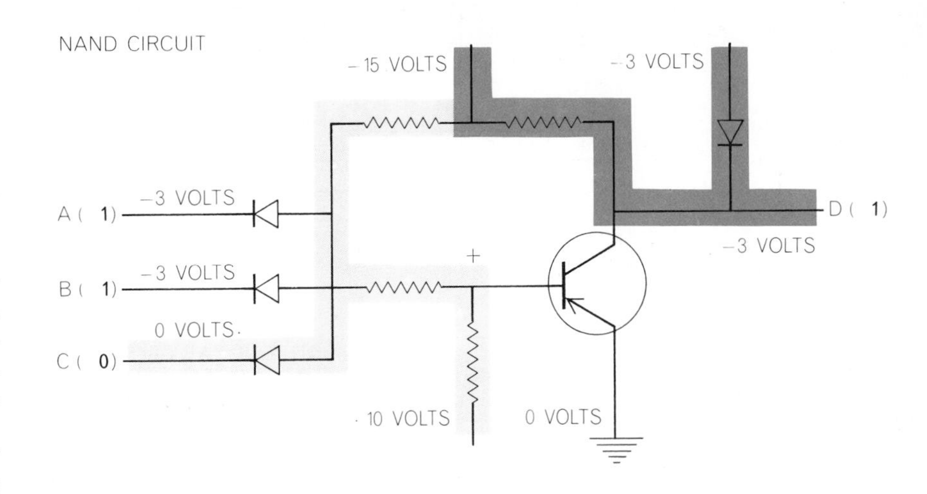

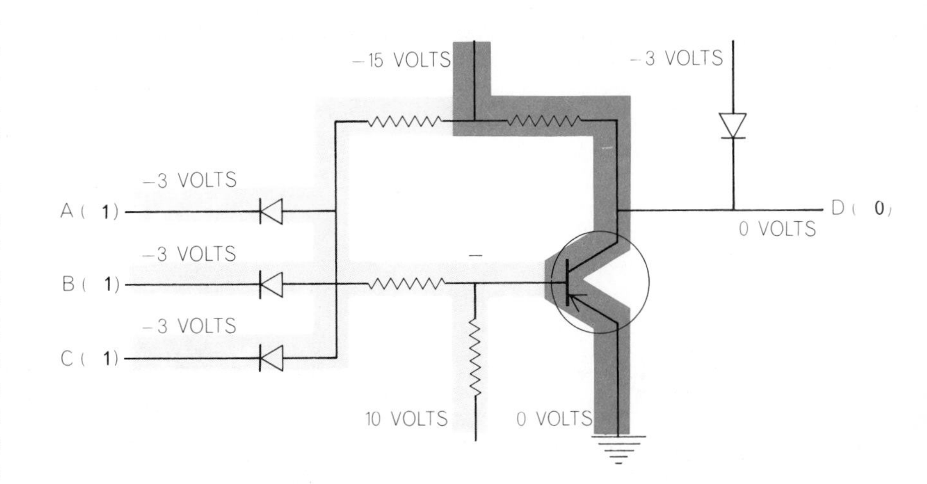

ELECTRONIC EMBODIMENT OF *NAND* CIRCUIT contains four diodes (*triangular shapes*), four resistors (*zigzags*) and one transistor (*inside circle*). The symbol for this three-input *nand* circuit and its Boolean statement appear at the top. In the circuit the dark color represents the flow of large current that is switched to produce the output, 0 or 1, depending on the flow of small current (*light color*), which is controlled by the input voltages. Current flow is shown for two different inputs: 1, 1, 0 and 1, 1, 1. By reversing the choice of voltage the *nand* circuit shown here acts as a *nor* circuit. Such circuits can be designed in many ways.

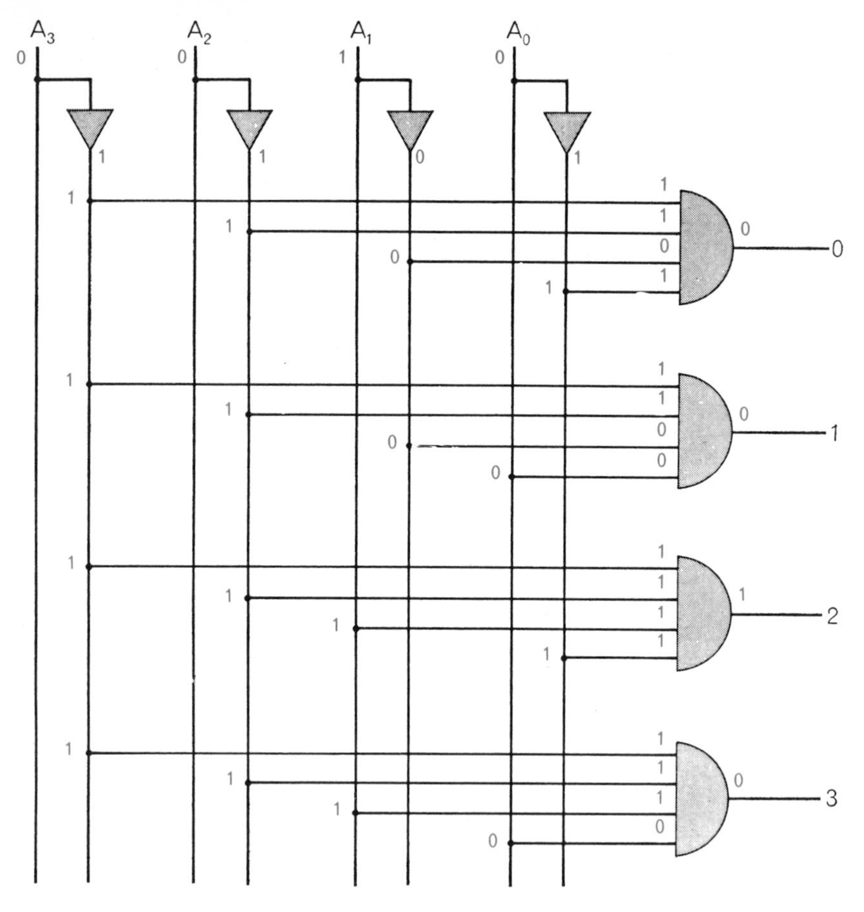

DECIMAL BINARY

	A_3	A_2	A_1	A_0
0	0	0	0	0
1	0	0	0	1
2	0	0	1	0
3	0	0	1	1
4	0	1	0	0
5	0	1	0	1
6	0	1	1	0
7	0	1	1	1
8	1	0	0	0
9	1	0	0	1

CONVERSION OF BINARY to decimal digits is accomplished by this circuit, made up of four *not* circuits and four *and* circuits. The truth table at left shows the binary equivalent for the decimal digits from 0 to 9. To show the principle involved in decoding binary digits, the circuit carries the decoding only as far as decimal digit 3. The signal at each of the numbered outputs is 0 unless all the inputs are 1. In the example this is true for the third *and* circuit from the top, labeled 2. Thus the binary digits 0010 are decoded to yield the decimal digit 2.

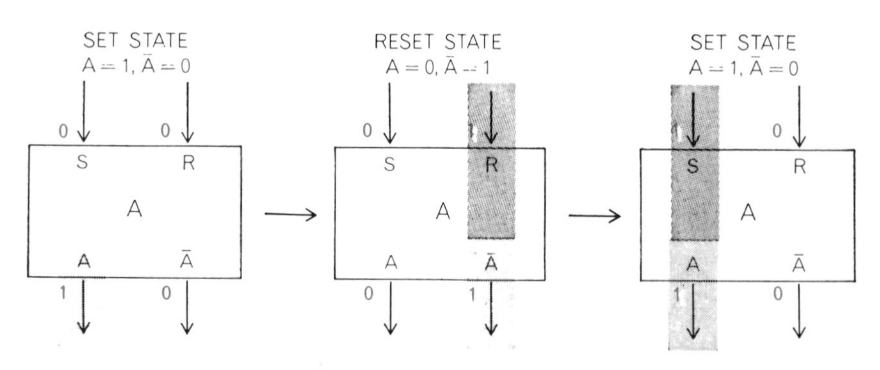

TYPICAL ONE-BIT STORAGE ELEMENT is represented by a "set-reset flip-flop." The one shown here is given the name *A* because it stores the variable *A*. A flip-flop "remembers" the most recent input state. If *A* has the value 1, it is in the set state; if *A* has the value 0, it is in the reset state. Applying a 1 to the *S* terminal yields the set state; applying a 1 to the *R* terminal yields the reset state. Flip-flops provide the transient memory in a computer.

ly the SHIFT signal has the value 0, but when it is desired to shift the three registers, the SHIFT signal is given the value 1 for a brief period, causing the registers to shift their contents one bit to the right. As in the case of the parallel adder, the serial adder can add one group of binary digits (such as X_3 X_2 X_1 X_0) to another group (such as Y_3 Y_2 Y_1 Y_0). At the first command to shift, the serial adder stores the sum of the first pair of digits (X_0 and Y_0); at the second command to shift, it stores the sum of the second pair of digits (X_1 and Y_1), and so on. The carry-out (C_i) of each addition is passed along at each command to shift.

Registers are needed for both serial and parallel adders. For the serial adder the registers must be shift registers and only a one-digit binary adder is required. For the parallel adder a binary adder is required for each bit of precision, that is, for each pair of *X* and *Y* inputs. The parallel adder is simply a large combinatorial circuit. The serial adder includes the binary adder, a flip-flop (known in this case as the *C* flip-flop) and associated circuitry. It is not a combinatorial circuit because its output (*S*) is not merely a function of the immediate inputs (*X* and *Y*); it is also a function of the internal state as represented by the value stored in the *C* flip-flop. Circuits in which the output is not only a function of the immediate inputs but also a function of the circuit's history as represented by its internal state are called sequential circuits. Such circuits are fundamental to the design of computers. Multiplication, for example, is usually implemented by a sequential circuit that repetitively uses an adder circuit.

For most of the period during which computers have evolved, the limiting factor in their design and cost has been memory. The speed of computers has been restricted by the time required to store and retrieve information. The cost of computers has been determined by the information-storage capacity of the memory. As a result much effort has been devoted to the development and improvement of memory devices.

A typical memory, which I have previously described as an array of registers of uniform size, is characterized by word length, storage capacity and access time. Each register in a memory is called a word; its size is expressed in bits and typically is in the range of 12 to 72 bits. The total storage capacity of a memory can be expressed in bits

C_{i-1}	X	Y	C_i	S
0	0	0	0	0
0	0	1	0	1
0	1	0	0	1
0	1	1	1	0
1	0	0	0	1
1	0	1	1	0
1	1	0	1	0
1	1	1	1	1

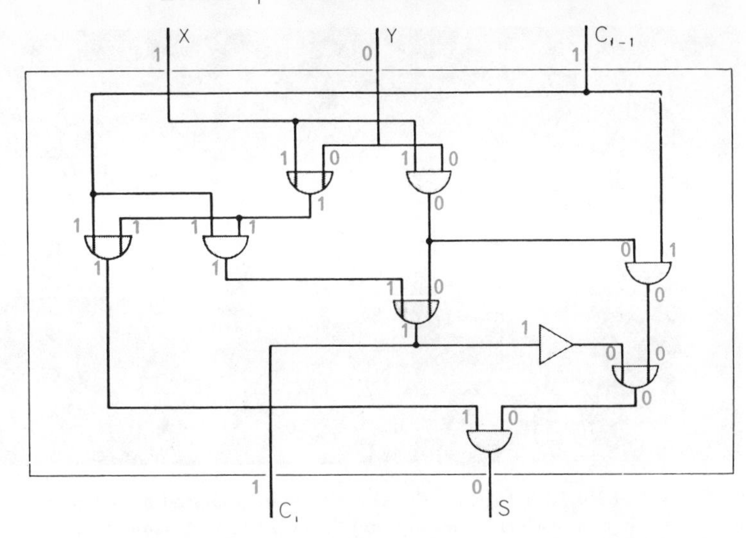

BINARY ADDER CIRCUIT *(right)* **can add two one-digit binary numbers. It is made up of** ***and, or*** **and** ***not*** **logic elements. Because the adder will usually be one of several linked in parallel** *(see illustration below)* **it must also be able to accept a digit known as the input carry (C_{i-1}) produced by an adder immediately to its right. The truth table** *(left)* **shows the "carry-out" (C_i) and the sum digit (S) for all combinations of three inputs. In the example the inputs are 1, 0 and 1. This is known as a combinatorial circuit.**

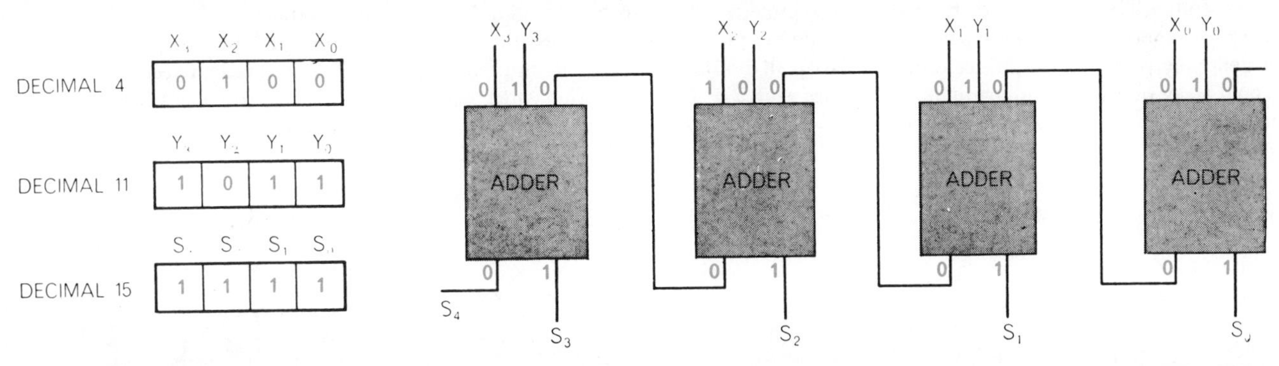

FOUR-DIGIT PARALLEL ADDER consists of four one-digit binary adders like the one shown at the top of the page. In a computer, registers *(not shown)* **would be needed to supply the input signals and to store the output signals. In the example illustrated here the binary number 0100 (decimal 4) is being added to 1011 (decimal 11). The sum is the binary number 1111 (decimal 15).**

but is more often expressed in words; depending on various factors, which will be examined below, the storage capacity can vary from 100 words to billions of words. The time required to store (write) or retrieve (read) a specified word of information is called the access time; it can range from a fraction of a microsecond to several seconds or minutes.

Access to a particular word in a memory is achieved by means of an addressing scheme. There are two classes of addressing schemes: "structure-addressing" and "content-addressing." In the first, which is the more common, each word is given a number by which it is identified; this number is called its address. Access to a particular word of a memory is achieved by specifying the address as a binary-coded number. In content-addressing, access is determined by the content of the word being sought. For example, each word of a content-addressed memory might contain a person's name and certain information about him (such as his bank balance or his airline reservation); access to that information would be achieved by presenting the person's name to the memory. The internal logic of the memory would locate the word containing the specified name and deliver the name and the associated information as an output. Since most memories are structurally addressed, no further consideration will be given to content-addressing.

Among the various memory designs there is a wide range of compromises among cost, capacity and access time [*see top illustration on page 186*]. Most memories fall into one of three access categories: random, periodic or sequential. In random-access memories the access time is independent of the sequence in which words are entered or extracted. Memories with short random-access times are the most desirable but also the most costly per bit of storage capacity. Magnetic-core devices are the most widely used random-access memories. An example of a memory device that provides periodic access is the magnetic drum, in which information is recorded on the circumference of a cylinder that rotates at a constant rate. Sequentially located words may be read at a high rate as they pass the sensing position. The maximum access time is one revolution of the drum, and the average access time to randomly selected words is half a drum revolution. The most common sequential memories—used when neither random nor periodic access is required—are provided by reels of magnetic tape. To run a typical 2,400-foot reel of tape containing 50 million bits of information past a reading head can take several minutes.

Since magnetic materials, in one form or another, supply the principal

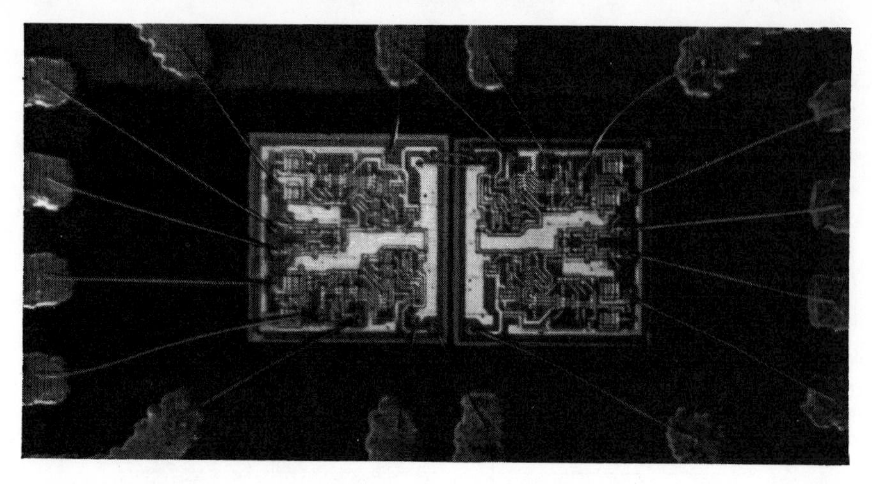

ACTUAL FOUR-DIGIT PARALLEL ADDER can be produced by linking two monolithic integrated circuits; each chip measures only 60 mils (.06 inch) on a side. This adder made by Texas Instruments Incorporated contains the equivalent of 166 discrete components.

storage medium in computers, I shall describe magnetic memories somewhat more fully. The high-speed random-access memory in a typical computer is generally provided by a three-dimensional array of about a million tiny magnetic cores, or rings, each of which can store one bit of information. The cores are threaded on a network of fine wires that provide the means for changing the magnetic polarity of the cores; the polarity determines whether a particular core stores a 1 or a 0. The cores are made of ferrite, a ferromagnetic ceramic. Highly automatic methods have been devised for forming, firing, testing and assembling the cores into memory arrays. In early magnetic-core memories the cores had an outside diameter of about a twelfth of an inch and cost about $1 per bit of storage capacity. The cycle time of these memories (the minimum time from the beginning of one access cycle to the beginning of the next) was in the range of 10 to 20 microseconds.

As the art has developed, the size of the cores has decreased, the cycle time has decreased and the maximum capacity has increased. The cores in most contemporary computers have a diameter of a twentieth of an inch; cycle times are between .75 microsecond and two microseconds. The fastest core memories have cores less than a fiftieth of an inch in diameter and cycle times of less than 500 nanoseconds (half a microsecond).

The essential requirement of a material for a random-access magnetic memory is a particular magnetic characteristic that allows a single element of

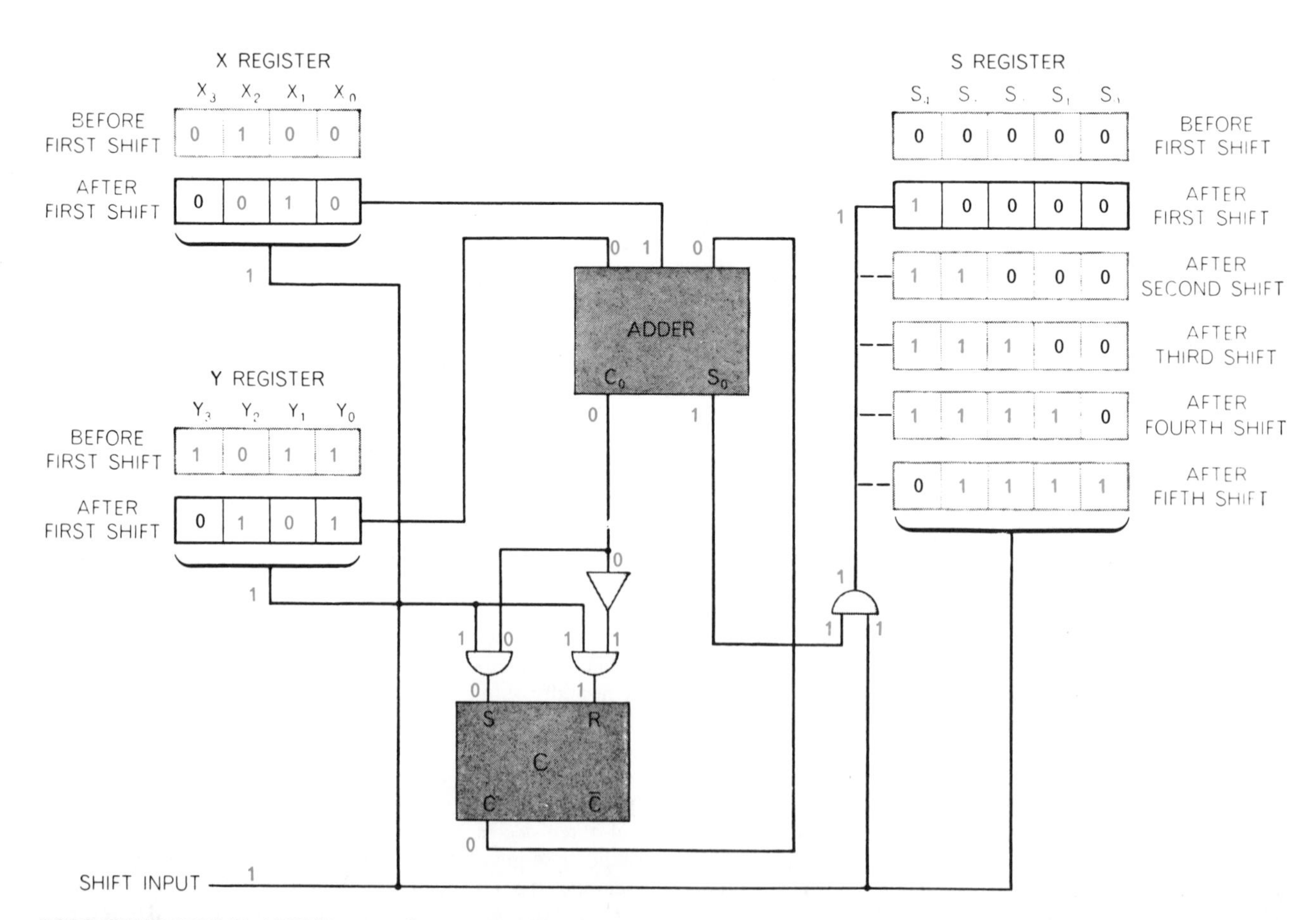

FOUR-DIGIT SERIAL ADDER uses only one adder like the one shown at the top of the preceding page but requires three shift registers and a flip-flop to pass along the carry-out of each addition. Each register has an input from the terminal called SHIFT. At the shift signal each register shifts its contents one bit to the right. Simultaneously the digits shifted out of the *X* and *Y* registers enter the adder, together with the input-carry from the *C* flip-flop. Five shift signals are needed to add two four-digit binary numbers.

a large array of elements of the material to be stably magnetized in either of two directions. Early in the 1950's it was discovered that certain thin metallic films also have this characteristic [*see illustration on page 185*]. The constant dream of computer designers since this discovery was made has been the development of a practical large-capacity memory that can be constructed directly from bulk materials without fabrication, test or assembly of discrete components for individual bits. Many geometries for thin-film memories, including flat films and films deposited on wires or glass rods, have been devised. Some film memories are in service and many more will be used in the future. It is anticipated that there will be dramatic reductions in the cost of random-access memories over the next few years.

In another widely used memory technology a thin film of magnetic material is deposited on some surface such as a plastic tape or card, or a metallic drum or disk. This magnetic surface is moved with respect to a head that can produce or detect patterns of magnetization in the magnetic film; the patterns are of course coded to represent the binary digits 1 and 0. The film for magnetic recording usually consists of finely ground iron oxides bonded together and to the surface by a small amount of organic binder. For magnetic drums and disks the magnetic medium often consists of a metallic film of a nickel-cobalt alloy.

Magnetic tape about a thousandth of an inch thick, half an inch wide and up to 2,400 feet long per reel has provided the main bulk information store for many years. Tape systems have reached a high state of development: they are able to transport the tape past the head at a rate of more than 100 inches per second and to start or stop the tape in a few milliseconds. Six or eight bits are usually written across the width of the tape; it is common for 800 of these six-bit or eight-bit groups to be written per inch along the tape. A current trend in information-processing systems is toward using tape for dead storage or for transporting data from one location to another. Magnetic recording devices with shorter random-access times are taking over the function of active file storage.

Storage devices with a capacity of a few hundred million words and an access time of a few seconds or less are just beginning to be delivered. These devices employ a number of magnetic

EVOLUTION OF CIRCUITS is reflected in these close-ups showing the central processing units in four generations of computers. UNIVAC I (*top*), the first large commercial electronic computer, used vacuum-tube logic circuits. The first model was delivered to the Bureau of the Census in 1951. International Business Machines' Model 704 (*second from top*) was a widely used large-scale vacuum-tube computer with a magnetic-core memory. The first 704 was installed in late 1955. In 1963 IBM delivered the first 7040 (*third from top*), a typical transistorized computer using discrete components. The Spectra 70/45 (*bottom*), recently delivered by the Radio Corporation of America, represents the latest generation. It uses monolithic integrated circuits similar to the one shown at the top of the opposite page.

TYPE OF MEMORY	RANDOM ACCESS TIME (MICROSECONDS)	INFORMATION TRANSFER RATE (BITS PER SECOND)	CAPACITY (BITS)	COST (DOLLARS PER BIT)
INTEGRATED CIRCUIT	$10^{-2} - 10^{-1}$	$10^{9} - 10^{10}$	$10^{3} - 10^{4}$	10
TYPICAL CORE OR FILM	1	10^{8}	10^{6}	10^{-1}
LARGE SLOW CORE	10	10^{7}	10^{7}	10^{-2}
MAGNETIC DRUM	10^{4}	10^{7}	10^{7}	10^{-3}
TAPE LOOP OR CARD	10^{6}	10^{6}	10^{9}	10^{-4}
PHOTOGRAPHIC	10^{7}	10^{6}	10^{12}	10^{-6}

COMPARISON OF MEMORY SYSTEMS shows a range of roughly a billion to one in access time and capacity and about 10 million to one in cost per bit. The spread in the rate of information transfer is smaller: about 10,000 to one from the fastest memories to the slowest. Integrated circuit memories (similar to logic circuits) and photographic memories (for digital storage) are just appearing.

cards or tape loops handled by various ingenious mechanisms [*see illustration on page 189*].

In memory systems that use magnetic drums and disks rotating at high speed, the heads for reading and writing information are spaced a fraction of a thousandth of an inch from the surface. The surface velocity is about 1,000 or 2,000 inches per second. In early drum systems severe mechanical and thermal problems were encountered in maintaining the spacing between the heads and the recording surface. In recent years a spectacular improvement in performance and reliability has been achieved by the use of flying heads, which maintain their spacing from the magnetic surface by "flying" on the boundary layer of air that rotates with the surface of the drum or disk. One modern drum memory has a capacity of 262,000 words and rotates at 7,200 revolutions per minute; it has a random-access time of about four milliseconds and an information-transfer rate of 11.2 million bits per second.

Magnetic information storage is meeting competition from other memory technologies in two areas: where fairly small stores of information must be accessible in the shortest possible time and where ultralarge stores must be accessible in a matter of seconds. For the first task, which today is usually performed by magnetic cores and thin

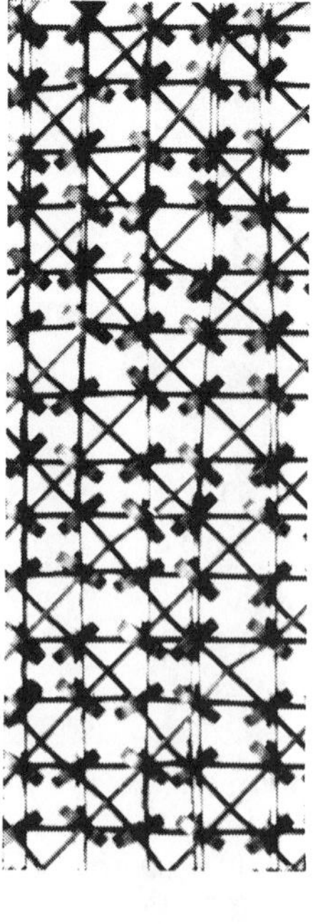

MAGNETIC-CORE MEMORY has been the standard high-speed memory in computers for many years. A typical core memory plane is shown two-thirds actual size at the left; a portion of the plane is enlarged about 10 diameters at the right. This example, made by Fabri-Tek Incorporated, contains 16,384 ferrite cores, each a fiftieth of an inch in diameter.

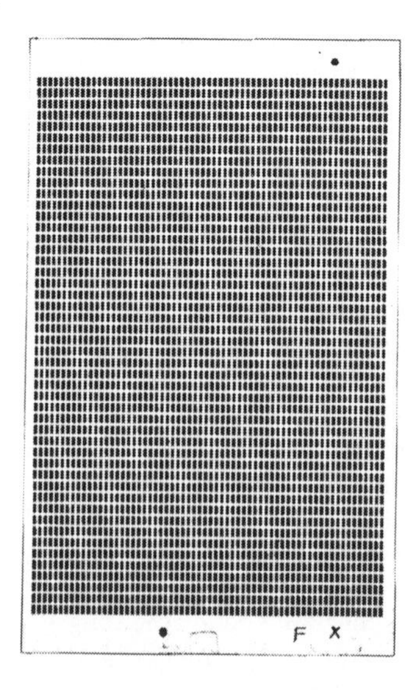

THIN-FILM MEMORY made by Burroughs, which operates even faster than magnetic-core memories, is shown here actual size. An enlargement in color appears on page 178.

films, one can now obtain memories fabricated by the same techniques used to produce monolithic integrated circuits [see "Microelectronics," by William C. Hittinger and Morgan Sparks; SCIENTIFIC AMERICAN, November, 1965]. Such circuits, resembling flip-flops, can be built up from tiny transistors and resistors; scores of such elements can be packed into an area no more than a tenth of an inch square [*see bottom left of illustration on page 189*]. A memory of this kind can store about 100 words and have a random-access time of 100 nanoseconds. Although the present cost of such memories is a few dollars per bit, the cost will probably decline to a few cents per bit by 1970. Integrated-circuit memories have the drawback that power is continuously dissipated by each element (unlike magnetic elements) whether it is actively being read (or altered) or not.

For very-high-volume storage and moderately fast access time, magnetic devices are being challenged by high-resolution photography. In these systems bits are recorded as densely packed dots on transparent cards or short strips of photographic film. During the next year or so several such systems will go into service; each will have a capacity of 10^{11} or 10^{12} bits and a maximum access time of a few seconds.

To combine rapid average access time and large storage capacity at a minimum cost to the user, computer designers have recently introduced the concept of the "virtual memory." Such a memory simulates a single large, fast random-access memory by providing a hierarchy of memories with a control mechanism that moves information up and down in the hierarchy, using a strategy designed to minimize average access time.

The logic and main memory of a very large modern computer contains nearly half a million transistors and a somewhat larger number of resistors and other electrical components, in addition to 10 million magnetic cores. In such a machine—or even in a smaller one with a tenth or a hundredth of this number of components—the matters of packaging, interconnection and reliability present very serious design problems.

The active circuit elements in early electronic computers were vacuum tubes. These computers encountered three major problems. First, the rate at which tubes failed was so high that in large computers the ratio of nonproduc-

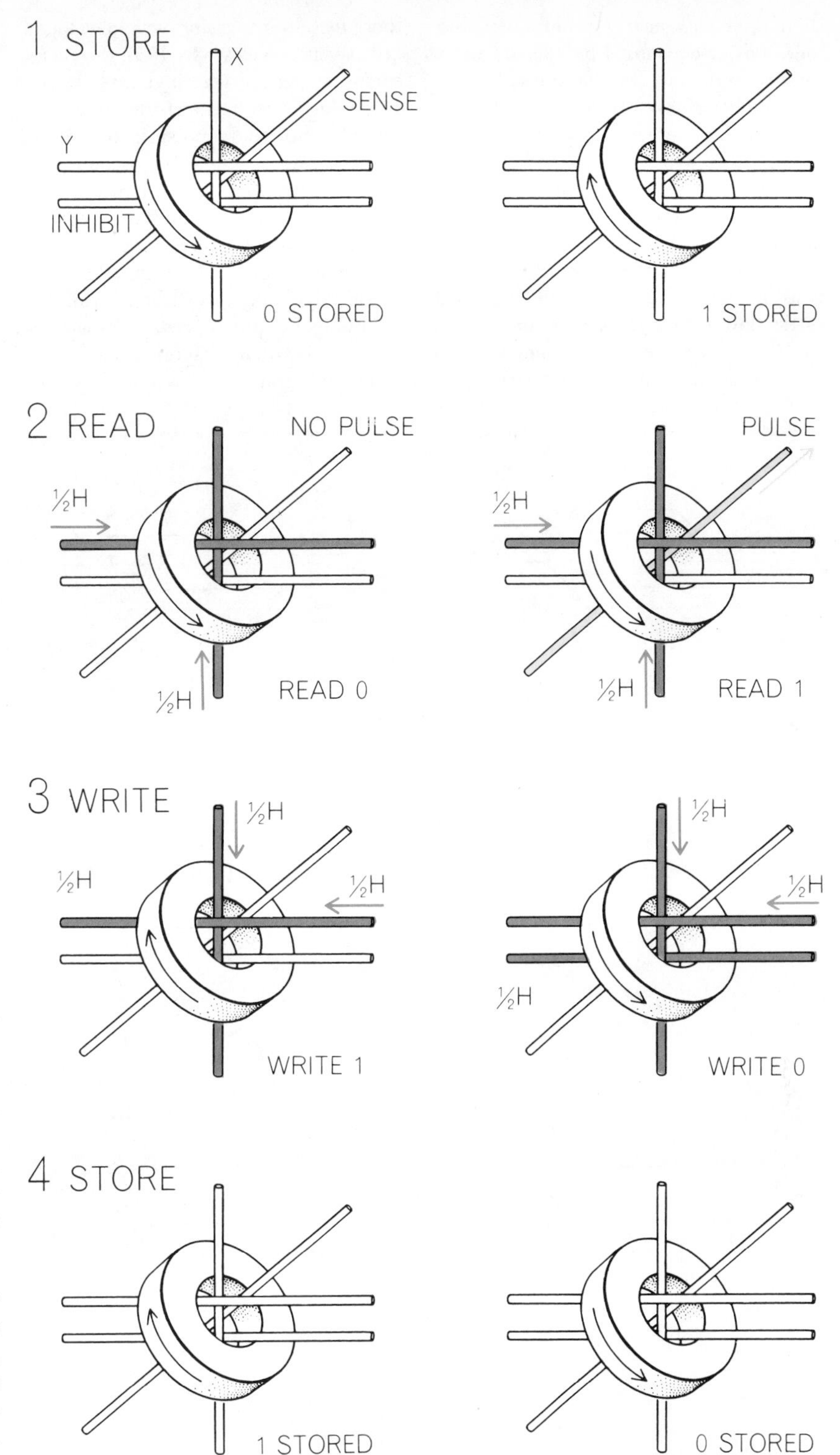

OPERATION OF MAGNETIC-CORE MEMORY involves switching the direction of magnetization, or polarity, of a ferrite core between two positions 180 degrees apart. One position is selected to represent 0, the other to represent 1. "Reading" and "writing" signals are carried on two wires (*X and Y*), each of which carries only half of the current (*½H*) needed to change the core's direction of polarization. During the reading cycle the direction of current flow is selected so that the pulses reverse the polarity of a core that is storing a 1, with the result that a voltage pulse signifying 1 (*light color*) is created in the "sense" wire. No pulse emanates from a core that is storing a 0. During the writing cycle the flow in the *X* and *Y* wires is reversed. This reverses the polarity of the core and writes 1 unless an opposing current is coincidentally passed through an "inhibit" wire, in which case the core polarity remains in the 0 position. A typical memory will contain a million cores.

tive time was nearly prohibitive. Second, power consumed by vacuum tubes was so large that adequate cooling was extremely difficult to achieve. Third, the components were so large that the distances over which signals had to travel would have limited computer speeds to levels that today would be regarded as slow.

In 1948 the point-contact transistor was invented. It was small and used little power, but it was too unstable a device to replace the vacuum tube in large-scale computers. A few years later the junction transistor was developed, but it was too slow. In 1957 the planar silicon transistor was invented. It provided high-speed transistors that were reliable and made possible the design of the present high-speed computers. Further development of the planar technology led to the monolithic integrated circuit, in which scores of components are created and linked together in a single tiny "chip" of silicon. A variation of this technique is used to create the integrated-circuit memories.

The integrated logic circuit, which is just beginning to make its way into large-scale use for computers, contributes substantially to the solution of the three problems that beset the vacuum-tube computer and that were only partially solved by discrete transistors. An integrated circuit on one chip of silicon can have the logic capacity of several of the logic circuits described earlier. It occupies far less space and consumes less power than an equivalent transistor circuit. Its small size makes possible systems with higher speeds because the interconnections of the circuits are shorter. Reliability is increased because the interconnections are themselves reliable. Indeed, the reliability of an entire integrated circuit is expected to approach that of an individual transistor. The latest integrated circuits have a signal delay of only a few nanoseconds, and still faster circuits are being developed. However, the physical size of a computer's components, together with their interconnections, remains a fundamental limitation on the complexity of the computer: an electrical signal can travel along a wire at the rate of only about eight inches per nanosecond (two-thirds the speed of light).

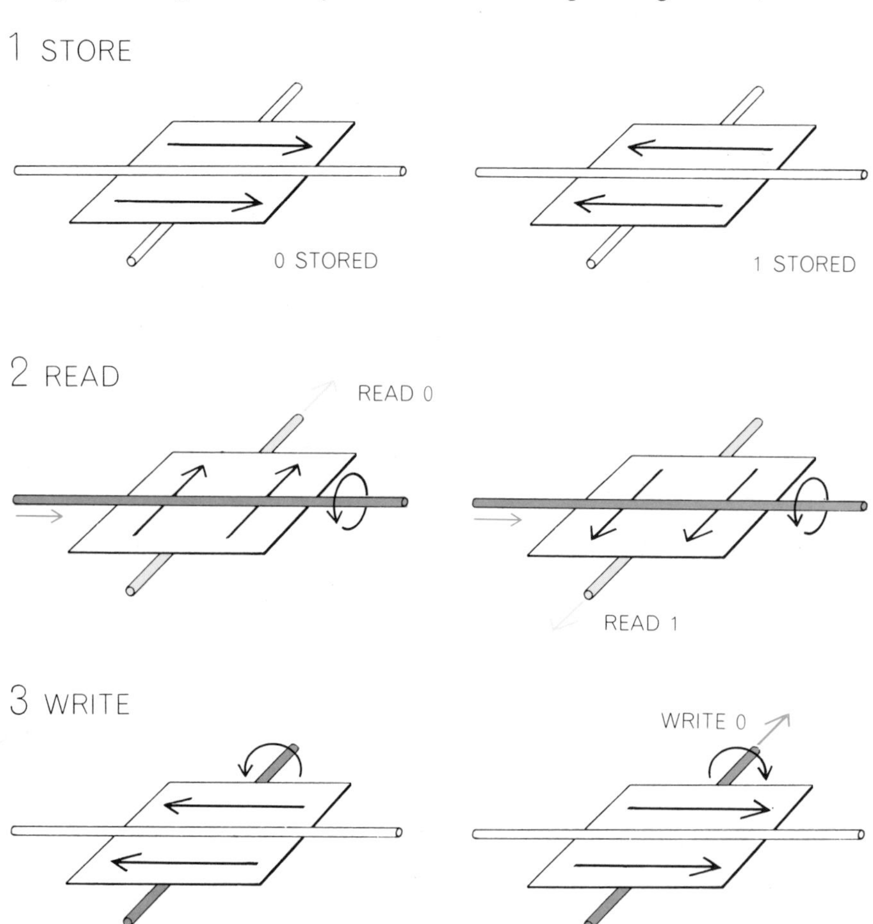

OPERATION OF THIN-FILM MEMORY differs from that of a magnetic-core memory, illustrated on the preceding page. One difference is that the read-out for a 0 or 1 is determined by the polarity of the voltage pulse in the sense wire rather than by the presence or absence of a voltage. Also, in the thin-film memory reading and writing are performed by passing current through different wires. Finally, the change in direction of magnetization that induces a read-out pulse involves a rotation of only 90 degrees rather than 180 degrees.

Computer technology has a way of confounding those who would predict its future. The thin-film memory, for example, has been "just around the corner" for more than 10 years, but the ferrite core is still the main element of random-access memories. Nevertheless, one can try to make certain predictions based on the situation at present. It now seems clear that integrated-circuit technology will soon produce circuits of great complexity at very low cost. These circuits will include high-speed memory circuits as well as logic circuits. Already one can get commercial delivery of a 100-bit register on a single chip of silicon that is a tenth of an inch in its largest dimension. It is my personal opinion that computer designers will be hard-pressed to develop concepts adequate to exploit the rapid advances in components.

Because computers built with integrated components promise to be much cheaper than present machines, one can expect significant changes in the comparative costs of information processing and information transmission. This in turn will influence the rate of growth of data-transmission facilities. Low-cost computers will also change the cost factors that help in deciding whether it is cheaper to do a job with human labor or to turn it over to a machine.

VARIETY OF MEMORY SYSTEMS are based on magnetism, electronic circuitry and photography. Magnetic-drum memory (*top left*), built by Univac Division of Sperry Rand Corporation, provides access in 17 milliseconds to any one of 786,432 36-bit words or some 4.7 million alphanumeric characters. "Random Access Computer Equipment" (*top right*), built by RCA, stores information on 2,048 flexible plastic cards. The basic unit holds 340 million alphanumeric characters; the average access time is 385 milliseconds. Magnetic-disk memory (*middle left*), made by Control Data Corporation, provides access in 34 to 110 milliseconds to any one of 131.9 million six-bit characters. "Data cell" system (*middle right*), offered by IBM, stores data on 2,000 narrow strips of magnetic film. It provides random access in 175 to 600 milliseconds to 800 million bits of information. Integrated-circuit memory (*bottom left*) provides access to 16 bits of information in about .01 microsecond. This example is made by Motorola Semiconductor Products Inc. A new photo-digital memory (*bottom right*) has been devised by IBM to provide rapid access to memory files containing a trillion bits. A single film chip, 1⅜ by 2¾ inches, can store several million bits of information; IBM is not yet ready to disclose the exact number.

33 The Role of The Computer

by Louis N. Ridenour
September 1952

The multifarious control loops of a fully automatic factory must be gathered into one big loop. This can best be done by means of a digital computing machine

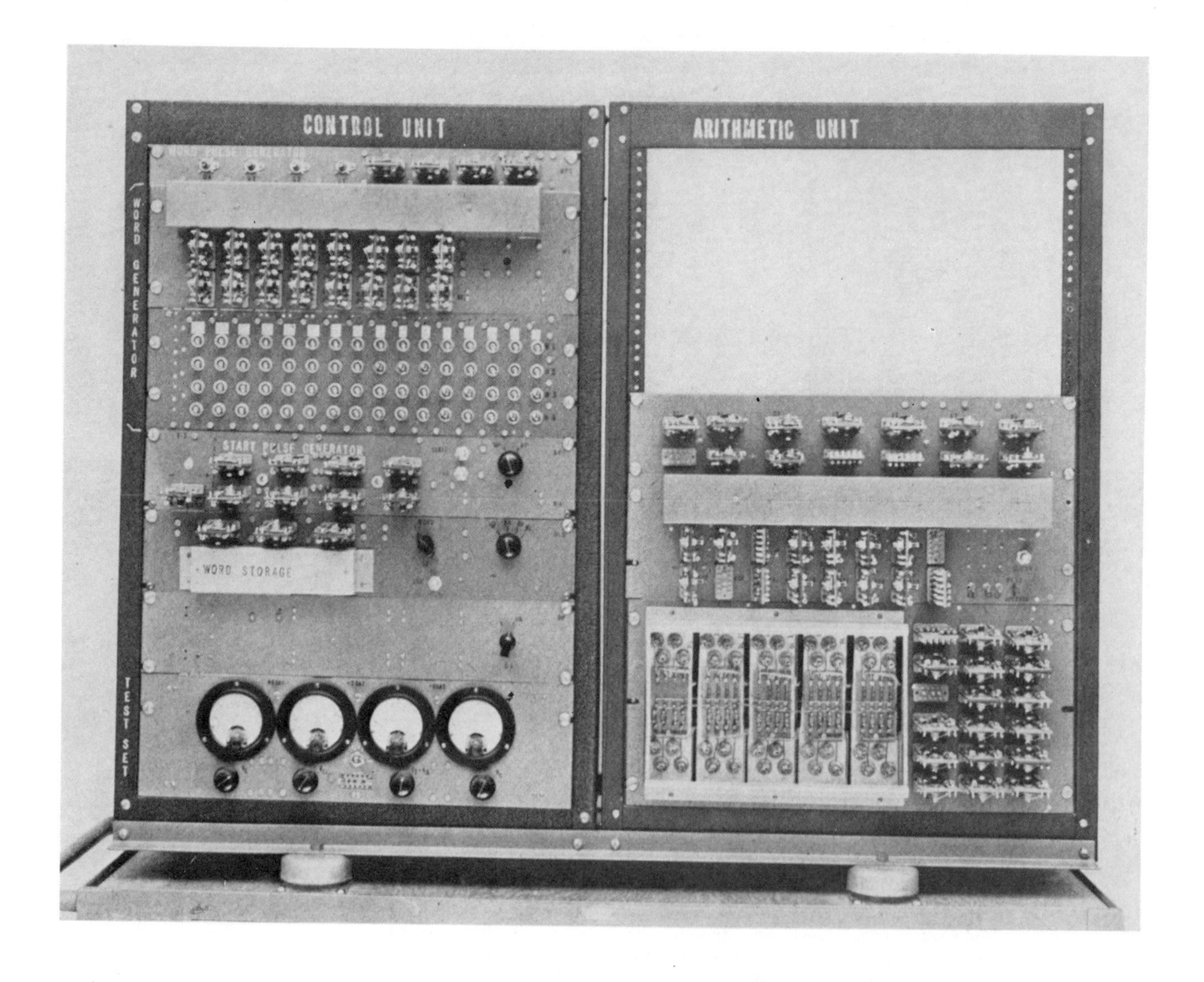

COMPUTER OF THE FUTURE is suggested by this experimental machine built by J. H. Felker of Bell Telephone Laboratories. Instead of vacuum tubes or relays it uses germanium diodes as its logic elements. It also uses the germanium triode, or transistor, as an amplifier. Because these germanium devices are about the size of a pea, a computer utilizing them is much smaller than an equivalent machine employing vacuum tubes or relays. The germanium diodes and triodes also use very little power; the entire computer draws only 5 watts. The machine is capable of multiplying 4,000 16-digit binary numbers a second. One of its interesting features is that each of its 80 transistors is part of an identical plug-in unit (*see illustrations, page 193*).

IF THE thermostat is a prime elementary example of the principle of automatic control, the computer is its most sophisticated expression. The thermostat and other simple control mechanisms, such as the automatic pilot and engine-governor, are specialized devices limited to a single function. An automatic pilot can control an airplane but would be helpless if faced with the problem of driving a car. Obviously for fully automatic control we must have mechanisms that simulate the generalized abilities of a human being, who can operate the damper on a furnace, drive a car or fly a plane, set a rheostat to control a voltage, work the throttle of an engine, and do many other things besides. The modern computer is the first machine to approach such general abilities.

Computer is really an inadequate name for these machines. They are called computers simply because computation is the only significant job that has so far been given to them. The name has somewhat obscured the fact that they are capable of much greater generality. When these machines are applied to automatic control, they will permit a vast extension of the control art—an extension from the use of rather simple specialized control mechanisms, which merely assist a human operator in doing a complicated task, to over-all controllers which will supervise a whole job. They will be able to do so more rapidly, more reliably, more cheaply and with just as much ingenuity as a human operator.

To describe its potentialities the computer needs a new name. Perhaps as good a name as any is "information machine." This term is intended to distinguish its function from that of a power machine, such as a loom. A loom performs the physical work of weaving a fabric; the information machine controls the pattern being woven. Its purpose is not the performance of work but the ordering and supervision of the way in which the work is done.

There are in current use two different kinds of information machine: the analogue computer and the digital computer. Several excellent popular articles have discussed the characteristics of these two types of computer; here we shall briefly recall their leading properties and then consider their respective possibilities as control mechanisms.

THE ANALOGUE machine is just what its name implies: a physical analogy to the type of problem its designer wishes it to solve. It is modeled on the simple, specialized type of controller, such as a steam-engine governor. Information is supplied to the machine in terms of the value of some physical quantity—an electrical voltage or current, the degree of angular rotation of a shaft or the amount of compression of a spring. The machine transforms this physical quantity into another physical quantity in accordance with the rules of its construction. And since these rules have been chosen to simulate the rules governing the problem, the resulting physical quantity is the answer desired. If the analogue machine is being used as a control device, the final physical quantity is applied to exercise the desired control.

Consider, as an example, the flyball-governor pictured on the cover, whose purpose is to hold a steam engine to a constant speed. We notice, first, that information on the engine speed reaches the governor in the form of the speed of rotation of a shaft, while the output of the governor is expressed as the motion of a throttle which is closed or opened as the whirling balls rise or fall. Second, we notice that the relation between these two physical quantities is determined by the actual construction of the governor. The design of the controller has been dictated by its function.

In contrast to the analogue machine, a digital machine works by counting. Data on the problem must be supplied in the form of numbers; the machine processes this information in accordance with the rules of arithmetic or other formal logic, and expresses the final result in numerical form. There are two major consequences of this manner of working. First, input and output equipment must be designed to make an appropriate connection between the log-

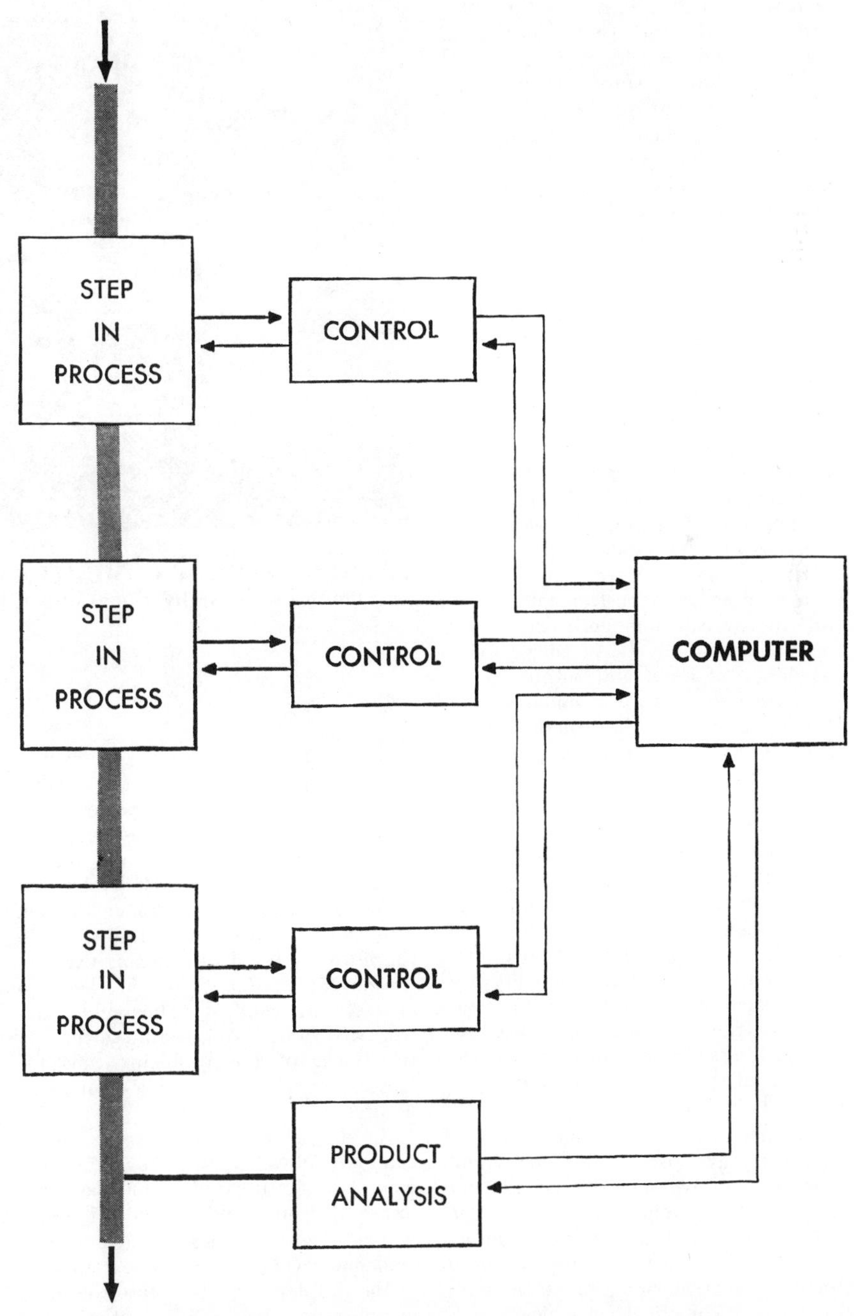

ROLE OF THE COMPUTER is shown in block diagram. Computer receives information from product analysis and feeds it into the various control loops.

ical world of the digital machine and the physical world of the problem being solved or the process being controlled. Second, the problem to be solved must be formulated explicitly for the digital machine. In the case of the analogue machine, the problem is implicit in the construction of the machine itself; construction of a digital machine is determined not by any particular problem or class of problems but by the logical rules which the machine must follow in the solution of *any* problem presented.

Thus far the need for specialized input and output equipment, more than any other factor, has restricted the role of digital information machines to computing. In a computation, both the input and the output quantities are numbers, so the most rudimentary equipment will suffice to introduce the problem and register the result. There is no need (as there would be in a control application) to transform various physical quantities into numerical form before submitting them to the machine, or to transform the results of the calculation into a control action, such as moving a throttle. To use a digital information machine as a computer it is necessary only to provide (1) an input device such as a teletypewriter, which with the help of a human operator can translate printed numbers into signals intelligible to the machine, and (2) an output device such as a page-printer or electric typewriter, which can translate the signals generated by the machine into the printed numbers intelligible to men. Even this simple requirement, however, has not always been well met by the designers of information machines.

When a digital information machine is to be used as an instrument of control—and we can confidently expect that this will eventually be its major role—the design of input and output equipment becames a more formidable task. While it is true that the structure of the machine itself depends on principles of logic rather than on the nature of its application, this is by no means true of the input and output elements. The input devices, or receptors, can use standard elements for receiving the program of instructions, but they must also receive data specifying the state of the particular process being controlled, and for this the detailed design will vary widely from one application to another. Similarly the effectors, which exercise the machine's control, must be designed in terms of the nature of the process or device being controlled.

In comparing digital and analogue machines as instruments for automatic control, we observe, first, that for simple control applications the analogue machine is almost always less elaborate than a digital machine would be. Even the most elementary digital machine requires an arithmetical (or logical) unit, a storage unit, a control unit, receptors and effectors. For simple problems, this array of equipment is wastefully elaborate. In contrast, an analogue machine need be no more complicated than the problem demands. A slide rule, for example, is a perfectly respectable information machine of the analogue type. The analogue machine's ability to do simple work by simple means explains its current predominance in the field of automatic control. The whole control art is so new and so little developed that most of the problems thus far tackled have been of a rather elementary nature.

As the control task becomes more complex, however, the analogue machine loses its advantage, and we begin to see a second fundamental difference between the two types of machine. The analogue machine is a physical analogy to the problem, and therefore the more complicated the problem, the more complicated the machine must be. If it is mechanical, longer and ever-longer trains of gears, ball-and-disk integrators and other devices must be connected together; if it is electrical, more and more amplifiers must be cascaded. In the mechanical case, the inevitable looseness in the gears and linkages, though tolerable in simple setups, will eventually add up to the point where the total "play" in the machine is bigger than the significant output quantities, and the device becomes useless. In the electrical case, the random electrical disturbances called "noise," which always occur in electrical circuits, will similarly build up until they overwhelm the desired signals. Since "noise" is far less obtrusive than "play," electrical analogue machines can be more complicated than their mechanical equivalents, but there is a limit. The great machine called Typhoon, built by the Radio Corporation of America for the simulation of flight performance in guided missiles, closely approaches

LARGE ANALOGUE COMPUTER **is exemplified by the machine of Project Typhoon, built by the Radio Corporation of America to simulate the**

performance of guided missiles, aircraft, ships, submarines and so on.

that limit. It is perhaps the most complicated analogue device ever built, and very possibly the most complicated that it will ever be rewarding to build.

The digital machine, on the other hand, is entirely free of the hazards of "play" and "noise." There is no intrinsic limit to the complexity of the problem or process that a digital machine can handle or control. The switching system of our national telephone network, which enables any one of 50 million phones to be connected to any other, is a digital machine of almost unimaginable complexity.

THE THIRD important difference between analogue and digital machines is in their accuracy potential. The precision of the analogue machine is restricted by the accuracy with which physical quantities can be handled and measured. In practice, the best such a machine can achieve is an accuracy of about one part in 10,000; many give results accurate to only one or two parts in 100. For some applications this range of precision is adequate; for others it is not. On the other hand, a digital machine, which deals only with numbers, can be as precise as we wish to make it. To increase accuracy we need only increase the number of significant figures carried by the machine to represent each quantity being handled. Of course in a control operation the machine's over-all precision is limited by possible errors in translating physical quantities into numbers and *vice versa,* but this does not alter the fact that where high precision is required, a digital machine is usually preferable to the analogue type.

There is a fourth respect in which the two machines differ. An analogue machine works in what is called "real time." That is, it continuously offers a solution of the problem it is solving, and this solution is appropriate at every instant to all the input information which has so far entered the machine. If the machine is doing a mathematical problem, for example, it need not formulate explicitly the equations to be solved and then go through the steps of solving them, as a digital machine would have to do. The equations are inherent in the very structure of the machine, and it solves them by doing just what it was built to do. It can thus respond promptly to changing input data, and offer an up-to-date solution at every moment. This property of working in "real time" is very important in most problems of automatic control. An autopilot flying a plane must respond at once to an attitude change resulting from a gust of wind; the most precise information on how to adjust the flight controls will be worthless if it comes 30 seconds too late.

Since a digital machine works by formulating and solving an explicit logical model of the problem, it can work in "real time" only if the time it requires to obtain a solution, given new input data, is short compared with the period in which significant changes can take place in the system being controlled. Present-day digital machines can achieve this speed for many important problems—flight control of aircraft, for example—but they are not yet fast enough to handle all the "real-time" problems that we should like to turn over to them. It has been estimated that the fastest existing digital machines are some 20 times too slow to deal with the problem of simulating the complete flight performance of a high-speed guided missile—the problem that Typhoon was built to handle. As development proceeds, the operating rates of digital machines can be expected to increase rapidly.

WE SEE, then, that both analogue and digital machines can be used for automatic control, and each has advantages in its own sphere. For simple applications in which no great precision is required, an analogue controller will usually be preferable. For complex problems, or problems in which high precision is required, a digital controller will be superior. Where "real-time" computations must be made, analogue machines are almost always used now, though digital machines are beginning to achieve speeds that fit them for this type of application.

All this refers to the present state of the art of automatic control. What can we guess about the developments to come?

The simple specialized analogue controllers already in use will surely be ex-

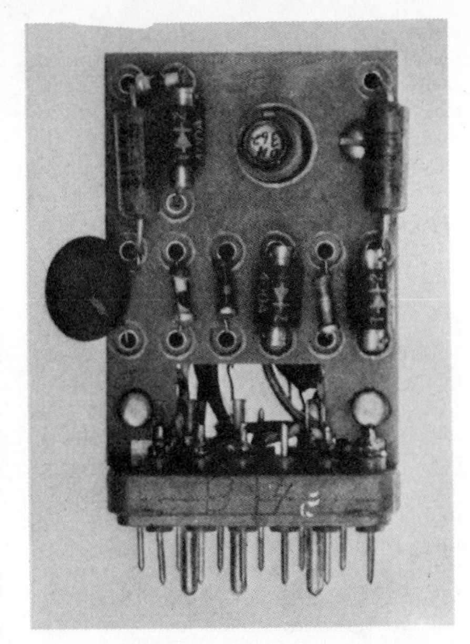

AMPLIFIER for the Bell Laboratories computer on page 190 is a standard unit an inch and a half wide.

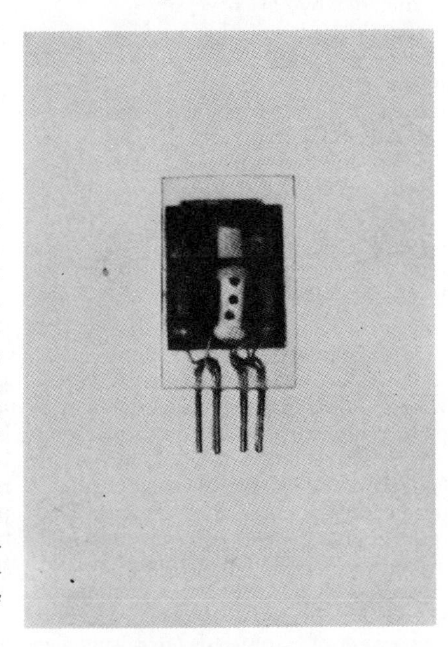

REFINEMENT of amplifier shown above is only ¾ inch wide, suggesting even smaller computers.

tended to wider application. But the most significant and exciting prospects reside in the digital machine. We can expect that it will soon open up a new dimension of control. The meaning of this prediction can be admirably illustrated in terms of the highly instrumented catalytic cracking plant which Eugene Ayres has described in a preceding article.

Mr. Ayres tells us of a plant in which there are some 150 different analogue controllers, each governing some aspect of the continuous process that the plant performs. Several hundred indicators on a central control panel offer the most detailed information on system performance. Many of these indicating instruments also provide continuous recordings. Manual controls which can override any automatic controller are present for use in emergency. The instruments and controls have been arranged on a flow diagram which simulates the organization of the plant and helps the human operator to find his way through the complexities of instrumentation. And the most important process-controls are adjusted manually according to the results of a periodic product analysis.

Clearly the human operator is still the master of this "automatic" plant. However elaborate the instrumentation, the readings of the instruments are still presented to men; however competent the automatic controllers, provision for human veto of their action is built into every one of them. Men are expected to meet emergencies, and to take control under "conditions of unstable equilibrium such as starting up or shutting down." The cracking plant is automatic only when the unexpected is not happening; in times of stress it falls back on human control, and its whole design is dictated by this necessity.

To this scheme there will soon be added end-point control—continuous adjustment of the main process-controls on the basis of a continuous product analysis within the system itself. This modification will improve performance, but it will leave the situation essentially as it was before: more routine responsibility will be given to machinery, but the human supervisor will still be vital to proper operation.

THE DIGITAL information machine, employed as an instrument of supervisory automatic control, can change this picture radically. Since such a machine can be instructed to perform any set of logical operations, however complicated, it can be programmed at the outset to react in emergencies precisely as would a well-instructed human operator—and it can react at least a thousand times faster. Further, the machine can be given a set of criteria for appraising the relative success of its various acts, and can be enabled to alter its own program of instructions in the light of experience on the job. Hence it will be capable of "learning" and of finding a better way to perform its operations than the one prescribed in the original instructions. And this universally adaptable machine can encompass the tremendous job of orchestrating the joint behavior of the hundreds of individual analogue controllers built into a modern cracking plant. The same machine can regulate the performance of the factory and keep the necessary accounting records.

The replacement of human operators in a refinery by a control machine would probably result in substantial economies, both in first cost and in operating cost per unit of product. Most of the saving in first cost would come from the elimination of the costly display and recording instruments that human operators require. In a machine-controlled plant, display would be unnecessary. The measurements vital to the process would be communicated directly to the control machine and processed there. The machine would issue the necessary commands to the specialized controllers which served it, and would print out in fully digested form the summary records of plant performance.

The saving in operating cost would come, not from eliminating the salaries of the few displaced operators, but simply from the fact that the machine could do a more efficient job. A human operator, even one of the greatest virtuosity, is a bottleneck in modern plant performance. Mr. Ayres has told us how the modern cracking plant simply cannot be operated, even by throngs of men, if its individual automatic controllers are left out. The cracking plant of tomorrow, controlled by a suitable information machine, will similarly be beyond the powers of human operators, even skillful ones equipped with all the control instrumentation—of the present variety—that can be devised.

The difficulty of designing a control room which will not baffle the operators is already substantial in present plants. This means that designers cannot increase the complexity of the plant, or its speed of operation, even though such changes might enhance efficiency. Removal of the limitations of human supervisors will open the way to vast design improvements. The information machine can remove them.

SOME CHEMISTS think that a big new development in industrial chemistry lies just ahead, a development based on exploiting certain new types of reactions. These are fast reactions which take place within microseconds, reactions of gases flowing at velocities above the speed of sound, and reactions that will make it possible to capture valuable but fleeting intermediate products in a chemical system by preventing the system from reaching equilibrium. The enthusiasts say that the jet engine is the model of the chemical plant of the future. A supersonic chemical plant of the kind envisioned cannot be operated by men in white overalls reading carefully arranged gauges in an elaborate control room; the speed of nerve impulses from eye to brain to muscle is just too slow for that. Reactions occurring in microseconds must be controlled by machines that can respond in microseconds. Men will design these machines, build them and give them instructions, but men will never be able to compete with their performance.

If this last assertion seems outrageous, it is not more outrageous than it once was to assert that a man could design and build a derrick which would lift a load no man could ever budge. We are familiar with power machinery, and we take for granted its su-

ANALOGUE DEVICE integrates variables with two disks. One variable

DIGITAL DEVICE such as the relay does not measure but counts. Shown

is given by position of small disk; the other, by angle of large disk.

periority to human muscles. We are not yet familiar with information machinery, and we are therefore not prepared to concede its superiority to the human nervous system. Nevertheless, a digital information machine can surpass human capabilities in any task that is governed by logical rules, no matter how complicated such rules may be.

Man's machines are beginning to operate at levels of speed, temperature, atomic radiation and complexity that make automatic control imperative. As an instrument of over-all automatic control the digital information machine has a great but as yet untouched potential. In the next few years this potential will begin to be realized, and the results are certain to be dramatic.

here is part of a panel of relays in a Bell Laboratories digital computer.

VI

APPLICATIONS

VI APPLICATIONS

INTRODUCTION

The major value of mathematics is that it can be used to obtain knowledge about our physical, economic, and political worlds. Much of this knowledge can be applied to the solution of practical problems. The articles in this section do not cover the full variety and depth of these applications; rather they are samples chosen from several areas. Before considering them, let us note that many of the preceding articles described several other types of applications relevant to their mathematical content. The articles on probability could be reread in this connection, as could those on computers.

The articles in this section begin with an application to the physical world. The most glorious achievement of applied mathematics is the science of motion—motion on the earth and in the heavens. Almost as impressive is the mathematical theory of sound and its uses. The first two articles treat the mathematics of sound.

What sounds should one bother to study? The sound of a tin can bouncing along the pavement can hardly be of interest. However, the sound of the human voice and the sound of musical instruments are certainly important for communication and for pleasure. Both of these are technically called musical sounds. One might think that there can be little in common in the variety of musical sounds, particularly if rock and roll is included. But mathematical analysis shows that they are all described by the same laws.

All musical sounds are periodic. That is, a musical sound is a motion of molecules of air that is repeated many times a second. What mathematical means do we have to study periodic motions? In trigonometry we learn about the function $y = \sin t$. The nature of this function is best understood from its graph. As Figure 1 shows, the function goes through what is called one cycle of values as t varies from 0 to 2π, and then the function repeats its behavior

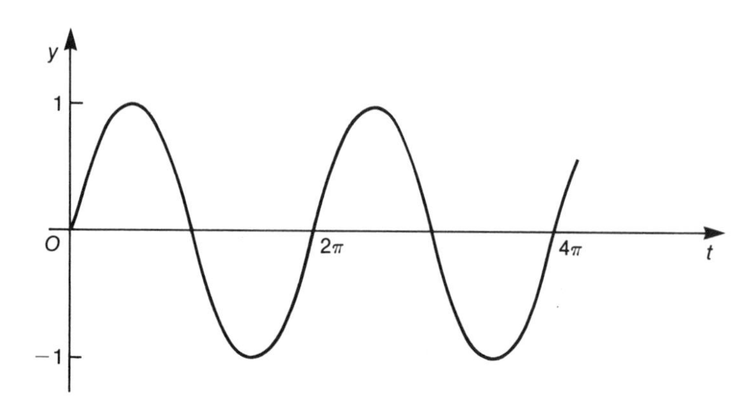

Figure 1.

in each interval of 2π along the t-axis. We say that the period of the function is 2π or, if t represents time in seconds, that the frequency with which the function repeats itself is 1 in 2π seconds.

The function $y = \sin 2t$ has a somewhat different behavior. The graph (Figure 2) shows that the function repeats its behavior in each t-interval of π. In other words, the period is π and the frequency is 2 in 2π seconds or 1 in π seconds. We can also say that the frequency is $1/\pi$ in 1 second. Correspondingly, the function $y = \sin 2\pi t$ has the period of 1 second and the frequency of 1 in 1 second. The function $y = \sin(256 \cdot 2\pi t)$ has the period of 1/256 of a second and the frequency of 256 per second. In other words, in each second, 256 full cycles or repetitions of the sine curve take place. This function represents a *pure* or *simple* sound that repeats itself 256 times a second. Such a sound is given off by a tuning fork that is designed to vibrate at this frequency. The y-values in the function represent the varying displacement of a typical air molecule from its rest or its undisturbed position.

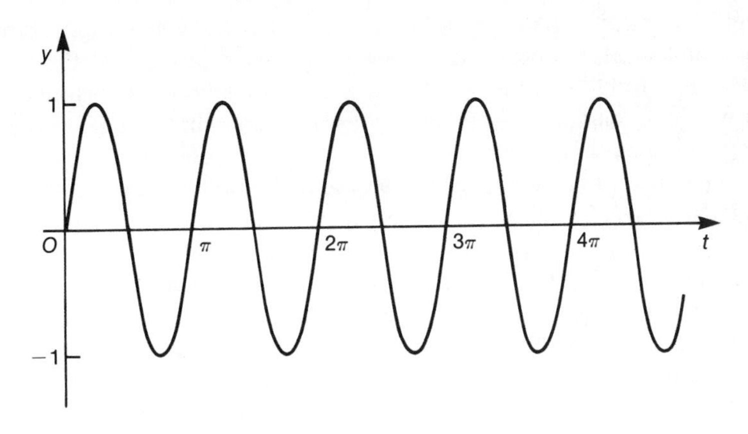

Figure 2.

Most musical sounds are not pure or simple. Though each strikes the ear as a distinctive sound, peculiar to the voice or instrument that emits it, the astonishing fact established through mathematics is that each musical sound is a combination of simple sounds. But not just any combination. All the simple sounds in any one combination have frequencies that are integral multiples, that is, twice, three times, etc., of the lowest frequency present in the sound. Musicians call the component of lowest frequency the first harmonic or fundamental, and the components of higher frequencies the second, third, etc., harmonics. The relevant mathematical theorem, created by Joseph Fourier, states that every musical sound can be represented as the sum of simple trigonometric functions. The form of such a sum would be

$$y = a_1 \sin b_1 x + a_2 \sin b_2 x + a_3 \sin b_3 x + \cdot \cdot \cdot ,$$

where $b_2, b_3, \ldots$, are integral multiples of b_1. The number of terms in the sum would, in a practical situation, depend upon the sound. Thus we can analyze musical sounds and account for their quality in terms of their component simple sounds.

Lineback's article gives pictures of the various harmonics and what the combination of two or more harmonics looks like on an oscilloscope or a cathode-ray tube (a simplified television tube). Saunders's article describes the kinds of sounds given off by musical instruments. (See also the article by David listed in the suggested readings.)

The subject of musical sounds is an excellent example of the values supplied by mathematics. Mathematicians and scientists primarily seek an understanding of natural phenomena, and in the case of sound we have the analysis just described. But understanding can be used to improve human life. For example, the telephone today is almost indispensable. The sound of the human voice may contain as many as twenty harmonics. However, only the first five to ten of these are needed to identify the sound. On the other hand, a high fidelity phonograph, which should reproduce the quality of the original sound as well as possible, may have to reproduce the first hundred harmonics. Far more sensitive electrical equipment and electrical circuits are required for other applications. Engineers who know the mathematical analysis of musical sounds can design the proper equipment for devices such as radio sets and amplifying systems. Though far more is involved in the transmission and reproduction of pictures, as in television, the basic principle is the same as that for musical sounds.

Does mathematics influence our culture? The use of amplifying systems in auditoriums and the use of radio and television for political speeches, advertising, and entertainment are familiar to all of us, and their effects and values need not be spelled out.

The applications of mathematics to physical phenomena have a power and sweep that stagger the imagination. On the other hand, the applications to social, biological, industrial, and economic problems are not at the present time as impressive. Yet they are by no means insignificant. The applications that we shall examine here are quality control, game theory, linear programming, and operations research.

The concept of quality control is easy to understand. Every manufactured article must meet specifications. If the article is a part in a delicate mechanism, these specifications must be met within tolerances that can be as small as one-thousandth of an inch. To test each of the thousands or millions of articles that a machine produces would be enormously expensive. Moreover, some testing procedures destroy the article. The alternative is to test samples. How big a sample should one use, and how often should such samples be tested? What variations in the samples can be accepted as indicative of satisfactory output of the entire lot? The techniques of quality control, which utilize mainly statistical theory, are quite varied, and Dalton's article describes a couple of them.

Game theory was introduced by the mathematician Émile Borel in 1921. Subsequently the mathematician John von Neumann and the economist Oskar Morgenstern collaborated in the early 1940s to develop the theory. The word *game* is used in a special sense. Throwing dice is a game, and probability theory alone can be used to predict the chance of, say, throwing a seven. Several people can compete in dice, but the outcome is independent of the number of competitors or of their thoughts or wishes. Poker, by contrast, is also a game, but beyond the element of chance in drawing cards to make a particular combination, the drawings of competitors and the responses that these competitors make to one's own bets have at least as much to do with success as the cards that one holds. The major theme of game theory is how to make rational decisions in the face of competition from opponents who are also free to make various possible decisions. Thus game theory as applied to economic competition and military maneuvers is a serious business and not a game. The articles by Hurwicz and Morgenstern describe the subject of game theory more fully and illustrate its uses.

Linear programming is today a very practical application of mathematics. This subject, to be distinguished from programming for computers, is a series of techniques used to solve what one might call the simpler algebraic problems of business. This type of problem is faced by a manufacturer who can produce different amounts of several items. The company knows the costs at which it can produce various amounts of each item and how many it can sell of each at a particular price. The company wishes to determine the quantity that it should produce of each item and the price at which to offer them on the market in order to maximize profits.

Linear programming can be used to determine the best locations of factories from the standpoint of accessibility to raw materials and closeness to the customers to whom the products will be shipped. Storage warehouses to serve customers in many different areas can be located so as to keep shipping costs down as much as possible by the methods of linear programming. The mathematics involved is algebra, coordinate geometry, a bit of calculus, matrix theory, and usually some statistics. The article by Cooper and Charnes presents the nature of linear programming.

Operations research is a conglomeration of conglomerations. It uses game theory, linear programming, probability, and indeed any mathematical technique that provides guidance in the operation of any large-scale organization. In applying these techniques to the management of corporations, tactics in warfare, the most effective operation of complex man-machine systems, or the organization of production, all relevant factors are taken into account. Since operations research encompasses game theory, quality control, and linear programming, one can say that it is also employed to solve the problems already described in connection with these individual topics. The virtue of operations research, where it is applicable, is that it replaces guesses by rational decisions and impulsive acts by strategy. Levinson and Brown describe the subject more fully in their article.

SUGGESTED READINGS

Benade, Arthur H. 1960. *Horns, Strings and Harmony.* Doubleday, Garden City, N.Y.

Bergeijk, William A. van. 1960. *Waves and the Ear.* Doubleday, Garden City, N.Y.

Blackwell, David, and M. A. Girshick. 1954. *Theory of Games and Statistical Decisions.* Wiley, New York.

Brams, Steven J. 1976. *Paradoxes in Politics.* Free Press, New York.

Charnes, A., W. W. Cooper, and A. Henderson. 1953. *An Introduction to Linear Programming.* Wiley, New York.

David, Edward E., Jr. 1961. "The Reproduction of Sound." *Scientific American,* August.

Ficken, F. 1961. *The Simplex Method of Linear Programming.* Holt, Rinehart and Winston, New York.

Grant, Eugene L. 1952. *Statistical Quality Control.* McGraw-Hill, New York.

Helm, E. Eugene. 1967. "The Vibrating String of the Pythagoreans." *Scientific American,* December.

Hillier, Frederick S., and Gerald J. Lieberman. 1974. *Operations Research,* second edition. Holden-Day, San Francisco.

Jeans, Sir James. 1968. *Science and Music.* Dover, New York.

Luce, R. D., and H. Raiffa. 1958. *Games and Decisions.* Wiley, New York.

McKinsey, J. C. C. 1952. *Introduction to the Theory of Games.* McGraw-Hill, New York.

Morse, Philip M., and George E. Kimball. 1950. *Methods of Operations Research.* M.I.T. Press, Cambridge, Mass.

Neumann, John von, and Oskar Morgenstern. 1955. *The Theory of Games and Economic Behavior.* Princeton University Press, Princeton, N.J.

Roederer, Juan G. 1974. *Introduction to the Physics and Psychophysics of Music.* Springer-Verlag, New York.

Schelling, Thomas C. 1960. *The Strategy of Conflict.* Harvard University Press, Cambridge, Mass.

Shubik, Martin (ed.). 1962. *Game Theory and Related Approaches to Social Behavior.* Prentice-Hall, New York.

Spiney, W. Allen, and Robert M. Thrall. 1970. *Linear Optimization.* Holt, Rinehart and Winston, New York.

Strum, Jay E. 1972. *Introduction to Linear Programming.* Holden-Day, San Francisco.

Williams, J. D. 1954. *The Compleat Strategyst: Being a Primer on the Theory of Games and Strategy.* McGraw-Hill, New York.

Wood, Alexander. 1962. *The Physics of Music.* Dover, New York.

34
Musical Tones

by Hugh Lineback
May 1951

The wave structure of music is made visible for the classroom

THE photographs on these two pages show oscilloscope recordings of musical notes and combinations of notes. They make visible the vibrations that the ear hears as sound, and they have proved to be very effective in teaching students the science of sound. The tones shown here were all made by a Hammond organ, which is especially suited to such a demonstration because the electrical signals that are converted to sound by the organ's loud-speaker can also be translated into patterns of light by a cathode-ray tube, so that the vibrations produced by the organ can be heard and seen at the same time. The visible wave patterns can be used to explain the three characteristics of a musical tone: pitch, loudness and quality.

The pitch of a note is determined by the number of complete vibrations, or cycles, per second—what is called the frequency. The note A above Middle C on the piano, for example, is produced by a sound vibration with a frequency of 440 cycles per second. Taking this note as a reference, we call it the fundamental tone and designate it as A-440. By doubling the frequency we produce a note an octave higher, or A-880, and doubling it again we get the second octave, A-1760. A musical instrument produces these various notes simultaneously as harmonics, which are simply multiples of the fundamental frequency. In this case the first octave could just as well be called the second harmonic of A-440, the next octave the fourth harmonic and the next the eighth harmonic. Other harmonics fall in the intervals between octaves: thus the third harmonic of Middle A falls on the second E above it, the fifth harmonic on the third C Sharp, and so on.

The loudness of a tone is measured by the amplitude, that is, the height of the wave peaks. The larger the vibration, or amplitude, the louder is the sound.

Even at the same pitch and loudness, one kind of instrument produces a markedly different tone from another. This third characteristic of a sound is known as its quality, or timbre. Timbre is determined by the blending of the fundamental and its harmonics in certain proportions. To obtain complex waves representing various instruments or tone qualities, the organ was set to produce combinations of harmonics of predetermined relative intensities, with the fundamental frequency at a certain strength, the second harmonic in a certain ratio to this, and so on.

These representations of harmonics make clear the qualities and limitations of various instruments. The fundamental tone alone would make rather dull listening. A predominance of high harmonics adds an effect of crispness, while the lower frequency components give power and dignity to music. Too much emphasis on the low harmonics may impart a muffled quality, and an absence of frequencies between the lower and upper harmonics contributes a weird and hollow effect.

Fundamental

Second harmonic

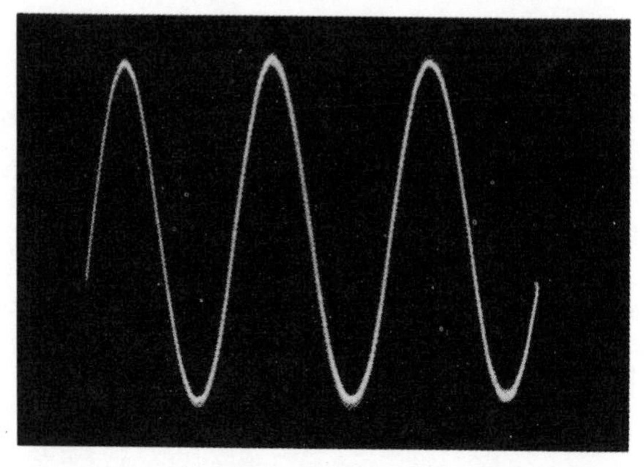

Third harmonic

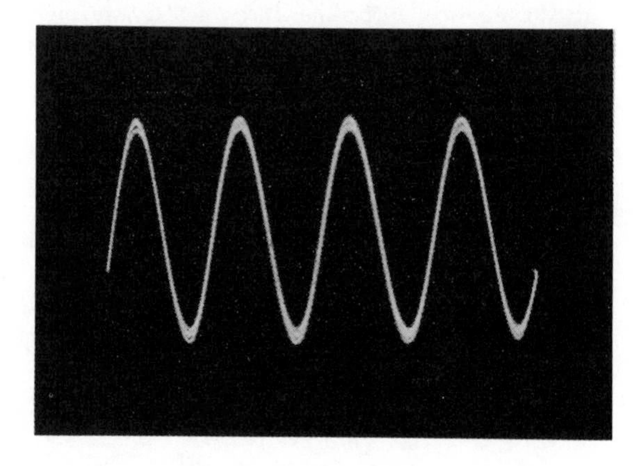

Fourth harmonic

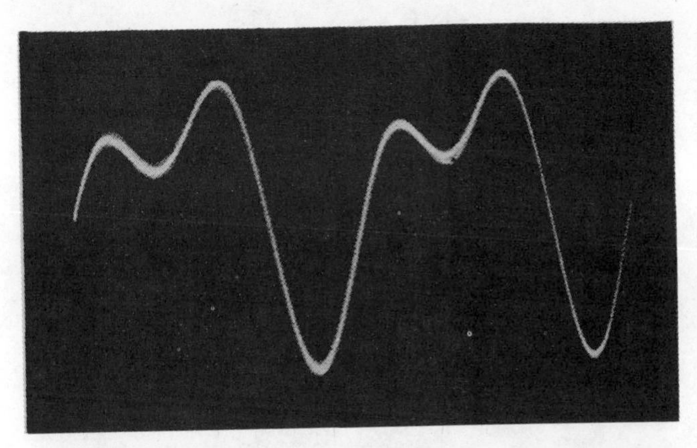

Combination of fundamental and second harmonic

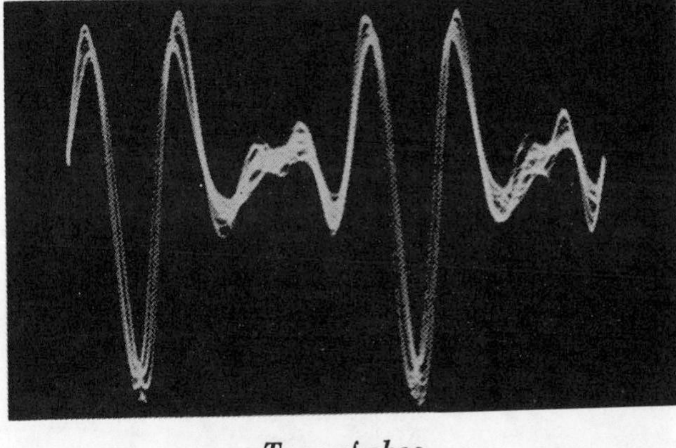

Tone of oboe

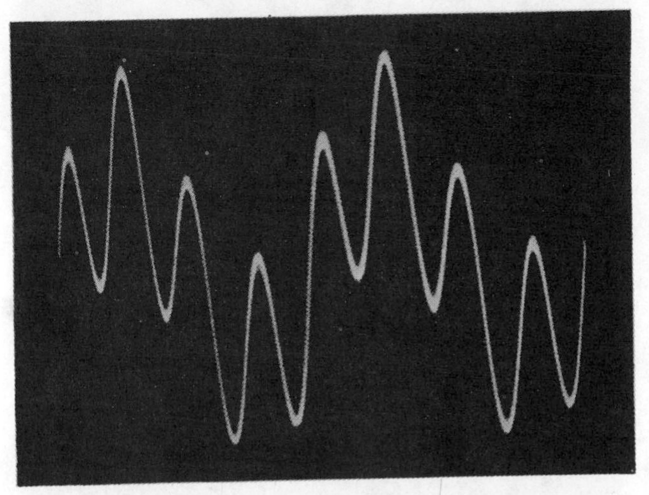

Combination of fundamental and fourth harmonic

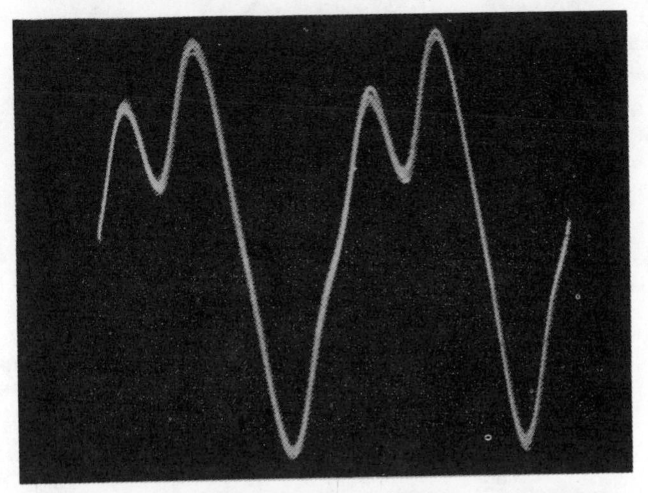

Tone of French horn

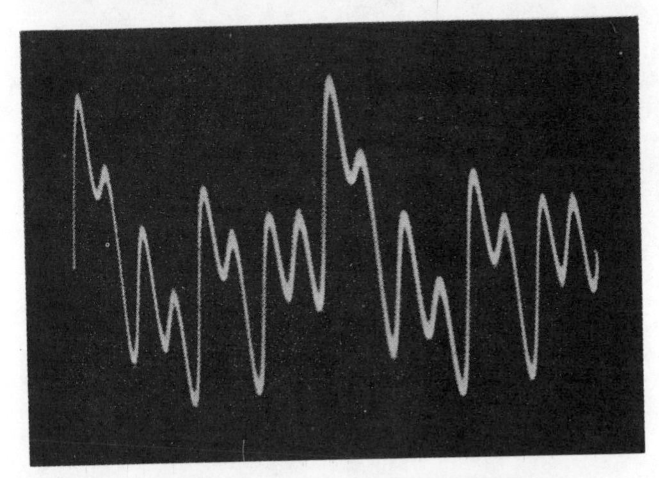

Combination of all even harmonics

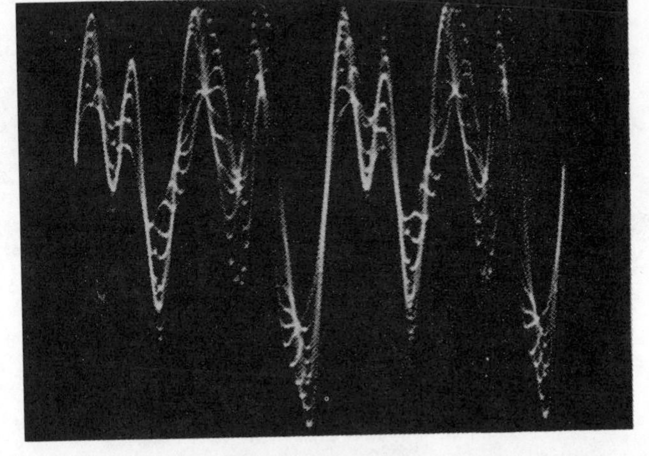

Tone of trumpet

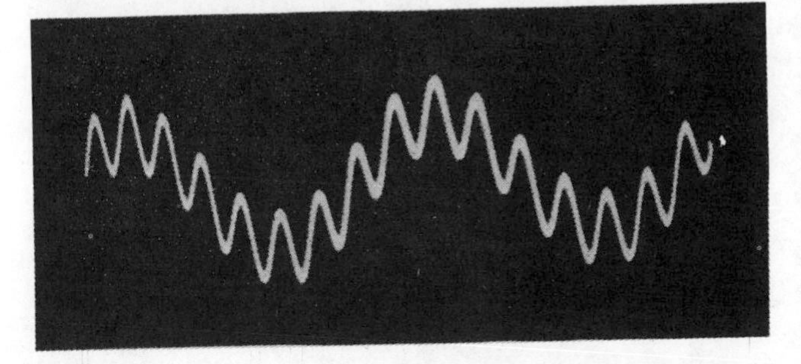

Combination of fundamental and eighth harmonic

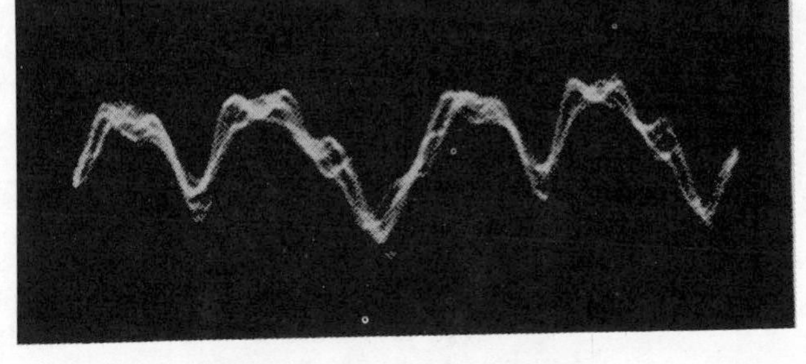

Tone of flute with strings

Physics and Music

by Frederick A. Saunders
July 1948

The agreeable sound of simple melodies and Beethoven symphonies is guided by physical rules, plus a little physiology and psychology. The understanding of these principles can enhance musical creation and enjoyment

ANYONE who looks upon a great bridge arching across a wide river is thrilled by its beauty, and aware at the same time that a great deal of measuring, testing and calculating must have gone into its planning to make the structure safe. A bridge is an obvious combination of art and science. Not so obvious is the physical architecture of great music. One who listens to a symphony at an orchestral concert may know that the composer drew on his inspiration to fill pages with symbols, and that the conductor and his musicians interpret these to help bring to life again what was in the composer's mind. The listener is intellectually and emotionally moved by the sequence of sounds coming to him from many different sorts of instruments. But what has this bewilderingly complex example of art to do with science?

The answer is simple enough. Music is based on harmony, and the laws of harmony rest on physics, together with a little psychology and physiology. The simplest and most pleasant intervals of music have always existed among the harmonics of pipes and strings. From them grew the study of harmony, and they have formed the basis of many noble melodies. A classic example is the opening melody of Beethoven's *Eroica* symphony, whose first part consists of the simplest possible intervals flowing one after the other. Such simple combinations do something to our ears which is fundamentally pleasant and satisfying. Some musical instruments were well developed long before the subject of musical acoustics was born. Today the physics of music helps to guide improvements in musical instruments, in the construction of buildings with good acoustics, in the reproduction of music for immense audiences, and in many other ways.

To examine the physical basis of music we begin by considering the nature of sound. Sound is a word used in at least two senses: (1) the sensation produced in the brain by messages from the ear, and (2) the physical events outside the ear.

SYMPHONY is a vast blend of frequencies from many instruments. At left: Leopold Stokowski conducts rehearsal of New York Philharmonic.

The context usually makes it plain which meaning is intended. Thus we avoid long arguments over whether a sound can exist if there is no one present to hear it. Sound has its origin in a vibrating body, and the vibration may be *simple* or *complex.* The motion of the pendulum of a clock represents a simple vibration, one which is not rapid enough to be audible. To be heard as a musical tone, a vibration must have a frequency of at least 25 cycles per second. A pure tone is represented by a smooth

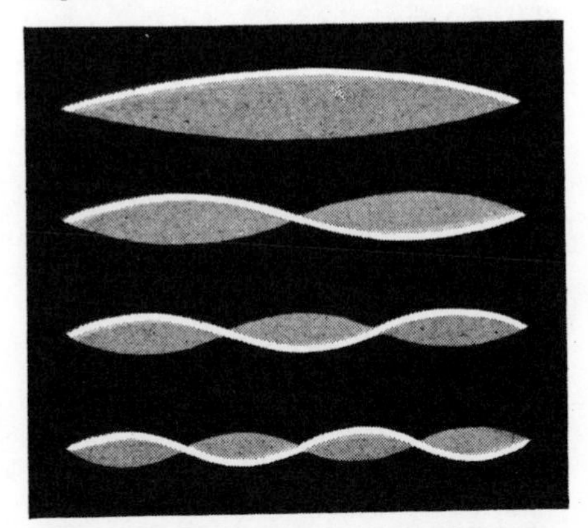

HARMONIC series is defined in various vibrations of a string. Harmonizing frequencies are two, three, four or more times simplest vibration (*top*).

curve in which distances to the right stand for time, and distances up and down correspond to the displacement of the vibrating body from its position of rest. A vibration of this sort is often called simple periodic motion because it repeats itself regularly with a constant period of time for each repetition. But pure musical tones are rare; the tones that are produced by musical instruments are almost always complex.

Complex vibrations can always be regarded as made up of a combination of simple vibrations of different frequencies. Their forms are very varied, as shown in the illustration on page 210. Sometime when you are out walking and have nothing better to do, try swinging your arms at different rates. The simplest case is easy: right arm going at twice the rate of the left. It is not quite so simple to make the right arm alone combine both of these motions, and it is still harder to combine rates whose ratio is one to three, two to three, and so on. One gives up before long; yet any violin string can do this easily without becoming confused. It can combine as many as 20 different rates at the same time into one complex vibration, which is caused in this case by the complicated motion of the string under the bow. These frequencies are simply related; their values are proportional to the integers 1, 2, 3, 4 and so on. They form a harmonic series. The vibration with the lowest frequency, corresponding to the number 1, is called the fundamental; the sound with double this frequency is the first harmonic, and the higher harmonics are calculated in like manner.

I. Harmonic Analyzers

The scientific study of musical instruments depends partly upon the resolution of complex tones into their harmonic elements, a process called harmonic analysis. It is often of practical importance to determine what components are present in a tone and how strong each one is. One old method of analyzing a musical tone is to study its wave form, as pictured by means of a microphone, an amplifier, and a cathode-ray oscilloscope. But the wave is frequently very complicated, and its analysis by mathematical methods into the simple waves of which it is built is very slow and tedious. In recent years instruments have been developed which analyze complex tones automatically, yielding rapid and accurate results.

Some of these harmonic analyzers make use of the physical effect called resonance, which is a response produced in one body from the vibration of another body. It is easily demonstrated on a piano. In piano strings the harmonics are strong. If you press gently on the key an octave below middle C, so as to free the string but not to strike it, and then strike the middle C key sharply, you will hear a continuing middle C tone coming from the lower string. The experiment succeeds only if the strings are in tune. The middle C frequency (about 260 cycles per second) is

FREQUENCY RANGE of some musical instruments and other producers of sound is tabulated in chart adapted from book *The Psychology of Music*, by C. E. Seashore. Frequencies, noted in scale at the bottom of page, are plotted horizontally. Range of scale is 40 to 20,000 cycles, as compared with the human ear's approximate range of 25 to 30,000 cycles. The thin line within each light horizontal bar indicates actual range of frequencies produced by each method. Circles on each line indicate effective range estimated by a group of expert musicians. Vertical lines at the right end of each frequency line indicate range of associated noise. The instruments in black panel are, from top to bottom, tympani, snare drum, cello, piano, bass tuba, French horn, bassoon, clarinet, male speech, female speech and jingling keys. In blue panel are cymbals, violin, trumpet, flute and clapping hands.

equal to that of the first harmonic of the lower string; hence the lower one can respond.

By a variation of the experiment, one can play a chord on a single string. Hold the lower string open as before, but now give a strong impulse to three keys at once—middle C, the C above and the G between. After the upper strings have been quieted, all three tones will be heard coming from the lower string alone, which is resonating to three frequencies at once. This works as well the other way around: hold the same three upper keys open with the right hand and give the lower C a sharp blow. The three upper tones will be heard, coming from the three untouched strings. Or again, try singing a tone into a piano with the loud pedal pressed down. (This frees the strings to vibrate in resonance with any tone with which they agree in frequency.) When you stop singing, you will hear a faint mixture of tones issuing from the piano.

If we had some kind of attachment to the strings by means of which the response of each could be recorded, we should have one type of harmonic analyzer, but not a very good one. It would be unable to respond properly to frequencies lying between those of the strings. A more useful type of analyzer would be a single string whose pitch we could change slowly and steadily throughout the whole range of the musical scale. This could be fitted with an attachment which would record the string's responses, whenever they occurred, to the tone being analyzed. Such a device would be like the tuning apparatus in a radio receiver, which picks up radio waves on each frequency over which they are being broadcast. The device would miss nothing, but it would not be capable of making analyses instantaneously. The same sort of plan, carried out electrically, gives more rapid results. With suitable equipment it is possible to obtain within a few seconds a complete photographic analysis of a sustained tone, yielding numerical values for the strength and frequency of all harmonics present in the frequency range from 60 to 10,000 cycles per second. This method has been applied to the study of the tones of many instruments.

A remarkable frequency analyzer recently developed by R. K. Potter of the Bell Telephone Laboratories gives a continuous analysis of speech; its result is appropriately called "visible speech." One speaks into a microphone and the oscillations of his speech are then passed through 12 electrical filters, each of which allows only a narrow range of frequency to pass. When amplified, each filtered set of oscillations lights a tiny "grain-of-wheat" lamp; there are 12 lamps, arranged vertically. The fundamental tone of the speech lights one lamp, the first harmonic another farther up, and so on. The lamps that light in response to the speaker indicate the frequencies present in his speech. To reproduce his speech pattern, the light from the lamps falls on a horizontal moving belt made of phosphorescent material, so arranged that each lighted lamp traces a separate luminous line on the belt. The result is a characteristic pattern for each vowel and consonant, defined by lines of varying frequency and duration. The accompanying illustration demonstrates how a phrase looks to the eye. A trained observer can read words and phrases at sight, and a person who has been deaf from birth may thus learn to read speech. He can also correct imperfections in his own speech by matching the patterns he produces against standard ones. This visible speech is exciting to watch, and it is likely to be of great help to the deaf.

II. The Violin

Now let us turn to the consideration of musical instruments, a subject in which harmonic analysis has been very useful. We may agree at the start that nothing deserves the name of musical instrument unless it can make a loud sound. Our greatest musical artists must fill large concert halls, and for this they need loud voices, violins, pianos or other instruments. Some musical instruments require a method of amplifying the vibrations created by the player to produce powerful tones. Consider the violin as an example.

A wire mounted on a bent iron rod, with no body or plate to shake, gives almost no sound when it is excited by bow or finger. The wire is too narrow to push the air about sufficiently to create a strong sound wave. Such a performance is analogous to trying to push a

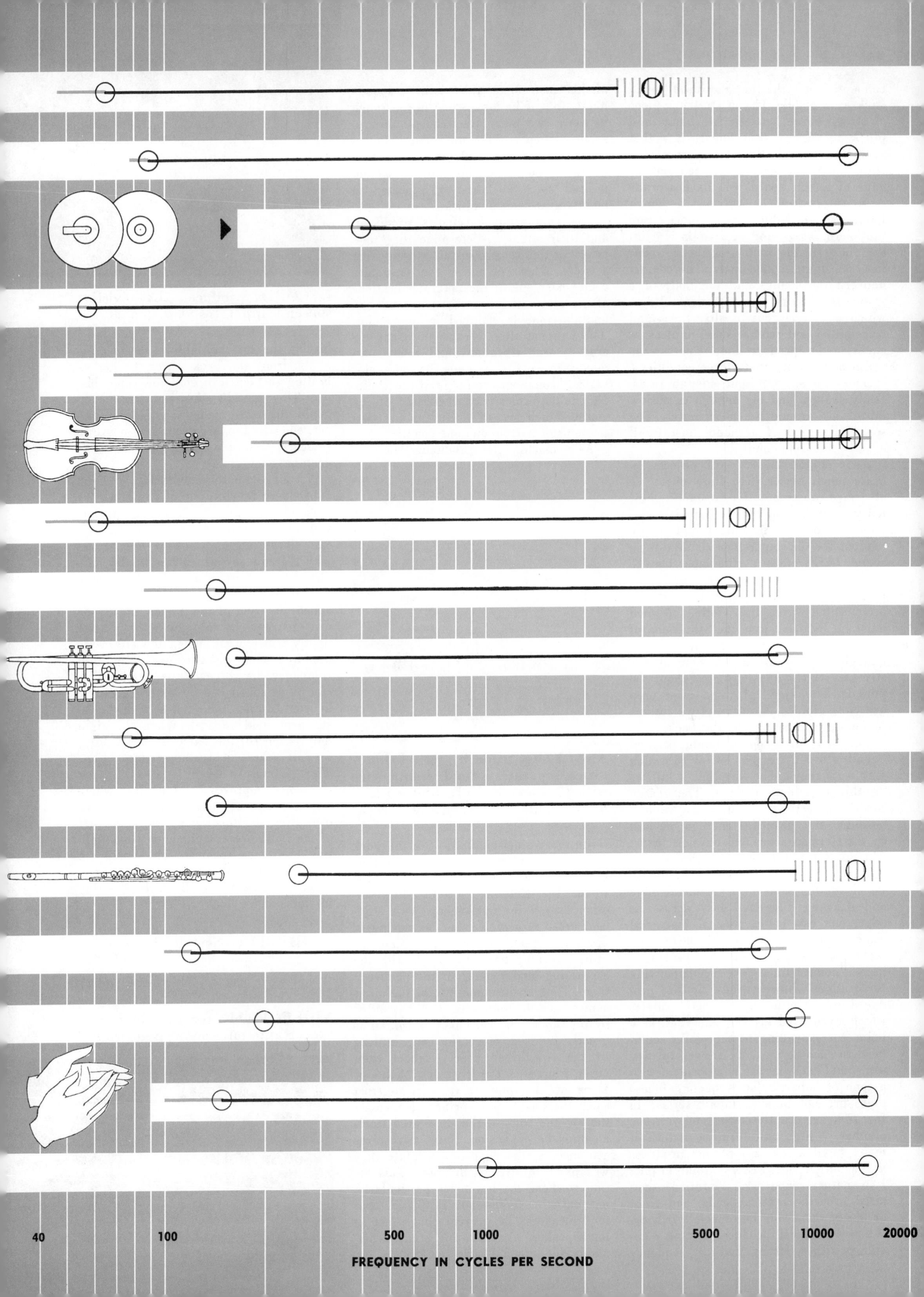

40
100
500
1000
5000
10000
20000
FREQUENCY IN CYCLES PER SECOND

canoe through the water with a round stick as a paddle. If you stretch a piece of strong twine between two hands and pluck it with a free finger, it makes very little sound. But if a part of the twine near one end is pressed on the edge of a thin board, you have a crude stringed instrument, giving a much louder sound, which now comes from the board. The sound of a violin is emitted not from the strings or the bow but from its light wooden body. The contact between the strings and the body of a violin is through the wooden bridge, which is cleverly cut to filter the sound transmitted and remove some unpleasant squeaks. To produce loud sounds, the violin body must satisfy three conditions. It must be strong, light enough to be easily shaken, and big enough to push a lot of air around when it moves. The sounding board of a piano must fulfill exactly the same conditions.

As everyone knows, stiff objects vibrate much better than limp ones; we all have observed, for instance, how noisy a job it is to wrap a parcel in stiff paper, whereas if a handkerchief is substituted for the paper there is almost no sound. Large areas of stiff paper tend to move together as one piece, and thus push on the air sufficiently to start vigorous sound waves. In a violin, the wood must be light, so that the vibrations of the strings can move it, and strong enough to sustain the tension of the strings, which adds up to about 50 pounds. The kinds of wood most used are close-grained Norway spruce for the top plate, and maple for the back.

The body of a violin should respond equally to all frequencies of vibration within its range. The fact that it fails to do this is seldom noticed. The reason for the defect will be clear if we first consider the beautiful method devised by the German acoustical physicist Ernst Chladni (at about 1800) which discloses the natural modes of vibration of plates. By sprinkling sand on a flat metal plate and drawing a rosined bow across its edge, one can get a musical tone, and some of the sand is seen to move from certain areas and some to rest along quiet "nodal" lines. The accompanying illustrations show various figures produced on violin-shaped metal plates which were fixed at both ends and at a point corresponding to the violin sound post. In each figure there are several patterns, and each pattern is associated with a tone of a particular frequency. These tones are not in a harmonic series; in fact they are usually discordant with one another. A high tone forms a pattern of many small areas; a low tone produces a few larger ones. Every violin has its own natural modes of vibration, scattered over the musical scale, and eight or ten of them may be especially strong. When a violinist produces a tone coinciding with a strong natural frequency of his instrument the violin responds loudly, but if he makes one in the range between two such frequencies, the response is poor. This unevenness in response occurs in the playing of the best artists on the best violins, but it is seldom noticed since no artist is expected to maintain an even loudness.

The number of harmonics produced, and their strength, determine the "tone color" or timbre of a sustained tone from a violin. Whenever one of the harmonics comes near one of the natural vibrations of the plates, it is increased in loudness, and the tone is changed in tone color. This happens often, because there are several natural vibrations and many harmonics in each tone. Thus the tone color varies throughout the range of the violin. No one tone color is characteristic of any violin—much less of violins of any particular age or from any one country.

As a machine for producing sound a violin is very inefficient. Most of the work done by the player in rubbing the bow against the strings is lost as heat in the wood. The Chladni patterns show another reason for inefficiency. Two adjacent areas in a plate must be moving in opposite directions when the plate vibrates, rocking back and forth with the separating nodal line at rest between them. Thus at the same instant the air is compressed by one area and expanded by the other. The net effect on the air is greatly reduced, since the contributions of the two areas nearly cancel each other. Moreover, the front and back surfaces of any plate may work against each other: while one surface compresses the air, the back of the same area starts an opposite expansion. If the two waves can meet at the edge of the plate they will partly destroy each other. This action weakens the low tones particularly, not only in violins but in pianos and loud-speakers. To prevent this effect in loud-speakers, the vibrating area is commonly set into a "baffle" which, by enlarging the surface, inhibits the meeting of the front and back waves. Larger vibrating surfaces can emit low tones better. This is why the violoncello and double bass are made progressively bigger, and why the large sounding board in a concert grand piano helps to improve its deep bass tones.

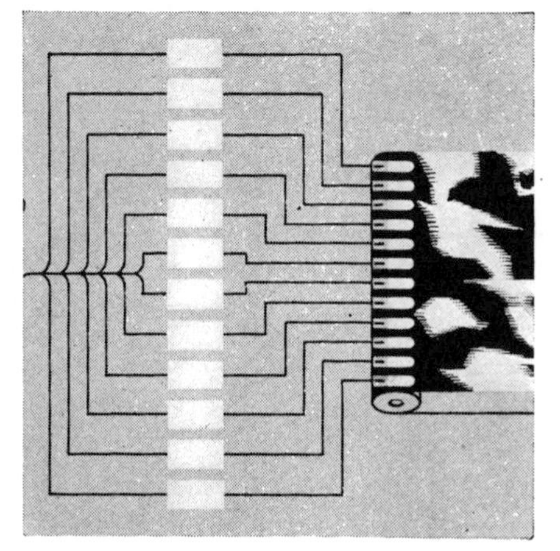

ANALYZER made by Bell Laboratories separates sound frequencies with 12 filters. Each regulates a tiny light. Lights make image on screen.

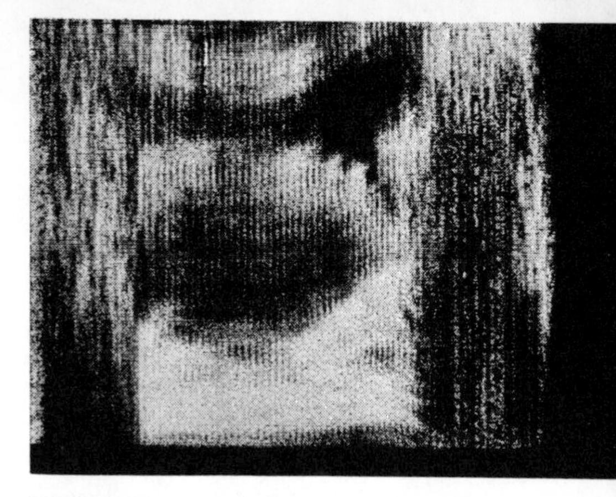

RECORD produced by "visible speech" apparatus depicted at left

Not all of the tone emitted by a violin is produced by vibration of its plates. We must also credit the air inside the box with an important contribution. This air can vibrate in and out of the *f*-holes with a frequency which lies in the middle of the lowest octave. The tone there would be mean and ugly without the added vibration of the innner air, as one can discover by plugging the *f*-holes lightly with cotton. When the air inside the box is vibrating at or near its natural frequency, its resonance is strong. This can be demonstrated by setting a candle in front of one of the holes, with the instrument held vertical. When the right note is bowed, the flame dances wildly; for all others it remains quiet. (The effect is most marked in a cello.) Air resonance improves the tone just where improvement is most needed; that is, over a few semitones where the small size of the violin prevents the body from emitting the tones strongly. The maximum effect is near C sharp on the G string in violins, and near A or B on the G string in violas and cellos.

III. Old *v.* New Instruments

Now what makes a superlative violin?

VIOLIN MUSIC recorded by visible speech apparatus shows a horizontal

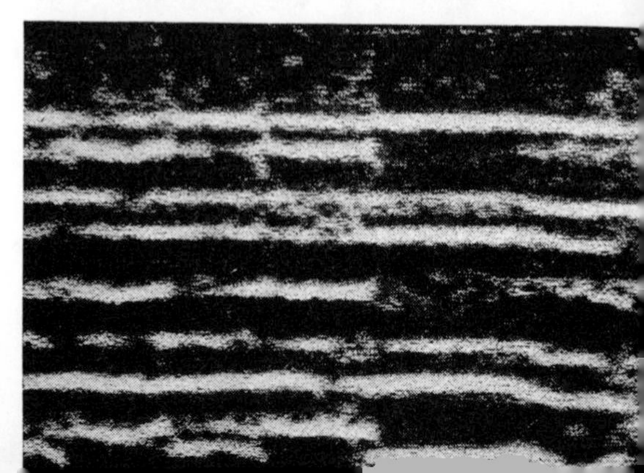

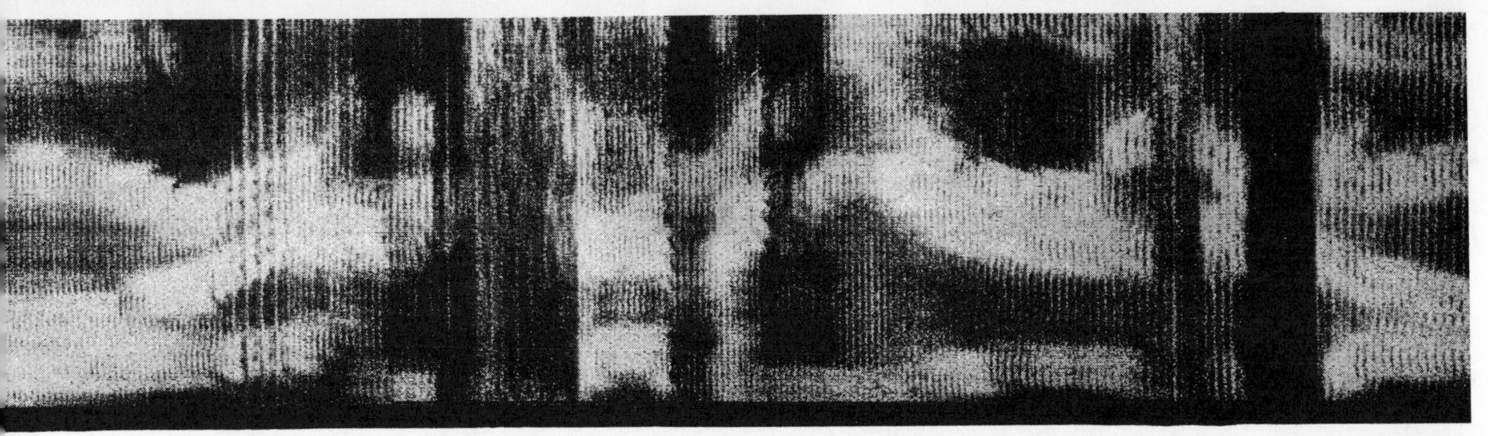

may be temporary image on a phosphorescent screen or, as in the illustration above, a pattern on a paper strip. This pattern, which may be read by a trained observer, represents phrase "Four score and seven years ago . . ."

This question is endlessly debated, but it cannot be settled by arguments. The most accurate and careful measurements in a laboratory with modern equipment are required, and a start has already been made. The impression made by a violin on a listener is due to many features: the quality or "tone color" of sustained tones, the ease with which the tones begin, the rate of decay of the sound, the loudness in different parts of the range. These items are often lumped together under the word "tone"; here we must separate them carefully. The tone color of sustained tones is probably the least important of the lot. The loudness in different ranges of pitch may be the most vital consideration in the judgment of a violin. A bad violin is weak in the low tones and too strong in the squeaky top frequencies.

Old violins are almost always thought to be better than new ones, and European better than American. This opinion may come in part from psychological causes —our admiration of old civilizations, the influence of tradition and so on—but part of it certainly comes from the beauty of workmanship characteristic of the best old instruments, and from their rarity. It is as difficult for most violinists to find any defect in a Stradivarius as it is easy for them to criticize the best-made American violin. In recent years careful experiments have been made with excellent modern apparatus, seeking to measure all the mechanical features mentioned in the preceding paragraph. A great variation in values was found among 12 Strads, many other old violins and a few dozen new ones, but the average values failed to show any consistent difference between old and new. This is not to say that there are no differences, but that the results were the same within the limits of error in the measurements, using very sensitive methods. These bold statements are supported by many "blindfold" audience tests, as well as by variations in professional opinions as to the merits of certain famous violins.

Violins seem to become lighter and better when played for a century or two. The effect of age on instruments which are not played appears to be small. Changes in the physical character of a violin can come about through vibration and also from contact with players. After a period of use a violin usually weighs more because it has absorbed water vapor from the air around the player. This makes the wood expand across the grain; when not in use it dries out again and contracts. These changes may alter both the physical and chemical properties of the wood. Some day it may be possible to attain the effects of years by a quick treatment of the wood; promising work along this line is now in progress.

There are methods of mapping out the natural vibrations of a violin by exciting it electrically and measuring its response at every frequency. This yields a curve, called the response curve, by means of which violins can be compared. The inequalities in response at various frequencies are remarkable, in both old and new instruments. All good violins should in the future have a certified response curve furnished with them when they are offered for sale.

IV. The Piano

Some of the statements which I have made about the violin apply equally to the piano. The piano's sounding board acts like the violin body. While the violin has not changed in the last century, the piano has seen constant improvements in the sounding board, the strings, the hammers and the key action. So loud has the instrument become that the vibrations now shake the floor and are sometimes transmitted through the solid structure of a building to unexpected distances. In apartment houses peace may sometimes be preserved with the neighbors by placing rubber pads between the piano legs and the floor.

New problems arose with the invention of the piano's key and hammer mechanism. The hammer must be light but strong, in order to act quickly and give powerful blows to the strings. The pads must be soft to avoid the production of strong high harmonics that a hard hammer creates. (One can almost convert a piano into a harpsichord by using a teaspoon for a hammer.) When a player hits a key on the piano, the action gives the hammer a throw; at the moment when the hammer strikes a string it is not connected with the key, but is flying freely. It is as if the player were throwing soft balls at the strings from a distance. Once the hammer is on its free way, the player can do nothing more to it. His only control is

band for each harmonic produced by the instrument. Large number of bands illustrates complex nature of musical sounds. Record shows passage from Glazounov's "Concerto in A Minor," played by Jascha Heifetz.

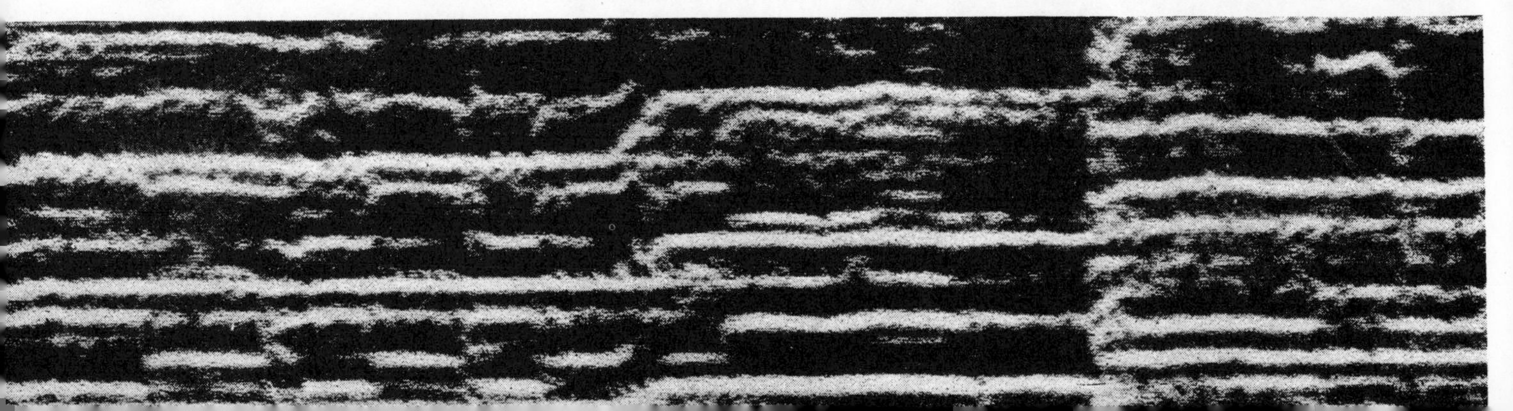

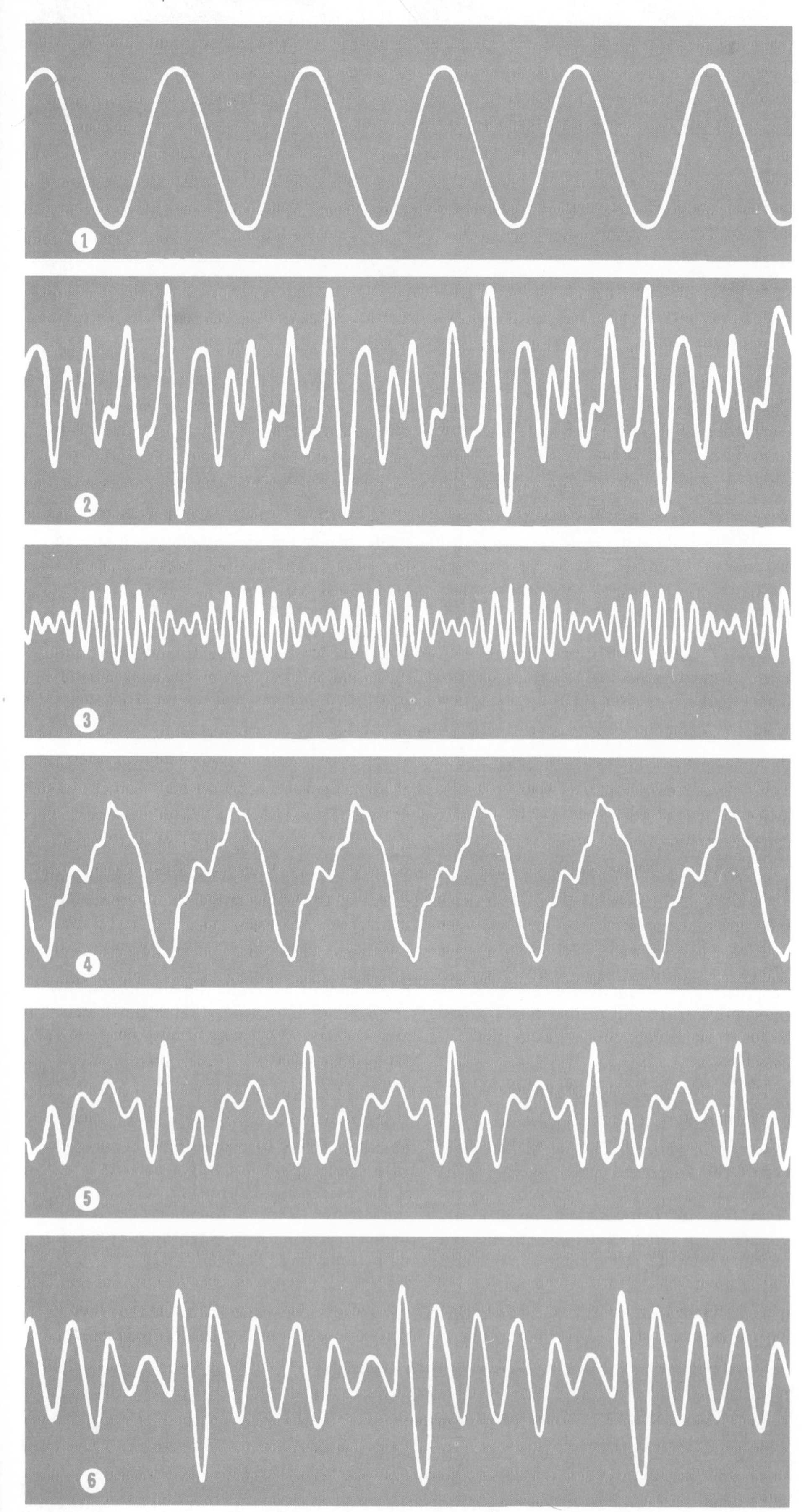

COMPLEX WAVE FORMS of musical sounds are the result of combining several simple forms. The forms above are (1) the simple tone of a tuning fork, (2) pure chord produced by four tuning forks struck together, (3) "beat" tone of two tuning forks with almost the same frequency. Characteristic instrumental forms were made by (4) violin, (5) oboe, (6) French horn.

through the initial speed he imparts to the hammer. Thus it is a fact that for a given hammer speed the tone is exactly the same whether the key is pressed by the finger of a great artist or by the tip of an umbrella. Any skeptic to whom this statement is repulsive should open up a piano and watch the motion of a hammer. Piano "touch" is of course a mixture of effects: besides hammer speed, which affects loudness and tone color, it depends on the sequence of tones, the length of time each key is held, the management of the pedals, the phrasing and so on. Of these the last three are perhaps the most important.

There are two subjects in musical acoustics, incidentally, which often arouse furious arguments. One is piano touch. The other is the alleged characteristic flavor of music in different keys. Pupils are often taught that D major is a martial key. Today military marches are played on a piano whose D is 294 cycles per second. A musician in Mozart's time would have had a D of about 278 cycles per second (our C sharp), since the pitch has risen about a semitone in this interval. If two performances of the same music in different pitches can produce the same impression, then the flavor of the key must come from its name and not from its pitch.

V. Wind Instruments

The wind instruments operate on a very different plan from the strings, and as sound-producers they are much more efficient. A stringed instrument loses considerable energy in transmitting its vibrations from the plate to the air; in a wind instrument the sound is emitted directly by vibrations of the air inside the pipe. Hence an instrument like the oboe or clarinet in the orchestra stands out against the string section, and two or three of them are considered sufficient to balance a much larger group of violins.

The sound waves in wind instruments are generated in a variety of ways: by thin streams of air issuing from slots (the organ) or from the player's lips (flute); by the vibrations of single reeds of cane (clarinet family), of double reeds of cane (oboe family), of metal (organ), or of the player's lips (cornet, horn). Except in the case of a metal reed, which has to be tuned to its pipe, the mechanism that excites a wind instrument has no very definite natural frequency but will accommodate itself to the rate of the vibration of the air in the pipe. This rate is determined by the time it takes an exciting air impulse, traveling with the speed of sound (about 1,100 feet a second), to go down the pipe and back. In instruments which have side holes, this wave is reflected not from the end of the pipe but from the first hole that is open. By this means the player controls the effective length of the pipe and the frequency or pitch of the sound produced. Shortening the pipe by opening successive holes makes it possible to produce the notes of the musical scale; the higher tones are obtained as harmonics of these fundamental vibrations. The sound of the flute comes from two holes, the one at the mouthpiece and the first open one lower down; the vibrating air dances in and out of these two holes simultaneously. The holes still lower down emit practically no sound. The same principles apply to the oboe or clarinet except that there is no hole in the mouthpiece. The lowest tone is the only one whose sound issues from the end of the instrument.

In the brass instruments, the length of the tube is governed either by a sliding piece (slide trombone) or by insertion of additional lengths of pipe by means of valves operated with the fingers. At each length a large series of harmonics can be blown, and with several lengths available all the notes of the scale can be played, many of them in more than one way. The fundamental tones are not often used.

The tone colors of wind instruments are not as variable as those of violins; hence the player's opportunities for virtuosity are more limited. On the organ, the only wind instrument that has separate pipes for each pitch, the organist can build up tone colors by combining pipes of the same pitch but different colors. In the brass instruments, a player produces a marked change in tone color when he puts his fist or some other object in the "bell" from which the sound comes. This muting of the tone corresponds to what happens when one loads the bridge of a violin with an extra weight.

VI. The Singing Voice

But none of these instruments has the variety of tone color available to a singer. The voice is the most versatile and expressive of all musical instruments.

The vocal cords vibrate somewhat as do the lips of a cornet player, that is, as a double reed. They produce a range of fundamental frequencies which is determined by the muscular tension that can be put on them and by their effective mass and length. The action of the cords has recently been photographed with a motion-picture camera, showing that they have a complicated, sinuous back-and-forth motion. Such a motion would be expected to generate a complex sound wave; voice sounds are indeed found to be rich in harmonics.

The throat and mouth space through which the sound passes on its way out can take the form of one chamber or be divided almost in two by the back of the tongue. A singer also varies the size of the mouth opening. These alterations enable the chamber to resonate to a variety of frequencies, some low as fundamentals, some high as harmonics. In singing, we presumably tune the chamber to resonate with the vocal cords at their fundamental

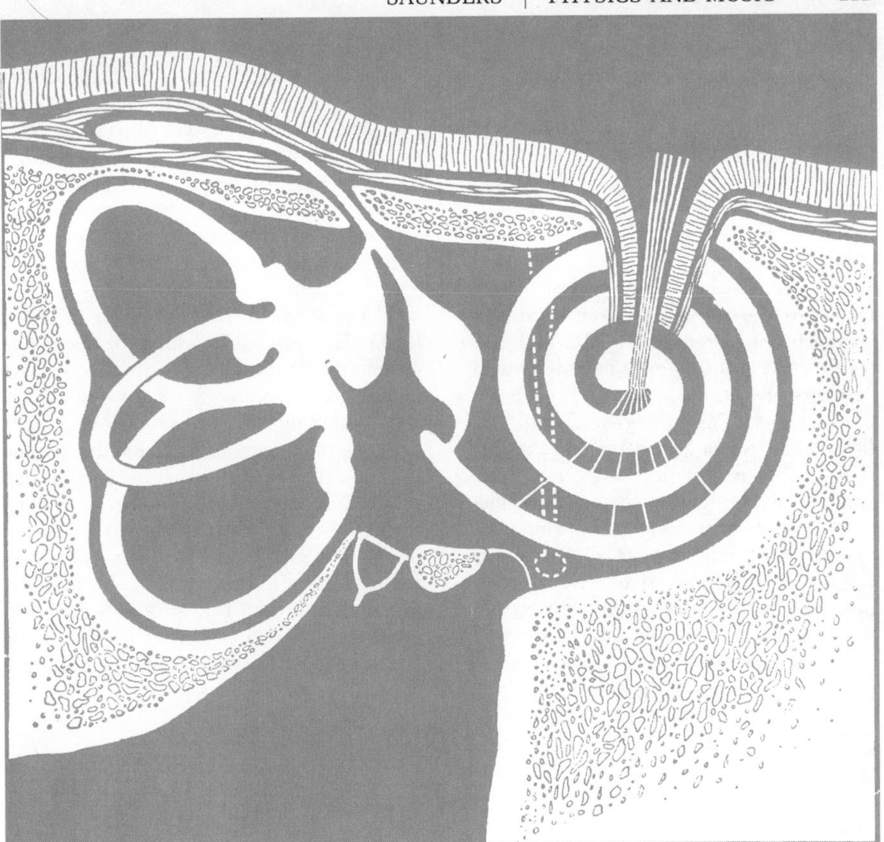

INNER EAR, here shown in highly diagrammatic drawing, is detector of sound. Spiral organ at right is the cochlea. From it run branches of the auditory nerve (*upper right*). These branches are attached to the basilar membrane, stretched across the cochlea's inside diameter along its full length. When sound vibrates membrane, nerve impulses are sent to the brain.

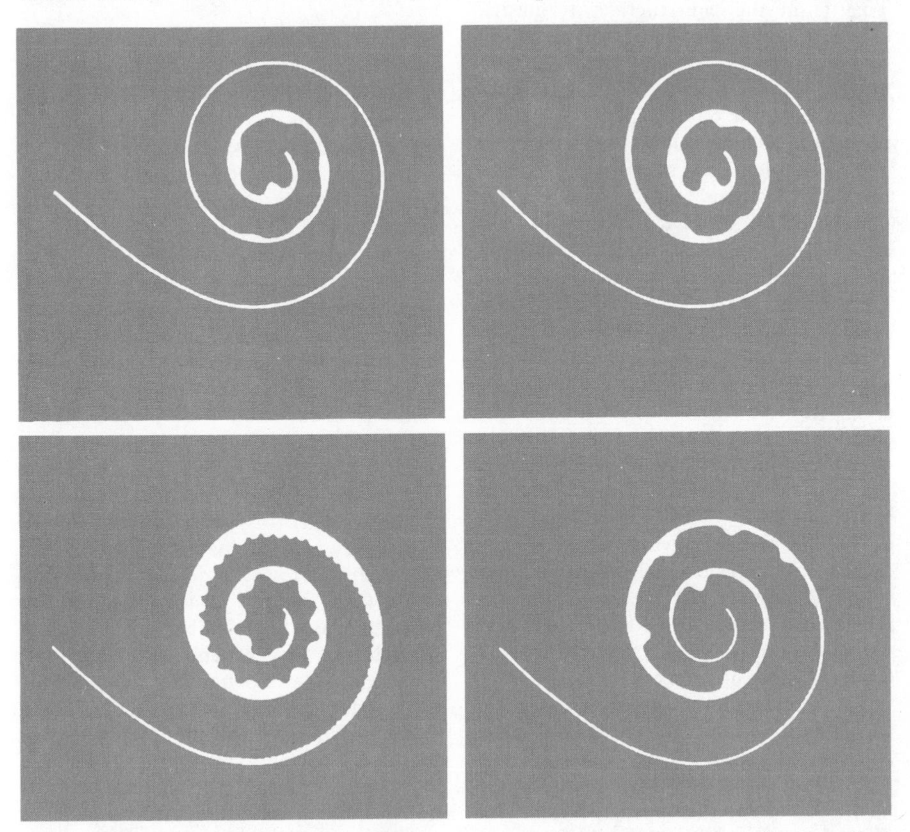

BASILAR MEMBRANE responds to various frequencies at various points along its length. Peaks on spiral diagrams show relative response. Two drawings at top show "false harmonics" of ear's response to a pure tone of increasing loudness. Two drawings below show membrane's accurate response to many harmonics of steamboat whistle (*left*) and note of a bugle (*right*).

or some harmonic. In general the pitch of the voice varies with the tension and length of the cords, its quality depends on the shape and size of the chamber, its loudness is determined by the amount of air pressure supplied by the lungs. The versatility of the voice comes from the ease and quickness with which all these changes can be made.

Singing teachers use certain special terms to describe all the processes involved in tone production. Although these terms have quite definite meanings to the teachers, to others such descriptions as "head tones," "chest register," and "tone placement" mean very little, and that little is probably misleading. One would suppose, for example, that the head and chest must vibrate somewhat at all times, and that the tone must always originate in the same place. One may also object to crediting the bony cavities in the head and the absorbent lung-space with helping to produce loud sounds, since these areas are powerless to contribute anything appreciable. It is to be hoped that before long there will be further experimental studies that will disclose the real behavior of the whole vocal apparatus, and that then such language can be used as will be understood by all.

VII. Musical Scales

There is one special study in which mathematics and music go hand in hand. This is in the construction of musical scales. People with unmusical ears sing up and down the range of pitches without hitting the same spot twice; but music cannot be built on this plan. The piano must have a pattern on its keyboard, and a fixed frequency for each key, as the flute has fixed positions for its side holes. The pattern of the keyboard repeats itself in each octave. An octave is measured by the first interval in the harmonic series. Two tones an octave apart have a frequency ratio of 2 to 1; they produce in our ears a simple motion and a pleasant impression. To produce a similarly pleasing effect within the octave, its intervals also should be simple, with ratios like the ones found in the harmonic series, such as the musical fifth (ratio 3 to 2) and the fourth (4 to 3). Thus the scale is built up on the plan of having as many pairs of tones as possible which please us when sounded together. At the same time the musician demands freedom to shift keys without running into any trouble with different sorts of intervals.

The final result is a scale of 12 notes with semitone intervals all exactly alike, and just filling an octave. The mathematician tells us that if we multiply the frequency of any starting note by the 12th root of two (or 1.05946), we obtain the frequency of the next higher note, and if we continue this process, after 12 multiplications we arrive at the beginning of the next octave. This scale does not give us perfect musical intervals inside the octave, but there seems to be no way in which we can get a better one to fit all the conditions stated. The purist objects; he has a wonderful ear and he says it hurts to hear these intervals the least bit off. So the mathematician writes another paper on a perfect—but unusable—scale.

Recently the physicist and the psychologist have joined in the discussion. A new measuring device has been invented by O. L. Railsback, which he calls the chromatic stroboscope. With this he can measure the frequency of any tone while it is sounding, with a precision greater than we may ever need. It has been used to check the tuning of pianos. The results show that expert tuners agree among themselves but they tune the low notes too low and the high ones too high to fit the scale. They do this because it actually sounds better, and the explanation of this odd fact is that the harmonics of a piano string are themselves out of tune, and are sharper than they should be. The 14th harmonic occurs about where the 15th belongs. The scale that results is no longer the exact scale of "equal temperament," which we have just considered, but a "spread" one whose octave ratio is slightly greater than two to one, while its fifths are almost true. All these years we have been using two scales without knowing it. To make matters worse, it has been shown that, in contrast to the piano, the harmonics of pipes and of bowed strings are not out of tune; so that the organ is presumably tuned in equal temperament. The violin is always tuned to perfect fifths, yet nobody minds when it is played with a piano tuned to a different scale. These strange differences seem to have escaped the notice even of our friend the purist.

Recently musical psychologists of the University of Iowa, under the leadership of Dr. C. E. Seashore, have measured the performance of a number of first-class professional singers and violinists, and found that they do not use the scale of equal temperament nor any other scale exactly. We must all be less sensitive to the refinements of tuning than was supposed. The scale (or scales) we now use is quite good enough for such ears as the best of us possess.

VIII. The Listening Ear

The ear, in fact, is a surprising organ; it does not always tell the strict truth. S. S. Stevens of Harvard University has shown that the pitch of a pure tone varies with its loudness. Low tones may drop a whole tone on the musical scale, while very high tones go the other way. If, while listening to a loud tone whose pitch is off, you cover your ears, the pitch goes back to where it belongs. Fortunately, since this effect is observed to an appreciable degree only for pure tones, it is of little importance in listening to most music, because the tones are complex. Moreover, at the pitch where the ear is most sensitive (2,000 cycles per second), the effect disappears.

The ear may even manufacture sounds that do not exist. Harvey Fletcher and his group at the Bell Telephone Laboratories have found that as the loudness of a pure tone increases, the ear begins to hear a change of tone color, seemingly caused by harmonics which appear in the tone in increasing number and strength. The tone increases in shrillness and harshness until it sounds like the blast of a cornet in one's ear. Yet an oscilloscope picture of the wave form of the sound shows no trace of these harmonics.

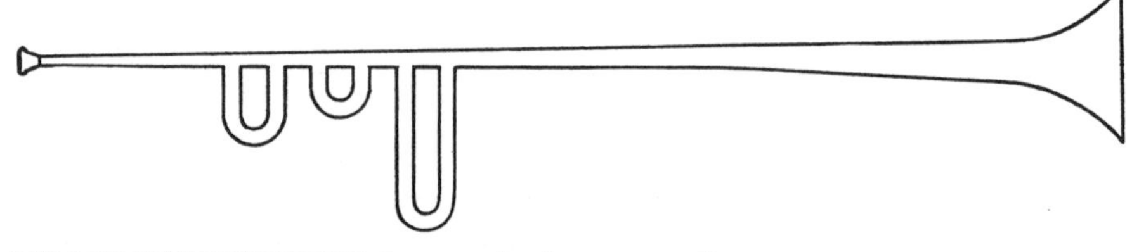

BRASS INSTRUMENT is stretched out to illustrate function of valves. Manipulating valves adds extra segments to effective length of pipe. This changes the rate of vibration of air in pipe and frequencies of its tones.

These ghostly harmonics arise somehow in the ear itself. The sensitive basilar membrane, where sound is detected by a series of nerve endings, has been proved to respond to different frequencies at different positions along its length. The membrane is spiral-shaped, and Fletcher pictures its "auditory patterns" by means of a set of spiral diagrams showing where disturbances occur in response to sounds of different pitch. In the case of a soft, pure tone, the membrane is disturbed only at the place appropriate to the frequency. But as the same tone grows louder, new disturbances mysteriously appear at the points where the harmonics of this tone would be recorded. The source of the false harmonics is probably traceable to a natural imperfection in the action of the mechanism of the middle ear.

A practical consequence of this quirk is that any tone, pure or complex, increases greatly in harshness as it becomes louder. Thus even a good radio gives a bad tone when turned up too loud; the ear is to be blamed, not the radio set. A violin has a harsher tone to the ear of the player than to a listener some distance away. A violin whose sound was amplified electrically to fill a large hall would sound quite unnatural.

IX. Room Acoustics

Science has made a very considerable contribution to music in connection with the acoustics of halls. To make clear the nature of this contribution we must consider some of the facts about sound in rooms. If the source of sound in a room is suddenly stopped, the sound lasts a little while; it dies down as it is absorbed or escapes through openings. The duration of this sound is long if the room is large or if the sound was a loud one; it is shortened if many absorbing substances are present. The absorptivity of a material is great if it is full of fine pores in which the regular vibrations that constitute sound are made irregular and thus turned into heat.

The best absorptive material is a closely packed audience. But porous plates of various sorts are available for covering walls or ceilings to cut down reflection of sound and increase absorption, in case the audience is not large enough. A bare room with hard walls reflects excellently, and this has two effects: the sound is made louder (just as white walls make a room lighter), and it is prolonged. Speech becomes hard to understand, because successive syllables overlap. Music usually benefits more by reflection than speech does: it has fewer short "syllables," and the reflections can make it loud enough to be heard well even in the rear seats of a very large hall.

Wallace Sabine of Harvard was the first to work out the proper way of correcting the acoustics of noisy halls by increasing their absorptivity. He founded architectural acoustics, which is fast becoming an exact science. It is now a simple matter to provide for good acoustics in a hall before it is built, and a bad hall can usually be made tolerable by treatment at any time.

One musical application of acoustics concerns the marked effect which the character of a room may have on the tone color and the loudness of a voice or other musical instrument. Most absorptive materials absorb more of the high tones than the low ones. When you select a piano in a bare showroom, it is likely to have a "brilliant" tone, meaning that it is strong in high frequencies. But if you place the same piano in a living room full of stuffed furniture, cushions and thick carpets, you may find its tone dull and weak. The high frequencies are still present, but they are quickly absorbed, and so you do not get the reinforcement of these tones that occurred in the showroom. A violin's tone color and power likewise depend on the sort of room in which it is played. On the other hand, a singer whose shrill high tones are hard to bear in an ordinary room should bring along a truckload of cushions, the presence of which would have the effect of greatly increasing the listeners' pleasure.

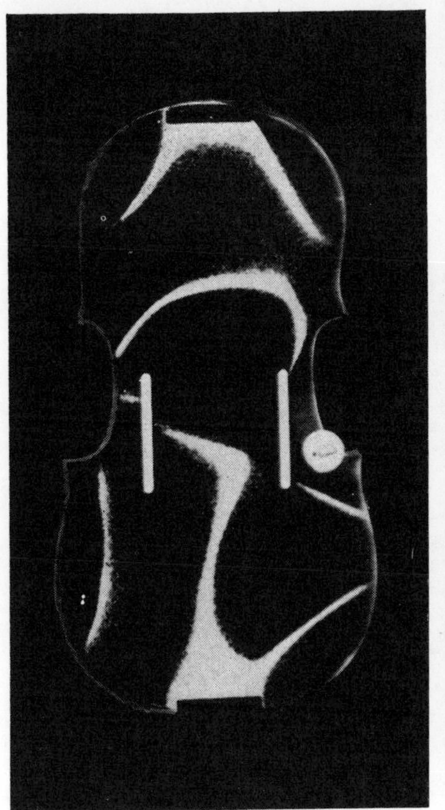
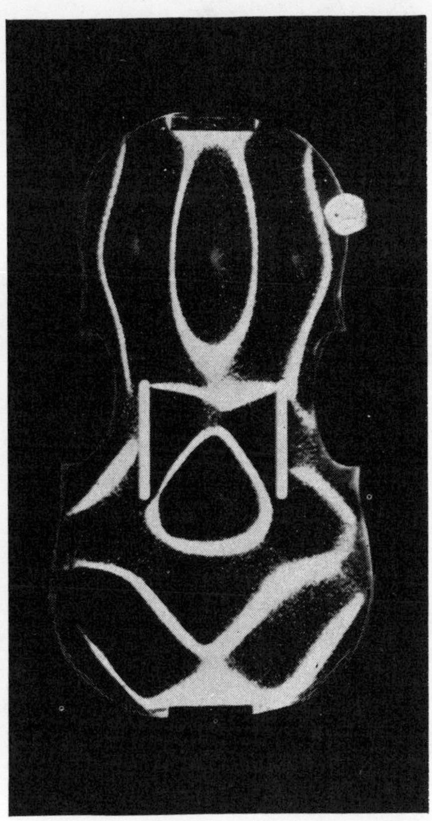

CHLADNI PLATES indicate the vibration of the body of a violin. These patterns were produced by covering a violin-shaped brass plate with sand and drawing a violin bow across its edge. When the bow caused the plate to vibrate, the sand concentrated along quiet nodes between the vibrating areas. Bowing the plate at various points, indicated by round white marker, produces different frequencies of vibration and different patterns. Low tones produce a pattern of a few large areas; high tones a pattern of many small areas. Violin bodies have a few such natural modes of vibration which tend to strengthen certain tones sounded by the strings. Poor violin bodies accentuate squeaky top notes. This sand-and-plate method of analysis was devised 150 years ago by the German acoustical physicist Ernst Chladni.

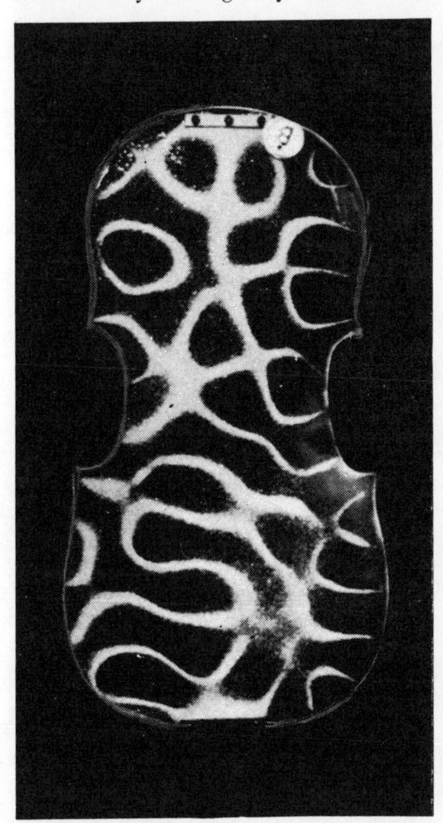
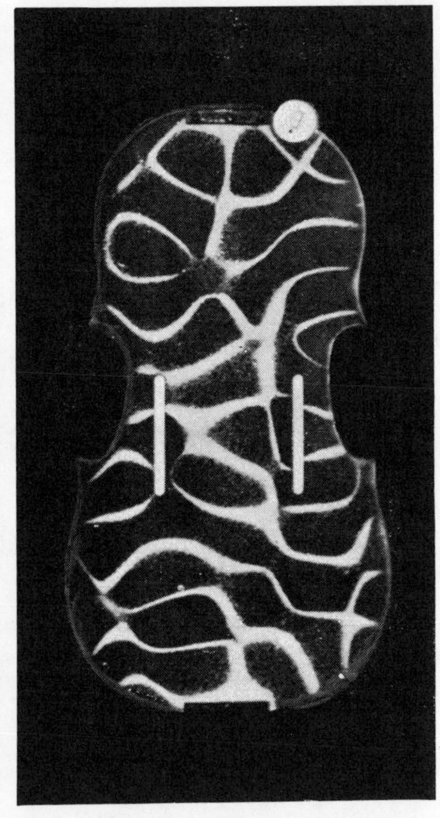

The Practice of Quality Control

by A. G. Dalton
March 1953

The statistical analysis of manufacturing processes has become a powerful tool of technology. An account of how its principles are now applied in the factory

STATISTICAL quality control has been employed to some extent for a quarter of a century, but only recently have the mists surrounding the rather formidable adjective "statistical" begun to clear. And only recently has industry generally come to see the potentialities of such methods for the solution of production problems.

What kind of problems? Let us consider a simple operation such as boiling eggs. Some people like their eggs boiled hard, some soft, some medium. Probably the most common desire is for medium-boiled eggs. Various mechanical egg-timing devices on the market indicate that the generally accepted average time for boiling an egg neither too hard nor too soft is three and a half minutes. However, even if we consistently boil our eggs for precisely that length of time, they don't always turn out medium boiled. Sometimes they are too congealed, at other times too runny.

Undoubtedly variations in the size or weight of eggs have much to do with this variability. Differences in the age of the eggs, in their original temperature, in the number of eggs put into the pot at one time, in the amount of water, in the atmospheric pressure, in the length of time they remain unopened after cooking—all these factors may influence whether they turn out medium boiled or not. If it seemed worth while, we could go to some lengths to find out how much these many variables might be controlled to produce more uniformly cooked eggs. We might establish a number of control points for individual variables: the size or weight of the eggs, their temperature immediately before boiling, and so forth. Some controls might have an important influence on the quality of the end product, others a negligible effect; all would cost something in time, effort and money. Through such experiments we can weigh the costs of controlling undesirable variables against the benefits. Will we enjoy our eggs more? Will we, if we are restaurateurs, sell more? Will we have fewer complaints and fewer eggs returned to the kitchen?

This problem of achieving a desired uniformity of product or service is a common one in industry, applying to baked plastics and machined gears, to toasters and bicycles, to railroad schedules and telephones. In industry, of course, the problem is much more complicated than in the home, but essentially it yields to the same methods we might have used to determine why eggs boiled for equal lengths of time don't always come out alike.

THE NEED FOR statistical quality control derives from the inability of hens and manufacturers to make two or more things exactly alike, however hard they may try. It is these differences among units of product that cause trouble. If large enough, particularly in materials or subassembly parts, the variations may make fabrication or assembly difficult, costly or impossible. Even if small enough to go undetected, they may still lead to customer dissatisfaction. This dissatisfaction may mean only that the customer, a creature of habit, is disturbed by changes in product. Such changes may have nothing to do with actual quality, but they are all part of the problem.

Differences in product may come from one or both of two general types of causes. One kind is the normal, chance variability of materials, machines, temperatures, atmospheric conditions, manual operations, measuring devices and other factors entering into the manufacture of an item. These variations are inherent to some degree in all processes converting raw materials into useful products, and it usually costs something to reduce their effects. The other category covers other-than-chance causes of variations, as opposed to expected random fluctuations. This group includes extraordinary variations in materials or machine operation, interruptions of the power supply, operators' carelessness or lack of skill, rough handling, poor organization of the work, and so on. Harmful combinations of chance causes also may be included in this category. Once identified, the other-than-chance causes can often be eliminated at little or no expense.

The first step in the problem of improving the uniformity of a product is to determine what kinds of differences are occurring and to what extent. The only way to find out is by inspection, which is a problem in itself. To inspect every single item produced is often impracticable, sometimes would be damaging to the product and in any case would still leave the reliability of the results open to question, because of inevitable errors of measurement and judgment. Hence inspection is often restricted to random samples, and the results are interpreted by application of the laws of probability. The effect of human error is reduced and reliability greatly improved by this method.

Further complicating the problem is the fact that every product has many characteristics, in each of which some differences will occur. Pencils, for instance, have weight, length, thickness, color, hardness of graphite, hardness of wood and other properties. The only

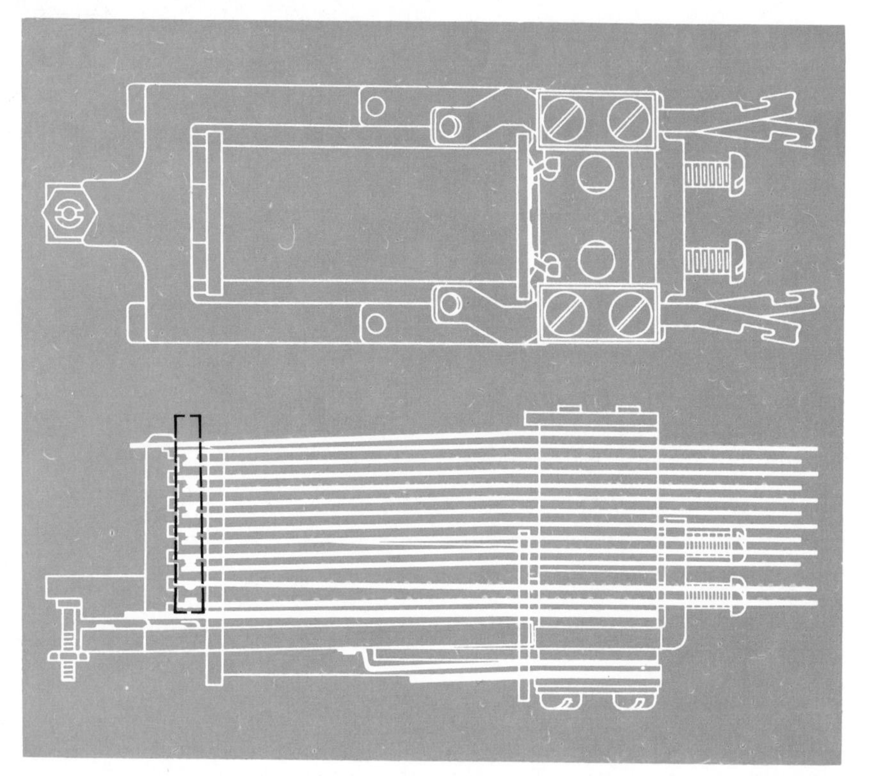

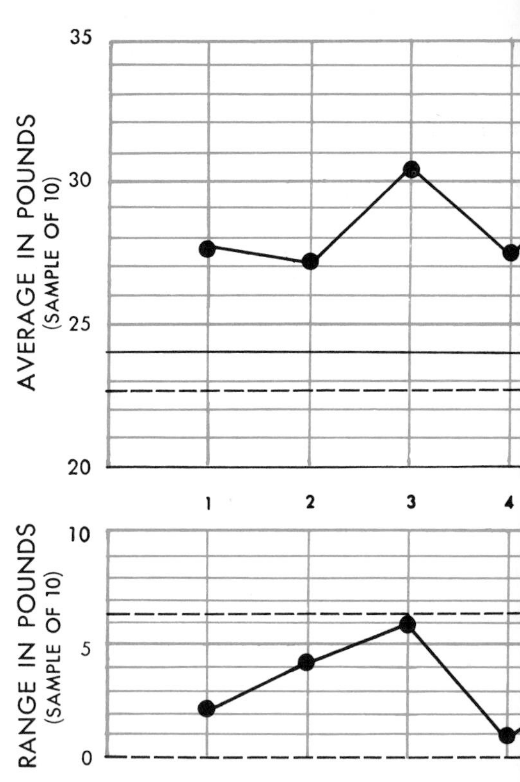

CHART is used to control the strength of the welds which fasten the contacts to the springs of an electric relay. At the left is a drawing of the relay from the top and side; the contacts are within the dotted black line at the far left. The curves at the right reflect the changes in the average strength of welds made by one machine over a period of 18 hours. Once an hour a sample of 10 welded contacts is taken from the output of each machine for testing. The test consists of measuring the number of pounds of force required to

variable of concern to consumers may be the hardness of the point. To the manufacturer, however, irregularities in length may be sufficient to disrupt packaging; variations in the wood may introduce tooling difficulties. Fortunately product characteristics are not all of equal importance, and their variabilities are not all likely to be of the same magnitude. With reliable inspection reports before us, we can proceed to sort them out and by statistical analysis find out what, if anything, we should do about the variations in our product.

IN ITS simplest form this analysis means plotting the inspection results on charts and comparing their distribution pattern with a normal law curve, in which the values are distributed uniformly around the average. The analyst selects upper and lower control limits, such that if any inspection results fall outside the limits, it is assumed that other-than-chance causes are affecting the process. Once in a great while this assumption may be incorrect, but the margin of error is comfortingly small.

Thus from the charts we can see whether variations in product derive from chance causes inherent in the process or from other-than-chance causes. If all are chance variations, but some are large enough to trouble us or our customers, we have only two choices of action. We can live with the condition so long as we can sell the product; or we can change the process, weighing the penalties of not changing it against the often considerable costs of improving it. If, as is more likely, many of the troublesome variations are due to other-than-chance causes, then a series of exploratory actions should be undertaken.

Control points, involving some kind of informative inspection and record keeping, can be set up at almost every step of production from raw materials to finished product, to find out at what stage things are going wrong. Since even the simplest product may undergo many stages, the cost of analyzing the product at every stage might well be prohibitive. Hence engineering judgment must step in to select the most promising points of control; *i.e.*, those which will result in the largest saving of waste. After experiments on the usefulness of various control points, the most effective are retained as routine inspection stations.

We shall consider here three important applications of statistical quality control: (1) to give warning of abnormal behavior in a process, (2) to diagnose the underlying causes of a wasteful process, and (3) to establish economical inspection plans.

AS AN EXAMPLE of the first, let me cite a process in the manufacture of a product used in the Bell Telephone System. The product is the electric relays, made in the millions by the Western Electric Company, whose function is to make, break or transfer electrical circuits in telephone offices. These relays have a number of contact springs, which are pulled together or separated as the relay is electrically energized or released. The contact points, often of semiprecious metal, are electrically welded to the ends of the relay springs and must withstand millions of shocks and slight rubbings as the relay releases and operates over a normal lifetime. If one of these points broke off, it might interrupt telephone service and mean a major job of replacement or repair. Specifications have been established defining the strength of a satisfactory weld in terms of the minimum pull required to break a contact point loose from the spring. The problem is, how can we be sure that the process is consistently producing welds that meet this specification?

At each welding machine we establish a control station which will let us know immediately when the machine or any other element in the process falters. The operations at a control station are as follows. A sample of 10 units is selected from each hour's production. The welded contacts are torn from the springs, and the amount of force required in each instance is recorded. The results are plotted on a chart to provide a continuing record of two things: (1) the average strength of each 10-unit sample,

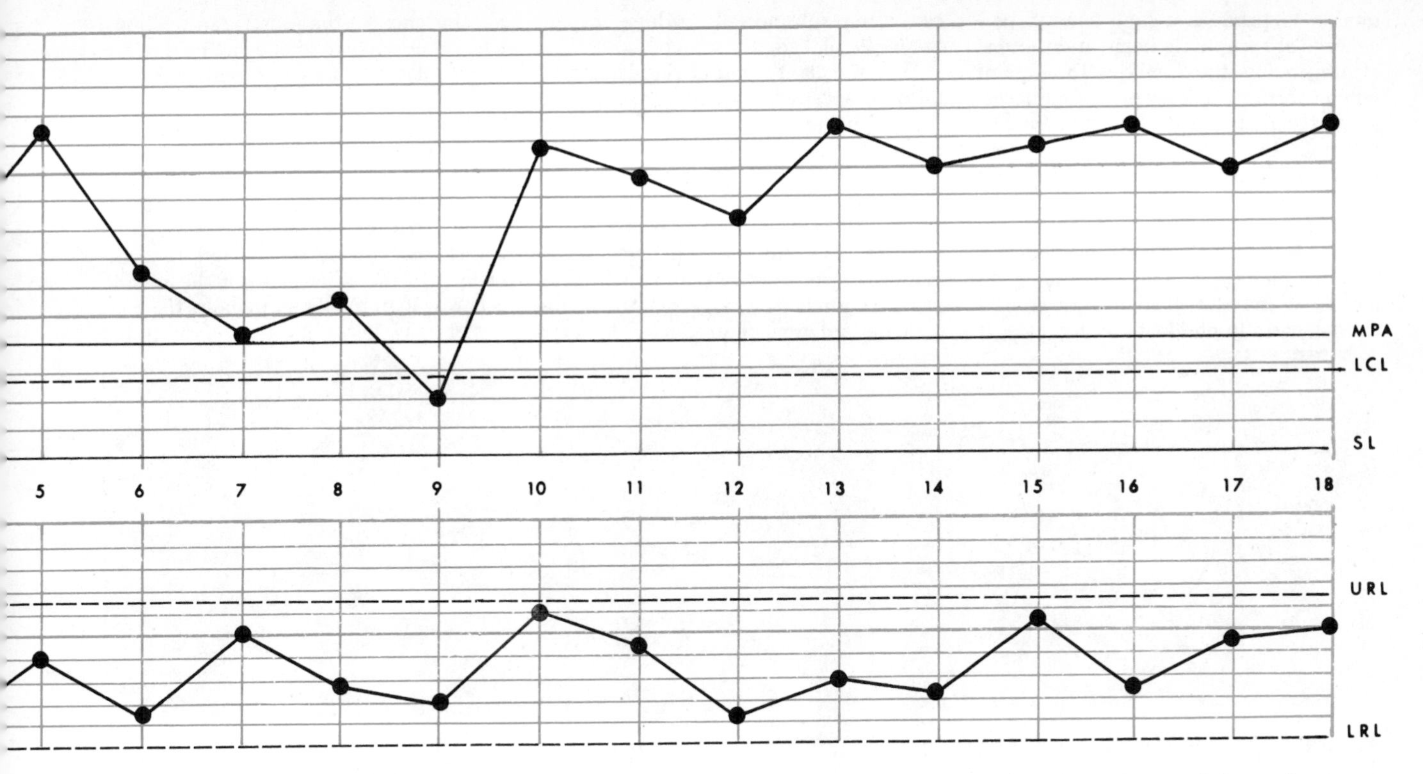

tear each contact from its spring. The curve at the top indicates how the average strength of the sample welds varied. The letters MPA at the far right stand for the minimum permissible average for the process; LCL, for the lower control limit; SL for the specification minimum limit for a single weld. At the ninth hour the average strength of the sample welds fell below the lower control limit; corrective action was then taken. The bottom curve shows range of the samples. URL stands for upper range control limit; LRL for lower range limit.

and (2) the range from the strongest to the weakest unit. Statistical control limits also are put on the chart. The control limits are so calculated that if any average value falls outside them, it indicates that the margin between the average strength and the minimum required is too narrow. The control limits for range are so calculated that when units begin to fall outside these limits, this also indicates that other-than-chance causes have intervened. Only when both the average and the range stay within their respective control limits is there adequate assurance that the welds made during the period in which the sample was taken are satisfactory. Whenever an average or range value of a sample falls outside the control limits, we know that we must stop the welding machine and look for the causes.

At this stage of manufacture the spring plus welded contact has some cash value, but the value is very small in comparison with the cost of the completed relay. Therefore, the cost of maintaining a system of control at each welding machine represents a very small insurance payment against the much greater penalties that would be paid if defective parts were not detected and went on into the finished product. Moreover, so long as the samples are statistically satisfactory, we need not interrupt operation of the process or reset the welding machines, which means a saving in time, labor and the useful life of the machines.

THIS EXAMPLE shows how statistical quality control can provide reliable warning signals on a production line. The following example will illustrate the second application: how statistical methods are used to diagnose and eliminate the causes of abnormally low production yields in a simple product—in this case carbon inserts for a tiny protector block. These blocks are used to protect home telephones from lightning discharge or an accidental power contact with the telephone lines. It was found that in the manufacture of these inserts the basic cause of rejections was irregularity in their dimensions. Several operations are involved, including molding the carbon, two separate firings, cementing in a porcelain cup and gauging for size. The major losses occurred during inspection after the second firing, in which about 60 per cent of the inserts were rejected for shrinkage and softness. A more informative inspection procedure was instituted, and statistical analyses of the results showed that the trouble was in the average value rather than in the range of values after this step in the process. To raise the average, changes were made in processing the carbon powder and in the molding dies. The result was a harder insert, with a more favorable average length and no harmful increase in the spread between the highest and lowest values. The changes reduced losses after the second firing from about 60 per cent to .2 per cent.

For the statistical analyses six experimental control points had been established. Four of these were kept, to give early warnings of any temporary hitch in the process. The total time spent in investigating and statistically analyzing the problem was less than three months. The results: a reduction of waste from 75 to 15 per cent, an increase in yield from 5,000 satisfactory units a day to 25,000, and reduction of the labor force by three operators and five inspectors.

Experience indicates that these spectacular results from a relatively inexpensive study are by no means exceptional. Almost limitless opportunities exist throughout industry for securing impressive benefits from the use of statistical quality control in diagnosing harmful irregularities.

A THIRD area for the use of statistical quality control is in planning economical inspection procedures. Inspection *per se* adds nothing to a product and is often regarded as a necessary evil. In many instances it gets this bad name from being used so extensively as a sorting operation to separate bad units from good. Occasionally this may be economical, but not often. It is economical only when a manufacturing plant is unable to produce a high enough percentage of acceptable items, and the cost of improving the facilities to a point where sorting operations are unnecessary exceeds the cost of sorting. In sorting operations as much or even more inspection time is usually expended on good items as on bad, giving rise to the age-old question, "Why don't we confine inspections to the bad units and stop wasting money looking at the good ones?" While statistical quality control cannot entirely eliminate such waste, it can often substantially reduce inspection costs on "good product."

When a plant's output regularly contains an acceptably small percentage of defects, inspection can safely be limited to those operations necessary to verify maintenance of this performance. Sampling inspections of this kind are essentially informative. They can be made very sensitive to any change in the manufacturing process and will thus provide timely warnings of any imminent degradation of product. They are also far cheaper than sorting inspections, since they involve looking at or measuring a relatively small percentage of the output.

It is important to recognize, however, that any sampling plan involves the risk that the sample may not be truly representative. If the quality of the sample is below that of the lot from which it comes, the lot may be rejected unnecessarily. On the other hand, if the sample is better than the lot as a whole, the consumer may get a product of marginal

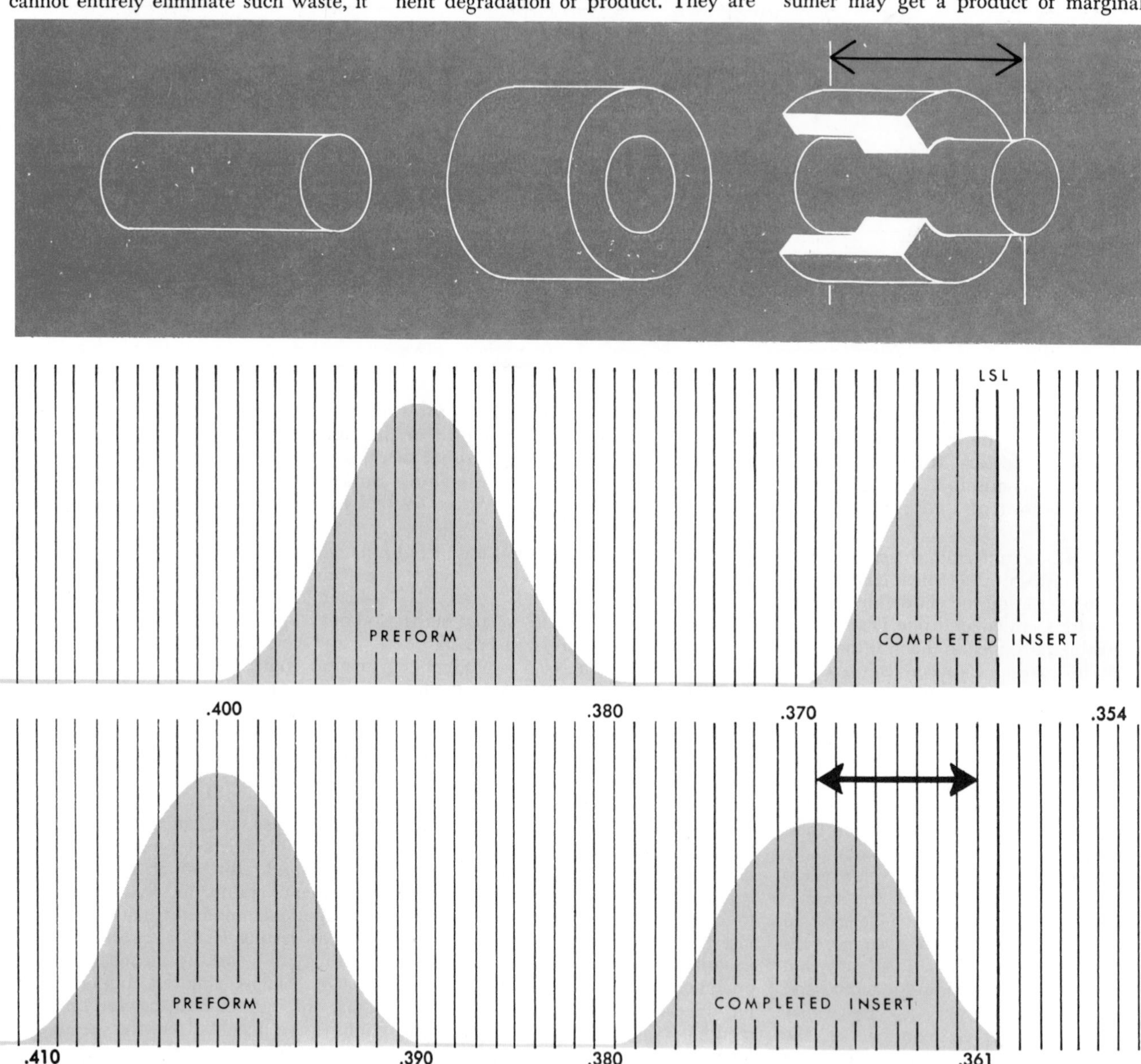

DISTRIBUTION CURVES were used to study the production of carbon inserts for protector blocks, used to protect telephones from lightning. At the upper right is a cutaway drawing of the insert; its length is indicated by the black arrow above it. The insert is made with a carbon rod (*top left*) which is given two firing operations and is then cemented into a porcelain cup (*top center*). The over-all length of the insert is given in fractions of an inch by the horizontal coordinate of the chart. The two curves in the top chart show the distribution of the lengths of sample inserts taken from one month's production. The curve labeled "preform" gives the distribution of the lengths before the inserts were fired; the curve labeled "completed insert" gives the distribution after firing. The letters LSL at the top of the chart stand for lower specification limit. It was found that a large number of completed inserts were rejected because their lengths were below this limit (*light blue area on curve at upper right of chart*). By altering the manufacturing process so that the distribution of the preform inserts was moved to the left, as shown at the bottom, the distribution of the completed inserts was also shifted to a more favorable position and the rejects were sharply curtailed. The black arrow above the curve at lower right shows the distance the distribution pattern of the completed inserts was shifted.

quality. Sampling inspections can be used to protect the "lot" quality or the "average" quality. In either case random samples are drawn from specific lots, and a lot is accepted when the number of defects in the sample is no greater than an allowable number. That number is based on a calculated risk chosen as tolerable. If the allowable number of defects is exceeded, each unit in the lot is then inspected individually.

Sometimes a system of double sampling is used. Instead of accepting or rejecting a lot on the basis of a single sample, the inspector first takes a smaller sample and applies a stricter test: *i.e.*, it must have fewer defective items. If this standard is met, the lot is accepted, and the inspection effort is minimal. If the first sample does not meet this strict test, a second sample is taken, usually of substantially larger size, and for both samples combined the allowable number of defective items is larger. Double sampling, properly applied, usually results in a net saving in the number of units that have to be inspected.

Double sampling in turn may be extended to multiple or sequential sampling, using a larger series of sampling trials. However, in many instances the greater complexity, the chances for misapplication and the extra bookkeeping tend to cancel the theoretical economies of sequential sampling.

In any of these sampling procedures the relationship between lot sizes and sample sizes is not constant. From a small lot one needs a relatively larger percentage sample than from a large one. When a process consistently produces well above the quality requirement, it is usually advantageous to sample large lots, because the chances of a lot's being rejected are small. When the process is less consistent or dependable it may be better to test small lots.

When sampling plans are properly applied, the economies in inspection time are not the only benefits. Sampling inspections also provide a continuing historical record of the capabilities of a process or a machine, which may be helpful to engineers planning the production of similar products. Such records can help consumers as well, for when a producer can furnish evidence that the quality of a product is consistently controlled at the specified level, his customers can reduce or eliminate their own inspections of incoming products.

STATISTICAL quality control is invaluable for giving warning of impending trouble, and its warnings must be heeded. However drastic it may seem to have to curtail or stop production and search for the remedy, it must be remembered that the cheapest thing to do about an unsatisfactory product is not to make it.

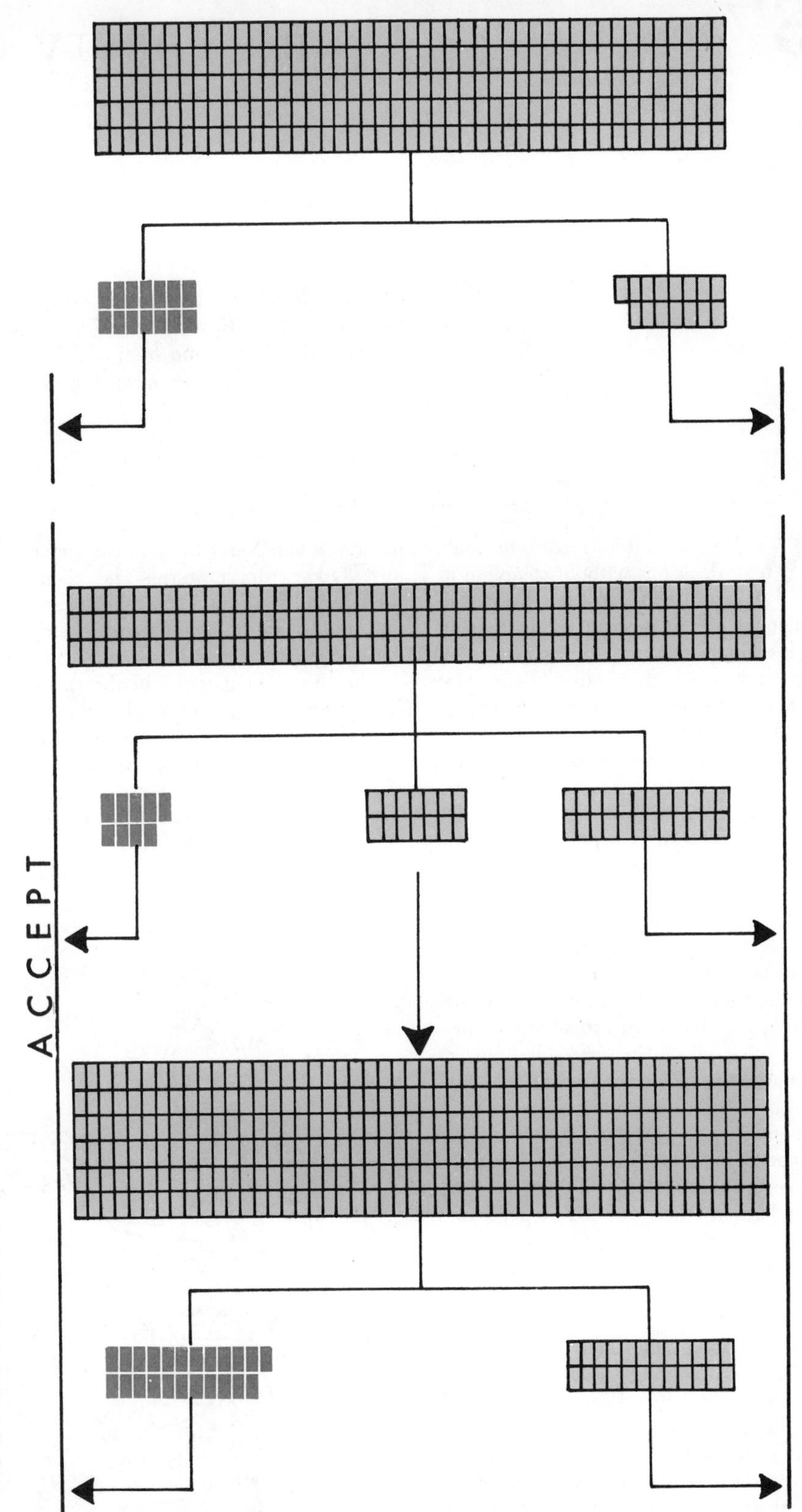

SINGLE AND DOUBLE SAMPLING are illustrated by this chart. In single sampling (*top*) a sample of 225 items might be taken from a much larger lot. If 14 or fewer were defective, the lot would be accepted; if 15 or more were defective, it would be rejected. In double sampling (*bottom*) the initial sample would be smaller, perhaps 150 items. If 9 or fewer were defective, the lot would be accepted; if 24 or more were defective, it would be rejected. If, however, from 10 to 23 were defective, a second sample, in this case one twice as large as the first, would be taken. Then if 23 or fewer defective units were found in the first and second samples combined, the lot would be accepted; if 24 or more were found, it would be rejected.

37

Game Theory and Decisions

by Leonid Hurwicz
February 1955

In which Smith plays a game with Jones and Columbus plays a game with nature to illustrate how this comparatively new mathematical tool can be used to grapple with problems involving uncertainties

We are often forced to make decisions without complete information as to the consequences of the possible alternative actions. Such is the case, for instance, when an individual must decide in May whether to take his vacation in July or in August, when a nation must decide on the size of its defense program though uncertain about other nations' intentions, when a scientist must decide on a plan for an experiment. Uncertainty is present in many decision problems, big and little, routine and unusual.

Some problems involving uncertainty can be treated scientifically by means of the mathematics of probability. The modern sciences of genetics and physics are largely based on probability theory. But what of the innumerable kinds of situations in which the probabilities cannot be computed? Think, for instance, of Columbus' problem when his crew demanded that he turn back. Could he have evaluated the probability of finding land to the west before food and water gave out?

Within the last few years mathematicians have begun to develop a systematic theory of "rational" decision-making in problems involving such uncertainties. Like the probability theory, originally developed in the 17th century from studies of simple games of chance (*e.g.*, dice), the new theory has grown out of studies of a "laboratory model"—in this case certain simple games of strategy against a thinking opponent (*e.g.*, chess and poker).

John von Neumann constructed the theory of games in the 1920s (earlier the mathematician Emile Borel had also had some ideas on the subject), but the subject did not achieve prominence until the publication in 1944 of the now classic *Theory of Games and Economic Behavior* by von Neumann and the economist Oskar Morgenstern. The theory then "caught on," and there has been a multitude of studies and papers developing it in a great many directions.

The theory of games and the theory of decision-making met on the territory of statistical inference. It had occurred to Abraham Wald, one of the founders of modern statistics, that statistical inference could be thought of as a game played against nature by the statistician attempting to uncover its secrets. Wald's principle of "minimizing the maximum risk," indeed, turned out to be equivalent to a principle of choosing a strategy in a game.

Game theory is so complex and heavily mathematical that it cannot be presented in a comprehensive fashion in one article. But many of us are not so much interested in the details of the theory as in its underlying logic, and of that one can get a rough idea from some simplified examples.

Among games of strategy it is convenient to distinguish between games of pure chance and what we shall call games with strategic uncertainty. In a game of pure chance (*e.g.*, dice) wheth-

Could Columbus have used the theory of games to

er a player wins or loses, and how much, depends only on his own choices and on luck. In a game with strategic uncertainty (*e.g.*, poker) he must think about an additional factor: What will the other fellow do? Our main interest is in games involving strategic uncertainty, but we shall find them easier to understand if we first devote some attention to how one might apply general principles of "rational" conduct to games of pure chance.

Suppose that I am invited to place a bet on the outcome of a simultaneous throw of two dice: I will be paid $10 if two aces (single dots) show, otherwise I shall have to pay $1. Should I accept the bet? To answer, we start by doing a little computing. On the average a double ace will appear once in 36 throws. Hence I can expect that in 36 throws I shall win $10 once and lose $1 35 times. The "mathematical expectation" would be a loss of $25, about 69 cents per throw. If all I cared about was the mathematical odds, I would obviously refuse to bet on such terms, since my expectation when not playing is zero—which is better than minus 69 cents! In fact, if I cared only about the mathematical expectation, I would insist that if I am to pay $1 whenever I lose, I ought to be paid at least $35 when the two aces come up; for only then would I be, in terms of my expectation, no worse off than if I refrained from betting.

But we know that people do make bets on a roulette wheel or in a lottery where their expectation is negative, *i.e.*, where, on the average, they must expect to lose. Of course, one could say that this only shows how irrational they are. Yet simple examples will show that a reasonable person will sometimes refuse a bet with a positive expectation and accept one with a negative expectation.

Imagine, for instance, a rich man who has walked far from his house, is tired and plans to take a bus home. The bus fare is 20 cents and it so happens he has only 20 cents in his pocket. At this point someone offers him the following bet: A coin will be tossed; if heads come up, he will be paid $1, if tails come up, he will have to pay 20 cents. In other words, he is offered five to one on what should be an even money bet. Yet we can be pretty sure that the rich man would not be lured into the game, for winning a dollar would mean very little to him, but having to walk home would be a darned nuisance.

Thus the amount of money one can expect to win or lose per throw is not all that matters. What does matter is the amount of satisfaction (or discomfort) associated with the possible outcome of a gamble. If one is willing to measure satisfaction in numerical units, there is a way to explain the rich man's decision in mathematical terms. Suppose that walking home would mean to him a loss of five units of satisfaction while winning a dollar would mean a gain of only three units of satisfaction. In units of satisfaction rather than in dollars his expectation on each toss of the coin would be negative.

On the other hand, the expectation in terms of satisfaction units may be positive when that in terms of dollars is negative. Imagine that it costs $2 to buy a ticket in a lottery where there is one chance in a million of winning a million dollars. Since one would have to bet $2 a million times in order to win a million dollars once, on the average, the expectation here is minus one million dollars, or minus $1 per drawing. But to a person with drab prospects in life the gain of one million dollars might mean, say, 10 million units of satisfaction as against only four units being lost when $2 is paid out. For such an individual the outcome in a million drawings, *in satisfaction units,* would be 10 million minus four times one million, which amounts to an expectation of gain of one and a half units per drawing.

Is it meaningful to speak of satisfac-

decide whether it was really worth while to sail on?

Rabelais's Judge Bridlegoose couldn't see the dice

tion units? Isn't satisfaction an inner psychological phenomenon that defies numerical measurement? It turns out that such measurement is possible if one is willing to postulate that the individual will always try to make his decision so as to maximize the expectation. Of course we have to construct a satisfaction scale, but, as in measuring temperature, we are free to select the zero point and the unit arbitrarily. Suppose, for instance, that I locate the zero of my scale at my present money holdings and decide that a $10 gain would mean one positive satisfaction unit. Imagine, further, that I am offered $10 for a correct call on the toss of a coin at various odds and that I am unwilling to bet $8, eager to bet $4 and more or less indifferent as to betting $7 against the $10. Assuming that my behavior is consistent with choosing the course of action leading to highest expectations, it must be that to me a loss of $8 means losing more than one unit of satisfaction, a loss of $4 means losing less than one unit of satisfaction and a loss of $7 is just about equivalent to one unit of satisfaction. Thus my satisfaction scale can be constructed by experimental methods.

In what follows the numbers in our examples can be interpreted as units of satisfaction. But readers who feel some reluctance to indulge in satisfaction measurement may prefer to think of the units as dollars.

The idea of computing expectations in terms of satisfaction units dates back at least to Daniel Bernoulli, who in the first half of the 18th century formulated a concept which he called the "moral expectation." Now the computation, with the new approach via maximizing expectations, has been put on a rigorous theoretical basis by the recent work of von Neumann and Morgenstern, Jacob Marschak, Milton Friedman, L. J. Savage and others, while Frederick Mosteller and others have done some interesting experiments.

Let us proceed to games possessing strategic uncertainty. If you knew the chances of the other fellow's playing one way or another in a poker game, you could determine the best strategy simply by computing expectations as in a game of chance. But in most social games peeking is frowned upon. It is precisely this lack of knowledge as to the opponent's probable strategy that gives poker its additional element of uncertainty and makes it so exciting.

In order to get a better picture of the problem, we shall consider an artificially simple game. Jones plays against Smith. Jones is to choose one of the three letters A, B or C; Smith, one of the four Roman numerals I, II, III or IV. Each writes his choice on a slip of paper and then the choices are compared. A payment is made according to the upper table on the opposite page. The figure zero means that neither pays; a positive number means that Smith pays that amount to Jones; a negative number, that Jones pays Smith. Thus if Jones chooses A and Smith chooses II, for example, Smith pays Jones $100.

Let us put ourselves in Jones's shoes and see how he might make his choice. If he peeked and knew what Smith had chosen, the answer would be simple; for instance, if he knew Smith had selected II, he would choose A, because C would get him only $2 and if he chose B he would have to pay Smith $1,000. Suppose that Jones happens to know only that Smith has eliminated III and IV and the chances are even as between I and II. If he played A, his expectation would then be minus 50 (dividing minus 200 plus 100 by 2); if he played B, it would be minus 500 (0 minus 1,000 divided by 2); if he played C, the expectation would be 1½ (1 plus 2 divided by 2). Thus in terms of the expectation C is the best choice.

But ordinarily Jones will have no such information. Nonetheless there are principles which can guide his play; we shall present a few of them. The first is "the principle of insufficient reason," associated with the names of the mathematicians Thomas Bayes and Pierre-Simon de Laplace. This principle would require that Jones behave as if Smith were equally likely to make any of his four choices. He would compute his expectations on that basis, and would find that if he chose A his expectation would be 49.5, for B it would be 0 and for C it would be 2.5. Thus A would be the best choice.

If Jones is an optimist, he might make his choice on the basis of another principle we shall call "visualize the best." In that case he would choose B, because it offers the opportunity for the largest pay-off ($1,000).

On the other hand, Jones may be a conservative man, even a pessimist. It would then be natural for him to follow the "visualize the worst" principle, named by mathematicians "minimax," because it amounts to minimizing the maximum possible loss—the principle suggested, as we have seen, by Wald.

Pascal applied mathematics to gambling

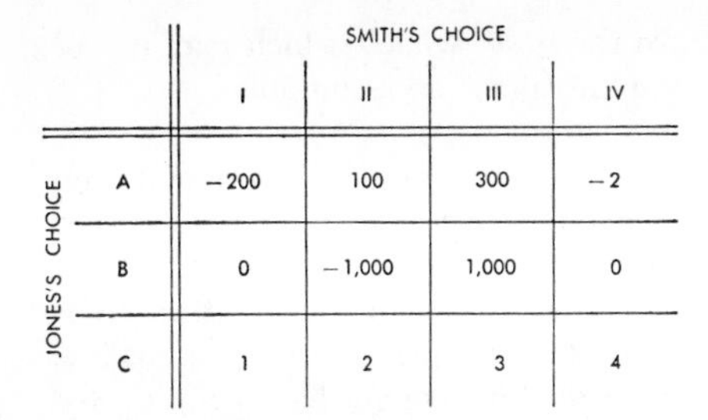

JONES'S CHOICE	SMITH'S CHOICE I	II	III	IV
A	−200	100	300	−2
B	0	−1,000	1,000	0
C	1	2	3	4

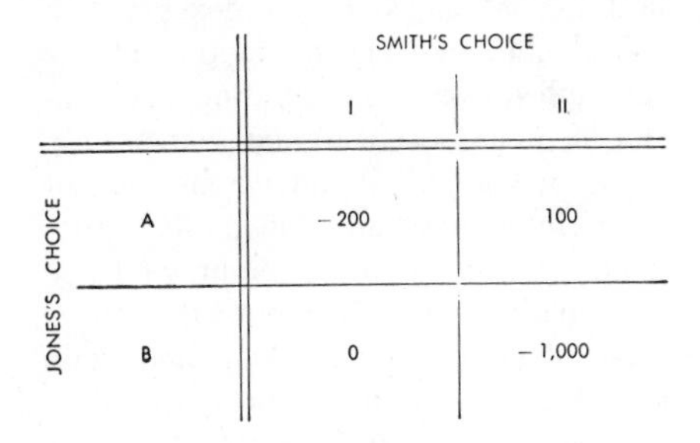

JONES'S CHOICE	SMITH'S CHOICE I	II
A	−200	100
B	0	−1,000

Smith v. *Jones*

Jones would then choose C, for while it affords no possibility of a large gain, its "worst" is a gain of 1.

Similar computations on Smith's behalf would show that the principle of insufficient reason and the "visualize the best" principle lead to the choice of II, while "visualize the worst" favors I. We should note that under no principle would it make sense for Smith to choose IV, because I is superior to IV if Jones chooses A or C and just as good as IV if Jones's choice is B. In the jargon of the decision theory, IV is "inadmissible." Similar comparison shows that III also is inadmissible. Thus the principle of insufficient reason, postulating that all four of Smith's choices are equally likely, is actually ruled out for Jones; he knows that Smith will never play III or IV.

Suppose that Smith knows Jones to be of the "visualize the best" school. He can collect $1,000 from Jones by playing II, anticipating that Jones will play B according to the optimistic principle. On the other hand, if Jones gets wind of this reasoning by Smith, he may switch to A and win $100. Thus a stable pattern of behavior is not likely to be established.

But things are strikingly different when both players visualize the worst, so that Jones plays C and Smith plays I. In this case it makes no difference whether the two players know each other's strategy; they can still do no better than play C and I, respectively. In other words, the "visualize the worst" principle apparently is spyproof—if either player had hired a spy to find out the other's strategy, he would have wasted his money.

Now it is easy to construct a game in which this principle seemingly is not spyproof. For instance, suppose we give each player only two choices—the first two choices of the preceding game, with the same pay-off schedule [*see lower table at left*]. In the new game if both players visualize the worst Jones will choose A and Smith I. But now if Jones knows that Smith is operating on this principle, he will switch to B, because he would lose $200 by playing A and break even by playing B. Certainly Smith has good reason to guard against espionage.

So it seems that the "visualize the worst" policy is not always spyproof after all. But at this point one of the most ingenious ideas of the theory of games enters the stage. The idea is to let chance play a role in the choice of strategy, that is, to use a randomized or "mixed" strategy.

Suppose that Jones marks A on 10 slips of paper and B on three slips, then mixes them up very thoroughly and proceeds to draw blindly to determine his play. What is his expectation? On the average he will play A 10 times and B three times in 13 games. If Smith were to play I all the time, Jones would lose 200 units 10 times and break even three times, thus losing 2,000. If Smith were to play II all the time, in 13 games Jones would, on the average, gain 100 units 10 times and lose 1,000 units three times; the total net loss again would be 2,000. Were Smith to alternate between I and II, whether according to a system or at random, Jones's expectation would still be minus 2,000 for 13 games. Thus his randomized strategy would yield the same result no matter what Smith did—and the result would be better than the worst he could expect (a loss of 200 per game) if he played A all the time, which, on the "visualize the worst" principle, is the best of the "pure" (nonrandomized) strategies.

This example shows that a mixed strategy may be better than the best pure strategy. It does not, of course, imply that any strategy using random choices has this property. The fact that the slips were marked A and B in the ratio 10 to 3 was of crucial importance. Had there been five As and five Bs to draw from, for instance, the outcome would have been inferior to playing "pure" A. It can be shown by algebraic computation that the 10-to-3 ratio yields the optimal strategy for Jones.

Let us now recall that what started us on the investigation of the mixed strategies was the fact that Smith's best "pure" strategy, namely I, was not spyproof. With mixed strategies in the picture, has the situation changed? To answer the question we must first find Smith's optimal strategy, which turns out, like Jones's, to be of the mixed variety; in his case he must play I and II in the ratio 11 to 2. On the assumption that Jones plays A, this mixture gives Smith the expectation of a gain of 2,000 units in 13 games (11 times 200 plus 2 times minus 100). And his expectation is exactly the same if he assumes that Jones will play B; Smith then wins 1,000 twice and breaks even 11 times for a total gain of 2,000 in 13 games. Indeed, it would make no difference if Jones were to alternate, in any manner whatsoever, between A and B. Thus Smith's strategy is spyproof in the sense that it would not help Jones to know that Smith was playing I and II in the ratio 11 to 2; Jones could still do no better than play 10 As to three Bs.

The preceding example illustrates a general phenomenon discovered and proved by von Neumann: in "zero-sum" two-person games (*i.e.*, in games where the amount lost by one player equals the amount gained by the other) the "visualize the worst" principle is spyproof provided mixed strategies are not disregarded.

Let us go back to Columbus and see whether the theory of games would have helped him in his dilemma, or at least how it might have formulated the problem for him. We start by setting up in table form Columbus' two possible choices (to turn back or keep going), the uncertain factual alternatives (that land was near or not near) and the probable consequences of Columbus' decisions in either case [*see top table, p. 224*]. Now as an experimental approach suppose we assign very hypothetical and preliminary values in satisfaction units to the various consequences [*middle table, p. 224*]. That is to say, let us assume that Columbus, attempting to envisage how disappointed he would feel if he later learned that he had turned back on the verge of discovering land, appraises this disappointment as a loss of 50 satisfaction units; that he values the saving of life by turning back from a hopeless quest as a gain of 20 satisfaction units, and so on. Let us also make one further assumption: that Columbus feels he can make some kind of estimate as to the probability of land being near.

If he supposed that the chances of

COLUMBUS' DECISION	ACTUAL LOCATION OF LAND: LAND NEAR	ACTUAL LOCATION OF LAND: NO LAND NEAR
TURN BACK	PROBABLE LATER DISAPPOINTMENT	LIFE SAVED
KEEP GOING	PROSPECT OF GLORY	PROSPECT OF DEATH

COLUMBUS' DECISION	ACTUAL LOCATION OF LAND: LAND NEAR	ACTUAL LOCATION OF LAND: NO LAND NEAR
TURN BACK	−50	20
KEEP GOING	100	−1,000

COLUMBUS' DECISION	ACTUAL LOCATION OF LAND: LAND NEAR	ACTUAL LOCATION OF LAND: NO LAND NEAR
TURN BACK	−1,000	20
KEEP GOING	500	−500

Columbus* v. *nature

land being near were 3 to 1, he would compute the expectation of "satisfaction" (actually dissatisfaction!) from turning back as follows: 3 times minus 50 added to 1 times 20 and the sum divided by 4—*i.e.*, minus 32.5. In other words, if he turns back, the net expectation is a loss of 32.5 satisfaction units. On the other hand, if he keeps going, the expectation is a loss of 175 satisfaction units (3 times 100 added to 1 times minus 1,000 and the sum divided by 4). Since the expectation of loss in going on is so much greater than that in turning back, Columbus' decision would be: better turn back. On the basis of the satisfaction values we have postulated, it would have taken a probability of 9 to 1 that land was near to induce Columbus to keep going.

Would he actually have insisted on such high odds in favor of success? If not, it must be that the satisfaction units we have assigned to the various possible consequences are unrealistic; perhaps we have overvalued Columbus' fear of death and undervalued his eagerness for the prize of discovery. We may therefore construct another table of values which might be considered more realistic [*see see the lowest table at the left*]. On this new basis a probability of 3 to 1 that land was near would have been sufficient to make Columbus decide to keep going.

But what if he had no idea as to the chances of land being near? The theory of games and decision-making would still have offered him several means of calculating his expectations. He might have followed the principle of insufficient reason, the strategy of "visualize the best" or the strategy of "visualize the worst." On the basis of the satisfaction figures in our last table Columbus would have found it worth while to keep going no matter which of these principles he applied. But on the basis of the first figures [*middle table*] he would have turned back unless he belonged to the "visualize the best" school—which may not be too unrealistic an assumption.

It may seem strange that principles for making decisions should be served cafeteria style—take your choice. Is there not some way of proving that only one of these principles is truly rational? A great deal of thought has been devoted to this problem, mainly via attempts to find logical flaws or paradoxes which would eliminate one or another of the principles from consideration. For instance, it has been argued that nature, being presumably nonmalicious and not out to inflict maximum loss on its "opponents" (investigators), might well use an "inadmissible" strategy though a smart player would not. Also, some argue that there is no need for spyproofing against nature, and this raises doubts as to whether a principle leading to the use of randomized strategies is reasonable. In defense of the rationality of randomized decision-making, one is tempted to recall Rabelais's Judge Bridlegoose, who decided lawsuits by the throw of dice and was known for his wisdom and fairness until his failing eyesight made him commit errors in reading the spots. (Less facetious arguments in favor of randomized decision-making also are available!)

The development of methods for rational decision-making where uncertainties exist certainly has a long way to go. The field is still rife with differences of opinion. Nevertheless, it is highly instructive to study the tools we have, and particularly to notice how often the various principles, despite the difference of their underlying assumptions, all lead to very similar if not identical conclusions as to the best decision to take in a given situation.

The Theory of Games

by Oskar Morgenstern
May 1949

From it is being forged a new tool for the analysis of social and economic behavior. The new approach already has shown its superiority to classical economic theory

THE analogy between games of strategy and economic and social behavior is so obvious that it finds wide expression in the thinking and even the language of business and politics. Phrases such as "a political deal" and "playing the stock market" are familiar reflections of this. The connection between games and these other activities is more than superficial. When they are examined by the methods of modern mathematics, it becomes evident that many of the forms of economic and social behavior are strictly identical with—not merely analogous to—games of strategy. Thus the mathematical study of games offers the possibility of new insights and precision in the study of economics.

The theory of probability arose from a study of lowly games of chance and from the desire of professional gamblers to find ways of taking advantage of the odds. Far more difficult problems are presented by games of strategy such as poker, bridge and chess. In these games, where the outcome no longer depends on chance alone but also on the acts of other players and on their expectations of one's own present and future acts, a player must choose among relatively complex strategies. Mathematically, these problems remained not only unsolved, but even untouched.

Gottfried Wilhelm Leibnitz, the German philosopher and mathematician, seems to have recognized that a study of games of strategy would form the basis of a theory of society. On the other hand, many efforts along quite different lines were made by philosophers and economists to provide a theory for "rational behavior" for individuals, business corporations, or even for entire communities.

Such a theory must be quantitative, which means that it must ultimately assume a mathematical character. A theory of games fulfilling these requirements would take into account that participants in a game vary in information and intelligence, that they have various expectations about the other players' behavior, and that different paths of reaching their goal may be open to them. The theory must also allow for the fact that the position of a player (or, equivalently, of an economic individual or a firm) is often adversely affected if his opponent finds out his intentions. The player has to take steps to protect himself against this contingency, and the theory must indicate how he should proceed most efficiently—and what his countermeasures would mean to the other players.

Why should such a theory be of interest to the sociologist and, in particular, to the economist? Does not the economics of today have an adequate model in mechanics, with its notions of forces, of equilibrium and stability? Physics is, indeed, at the bottom of current efforts to provide a statement of rational economic behavior, whether it is mathematically formulated or not. But many important situations that arise at all levels in economics find no counterpart whatever in physics.

A typical example is the fixing of wage rates between workers and employers when both groups have found it to their advantage to combine into unions and associations. Current economics cannot tell us, except in a general manner, under what circumstances such combinations will arise, who will profit, and by how much. The two groups have opposing interests, but do not have separate means to pursue their contrary aims. They must finally come to some agreement, which may turn out to be more advantageous to one side than to the other. In settling their differences they will feint, bluff, use persuasion; they will try to discover each other's strategies and prevent discovery of their own. Under such circumstances a theory of rational behavior will have to tell a participant how much a given effort will be worth in view of the obstacles encountered, the obstacles being the behavior of his opponents and the influence of the chance factor.

Monopoly and monopolistic market forms—that is, trading among only a few individuals or firms on one side of the market at least—are characteristic of all social economies. They involve serious feuds and fights, a very different picture from the general, "free" competition with which classical economic theory usually deals. On the orthodox theory, the individual is supposed to face prices and other conditions that are fixed, and is supposed to be in a position to control all the variables, so that his profit or utility depends only on his own actions. Actually, however, when there are only a few individuals, or many individuals organized into a few combinations, the outcome never depends on the actions of the individual alone. No single person has control of all the variables, but only of a few.

The case of an individual acting in strict isolation can be described mathematically as a simple maximum problem—that is, finding the behavior formula that will yield the maximum value or return. The cases involving combinations are of an entirely different mathematical and logical structure. Indeed, they present a peculiar mixture of maximum problems, creating a profound mathematical question for which there is no parallel in physical science or even in classical mathematics.

Yet this is the level at which the problem of economic behavior needs to be attacked. Clearly it is far more realistic to investigate from the outset the nature of the all-pervading struggles and fights in economic and social life, rather than to deal with an essentially artificial, atomistic, "free" competition where men are supposed to act like automatons confronted by rigidly given conditions.

THE theory of games defines the solution of each game of strategy as the distribution or distributions of payments to be made by every player as a function of all other individuals' behavior. The solution thus has to tell each player, striving for his maximum advantage, how to behave in all conceivable circumstances, allowing for all and any behavior of all the other players. Obviously this concept of a solution is very comprehensive, and finding such a solution for each type of game, as well as computing it numerically for each particular instance, poses enormous mathematical difficulties. The theory makes important use of mathematical logic, as well as combinatorics (the study of possible ways of combining and ordering objects) and set theory (the techniques for dealing with any collection of objects which have one or more exactly specified properties in common). This domain of modern mathematics is one of exceptional rigor. But it is believed that great mathematical discoveries are required to make a break-through into the field of social phenomena.

A single individual, playing alone, faces the simplest maximum problem; his best strategy is the one that brings him the predetermined maximum gain. Consider a two-person game: Each player wishes to win a maximum, but he can do this only at the expense of the other. This situation results in a zero-sum game, since the sum of one player's gains and the other's losses (a negative number) is zero. One player has to design a strategy that will assure him of the maximum advantage. But the same is true of the other, who naturally wishes to minimize the first player's gain, thereby maximizing his own. This clear-cut opposition of interest introduces an entirely new concept, the so-called "minimax" problem.

SOME games have an optimal "pure" strategy. In other words, there is a sequence of moves such that the player using it will have the safest strategy possible, whatever his opponent does. His position will not deteriorate even if his strategy is found out. In such "strictly determined" games, every move—and hence every position resulting from a series of moves— is out in the open. Both players have complete information. The mathematical expression of this condition is that the function describing the outcome of a game has a "saddle point." This mathematical term is based on an analogy with the shape of a saddle, which can be regarded as the intersection of two curves at a single point. One curve in a saddle is the one in which the rider sits; the other is the one that fits over the horse's back and slopes down over its sides. The seat of the saddle represents the "maximum" curve, and its low point is the "maximin." The curve that straddles the horse's back is the "minimum" curve, and its high point is the "minimax." The point at which the two curves meet at the center of the saddle is the "saddle point." In the theory of games, the somewhat more special saddle point is the intersection of two particular strategies.

The mathematical values of the strategies involved in a hypothetical game of this kind are represented in the diagram on this page. This shows a simple game between two players, A and B, each of whom has available three possible strategies. There are nine possible combinations of moves by A and B. The numbers in the boxes represent A's gains or losses for all combined strategies and, since this is a zero-sum game, their negatives represent B's losses or gains. A's minimax strategy is A-2, because if he follows that sequence of moves, he is sure to win at least two units no matter what B does. Similarly, B's minimax strategy is B-1, because then he cannot possibly lose more than two units whatever A's plan of action. If a spy informed A that B was planning to use B-1, A could make no profit from that information. The point where the A-2 row intersects the B-1 column is the saddle point for this game.

It may seem that B has no business playing such a game, since he must lose two units even with his best strategy, and any other strategy exposes him to even heavier loss. At best he can win only a single unit, and then only if A makes a mistake. Yet all strictly determined games are of this nature. A simple example is ticktacktoe. In perfectly played ticktacktoe every game would result in a tie. A more complex example is chess, which has a saddle point and a pure strategy. Chess is exciting because the number of possible moves and positions is so great that the finding of that strategy is beyond the powers of even the best calculating machines.

A \ B	B-1	B-2	B-3
A-1	2	1	4
A-2	2	3	2
A-3	2	-1	1

GAME OF STRATEGY between two players, each with three possible strategies, has nine possible results. Numbers in boxes represent A's gains or losses for each combination of plays by both players.

Other two-person, zero-sum games, however, have no single best possible strategy. This group includes games ranging from matching pennies to bridge and poker—and most military situations. These games, in which it would be disastrous if a player's strategy were discovered by his opponent, are not strictly determined. The player's principal concern is to protect his strategy from discovery. Do safe and good strategies exist for "not strictly determined" games, so that their choice would make the games again strictly determined? Can a player in such a game find strategies other than "pure" strategies which would make his behavior completely "rational"? Mathematically speaking, does a saddle point always exist?

It does, and the proof was originally established in 1927 by the mathematician John von Neumann, the originator of the theory of games, now at the Institute for Advanced Study in Princeton. He used various basic tools of modern mathematics, including the so-called fixed-point theorem of the Dutch mathematician L. E. J. Brouwer. Von Neumann proved, by a complex but rigorous application of this theorem to the theory of games, that there is a single "stable" or rational course of action that represents the best strategy or saddle point even in not strictly determined games.

This principle can also be demonstrated in practical terms. Observation shows that in games where the discovery of a player's plan of action would have dangerous consequences, he can protect himself by avoiding the consistent use of a pure strategy and choosing it with a certain probability only. This substitution of a statistical strategy makes discovery by the opponent impossible. Since the player's chief aim must be to prevent any leakage of information from himself to the other player, the best way to accomplish this is not to have the information oneself. Thus, instead of choosing a precise course of action, the various possible alternatives are considered with different probabilities.

It is in the nature of probability that individual events cannot be predicted, so that the strategy actually used will remain a secret up to the decisive moment, even to the player himself, and necessarily to his opponent as well. This type of indecision is a well-known empirical fact. Wherever there is an advantage in not having one's intentions found out—obviously a very common occurrence—people will be evasive, try to create uncertainty in the minds of others, produce doubts, and at the same time try to pierce the veil of secrecy thrown over their opponents' operations.

The example *par excellence* is poker. In a much simpler form, this type of behavior is illustrated in the game of matching pennies. Here the best strategy is to show heads or tails at random, taking care only to play each half the time. Since the same strategy is available to the opponent, both players will break even if they play long enough and both know this principle. The calculation of the best strategy grows in difficulty as the number of possible moves increases: *e.g.*, in the Italian game called morra, in which each player shows one, two or three fingers and simultaneously calls out his guess as to the sum of fingers shown by himself and his opponent, a player has nine possible strategies. His safest course is to guess a total of four fingers every time, and to vary his own moves so that out of every 12 games he shows one finger five times, two fingers four times and three fingers three times. If he plays according to this mixture of strategies, he will at least break even, no matter what his opponent does.

LET us apply these principles to a simple economic problem. Suppose that two manufacturers are competing for a given consumer market, and that each is considering three different sales strategies. The matrix on this page specifies the possible values of the respective strategies to manufacturer A: This situation does not have a single best strategy. If A chooses strategy A-1, B can limit his profit to one unit by using strategy B-2 or B-3; if A chooses strategy A-2 or A-3, B can deprive him of any profit by choosing strategy B-1. Thus each manufacturer stands to lose if he concentrates on a single sales technique and his rival discovers his plan. Analysis shows that A will lose unless he uses a combination of A-1, A-2 and A-3, each a third of the time. On the other hand, if manufacturer B fails to employ his best mixed strategy—B-1 a ninth of the time, B-2 two ninths of the time, and B-3 two thirds of the time—his competitor will gain. These mixed strategies are the safest strategies. They should be used whenever each manufacturer does not know what the other will do.

An example which illustrates in statistical terms many of the conflicts of choices involved in everyday life is the famous story of Sherlock Holmes' pursuit by his archenemy, Professor Moriarty, in Conan Doyle's story, "The Final Problem." Holmes has planned to take a train from London to Dover and thence make his escape to the Continent. Just as the Dover train is pulling out of Victoria Station, Moriarty rushes on the platform and the two men see each other. Moriarty is left at the station. He charters a special train to continue the chase. The detective is faced with the problem of outguessing his pursuer. Should he get off at Canterbury—the only intermediate stop—or go all the way to Dover? And what should Moriarty do? In effect, this situation can be treated as a rather unusual version of matching pennies—a "match" occurring if the two men decide to go to the same place and meet there. It is assumed that such a meeting would mean the death of Sherlock Holmes; therefore it has an arbitrarily assigned value of 100 to Moriarty. If Holmes goes to Dover and makes his way to the Continent, it is obviously a defeat for the professor, but—also obviously—not as great a defeat as death would be for the detective. Hence, a value of minus 50 to Moriarty is given to this eventuality. Finally, if Holmes leaves the train at Canterbury and Moriarty goes on to Dover, the chase is not over and the temporary outcome can be considered a draw. According to the theory of games, the odds are 60 to 40 in favor of the professor.

In the story, of course, this game is played only once: Sherlock Holmes, deducing that Moriarty will go to Dover, gets off at Canterbury and watches triumphantly as the professor's pursuing train speeds past the intermediate station. If the game were continued, however, Holmes' look of triumph would hardly be justified. On the assumption that Moriarty persisted in the chase, calculations indicate that the great detective was actually as good as 40 per cent dead when his train left Victoria Station!

The theory of games has already been applied to a number of practical problems. Situations similar to that of Holmes are being analyzed in that branch of operational research which deals with military tactics, the possible courses of action being various dispositions of troops or combinations of measures and countermeasures. The handling of the more complex situations that exist in economics is expected to require the aid of calculating machines. For example, two competing automobile manufacturers may each have a large number of strategies involving the choice of various body designs, the addition of new accessories, the best times to announce new models

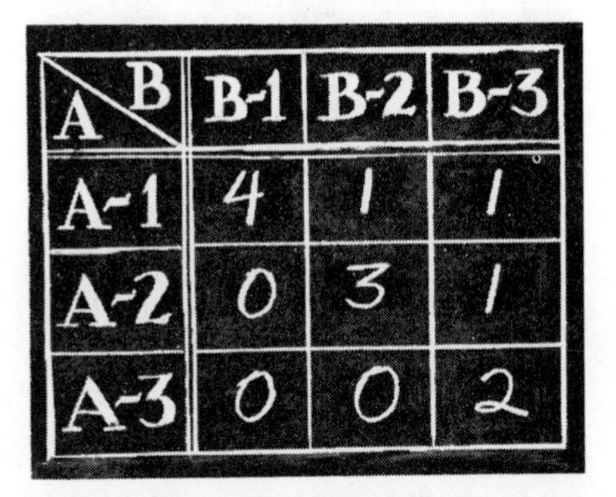

A \ B	B-1	B-2	B-3
A-1	4	1	1
A-2	0	3	1
A-3	0	0	2

BUSINESS RIVALRY between two firms with three strategies each again diagrams A's possible gains. No single strategy is best if the opponent discovers it; hence the rivals must use a mixture of all three.

and price changes, and so on. It has been estimated that the calculations for a game in which one manufacturer had 100 possible strategies and his competitor had 200 (a not uncommon situation) would take about a year on an electronic computer.

If we now make the transition to games involving three or more persons, a fundamentally new phenomenon emerges—namely, the tendency among some players to combine against others, or equivalently in markets to form trade unions, cartels and trusts. Such coalitions will be successful only if they offer the individual members more than they could get acting separately. Games where that is the case are called essential. Coalitions will then oppose each other in the manner of individual players in a two-person game. A coalition will have a value for the players who form it, and they may therefore require payments or "compensations" from newcomers who want to enter the coalition and share in its proceeds. As a rule a great deal of bargaining will precede the determination of the system of distribution of gains or profits among the members of the coalition.

Basically, the formation of a coalition expresses the fundamental tendency toward monopoly, which is thus found to be deeply characteristic of social and economic life. Indeed, Adam Smith already had noted the tendency of businessmen to "conspire" against the common welfare, as he stated it, by getting together into groups for better exploitation. Important chapters of American economic history deal with the efforts of government to break conspiracies of various kinds in order to limit the power of trusts and other amalgamations. When these are broken—if at all—they tend to arise again, so a continuous watchfulness is necessary.

The powerful forces working toward monopoly ought therefore to be at the very center of economic studies. They should replace the preoccupation with a nonexistent pure or free competition where nobody has any perceptible influence on anything, and where all data are assumed to be immutably given. Since this is the imaginary setup from which current economic theory starts, it encounters insuperable difficulties when it enters the realm of monopolistic competition. It is not surprising, therefore, that classical economics has failed to yield a general theory that embraces all economic situations.

THE approach to the coalition problem in the theory of games can be shown by a three-person situation in which it is assumed that a player can achieve a gain in any given play only if he joins with one other player. The gains and losses that would result for the individual players in the case of each possible coalition are shown in the diagram on page 303. Thus if A and B form a coalition, each gains a half unit and C loses one unit. What keeps the players in the game is that they all stand a chance of profit; each player's problem is to succeed in forming a coalition with one of the other two on any given deal. This simplified situation illustrates in essence much of the conflict that occurs in modern economic life.

Now the important characteristic of this type of game is that there is no single "best" solution for any individual player. A, for example, can gain as much by forming a coalition with C as with B. Therefore all three of the possible distributions of payments, taken together, must be viewed as the solution of this three-person game.

There are, of course, many other distribution schemes that might be considered by the players. For example, one of the partners in a coalition could make a deal with the third player whereby both improved their positions (the third player reducing his losses) at the expense of the other partner. What is to prevent the participants in the game from considering all these other possibilities?

The question can be answered by introducing the concept of "domination." In mathematical terminology the various possible schemes for distribution of payments are called "imputations." One imputation is said to dominate another if it is clearly more advantageous to all the players in a given coalition. It is found, as shown in the three-person game described above, that the imputations belonging to a solution do not dominate each other: in this case all three imputations have an equal chance of being chosen; none is most advantageous to the players in each coalition. While it is extremely difficult to prove mathematically that such a solution would exist for every game with arbitrarily many players, the principle can be expected to hold true.

Now it is also found that while the imputations belonging to the solution do not dominate each other, individually they are not free from domination by imputations outside the solution. In other words, there are always outside schemes from which some of the players could profit. But any and every imputation outside the solution is dominated by one belonging to the solution, so that it will be rejected as too risky. It will be considered unsafe not to conform to the accepted standard of behavior, and only one of the imputations which are part of the solution will materialize.

These examples give an idea of the great complexity of social and economic organization. In this realm "stability" is far more involved than it is in the physical sciences, where a solution is usually given by a number or a set of numbers. In essential games, in economics and in warfare, there is instead a set of alternatives, none of which is clearly better than another or all others. One imputation in a set is not more stable than any other, because every one may be threatened by one outside the solution. But each has a certain stability because it is protected by other potential imputations in the solution against upsets from outside. Collectively they eliminate the danger of revolutions. The balance is most delicate, however, and it becomes more sensitive as the number of players increases. These higher-order games may have many solutions instead of a single one, and while there is no conflict within an individual solution, the various solutions or standards of behavior may well conflict with one another.

This multiplicity of solutions may be interpreted as a mathematical formulation of the undisputed fact that on the same physical background of economic and social culture utterly different types of society can be established. Within each society, in turn, there is possible considerable variation in the distribution of income, privileges and other advantages—which corresponds to the multiplicity of imputations or distribution schemes in a single solution in a game.

The theory also yields insight into even more delicate social phenomena. Although it assumes that every player has full information, discrimination may exist: two players may make a third player "tabu," assigning him a fixed payment and excluding him from all negotiations and coalitions. Yet this arrangement need not lead to complete exploitation of the third player. In practical economic life, for example, cartels do not annihilate all outside firms, although it would not be a technically difficult operation. Rather, in deference to socially accepted standards of behavior they allow certain outsiders a share in the industry, so as not to attract undue attention—and to be able to point out to the government and the public that "competition" exists in the particular industry.

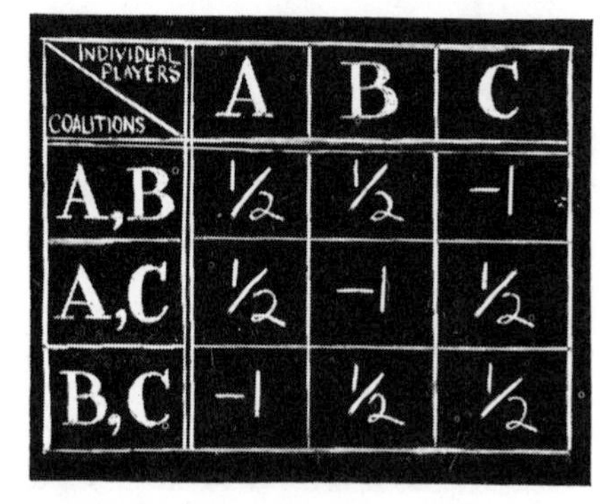

INDIVIDUAL PLAYERS / COALITIONS	A	B	C
A,B	½	½	−1
A,C	½	−1	½
B,C	−1	½	½

COALITION GAME with three players produces still another matrix. Here gains or losses to players resulting from various possible coalitions are shown in vertical columns. Player must form partnership to win.

It is surprising and extremely significant that, although the theory of games was developed without any specific consideration of such situations, the fact that they exist was derived from general theorems by purely mathematical methods. Furthermore, the theory shows—again purely mathematically—that certain privileges, even if anchored in the rules of a game (or of a society), cannot always be maintained by the privileged if they come into conflict with the accepted standard of behavior. A privileged person or group may have to give up his entire "bonus" in order to survive economically.

These and many other implications can be derived from the study of simple three-person games. Games of more than three players provide further interesting insights—but at the price of great and, in many cases, still insuperable mathematical difficulties. The almost unimaginable complexity involved may be illustrated by poker, the game which, above all others, furnishes a model for economic and social situations. The subtleties of poker and the countless number of available strategies—*e.g.*, the technique of purposely being caught bluffing now and then so that future bluffs may be successful—prevent the thorough analysis that would be necessary to throw light on corresponding problems in practical everyday affairs. The matrix of possible strategies for poker is so large that it has not even been calculated, much less drawn. Consider a radically simplified version of the game which assumes a deck of only three cards, a one-card, no-draw hand, only two players, three bids between them (the first player gets two, the second one), and no overbetting. Even this watered-down version of poker involves a matrix of 1,728 boxes, and computing a single best possible strategy for each player to an accuracy of about 10 per cent might require almost two billion multiplications and additions.

BUT even with its present limitations the theory of games has made it possible to analyze problems beyond the scope of previous economic theory. Besides those already indicated, the problems now being explored include the application of the mathematics for a game involving seven persons to the best location of plants in a particular industry, the relation between labor unions and management, the nature of monopoly.

The initial problem in the theory of games was to give precision to the notion of "rational behavior." Qualitative or philosophical arguments have led nowhere; the new quantitative approach may point in the right direction. Applications are still limited, but the approach is in the scientific tradition of proceeding step by step, instead of attempting to include all phenomena in a great general solution. We all hope eventually to discover truly scientific theories that will tell us exactly how to stabilize employment, increase national income and distribute it adequately. But we must first obtain precision and mastery in a limited field, and then proceed to increasingly greater problems. There are no short cuts in economics.

Linear Programming

by William W. Cooper and Abraham Charnes
August 1954

Like the theory of games, it is a method of pure mathematics that can be applied to human affairs. It is used to calculate the best possible solution to a problem that involves a number of variables

Imagine that you are manufacturing a product at a number of factories and must freight it to markets in many different parts of the country. How would you go about calculating the pattern of shipments that would deliver the goods from your many warehouses to the many markets at the lowest possible freight cost?

By common sense and trial and error you might readily work out a reasonable schedule. But even a non-mathematician can see that to find the best solution among the infinite number of possible solutions would be a far more formidable problem.

We shall describe in this article a recently developed technique in applied mathematics which makes it possible to solve such problems in a relatively short time by means of simple computations. The theory of linear programming was developed by John von Neumann, G. B. Dantzig, T. C. Koopmans and a few other mathematicians, statisticians and economists. It was first applied as an operating tool by Marshall Wood and his staff in the Air Force's Project SCOOP (Scientific Computation of Optimum Programs). One of its applications was in the Berlin air lift. As a result of work by the Air Force group and others in linear programming and related developments, such as the theory of games, statistical decision theory and input-output analysis, truly scientific methods of analysis are now being applied to many problems in business and logistics which used to be considered beyond the scope of such analyses. In this article we shall confine ourselves to linear programming and explain the principle with a sample problem.

Linear programming derived its name from the fact that the typical problems with which it deals are stated mathematically in the form of linear equations. (Actually "linear" is too narrow a name for the technique, for it may be applied to nonlinear problems as well.) In essence it is a method for considering a number of variables simultaneously and calculating the best possible solution of a given problem within the stated limitations. Any manufacturer will at once appreciate that this is a precise statement of his own problem. In deciding what particular items to manufacture, and in what quantities, he must take into account a great complex of factors: the capacities of his machines, the cost and salability of the various items, and so on. To make matters worse, each subdivision of his problem has its own complexities; for instance, he may have to choose among several possible processes for making a particular item. And all the factors and decisions may interlock and react upon one another in unexpected ways. In the circumstances, the best that any management can hope to achieve is a reasonably workable compromise. With linear programming, however, it becomes possible to locate definitely the optimum solutions among all the available ones, both in the realm of over-all policy and in departmental detail.

To illustrate the method let us take a highly simplified hypothetical case. We have a factory that can make two products, which for simplicity's sake we shall name "widgets" and "gadgets." The factory has three machines—one "bounder" and two "rounders." The same machines can be used to make widgets or gadgets. Each product must first be roughed out on the bounder and then rounded on one of the rounders. There are two possible processes for making each product: We can use the bounder and rounder No. 1 or the bounder and rounder No. 2 for either a widget or a gadget. Let us name the respective processes for the widget One and Two, and for the gadget Three and Four. The key variables are the times involved. To make a widget by Process One requires .002 of an hour on the bounder and .003 of an hour on rounder 1; by Process Two, .002 of an hour on the bounder and .004 of an hour on rounder 2. A gadget by Process Three takes .005 of an hour on the bounder and .008 of an hour on rounder 1; by Process Four, .005 of an hour on the bounder and .010 of an hour on rounder 2. Finally, we know that the capacities of the machines for the period we are considering (say six months) are 1,000 hours of operation on the bounder, 600 hours on rounder 1 and 800 hours on rounder 2.

All this information is summarized at the top of the opposite page. A production superintendent might call this a flow chart; we can think of it as a model which specifies conditions, or constraints, that will govern any production decision we must make. Now it is readily apparent that we can translate these facts into an algebraic model. If we let x_1 represent the unknown number of widgets to be made by Process One, x_2 the number of widgets by Process Two, and x_3 and x_4 the numbers of gadgets to be made by Processes Three and Four, we can write all the information in an algebraic table [*middle of following page*]. The inequality sign before the numbers representing hours of capacity on the machines is the well-known symbol meaning "no more than." What this table means is simply that we can make no more widgets and/or gadgets than the capacities of the respective machines will allow. But with the conditions stated in this form, we are now in

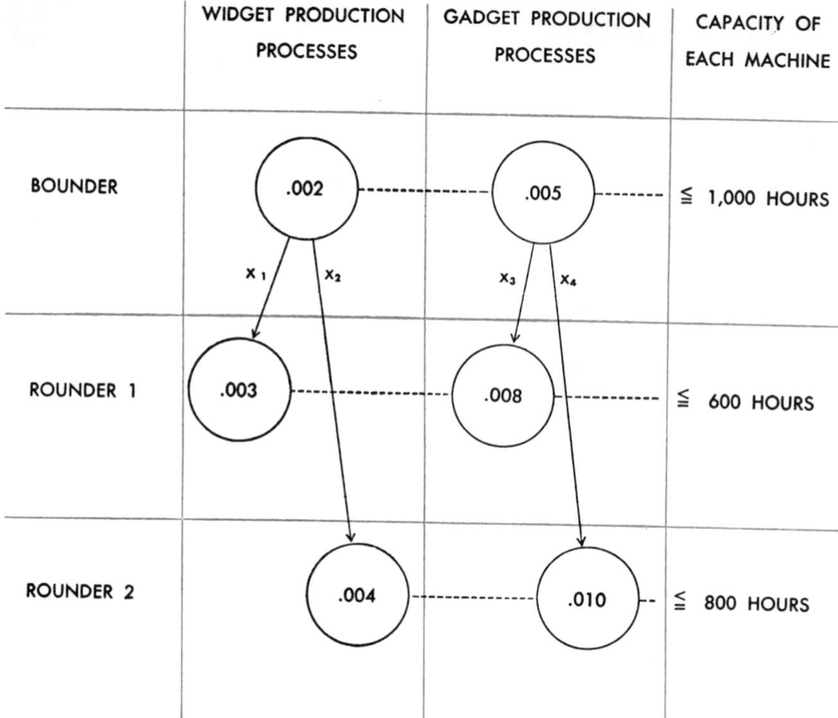

GRAPHIC MODEL depicts the restrictions on production of "widgets" and "gadgets" by three machines: one "bounder" and two "rounders." The numbers in the circles indicate the fraction of an hour required for each machine to perform its function on each part.

BOUNDER	$.002x_1 + .002x_2 + .005x_3 + .005x_4$	$\leqq$ 1,000 HOURS
ROUNDER 1	$.003x_1 + .008x_3$	$\leqq$ 600 HOURS
ROUNDER 2	$.004x_2 + .010x_4$	$\leqq$ 800 HOURS

ALGEBRAIC MODEL of the same conditions represents as x_1, x_2, x_3 and x_4 the numbers of items produced by the processes indicated by arrows in the graphic model. The blank spaces in each column can be disregarded because they represent zero in each case.

BOUNDER	$.002x_1 + .002x_2 + .002x'_1 + .005x_3 + .005x_4 + .005x'_3$	$\leqq$ 1,000 HOURS
ROUNDER 1	$.003x_1 + .008x_3$	$\leqq$ 600 HOURS
ROUNDER 2	$.004x_2 + .010x_4$	$\leqq$ 800 HOURS
ROUNDER 1 (OVERTIME)	$.003x'_1 + .008x'_3$	$\leqq$ 200 HOURS
CONTRACT	$x_1 + x_2 + x'_1$	$\geqq$ 450,000 WIDGETS

COMPLETED MODEL is based on the assumption of additional capacity for one of the machines. This is overtime on rounder 1. Now two new processes are introduced: x'_1 and x'_3. The sign before 450,000 widgets means at least this number must be produced.

a position to consider the variables simultaneously and to calculate solutions which will satisfy the constraints. A solution will be called a linear program; linearity here refers to the fact that the available capacity on each machine is used up *in proportion to* the number of items run through it.

It is important to note here that the unknowns x_1, x_2, x_3 or x_4 may be zero but none of them can be a negative number. Of course it is obvious that we cannot produce a minus number of products. But in mathematics the exclusion of negative values must be carefully noted. In fact, the successful development of the theory of linear programming required extensive study of the effects that this restriction would have on traditional methods of solving and analyzing equations.

Having stated the constraints, we can proceed to find the best production schedule attainable within these limitations. What we mean by "best" will of course depend on what criterion we choose to apply. We might decide to seek the schedule that would produce the largest possible number of items, or the one that would use the greatest possible amount of the machines' available running time. But ordinarily the objective would be the greatest possible profit. Let us assume that the profit on each widget produced by Process One is 85 cents, on each widget by Process Two 70 cents, on each gadget by Process Three $1.60, and on each gadget by Process Four $1.30. We then get this equation: Total Profit $= .85x_1 + .70x_2 + 1.60x_3 + 1.30x_4$.

From this information we could calculate the number of each item we should produce to realize the largest possible total profit within the machines' capacity. (Be it noted that gadgets, though yielding a larger per unit profit, should not necessarily pre-empt the machines, for widgets take less time to produce.)

The problem as so far outlined, however, is much too simple to represent an actual situation. To come closer to a real problem we should at least introduce a sales factor. Let us suppose, therefore, that our factory has orders for 450,000 widgets. Elementary arithmetic will show that our present machine capacity cannot turn them out within the time limit. We have enough capacity on the bounder for 500,000 widgets (1,000 hours divided by .002 of an hour per widget) but our two rounders combined could finish no more than 400,000. We must obtain more rounder capacity, and this may be done by authorizing overtime on one or both of the rounders. Suppose, then, we arrange for 200 hours of overtime on the faster of our two rounders (.003 of an hour per widget). Because of higher labor costs for overtime, the profit per widget will be reduced from the usual 85 cents to 60 cents during the overtime period, and if we should use any of the overtime capacity for making gadgets, the unit profit on them will drop from $1.60 to $1.40.

The constraints governing this expansion of the problem are summarized in the algebraic table above (the new symbols x'_1 and x'_3 represent the number of widgets and of gadgets, respectively, to be turned out on rounder 1 during the overtime period). Given the unit profit figures, we are now prepared to calculate the most profitable possible employment of the machines.

The answer can be computed by one of several methods. The most general, devised by Dantzig, is called the simplex method. Certain special methods which are more efficient for the kind of problem being considered here were developed by the authors of this article. These systems, though too involved and lengthy to be explained in detail here, require only simple arithmetical operations which can easily be carried out by clerks or commercially available computers. The simplex method starts from zero use of the machines' capacity, and the computation proceeds by a series of specified steps, each of which advances closer to the ultimate answer.

The answer in this case is that we should produce no gadgets but should make 466,667 widgets, finishing 200,000 on rounder 1 at straight time, another 200,000 on rounder 2 at straight time and the remaining 66,667 at overtime on rounder 1. The total profit will be $350,000. Our system of calculation tells us that it is not possible to devise a production schedule which will yield a larger total profit within our restrictions.

It is possible that there may be other programs which would provide as much (but no more) profit. If there are, the analyst can quickly find them. He can also determine the second best or third best program, and so on. Thus the method is not only powerful but also flexible; it can offer management a range of choices based on different considerations. Furthermore, linear programming methods can be extended to analyze the effects of any change in the restrictions—an improvement in efficiency, an increase or reduction in cost, an increase in capacity. These methods employ a "dual theorem," whereby a maximizing problem (such as the maximization of profits in the case we have been considering) is viewed as the reverse of a related minimizing problem. In this case the minimizing problem is concerned with the worth of the machine capacities. Using the same set of facts and calculations, it is possible to show precisely how much more profitable it is to employ overtime on rounder 1 than on rounder 2. If the overtime on rounder 1 were increased to 300 hours instead of 200 hours, the maximum profit would be $370,000 instead of $350,000. But an increase of overtime from 300 to 301 hours would be worthless, for at that point the possible rounder output would exceed the capacity of the bounder.

In short, linear programming may be applied not only to finding the best program within given restrictions but also to assessing the advisability of changing the restrictions themselves.

The hypothetical problem outlined in this article was simplified to illustrate some of the basic elements of the technique. In any real-life problem the factors at play are both more numerous and more difficult to identify. It is as important to locate the truly pertinent factors as it is to construct the correct mathematical model for dealing with them. The application of linear programming is full of pitfalls. To evaluate the various features of a problem and determine which should be included in the model requires understanding collaboration between the mathematical analyst and the operations people actually working at the job.

The range of problems to which linear programming may be applied is very wide. As we have already indicated, the Air Force has employed it in problems of logistics. In industry the method is solving problems not only in production but even in such matters as devising the most effective salary pattern for executives—a pattern which will not only meet competition for their services from outside but will also avoid inconsistencies within the company.

Through the dual theorem, linear programming has been related to the theory of games; it is thereby enabled to take probabilities as well as known restrictions into account. It also has fundamental, though indirect, connections with statistical decision theory, which the late Abraham Wald related to the theory of games shortly before his recent death in an airplane crash. All three of these disciplines are contributing to one another's progress. Indeed, our own chief interest in linear programming is to develop generalizations which will enlarge the scope of the technique.

It has been highly satisfying to us to see how often research on a particular problem has led to methods of much more general application, sometimes in altogether unexpected fields. For instance, the work we did in adapting linear programming to the problem of the executives' salary schedule opened a new path for studying the field of statistical regression and correlation analysis. Similarly, an investigation of certain problems in economic measurement and management science has paved the way for new approaches in totally unrelated fields of work in engineering and physics, such as plasticity and elasticity.

As research on these new tools of scientific analysis continues, we can expect to find many new uses for them. But what is perhaps most remarkable is the great and continuing revolution that science and technology have wrought in mathematics. As the mathematician and writer Eric Temple Bell has said in his book *The Development of Mathematics*:

"As the sciences . . . became more and more exact, they made constantly increasing demands on mathematical inventiveness, and were mainly responsible for a large part of the enormous expansion of all mathematics since 1637. Again, as industry and invention became increasingly scientific after the industrial revolution of the late 18th and early 19th centuries, they too stimulated mathematical creation. . . . The time curve of mathematical productivity [shoots up] with ever greater rapidity. . . ."

40 Operations Research

by Horace C. Levinson and Arthur A. Brown
March 1951

It is the application of scientific method to broad problems in war, government and business. First developed by the British during World War II, it is now beginning to flower in the U.S.

WORLD WAR II developed a new military use of science, christened operations research (in the U. S.) or operational research (in Britain). Unlike all previous applications of science to warfare, which were concerned almost entirely with the development of weapons, operations research is concerned with the use of weapons. Its province is the tactical and strategical aspects of military operations: it deals with methods of locating enemy submarines, for example, rather than with the technology of the torpedo or the bomb. What it does is to apply scientific method, including mathematical techniques, to the analysis of situations and of the efficiency of various systems of organization for coping with them.

Obviously this approach can be applied to business and governmental organizations just as well as to military ones, and since the war it has been used fruitfully to solve some business and industrial problems. In Britain the Government has made considerable use of it, and workers in the field have formed an Operational Research Club. In the U. S. there is now a Committee on Operations Research, organized by the National Research Council to further the development and use of operations research for nonmilitary purposes. Since 1948 a course on nonmilitary uses of it has been given at the Massachusetts Institute of Technology in collaboration with the Navy. On the military side, the Army, the Navy, the Air Force and the Joint Chiefs of Staff all have operations research organizations.

OPERATIONS research was born in the Battle of Britain. The British Government was exploring every available means to defend the country against the disastrous German bombings. The British had a skillful but small air force and they had radar. Could radar make up for the smallness of the Royal Air Force? How could the radar interception system be used to maximum advantage; how should the antennas be distributed, the signals organized, and so on? The Government called in half a dozen scientists of various disciplines to answer these questions. By collecting the relevant facts and analyzing them with the general methodology of science, these men devised a new operating technique that doubled the effectiveness of the air defense system.

Impressed with this spectacular success, Britain organized similar teams to tackle many other military problems. The U. S. armed forces likewise put operations research groups to work soon after this nation entered the war. The work of these teams in both countries paid high dividends in deciding such questions as the most effective altitudes at which planes should fly in hunting submarines, the best payload division between fuel, instruments and armament, the best search pattern. One short operations research study showed that planes attacking submarines could increase their effectiveness fivefold by changing the depth at which depth charges were set to go off. In the famous Allied anti-submarine campaign in the Bay of Biscay, British and U. S. operations research teams working together designed a patrol system which succeeded in sighting practically every submarine that came in or out of the Bay and sank about a quarter of those it attacked.

Operations research can handle very diverse problems, but to be eligible for solution the problem must satisfy two conditions: 1) it must be expressible in numbers or quantities, and 2) the data must be adaptable to the available techniques. If the data are statistical, for example, they must come from operations that are roughly similar and must be extensive enough to permit some sort of statistical regularity or law to show itself. Let us consider a couple of actual military examples.

Early in the war it was the conventional military view that in an airplane squadron it was most important to have as many planes as possible in fit condition to take to the air at all times. The RAF had set as a standard that no less than 70 per cent of a squadron's planes should be fit to fly; this percentage was called the maintenance efficiency. Since it was very difficult to keep up to this standard, the problem was given for study to Cecil Gordon, a member of an operations research team.

Gordon took a completely fresh approach. He was a biologist, and he decided there was a usable analogy between the life cycles of human beings and of aircraft. Flying, he reasoned, breeds disrepair and the necessity for repair; repair breeds readiness for flight; readiness for flight, given the opportunity, breeds flying, and so the cycle starts again. But in a military squadron it is not readiness for flight that you want: you want flying. An airplane on the ground is only potentially valuable—useless until you need to fly it.

Gordon lived with the squadron for a while, determined the rate at which flying time generated repair time and assimilated every significant feature of the squadron's operations and the life cycle of a plane. As the result of all this he came to a startling conclusion: the old criterion of maintenance efficiency was wrong. What counted was the percentage of demand for flying met by the planes and the amount of flying accomplished per maintenance man-hour. The upshot was that the target percentage of aircraft kept fit for flying was cut from 70 per cent to around 35 per cent, with a large increase in battle time and squadron efficiency.

The other example concerns the problem of the Japanese Kamikazes (suicide planes) that threatened our ships in the Pacific. The question was whether a ship should maneuver violently to spoil the aim of the diving Kamikaze or keep steady to improve the aim of its own defensive anti-aircraft fire. The operations research group that undertook to find the answer had the records of 477 attacks to study. In 172 cases the suicide plane had succeeded in hitting the ship, and in 27 cases the ship had been sunk. The scientists discovered that the effectiveness of the two types of defensive tactics depended on the size of the ship. Violent evasive maneuvers obviously have only a slight effect on the aim of anti-aircraft gunners in the case of a large ship but a very pronounced effect on their aim in a small ship. What is the net result when the effects on the aim of attacking Kamikazes are taken into account? After considerable study the group concluded that a large ship, when attacked by a diving Kamikaze, should change course violently and a small ship slowly. It also found that a ship should present its beam to an attacking plane that came in from a high dive and turn its bow or stern to a plane diving from a low altitude. In presenting its beam to a Kamikaze a ship could concentrate more anti-aircraft fire on it, but in the case of a low-diving plane the effect of the increased fire power was more than offset, the study showed, by the larger target offered by the ship. The operations research team's recommendations later proved their value in battle: ships that followed its suggestions were hit about 29 per cent of the time when attacked and those that did not observe these rules were hit about 47 per cent of the time.

OPERATIONS research makes use of such mathematical and statistical concepts as the variable, the statistical constant, the function and probability. Its military applications also introduced a number of new concepts, examples of which are exchange rate and sweep rate. The exchange rate of course is the ratio of one's own losses to those of the enemy, but when the losses are not in directly comparable units, a complex analysis may be necessary to determine the true value of an operation in the over-all picture.

The notion of sweep rate grew out of the problem of searching for enemy ships or submarines. Any search tool (a broom, a rake, a flashlight beam, a radar beam) essentially explores a field of targets whose exact locations are unknown. Its effectiveness is measured by its sweep width and its sweep rate. It has a certain probability, depending on these factors, of picking up a target at any given point. A new broom, which "sweeps clean," has a sweep width equal to its actual width, and the probability that it will pick up an object within these limits is 1, representing certainty. A rake swept over an area of small leaves has an irregular detection probability curve, and a man hunting a target at night with sweeps of a flashlight may have still another probability pattern.

Suppose that a plane, flying back and forth at a given speed without crossing its own path, is searching an area of ocean for submarines. To compute the sweep width you multiply the number of submarines assumed to be in the area by the distance flown by the plane, and then divide this product into the product of the number of submarines actually detected and the total number of square miles searched. This number, the sweep width, multiplied by the speed of the plane, gives the sweep rate.

This seems a lot of computing, and one is inclined to ask what is accomplished by it. The answer is that sweep rates are sensitive measures of the efficiency of the searching operation. If the sweep rate of a command of airplanes searching an area for submarines drops sharply, it is a signal to the command that something has gone wrong.

These concepts can also be applied to some business problems, for there are close analogies between military and business operations. In fact, a rough dictionary can be constructed that will translate one into the other. For "weapons" read "materials"; for "command" read "management" or "executive"; for "enemy" read "competitor" or, curiously enough, "customer"; for "destroy" read "out-compete" or "acquire"; for "enemy losses" read "own gains," and so on. The notion of exchange rate becomes, in business, the ratio of gains to expenditures. The notion of sweep rate is directly applicable to the search for customers.

THE use of operations research in business is not entirely new. Under such names as "business analysis" or "business research" sporadic activities of this kind have been conducted for a long time by some firms. What the operations research of World War II introduced was the development of systematic techniques and the enlistment of groups of trained scientists in such work.

To illustrate how operations research may be used in business let us take a study that was actually made of the effectiveness of department-store newspaper advertising. The particular type of advertising analyzed was the kind designed to produce immediate sales—what may be called "quick response" advertising. Since department stores spend millions of dollars each year on this "QR" advertising, the scientific measurement of its results is evidently a worth-while project.

The success of such advertising cannot be measured simply by the amount of extra sales of the particular goods advertised. The advertising is intended to attract customers to the store and increase the total sales. It is not effective if it merely increases the sales of one department at the expense of other departments in the store or of future sales of this department itself. For example, suppose a store advertises a coat that has been selling for $30 at a reduced price of $24. This will naturally attract a lot of quick sales. But how many of these sales are to customers who would have bought the coat anyway at its normal price of $30, either at the time or later on, if the advertisement had not been run? In how many cases did this $24 purchase take the place of purchases that would have been made in other departments of the store? How many customers attracted by the coat bargain also made spur-of-the-moment purchases elsewhere in the store? And finally, how much business did this advertisement bring to the store that would otherwise have gone to competing stores?

It is clear that to assess the net returns from this type of advertising the study must cover the store's total sales and all QR advertising over a considerable period of time. Its goal must be to determine the "true plus volume," that is, the extra volume of sales for the store as a whole which would not have been obtained if the advertising had not been run. Such a study was conducted some years ago by the research director of a certain department store. He made a week-by-week comparison of the QR advertising and total sales of this store with those of the competing department stores in the same city. His analysis of these data was then based on the following reasoning: Of the total sales in a given week a certain percentage is due to QR advertising. This fraction changes from week to week, both for the store and for the competing group, due to fluctuations in their respective QR expenditures. Moreover, the fluctuations in advertising by this store do not parallel those for the competing group: in some weeks it does relatively more advertising, in others relatively less. Consequently there should be corresponding fluctuations in the ratio of total sales by the store to total sales by the competing group, reflecting the variations in advertising ratio. The problem was to isolate these fluctuations.

In order to do so the research director

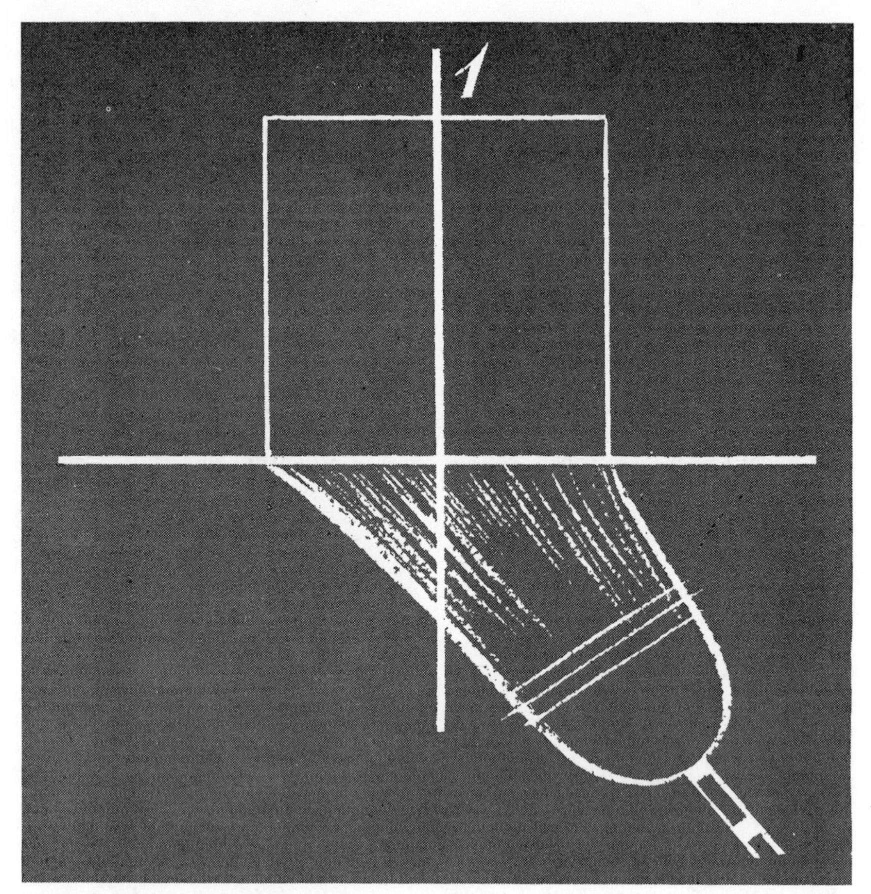

NEW BROOM has an effectiveness measured by its sweep width (*horizontal coordinate*) and its sweep rate (*vertical coordinate*). The probability that the broom will encounter an object in its path is 1, or certainty.

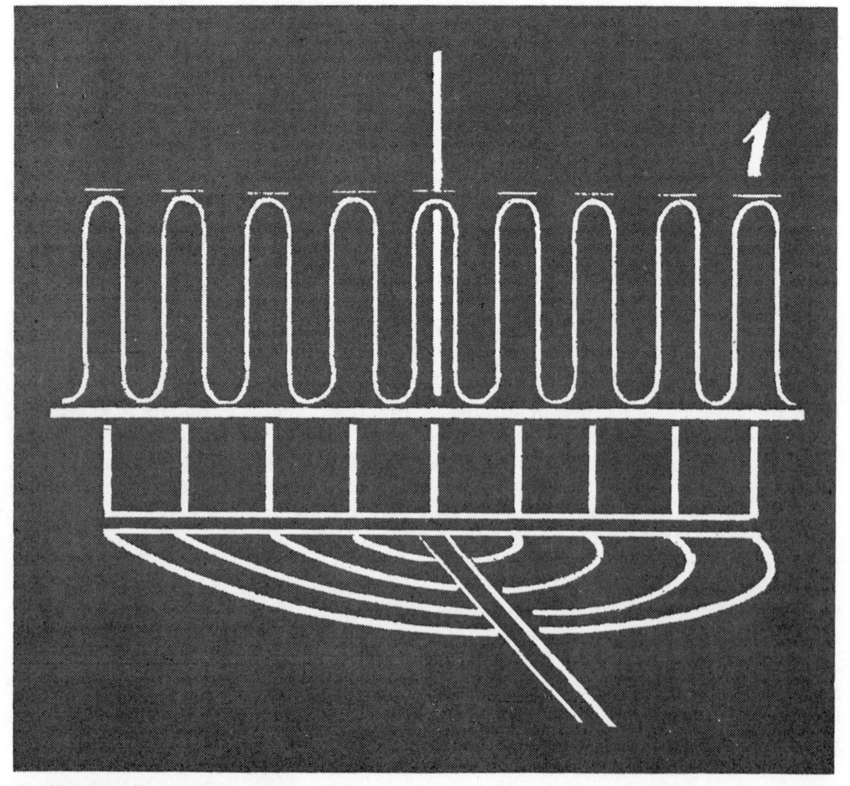

GARDEN RAKE obviously has less effectiveness than a broom. The probability that a tooth will encounter an object in its path is 1, but the probability that an object between the teeth will be encountered is 0.

resorted to mathematics. Let S represent the portion of the store's sales volume that does not include the sales attributable to QR advertising, in other words, the total sales minus the QR sales for a week. Let O represent the corresponding sales of the competing group. It is reasonable to assume that the ratio S/O is a statistical constant over an extended period, since sales due to the variable factor of QR advertising are eliminated and the effects of weather and of economic changes, if not too violent, will be uniform for all the stores. This assumption is the key to the solution, for it leads to a set of equations which, using the known figures on total sales and QR advertising expenditures, can be solved numerically to determine the unknown statistical constant and the pulling power of the advertising.

This study yielded the following conclusions, among others: 1) the QR advertising produced a large true plus volume; 2) the amount of true plus volume depended sensitively on price reductions of the advertised goods; 3) the average pulling power of the advertising remained practically constant over a period of several months, although individual advertisements varied greatly in effectiveness. The analysis also had important by-products not involving advertising: it threw much light on some phases of the store's operations and led to improvements in efficiency.

THE peacetime applications of operations research have included the analysis of such problems as the proper use of equipment and manpower, operating procedures in factories and public utilities, the planning of government projects. One operations research study of the laying of road surface materials in Britain, for example, resulted in an annual saving of a million pounds sterling.

Operations research is already a machine of great power. Like a farm tractor it must be expertly manned. One of the main objectives of the Committee on Operations Research in the U. S. is to create a supply of trained workers in this field. The Committee believes that operations research is particularly important and urgent during the present national emergency. To the extent that nonmilitary operations research is successful in increasing the efficiency of U. S. industry, it will contribute to reducing the critical shortage of manpower. And the more young scientists are trained in operations research, the greater will be the supply available to the armed forces in case of necessity.

SOLUTIONS TO PROBLEMS IN MARTIN GARDNER'S ARTICLES

8. The Remarkable Lore of the Prime Numbers

1. The two composite numbers are 10,001 (the product of primes 73 and 137) and 123,456,789, which is evenly divisible by 3. The other numbers are primes.

2. Two meshed gear wheels of different sizes cannot return to the same position until a certain number of teeth, k, have passed the point of contact on both wheels. The number k is the lowest common multiple of the number of teeth on each wheel. Let n be the number of teeth on the small wheel. We are told that the large wheel has 181 teeth. Since 181 is a prime number, the lowest common multiple of n and 181 is $181n$. Therefore the small wheel will have to make 181 rotations before the two wheels will return to their former position.

3. How can the nine digits be arranged to make three primes with the lowest possible sum? We first try numbers of three digits each. The end digits must be 1, 3, 7 or 9 (this is true of all primes greater than 5). We choose the last three, freeing 1 for a first digit. The lowest possible first digits of each number are 1, 2 and 4, which leaves 5, 6 and 8 for the middle digits. Among the 11 three-digit primes that fit these specifications it is not possible to find three that do not duplicate a digit. We turn next to first digits of 1, 2 and 5. This yields the unique answer

$$\begin{array}{r} 149 \\ 263 \\ 587 \\ \hline 999 \end{array}$$

4. The last number, 333333331, has a factor of 17. (The problem is based on a result obtained by Andrzej Makowski of Poland, which was reported in *Recreational Mathematics Magazine* for February, 1962.)

5. It is easy to find as large an interval as we please of consecutive integers that are not prime. For an interval of a million integers, consider first the number 1,000,001! The exclamation mark means that the number is "factorial 1,000,001," or the product of $1 \times 2 \times 3 \times 4 \cdot \cdot \cdot \times 1{,}000{,}001$. The first number of the interval we seek is $1{,}000{,}001! + 2$. We know that 1,000,001! is divisible by 2 (one of its factors), so that if we add another 2 to it, the resulting integer must also be divisible by 2. The second number of the interval is $1{,}000{,}001! + 3$. Again, because 1,000,001! has a factor of 3, it must be divisible by 3 after we add 3 to it. Similarly for $1{,}000{,}001! + 4$, and so on up to $1{,}000{,}001! + 1{,}000{,}001$. This gives a consecutive sequence of one million composite numbers. Are these the smallest integers that form a sequence of one million nonprimes? No, as Ted L. Powell pointed out in *The Graham Dial* for April, 1960; we can obtain a lower sequence just as easily by *subtracting*: $1{,}000{,}001! - 2$; $1{,}000{,}001! - 3$; and so on to $1{,}000{,}001! - 1{,}000{,}001$.

10. A Short Treatise on the Useless Elegance of Perfect Numbers and Amicable Pairs

The article's first problem was to find a rule, given a perfect number's Euclidean formula, for writing that number in binary form. The formula: $2^{n-1}(2^n - 1)$. The rule: Put down n ones followed by $n - 1$ zeros. Example: Perfect number $2^{5-1}(2^5 - 1) = 496$ has the binary form 111110000.

The rule is easily understood. In binary form 2^n is always 1 followed by n zeros. The expression on the left side of Euclid's formula, 2^{n-1}, therefore has the binary form of 1 followed by $n - 1$ zeros. The parenthetical expression $(2^n - 1)$, or one less than the nth power of 2, has the binary form of n ones. The product of these two binary numbers obviously will be n ones followed by $n - 1$ zeros.

Readers will find it amusing to test the theorem that the sum of the reciprocals of the divisors of any perfect number (including the number itself as a divisor) is 2, by writing the reciprocals in binary form and then adding.

There are several ways to state rules for determining the final digit of a perfect number by inspecting its Euclidean formula, but the following rule seems the simplest. It applies to all perfect numbers except 6. Halve the first exponent and note whether the result is even or odd. (If the exponent has more than two digits, only the last two must be halved to obtain this result.) If even, the perfect number ends in 6; if odd, it ends in 8. Example: The 23rd perfect has the formula $2^{11,212}(2^{11,213} - 1)$. Half of 12 is 6, an even number, and so the 23rd perfect ends in 6.

It is interesting to study the endings of perfect numbers in systems other than the binary and the decimal system. If the base is a multiple of 3, all perfects except 6 have terminal digits of 1. If the base is a multiple of 6, as in the duodecimal (base 12) system, all perfects except 6 end in 4.

11. Diophantine Analysis and the Problem of Fermat's Legendary "Last Theorem"

The Diophantine problems have the following answers.
following answers.

(1) The problem about the farmer and the animals reduces to the Diophantine equation $11x + 5y = 200$. Applying the method of continued fraction, three solutions in positive integers can be found:

Cows	*Pigs*	*Sheep*
5	29	66
10	18	72
15	7	78

(2) L. H. Longley-Cook, in *Fun with Brain Puzzlers* (Fawcett, 1965), Problem 87, solves the rectangle problem as follows. Let x and y be the sides of the large rectangle. The total number of cells it contains is xy. The border, one cell wide, contains $2x + 2y - 4$ cells. Since we are told that the border contains $xy/2$ cells, we can write the equation:

$$xy/2 = 2x + 2y - 4.$$

Double both sides and rearrange the terms:

$$xy - 4x - 4y = -8.$$

Add 16 to each side:

$$xy - 4x - 4y + 16 = 8.$$

The left side can be factored:

$$(x - 4)(y - 4) = 8.$$

It is clear that $(x - 4)$ and $(x - y)$ must be positive integral factors of 8. The only pairs of such factors are 8, 1 and 4, 2. They provide two solutions: $x = 12$, $y = 5$, and $x = 8$, $y = 6$.

The problem is closely related to integral-sided right triangles. The width of the border is an integer only when the diagonal of the large rectangle cuts it into two such "Pythagorean triangles."

If we generalize the problem to allow nonintegral solutions for borders of any uniform width, keeping only the proviso that the area of the border be equal to the area of the rectangle within it, there is an unusually simple formula for the width of the border. (I am indebted to S. L. Porter of Davis, Calif., for it.) Merely add two adjacent sides of the border, subtract the diagonal of the large rectangle and divide the result by four. This procedure gives the width of the border.

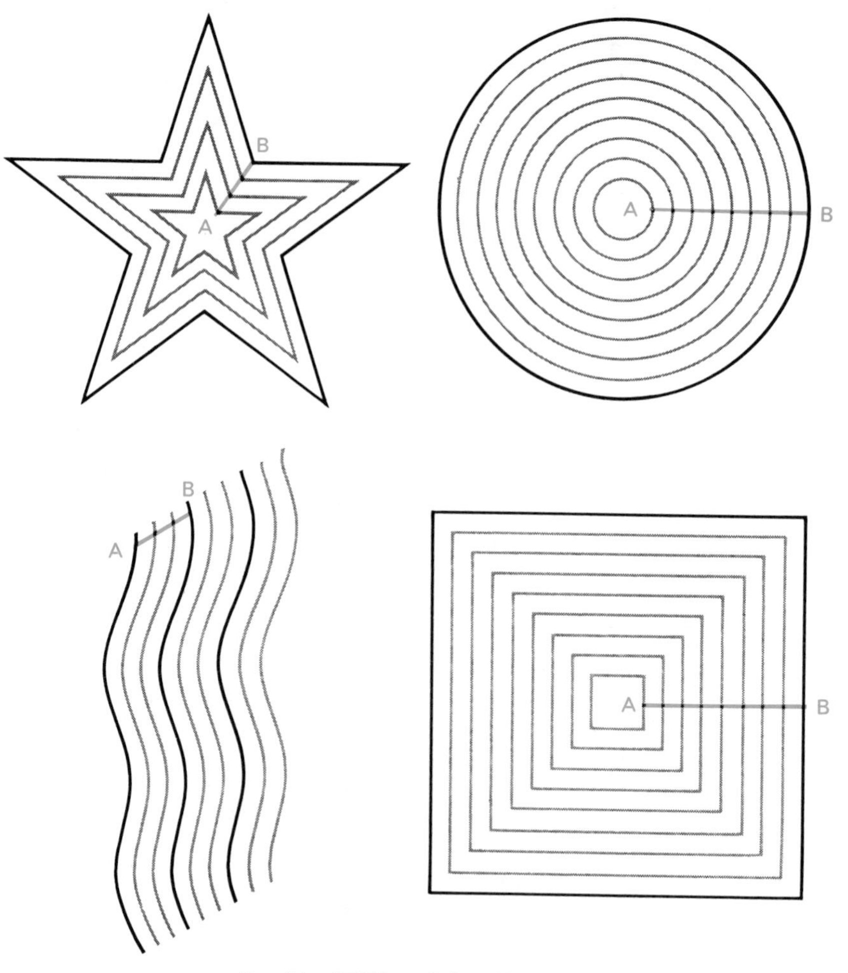

Proof for "ESP" symbol problem

13. The Hierarchy of Infinites and the Problems It Spawns

Which of the five "ESP" symbols cannot be drawn an aleph-one number of times on a sheet of paper, assuming ideal lines that do not overlap or intersect, and replicas that may vary in size but must be similar in the strict geometric sense?

Only the plus symbol cannot be aleph-one replicated. The illustration above shows how each of the other four can be drawn an aleph-one number of times. In each case points on line segment AB form an aleph-one continuum. Clearly a set of nested or side-by-side figures can be drawn so that a different replica passes through each of these points, thus putting the continuum of points into one-to-one correspondence with a set of nonintersecting replicas. There is no comparable way to place replicas of the plus symbol so that they fit snugly against each other. The centers of any pair of crosses must be a finite distance apart (although this distance can be made as small as one pleases), forming a countable (aleph-null) set of points.

The reader may enjoy devising a formal proof that aleph-one plus symbols cannot be drawn on a page. The problem is similar to one involving alphabet letters that can be found in Leo Zippin's *Uses of Infinity* (Random House, 1962), page 57. So far as I know, no one has yet specified precisely what conditions must be met for a linear figure to be aleph-one replicable. Some figures are aleph-one replicable by translation or rotation, some by shrinkage, some by translation plus shrinkage, some by rotation plus shrinkage. I rashly reported in my article that all figures topologically equivalent to a line segment or a simple closed curve were aleph-one replicable, but Robert Mack, then a high school student in Concord, Mass., found a simple counterexample. Consider two unit squares, joined like a vertical domino, then eliminate two unit segments so that the remaining segments form the numeral 5. It is not aleph-one replicable.

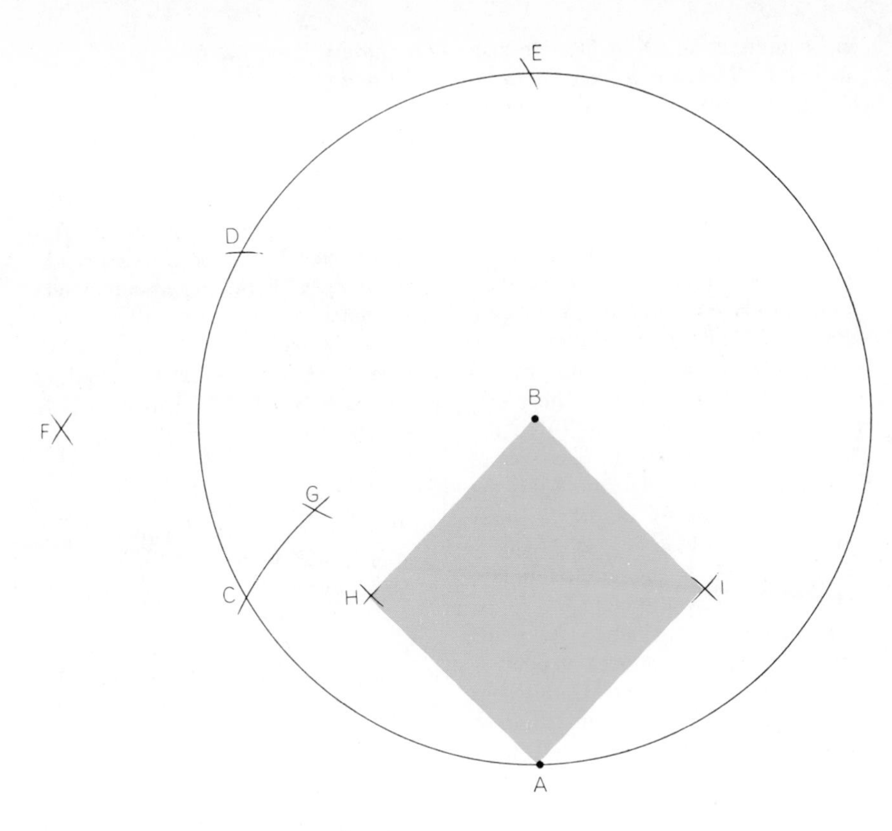

Construction of a square, given diagonal corners A and B

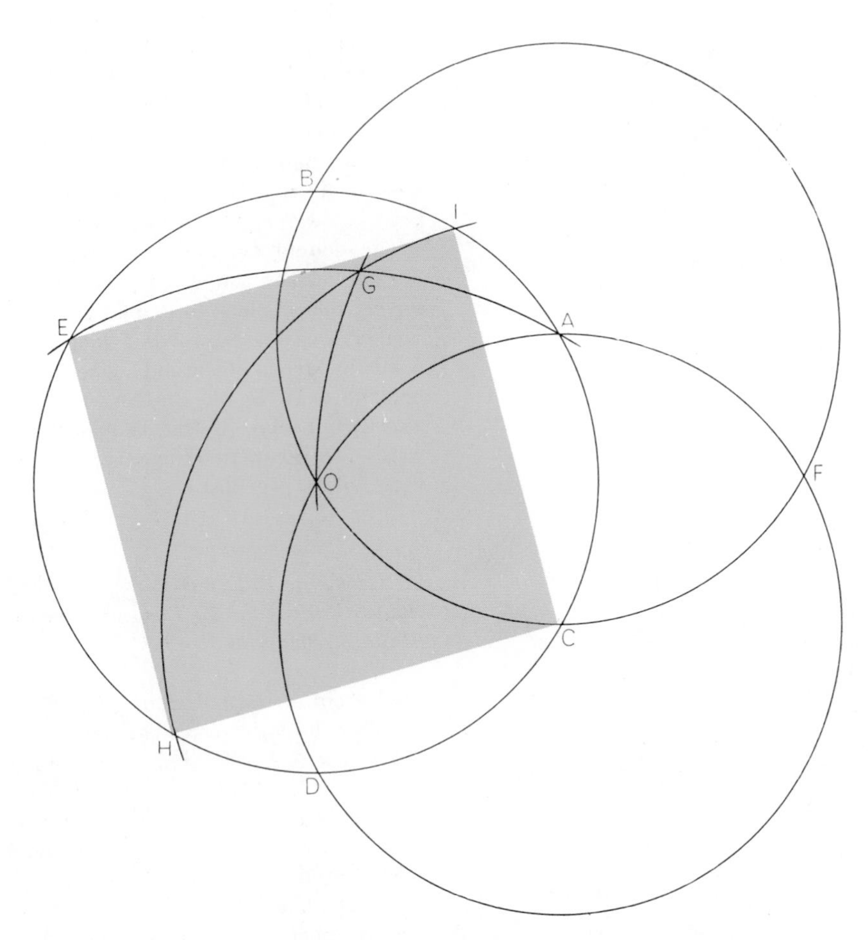

Simpler solution for "Napoleon's problem"

15. Geometric Constructions with a Compass and a Straightedge, and Also with a Compass Alone

The top illustration at the left shows a nine-arc method of solving the Mascheroni problem: Given two diagonally opposite corners of a square, find the other two corners using only a compass. *A* and *B* are the given corners. Draw the circle with radius *AB* and *B* as center. Keeping the compass at the same opening, draw arcs *C*, *D* and *E* (centers at *A*, *C* and *D*). With radius *CE* and centers at *A* and *E*, draw the two arcs that intersect at *F*. With radius *BF* and center at *E*, draw the arc that intersects a previous arc at *G*. With radius *BG* and centers at *A* and *B*, draw the arcs intersecting at *H* and *I*. *A*, *H*, *B* and *I* are the corners of the desired square. Philip G. Smith, Jr., of Hastings-on-Hudson, N.Y., sent a simple proof of the construction, based on right triangles and the Pythagorean theorem, but I leave it to interested readers to work out such proofs for themselves.

Since writing the article on Mascheroni constructions I have learned that the six-arc solution to "Napoleon's problem" is indeed Mascheroni's. Fitch Cheney sent me his paper "Can We Outdo Mascheroni?" (*The Mathematics Teacher*, Vol. 46, March, 1953, pages 152–156), in which he gives Mascheroni's solution followed by his own simpler solution using only five arcs.

Cheney's solution is shown at the bottom left. Pick any point *A* on the given circle and draw a second circle with radius *AO*. With *C* as center and the same radius, draw a third circle. With *D* as center and radius *DA*, draw the arc intersecting the original circle at *E*. With *F* as center and radius *FO*, draw an arc crossing the preceding arc at *G*. With *C* as center and radius *CG*, draw the arc intersecting the original circle at *H* and *I*. *E*, *I*, *C* and *H* mark the corners of the desired square.

Cheney calls attention in his article to the difference between a "modern compass," which retains its opening like a divider, and the "classical compass" of Euclid, which closes as soon as either leg is removed from the plane. Cheney's five-arc solution uses only classical arcs, in contrast to Mascheroni's. Cheney also gives in his article a seven-step classical method of inscribing a pentagon in a circle, two steps fewer than Mascheroni's modern-compass method.

16. Elegant Triangle Theorems Not to Be Found in Euclid

The problem is to find the side of an equilateral triangle containing a point p that is three, four and five units from the triangle's corners. The following solution is from Charles W. Trigg, *Mathematical Quickies* (McGraw-Hill, 1967), answer to Problem 201. He credits the answer to a 1933 source. The broken lines of the illustration below are constructed so that PCF is an equilateral triangle and AE is perpendicular to PC extended left to E. Angle $PCB = 60$ degrees minus angle PCA = angle ACF. Triangles PCB and FCA are therefore congruent and $AF = BP = 5$. Because APF is a right triangle, angle $APE = 180 - 60 - 90 = 30$ degrees. From this we conclude that AE is 2 and EP is twice the square root of 3. This permits the equation

$$AC = \sqrt{2^2 + (3 + 2\sqrt{3})^2}$$
$$= \sqrt{25 + 12\sqrt{3}},$$

which gives AC, a side of the original triangle, a value of 6.766+. For two ways to solve the general problem see L. A. Graham, *Ingenious Mathematical Problems and Methods* (Dover, 1959), answer to Problem 55.

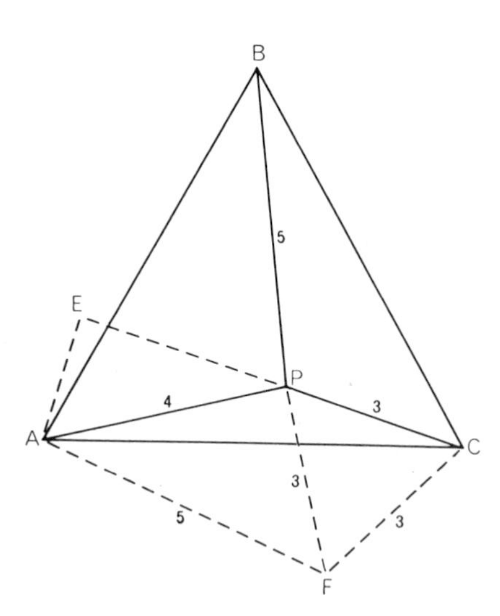

Solution to triangle problem

17. Diversions That Involve One of the Classic Conic Sections: The Ellipse

1. No regular polygon with more sides than a square can be inscribed in an ellipse for this reason: the corners of all regular polygons lie on a circle. A circle cannot intersect an ellipse at more than four points. Therefore no regular polygon with more than four corners can be placed with all its corners on an ellipse. This problem was contributed by M. S. Klamkin to *Mathematics Magazine* for September-October, 1960.

2. The proof that the paper-folding method of constructing an ellipse actually does produce an ellipse is as follows. Let point A in the illustration below be any point on a paper circle that is not the circle's center (O). The paper is folded so that any point (B) on the circumference falls on A. This creases the paper along XY. Because XY is the perpendicular bisector of AB, BC must equal AC. Clearly $OC + AC = OC + CB$. $OC + CB$ is the circle's radius, which cannot vary, therefore $OC + AC$ must also be constant. Since $OC + AC$ is the sum of the distances of point C from two fixed points A and O, the locus of C (as point B moves around the circumference) must be an ellipse with A and O as the two foci.

The crease XY is tangent to the ellipse at point C because it makes equal angles with the lines joining C to the foci. This is easily established by noting that angle XCA equals angle XCB, which in turn equals angle YCO. Since the creases are always tangent to the ellipse, the ellipse becomes the envelope of the infinite set of creases that can be produced by repeated folding of the paper. This proof is taken from Donovan A. Johnson's booklet *Paper Folding for the Mathematics Class*, published in 1957 by the National Council of Teachers of Mathematics.

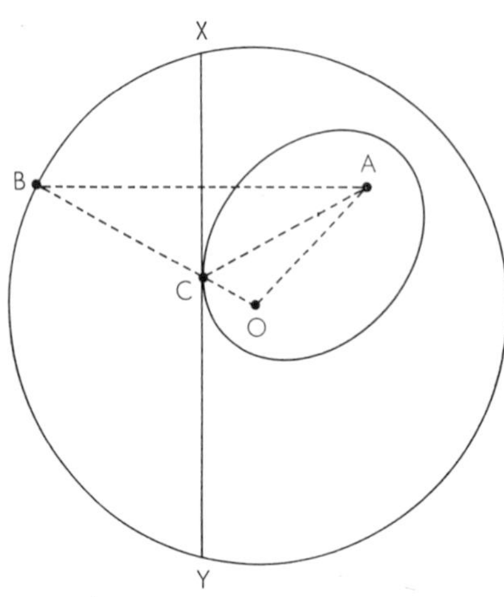

Answer to the paper-folding problem

18. Curious Properties of a Cycloid Curve

What kind of track will enable the car to travel without bobbing up and down? The illustration at the top of page 239 supplies the surprising answer: a series of semicircles! If a circle is rolled inside a circular arc, points on its circumference generate what are called hypocycloids. When the radius of a semicircular track is twice that of the rolling circle, as it is here, the hypocycloid is a straight line.

19. Curves of Constant Width, One of Which Makes It Possible to Drill Square Holes

What is the smallest convex area in which a line segment of length 1 can be rotated 360 degrees? The answer: An equilateral triangle with an altitude of 1. (The area is one-third the square root of 3.)

Any figure in which the line segment can be rotated obviously must have a width at least equal to 1. Of all convex figures with a width of 1, the equilateral triangle of altitude 1 has the smallest area. (For a proof of this the reader is referred to *Convex Figures*, by I. M. Yaglom and V. G. Boltyanskii, pages 221 and 222.) It is easy to see that a line segment of length 1 can in fact be rotated in such a triangle [*see illustration on p. 239*].

The deltoid curve was believed to be the smallest simply-connected area solving the problem until 1963 when a smaller area was discovered independently by Melvin Bloom and I. J. Schoenberg. (See H. S. M. Coxeter, *Twelve Geometric Essays*, [Carbondale and Edwardsville: Southern Illinois University Press, 1968], page 231.)

21. Geometric Fallacies: Hidden Errors Pave the Road to Absurd Conclusions

The errors in the fallacious geometric proofs are briefly explained as follows:

Theorem 1. An obtuse angle is sometimes equal to a right angle. The mistake lies in the location of point K. When the figure is accurately drawn, K is so far below line DC that, when G and K are joined, the line falls entirely outside the original square $ABCD$. This renders the proof totally inapplicable.

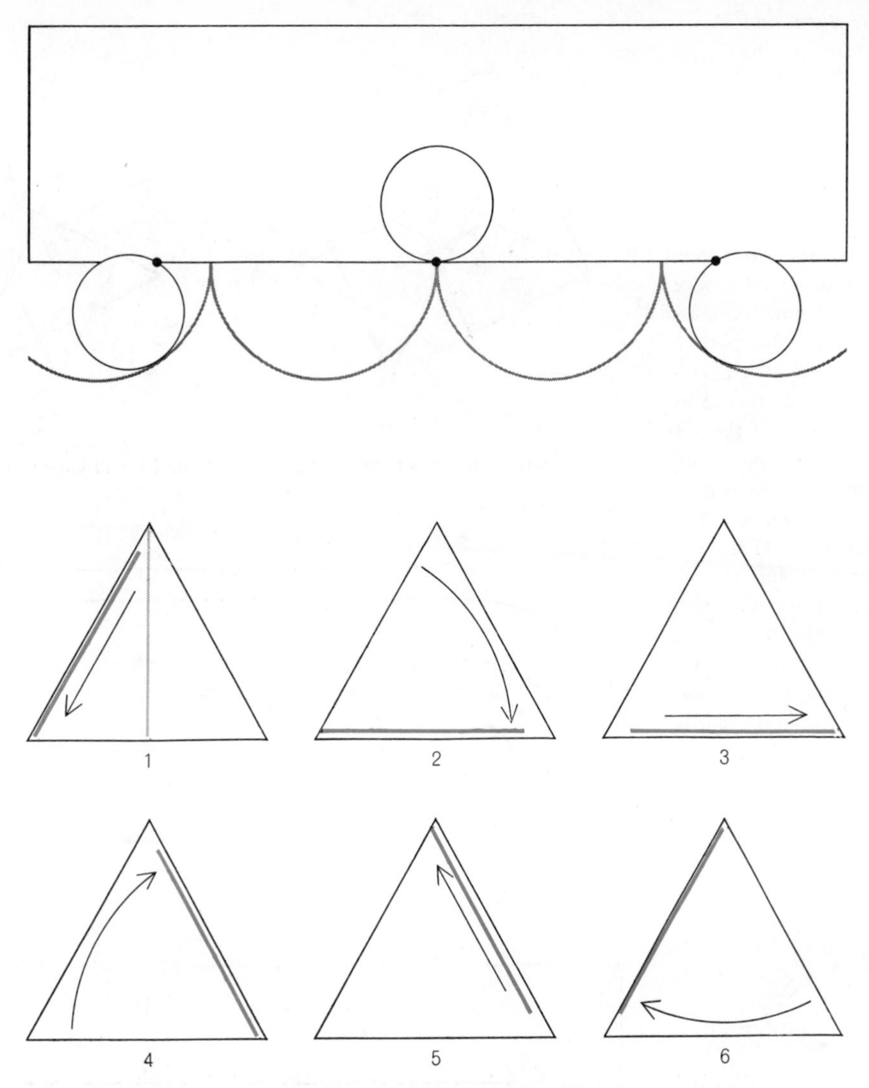

Answer to the needle-turning problem

Theorem 2. Every triangle is isosceles. Again the error is one of construction. *F* is always outside the triangle and at a point such that, when perpendiculars are drawn from *F* to sides *AB* and *AC*, one perpendicular will intersect one side of the triangle but the other will intersect an extension of the other side. A detailed analysis of this fallacy can be found in Eugene P. Northrop's *Riddles in Mathematics* (1944), Chapter 6.

Theorem 3. If a quadrilateral ABCD has angle A equal to angle C, and AB equals CD, the quadrilateral is a parallelogram. The proof is correct if *X* and *Y* are each on a side of the quadrilateral or if both *X* and *Y* are on projections of the sides. It fails if one is on a side and the other is on an extension of a side, as shown in the illustration at the right. This figure meets the theorem's conditions but obviously is not a parallelogram.

Theorem 4. Pi equals 2. It is true that as the semicircles are made smaller their radii approach zero as a limit and therefore the wavy line can be made as close to the diameter of the large circle as one pleases. At no step, however, do the semicircles alter their *shape.* Since they always remain semicircles, no matter how small, their total length always remains pi. The fallacy is an excellent example of the fact that the elements of a converging infinite series may retain properties quite distinct from those of the limit itself.

Theorem 5. Euclid's parallel postulate can be proved by Euclid's other axioms. The proof is valid in showing that one line can be constructed through *C* that is parallel to *AB,* but it fails to prove that there is only one such parallel. There are many other methods of constructing a parallel line through *C;* the proof does not guarantee that all these parallels are the *same* line. Indeed, in hyperbolic non-Euclidean geometry an infinity of such parallels can be drawn through *C,* a possibility that can be excluded only by adopting Euclid's fifth postulate or one equivalent to it. Elliptic non-Euclidean geometry, in which *no* parallel can be drawn through *C,* is made possible by discarding, along with the fifth postulate, certain other Euclidean assumptions.

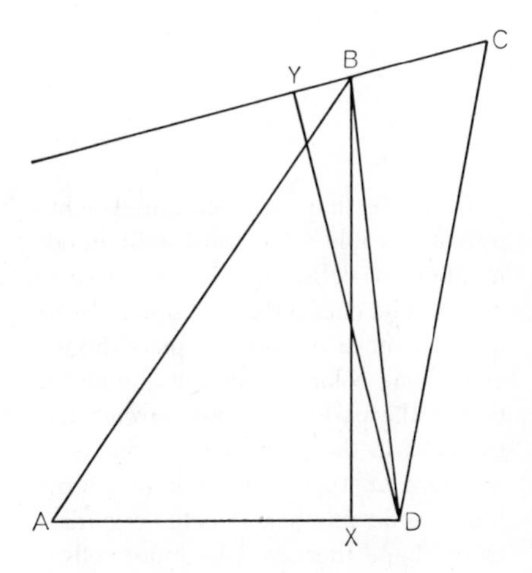

Quadrilateral-theorem counterexample

23. Various Problems Based on Planar Graphs, or Sets of "Vertices" Connected by "Edges"

The two printed-circuit problems are solved in the manner shown in the illustration at right. A symmetrical, non-self-intersecting Euler line for the four-circle puzzle is shown below, obtained by the coloring method. The path in the illustration at the right traces a re-entrant knight's tour on the cross-shaped board. To determine if there is a single path that will go over every possible knight's move, we first draw a graph [*at right in illustration*] showing every move. Note that eight of the vertices are meeting points for an odd number of edges. In accordance with the Euler theorem given in the article, a minimum of 8/2, or 4, paths are required to trace every edge once and only once. Each path must begin at one odd vertex and end at another.

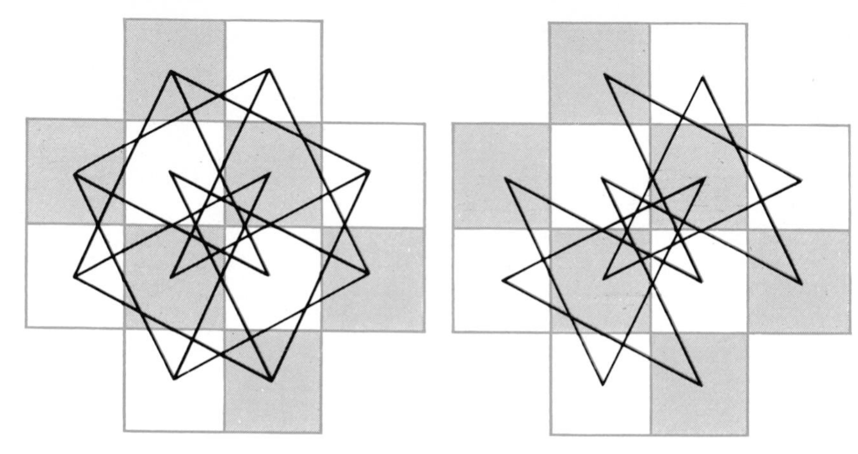

Graphs for re-entrant knight's tour (left) and for all knight's moves (right)

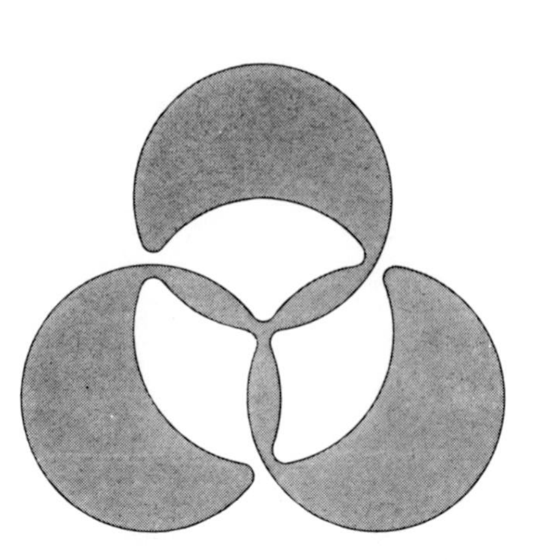

Solution to four-circle problem

To prove that no re-entrant knight's tour is possible on a board with an odd number of cells, first color the cells alternately, checkerboard fashion. Every knight's move carries the piece from a cell of one color to a cell of another, so that if the path is a closed circuit, half the cells in the path must be one color and half another color. But if a board has an odd number of cells, regardless of its shape there will be more cells of one color than of the other.

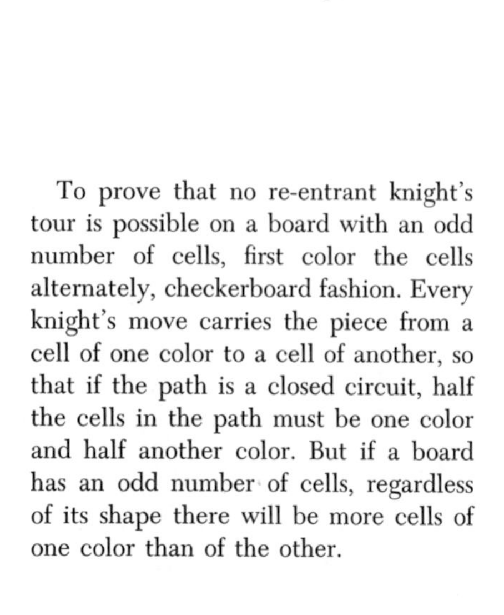

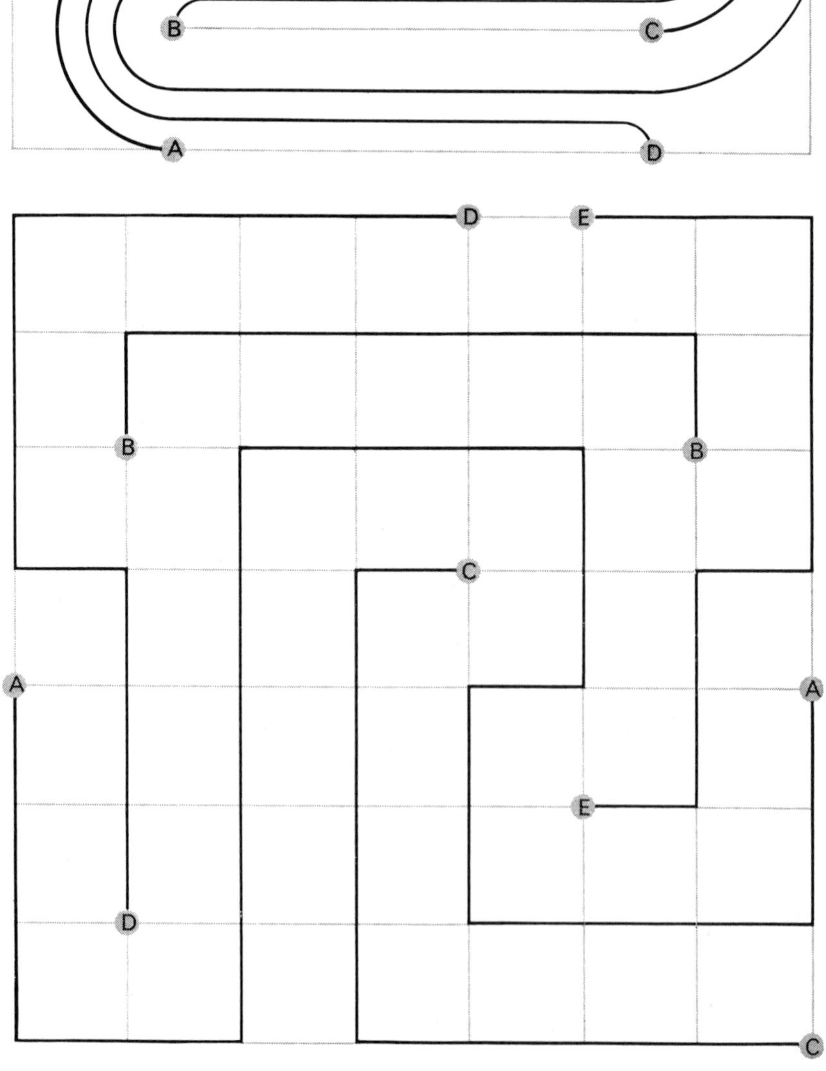

Solutions to printed-circuit problems

24. Topological Diversions, Including a Bottle with No Inside or Outside

The torus-cutting problem is solved by first ruling three parallel lines on the unfolded square [*see figure below*]. When the square is folded into a torus, as explained, the lines make two closed loops. Cutting these loops produces two interlocked bands, each two-sided with two half-twists.

How does one find a loop cut on the Klein bottle that will change the surface to a single Möbius strip? On both left and right sides of the narrow rectangular model described you will note that the paper is creased along a fold that forms a figure-eight loop. Cutting only the left loop transforms the model into a Möbius band; cutting only the right loop produces an identical band of opposite handedness.

What happens if both loops are cut? The result is a two-sided, two-edged band with four half-twists. Because of the slot the band is cut apart at one point, so that you must imagine the slot is not there. This self-intersecting band is mirror-symmetrical, neither right- nor left-handed. You can free the band of self-intersection by sliding it carefully out of the slot and taping the slot together. The handedness of the resulting band (that is, the direction of the helices formed by its edges) depends on whether you slide it out to the right or the left. This and the previous cutting problems are based on paper models that were invented by Stephen Barr and are described in his *Experiments in Topology*.

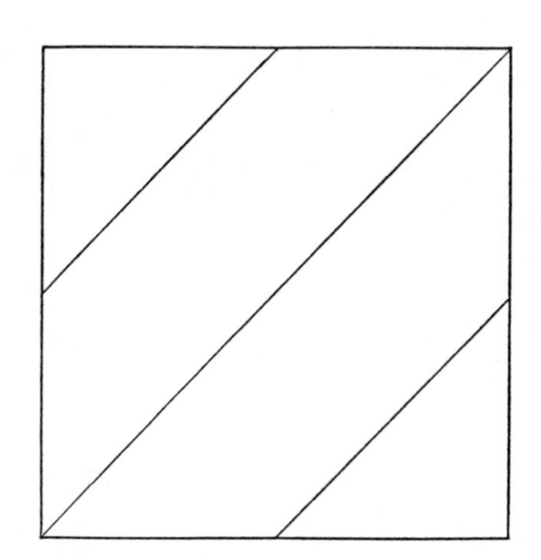

Solution to the torus-cutting problem

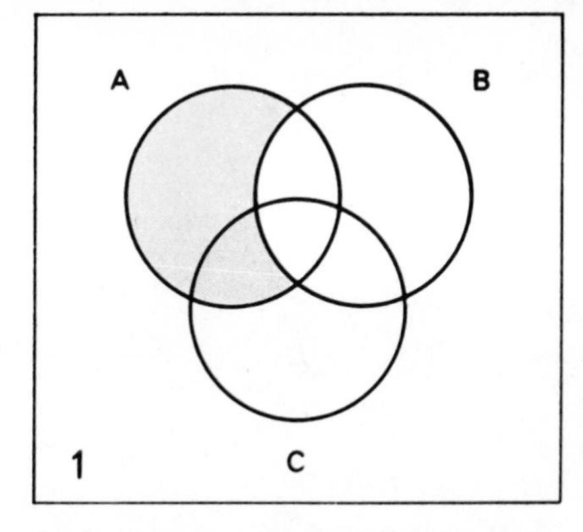

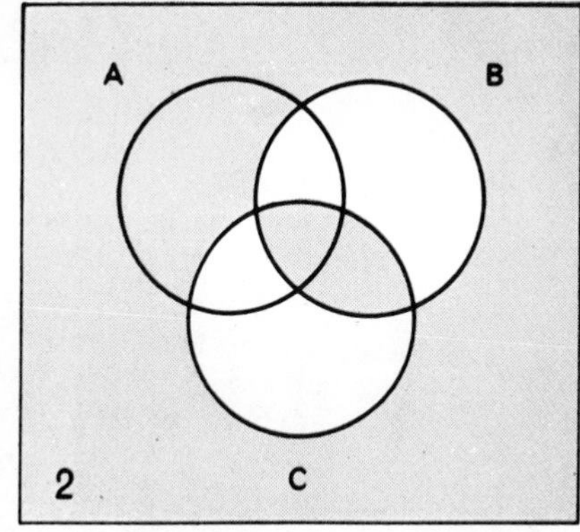

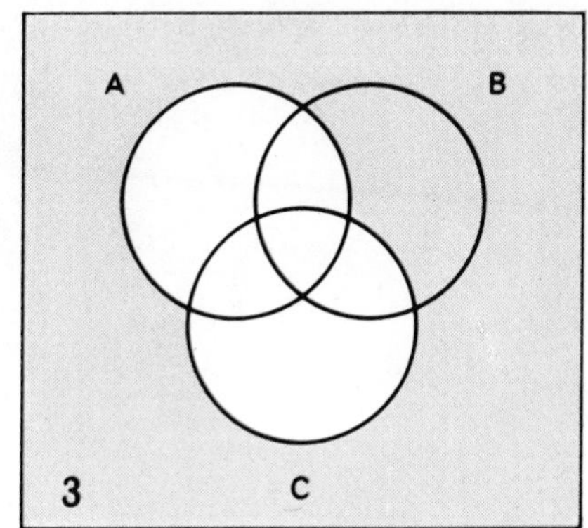

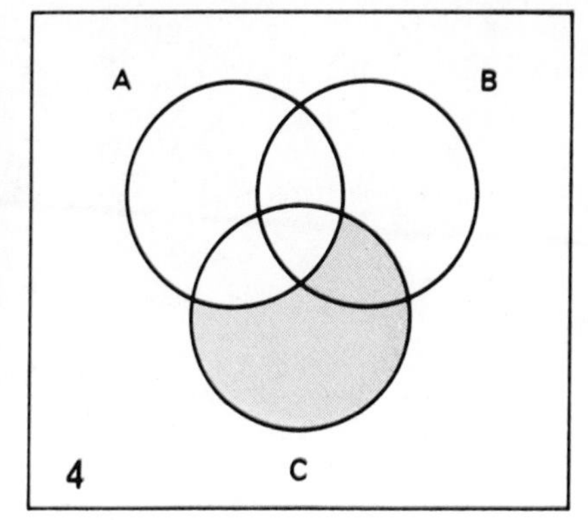

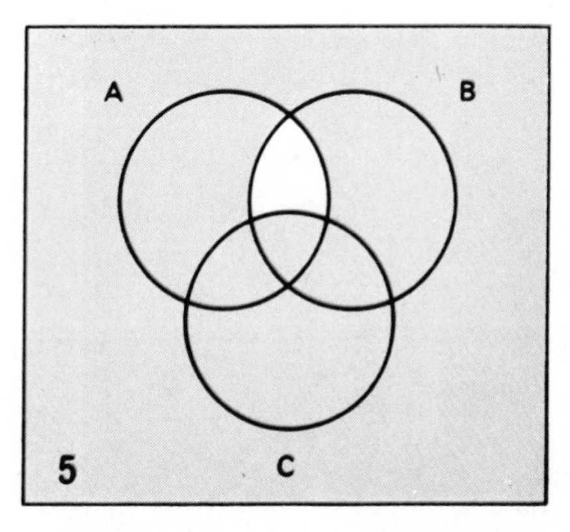

Venn-diagram solution to the martini problem

30. Boolean Algebra, Venn Diagrams and the Propositional Calculus

Three Venn circles are shaded as in the illustration above to solve the problem about the three men who lunch together. Each of the first four diagrams is shaded to represent one of the four premises of the problem. Superimposing the four to form the last diagram shows that if the four premises are true, the only possible combination of truth values is a, b, $\sim c$, or true a, true b and false c. Since we are identifying truth with ordering a martini, this means that Abner and Bill always order martinis, whereas Charley never does.

The method of generating 2^n integers to form Boolean algebras, as already explained, was given by Francis D. Parker in *The American Mathematical Monthly* for March, 1960, page 268. Consider a set of any number of distinct primes, say 2, 3, 5. Write down the multiples of all the subsets of these three primes, which include 0 (the null set) and the original set of three primes. Change 0 to 1. This produces the set 1, 2, 3, 5, 6, 10, 15, 30, the first of the examples given. In a similar way the four primes 2, 3, 5, 7 will generate the second example, the $2^4 = 16$ factors of 210. A proof that all such sets provide Boolean algebras, when the algebra is interpreted as explained earlier, can be found in *Boolean Algebra*, by R. L. Goodstein (Pergamon Press, 1963), page 126, as the answer to problem No. 10.

BIBLIOGRAPHIES

I. HISTORY

1. The Rhind Papyrus

THE RHIND MATHEMATICAL PAPYRUS, British Museum 10057 and 10058. Vol. I: Free Translation and Commentary by Arnold Buffum Chace. Vol. II: Photographs, Transcription, Transliteration, Literal Translation by Arnold Buffum Chace, Ludlow Bull and Henry Parker Manning. Mathematical Association of America, Oberlin, Ohio; 1927 and 1929.

4. Descartes

CORRESPONDANCE: RENÉ DESCARTES. Edited by C. Adam and G. Milhaud. Presses Universitaire de France, 1936–56.

DESCARTES' DISCOURSE ON METHOD. Leon Roth. The Clarendon Press, 1937.

EXPERIENCE AND THE NON-MATHEMATICAL IN THE CARTESIAN METHOD. Alan Gewirtz in *Journal of the History of Ideas,* Vol. II, No. 2, pages 183–210; April, 1941.

NEW STUDIES IN THE PHILOSOPHY OF DESCARTES. Norman Kemp Smith. The Macmillan Co., 1952.

THE PHILOSOPHIC WORKS OF DESCARTES. Elizabeth S. Haldane and G. R. T. Ross. Cambridge University Press, 1934.

THE SCIENTIFIC WORK OF RENÉ DESCARTES (1596–1650). Joseph Frederick Scott. Taylor & Francis, 1952.

6. Leibniz

CONCEPTS OF THE CALCULUS: A CRITICAL AND HISTORICAL DISCUSSION OF THE DERIVATIVE AND THE INTEGRAL. C. B. Boyer. Hafner Publishing Company, 1949.

A CRITICAL EXPOSITION OF THE PHILOSOPHY OF LEIBNIZ, WITH AN APPENDIX OF LEADING PASSAGES. Bertrand Russell. George Allen & Unwin Ltd, 1949.

GOTTFRIED WILHELM LEIBNIZ: PHILOSOPHICAL PAPERS AND LETTERS, VOLS. I AND II. Edited by Leroy E. Loemker. The University of Chicago Press, 1956.

A HISTORY OF MATHEMATICAL NOTATIONS. Florian Cajori. The Open Court Publishing Company, 1928, 1929.

THE LEIBNIZ-CLARKE CORRESPONDENCE. H. G. Alexander. Manchester University Press, 1956.

7. The Invention of Analytic Geometry

HISTORY OF GEOMETRICAL METHODS. J. L. Coôlidge. Oxford University Press, 1940.

A HISTORY OF THE CONIC SECTIONS AND QUADRIC SURFACES. J. L. Coolidge. Oxford University Press, 1945.

8. The Remarkable Lore of the Prime Numbers

MAGIC SQUARES MADE WITH PRIME NUMBERS TO HAVE THE LOWEST POSSIBLE SUMMATIONS. W. S. Andrews and Harry A. Sayles. *The Monist,* Vol. 23, No. 4; October, 1913. Pages 623–630.

HISTORY OF THE THEORY OF NUMBERS: VOLUME I. Leonard Eugene Dickson. Carnegie Institution, 1919. (Reprint. Bronx, N.Y.: Chelsea Publishing Co., 1952.)

THE FACTORGRAM. Kenneth P. Swallow. *The Mathematics Teacher,* Vol. 48, No. 1; January, 1955. Pages 13–17.

THE FIRST SIX MILLION PRIME NUMBERS. C. L. Baker and F. J. Gruenberger. Madison, Wis.: Microcard Foundation, 1959.

A Visual Display of Some Properties of the Distribution of Primes. M. L. Stein, S. M. Ulam, and M. B. Wells. *The American Mathematical Monthly*, Vol. 71, No. 5; May, 1964. Pages 516–520.

Peculiar Properties of Repunits. Samuel Yates. *Journal of Recreational Mathematics*, Vol. 2, No. 3; July, 1969. Pages 139–146.

II. NUMBER AND ALGEBRA

9. Mathematical Sieves

Gödel's Proof. Ernest Nagel and James R. Newman. New York University Press, 1958.

On Certain Sequences of Integers Defined by Sieves. Verna Gardiner, R. Lazarus, N. Metropolis and S. Ulam in *Mathematics Magazine*, Vol. 29, No. 3, pages 117–122; January–February, 1958.

The Random Sieve. David Hawkins in *Mathematics Magazine*, Vol. 31, No. 1, pages 1–3; September–October, 1957.

12. The Theory of Numbers

Number Theory and its History. Oystein Ore. McGraw-Hill Book Company, Inc., 1948.

First Course in Theory of Numbers. Harry N. Wright. John Wiley & Sons, Inc., 1939.

14. Number

The Development of Mathematics. Eric T. Bell. McGraw-Hill Book Company, Inc., 1945.

Episodes from the Early History of Mathematics. Asger Aaboe. Random House, 1964.

History of Mathematics, Vol. II: Special Topics of Elementary Mathematics. David Eugene Smith. Dover Publications, Inc., 1958.

The Lore of Large Numbers. Philip J. Davis. Random House, 1961.

Mathematics and the Physical World. Morris Kline. Thomas Y. Crowell Company, 1959.

Number: The Language of Science. Tobias Dantzig. The Macmillan Company, 1954.

Philosophy of Mathematics. Stephen F. Barker. Prentice-Hall, Inc., 1964.

III. GEOMETRY

18. Curious Properties of a Cycloid Curve

A Treatise on the Cycloid. Richard Anthony Proctor. London: Longmans, Green and Co., 1878.

Some Historical Notes on the Cycloid. E. A. Whitman. *The American Mathematical Monthly*, Vol. 50, No. 5; May, 1943. Pages 309–315.

A Book of Curves. E. H. Lockwood. Cambridge: Cambridge University Press, 1961.

Brachistochrone, Tautochrone, Cycloid—Apple of Discord. J. P. Phillips. *The Mathematics Teacher*, Vol. 60, No. 5; May, 1967. Pages 506–508.

20. Projective Geometry

Art and Geometry. William M. Ivins, Jr. Dover Publications, 1964. [Reprint.]

Mathematics in Western Culture. Morris Kline. Oxford University Press, 1953.

Projective Geometry. John Wesley Young. The Open Court Publishing Company, 1930.

Projective Geometry. Oswald Veblen and John Wesley Young. Ginn and Company, 1910–1918.

22. The Koenigsberg Bridges

Men of Mathematics. Eric Temple Bell. Simon and Schuster, 1937.

23. Various Problems Based on Planar Graphs or Sets of "Vertices" Connected by "Edges"

On the Traversing of Geometrical Figures. J. C. Wilson. Oxford: Oxford University Press, 1905.

Economic Applications of the Theory of Graphs. Giuseppe Avondo-Bodino. New York: Gordon and Breach, 1962.

The Theory of Graphs and Its Applications. Claude Berge. New York: Barnes and Noble, 1962.

Flows in Networks. L. R. Ford and D. R. Fulkerson. Princeton, N.J.: Princeton University Press, 1962.

Graphs and Their Uses. Oystein Ore. New York: Random House, 1963.

Groups and Their Graphs. Israel Grossman and Wilhelm Magnus. New York: Random House, 1964.

Finite Graphs and Networks: An Introduction with Applications. Robert G. Busaker and Thomas L. Sarty. New York: McGraw-Hill, 1965.

STRUCTURAL MODELS: AN INTRODUCTION TO THE THEORY OF DIRECTED GRAPHS. Frank Harary, Robert Z. Norman, and Dorwin Cartwright. New York: John Wiley and Sons, 1965.

CONNECTIVITY IN GRAPHS. W. H. Tutte. Toronto: University of Toronto Press, 1967.

A SEMINAR ON GRAPH THEORY. Edited by Frank Harary. New York: Holt, Rinehart and Winston, 1967.

24. Topological Diversions, Including a Bottle with No Inside or Outside

TOPOLOGY. Albert W. Tucker and Herbert S. Bailey, Jr. *Scientific American,* Vol. 182, No. 1; January, 1950. Pages 18–24.

ELEMENTARY POINT SET TOPOLOGY. R. H. Bing. *The American Mathematical Monthly,* Vol. 67, No. 7; August–September, 1960. Special Supplement.

INTUITIVE CONCEPTS IN ELEMENTARY TOPOLOGY. Bradford Henry Arnold. New York: Prentice-Hall, 1962.

EXPERIMENTS IN TOPOLOGY. Stephen Barr. New York: Thomas Y. Crowell, 1964.

VISUAL TOPOLOGY. W. Lietzmann. London: Chatto and Windus, 1965.

THE FOUR-COLOR PROBLEM. Oystein Ore. New York: Academic Press, 1967.

25. Geometry

GREAT IDEAS OF MODERN MATHEMATICS: THEIR NATURE AND USE. Jagjit Singh. Dover Publications, Inc., 1959.

A LONG WAY FROM EUCLID. Constance Reid. Thomas Y. Crowell Company, 1963.

PRELUDE TO MATHEMATICS. W. W. Sawyer. Penguin Books Inc., 1955.

WHAT IS MATHEMATICS? AN ELEMENTARY APPROACH TO IDEAS AND METHODS. Richard Courant and Herbert Robbins. Oxford University Press, 1941.

IV. STATISTICS AND PROBABILITY

26. Statistics

ELEMENTARY STATISTICAL ANALYSIS. Samuel S. Wilks. Princeton University Press, 1948.

27. What Is Probability?

A TREATISE ON PROBABILITY. John Maynard Keynes. Macmillan and Company, Ltd., 1921.

THE NATURE AND APPLICATION OF INDUCTIVE LOGIC. Rudolf Carnap. The University of Chicago Press, 1951.

PROBABILITY, STATISTICS, AND TRUTH. Richard von Mises. W. Hodge and Company, Ltd., 1939.

28. Probability

THE SCIENCE OF CHANCE. Horace C. Levinson. Rinehart and Company, Inc., 1950.

PROBABILITY AND ITS ENGINEERING USES. Thornton C. Fry. D. Van Nostrand Company, Inc., 1928.

AN INTRODUCTION TO PROBABILITY THEORY AND ITS APPLICATIONS. William Feller. John Wiley and Sons., Inc., 1950.

V. SYMBOLIC LOGIC AND COMPUTERS

31. Symbolic Logic

SYMBOLIC LOGIC. C. I. Lewis and C. H. Langford. The Century Company, 1932.

ELEMENTS OF SYMBOLIC LOGIC. Hans Reichenbach. The Macmillan Company, 1947.

32. Computer Logic and Memory

DESIGN OF TRANSISTORIZED CIRCUITS FOR DIGITAL COMPUTERS. Abraham I. Pressman. John F. Rider Publisher, Inc., 1959.

LOGICAL DESIGN OF DIGITAL COMPUTERS. Montgomery Phister. John Wiley & Sons, Inc., 1958.

SQUARE-LOOP FERRITE CIRCUITRY: STORAGE AND LOGIC TECHNIQUES. C. J. Quartly. Prentice-Hall, Inc., 1962.

SWITCHING CIRCUITS AND LOGICAL DESIGN. S. H. Caldwell. John Wiley & Sons, Inc., 1958.

THEORY AND DESIGN OF DIGITAL MACHINES. Thomas C. Bartee, Irwin L. Lebow and Irving S. Reed. McGraw-Hill Book Company, Inc., 1962.

33. The Role of the Computer

MATHEMATICAL MACHINES. Harry M. Davis in *Scientific American*, Vol. 180, No. 4, pages 28–39; April, 1949.

CALCULATING INSTRUMENTS AND MACHINES. Douglas R. Hartree. University of Illinois Press, 1949.

ELECTRONIC ANALOG COMPUTERS. Granino A. and Theresa M. Korn. McGraw-Hill Book Company, Inc., 1952.

MECHANICAL BRAINS. Louis N. Ridenour in *Fortune*, Vol. 39, No. 5, pages 108–118; May, 1949.

VI. APPLICATIONS

34. Musical Tones

THE PHYSICS OF MUSIC. Alexander Wood. Methuen & Co. Ltd., London, 1947.

35. Physics and Music

THE PHYSICS OF MUSIC. Alexander Wood. Methuen, 1947.

SOUND WAVES: THEIR SHAPE AND SPEED. D. C. Miller. Macmillan, 1937.

THE PSYCHOLOGY OF MUSIC. C. E. Seashore. McGraw-Hill, 1938.

THE ACOUSTICS OF MUSIC. W. T. Bartholomew. Prentice Hall, 1942.

VISIBLE SPEECH. R. K. Potter and others. Van Nostrand, 1948.

36. The Practice of Quality Control

ECONOMIC CONTROL OF QUALITY OF MANUFACTURED PRODUCT. W. A. Shewhart. D. Van Nostrand Co., Inc., 1931.

STATISTICAL QUALITY CONTROL. Eugene L. Grant, McGraw-Hill Book Company, 1952.

37. Game Theory and Decisions

THEORY OF GAMES AND STATISTICAL DECISIONS. D. H. Blackwell and M. A. Girshick. John Wiley & Sons, Inc., 1954.

FOUNDATIONS OF STATISTICS. L. J. Savage. John Wiley & Sons, Inc., 1954.

THEORY OF GAMES AND ECONOMIC BEHAVIOR. John von Neumann and Oskar Morgenstern. Princeton University Press, 1947.

38. The Theory of Games

THE THEORY OF GAMES AND ECONOMIC BEHAVIOR. John von Neumann and Oskar Morgenstern. Princeton University Press, 1955.

39. Linear Programming

AN INTRODUCTION TO LINEAR PROGRAMMING. A. Charnes, W. W. Cooper and A. Henderson. John Wiley & Sons, Inc., 1953.

ACTIVITY ANALYSIS OF PRODUCTION AND ALLOCATION. Edited by T. C. Koopmans. John Wiley & Sons, Inc., 1951.

BLENDING AVIATION GASOLINES: A STUDY IN PROGRAMMING INTERDEPENDENT ACTIVITIES IN AN INTEGRATED OIL COMPANY. A. Charnes, W. W. Cooper and B. Mellon in *Econometrica*, Vol. 20, No. 2; April, 1952.

40. Operations Research

METHODS OF OPERATIONS RESEARCH. Philip M. Morse and George E. Kimball. The Technology Press of Massachusetts Institute of Technology and John Wiley & Sons, Inc., 1950.

INDEX